Probability and Mathematical Statistics (Continued)

SCHEFFE · The Analysis of Variance
SEBER · Linear Regression Analysis
WILKS · Mathematical Statistics
WILLIAMS · Diffusions, Markov Processes, and Martingales, Volume I: Foundations
ZACKS · The Theory of Statistical Inference

Applied Probability and Statistics

BAILEY · The Elements of Stochastic Processes with Applications to the Natural Sciences
BAILEY · Mathematics, Statistics and Systems for Health
BARNETT and LEWIS · Outliers in Statistical Data
BARTHOLOMEW · Stochastic Models for Social Processes, *Second Edition*
BARTHOLOMEW and FORBES · Statistical Techniques in Manpower Planning
BECK and ARNOLD · Parameter Estimation in Engineering and Science
BELSLEY, KUH, and WELSCH · Regression Diagnostics: Identifying Influential Data and Sources of Collinearity
BENNETT and FRANKLIN · Statistical Analysis in Chemistry and the Chemical Industry
BHAT · Elements of Applied Stochastic Processes
BLOOMFIELD · Fourier Analysis of Time Series: An Introduction
BOX · R. A. Fisher, The Life of a Scientist
BOX and DRAPER · Evolutionary Operation: A Statistical Method for Process Improvement
BOX, HUNTER, and HUNTER · Statistics for Experimenters: An Introduction to Design, Data Analysis, and Model Building
BROWN and HOLLANDER · Statistics: A Biomedical Introduction
BROWNLEE · Statistical Theory and Methodology in Science and Engineering, *Second Edition*
BURY · Statistical Models in Applied Science
CHAMBERS · Computational Methods for Data Analysis
CHATTERJEE and PRICE · Regression Analysis by Example
CHERNOFF and MOSES · Elementary Decision Theory
CHOW · Analysis and Control of Dynamic Economic Systems
CLELLAND, deCANI, BROWN, BURSK, and MURRAY · Basic Statistics with Business Applications, *Second Edition*
COCHRAN · Sampling Techniques, *Third Edition*
COCHRAN and COX · Experimental Designs, *Second Edition*
COX · Planning of Experiments
COX and MILLER · The Theory of Stochastic Processes, *Second Edition*
DANIEL · Biostatistics: A Foundation for Analysis in the Health Sciences, *Second Edition*
DANIEL · Applications of Statistics to Industrial Experimentation
DANIEL and Wood · Fitting Equations to Data: Computer Analysis of Multifactor Data, *Second Edition*
DAVID · Order Statistics
DEMING · Sample Design in Business Research
DODGE and ROMIG · Sampling Inspection Tables, *Second Edition*
DRAPER and SMITH · Applied Regression Analysis

continued on back

Statistical Concepts and Methods

Statistical Concepts and Methods

GOURI K. BHATTACHARYYA
RICHARD A. JOHNSON
University of Wisconsin

John Wiley & Sons
New York • Santa Barbara • London • Sydney • Toronto

To JOLLY G.K.B.

To BOBBIE, ERIK and TOM R.A.J.

Library of Congress Cataloging in Publication Data:

Bhattacharyya, Gouri K 1940–
Statistical concepts and methods.
(Wiley series in probability and mathematical statistics)
Includes bibliographical references and index.
1. Statistics. I. Johnson, Richard Arnold, joint author. II. Title.
HA29.B46 1977 519.5 76–53783
ISBN 0-471-07204-4

Printed in the United States of America

10 9 8 7

Preface

The twentieth century has witnessed a progressively increasing use of statistical reasoning in all fields of science and even in the humanities. Consequently, it is now essential for students in most academic and professional programs to be acquainted with the basic principles and techniques of statistical analysis. Fortunately, comprehension of many of the core statistical concepts does not require knowledge of advanced mathematics. Our purpose in this book is to build both an understanding of these concepts and an awareness of the limitations of the methods while introducing a broad spectrum of widely used statistical techniques. It is designed for introductory statistics courses for sophomore-senior or first year graduate students from a wide range of disciplines.

Our goal throughout is to explain the concepts in the spirit of current statistical thinking and practice. These discussions rely mainly on intuitive reasoning, although, occasionally, a few steps of algebra are added. We believe that students will better appreciate statistical methods if they are provided with adequate explanations of the reasoning behind these methods. Without this understanding, students often lose interest in the subject and the cookbook recipe presentation may lead them into gross misuses of statistics. Moreover, before formally introducing a statistical technique we present illustrations to indicate its scope and motivate its study. A majority of teachers we know have found this motivation-reasoning-statistical method-example sequence to be very effective for introductory courses.

Drawing inferences from an analysis of data, the central theme of statistics, is introduced early in the book, as are the ideas of statistical models and the amount of uncertainty regarding conclusions. This initial exposure prevents students from becoming bored with preparatory material before obtaining an appreciation of its usefulness. The role of probability in making inferences from data is also explained at an early stage. Our treatment of probability in Chapter 3 is quite comprehensive even though elementary. In developing these concepts we have avoided overworked examples of die or coin tossing and have instead used examples of practical interest.

When describing the various methods of statistical analysis, the reader is continually reminded that the validity of a statistical inference is contingent upon certain model assumptions. Misleading conclusions may result when these assumptions are violated. We feel that the teaching of statistics, even at an introductory level, should not be limited to the prescription of methods. Students should be encouraged to develop a critical attitude in applying the methods and to be cautious when interpreting the results. This attitude is especially important in the study of relationship among variables, which is perhaps the most widely used (and also abused) area of statistics. In addition to discussing inference procedures in this context, we have particularly stressed critical examination of the model assumptions and careful interpretation of the conclusions.

Considerable effort was expended to assemble a wealth of examples and exercises drawn from a wide range of real life settings. Many of the illustrations contain data gathered from experimenters at the Madison campus and also from published reports in professional journals and government publications. We have, however, limited our attention to the aspect in each data set, that is relevant to the current discussion. We feel that a complete discussion of all the issues raised by an investigation or case study approach is not appropriate in a first course in statistics.

This book contains ample material for a one-semester or two-quarter course. The first nine chapters contain the core statistical concepts covering the descriptive summary of data, the idea of probability and its role in statistical inference, and the two basic processes of inference—testing hypotheses and confidence intervals. This material can be covered in nearly two-thirds of a semester leaving some time for topics selected from subsequent chapters. Since regression techniques are extensively used in most statistical analyses, we recommend that Chapter 10 and the first three sections of Chapter 11 follow the core material. Of course, individual instructors may give priority to other topics. Additional chapters that cover useful techniques in detail can be used for a second-semester or third-quarter continuation course. The portions of the book that involve either algebraic derivations or not so elementary concepts are marked by asterisks as are the corresponding exercises. In each chapter, mathematical exercises have been separated from the others.

Chapters 12 through 16 are essentially self-contained and the topics can be chosen in any sequence following the core material. Chapters 12, 13, 15, and 16 are likely to be of prime interest to social science majors. Courses given mainly to physical and biological science students may include the remainder of Chapter 11 and Chapters 13 to 15.

For a more elementary freshman-sophomore level course that is limited to one semester, the instructor may wish to omit some portions of the earlier chapters including the starred sections and Sections 3.6, 4.5, 4.6, 4.7, 5.6, 5.7, 5.8, and 6.6. Doing so amounts to omitting rigorous explanations concerning joint distributions of random variables and the properties of expectation and variance of a sum as well as a few discrete probability models. This approach allows an instructor to proceed at a slower pace or leaves more time for the coverage of applied topics in later chapters.

We would like to thank all our colleagues in the department of statistics for creating a stimulating environment, with a constant interplay of theory and practice, in which statistics becomes an exciting discipline. In particular we would like to express our sincere appreciation to Professor Norman Draper for many helpful comments on Chapters 10 and 11. We record our thanks to many of our graduate students, especially, Said Infante Gil, Fabian Hernandez, Tom Wehrly, and Shin Ta Liu for reading the manuscript and correcting numerical errors, and to Mary Esser for her careful typing of the manuscript. Also, we are indebted to several faculty members on the Madison campus, especially Professors Alan Ek and Carter Denniston, for providing us with data sets and examples. Finally, but not least, we are grateful to our wives who had to shoulder extra responsibilities during the writing of this book.

G. K. Bhattacharyya
Richard A. Johnson

Contents

CHAPTER 1

Introduction

1.1 WHAT IS STATISTICS?

Most people are familiar with the term statistics as it is used to denote and record numerical facts and figures: for example, the heights of skyscrapers in the country, the daily prices of selected stocks on a stock exchange, the tonnage of cargo vessels commissioned over the last 15 years, or even the number of yards gained by the winning team in a football game. However, this usage of the term is not the central focus of the subject. Statistics primarily deals with situations in which the occurrence of some event cannot be predicted with certainty. Our conclusions are often uncertain because we base them on incomplete data —assessing the current rate of unemployment in a state based on a survey of a few thousand people is an example. Uncertainty also arises when repeated observations on a phenomenon yield variable results even though attempts are made to control the factors governing the event being observed. For instance, one-year-old pine trees are not all the same height, even if they have grown from the same batch of seedlings under identical soil and weather conditions. The time it takes to mow a lawn, the weights of six-week-old chickens raised on a poultry farm, and the period of relief from hay fever symptoms after taking a certain medicine are other examples of situations in which variability in repeated observations occurs. Statistics is a body of *concepts and methods* used to collect and interpret data concerning a particular area of investigation and to draw conclusions in situations where uncertainty and variation are present.

Historically, the word "statistics" is derived from the Latin word "status" meaning "state." For several decades, statistics was associated solely with the display of facts and figures pertaining to the economic, demographic, and political situations prevailing in a country. Even today, a plethora of government reports that contain massive numerical documentation and bear such titles as "Statistics of Farm Production" and "Labor Statistics" are reminders of the origin of the word "statistics." A major segment of the general public still has the misconception that statistics is exclusively associated with traumatic arrays of numbers and sometimes disconcerting series of graphs. Therefore, it is essential to remember that modern statistical theory and methodology have made giant strides beyond the mere compilation of numerical tables and graphs. As a subject, statistics now encompasses concepts and methods that are of far-reaching importance in all inquiries that involve gathering of data, by a process of experimentation or observation, and making inferences or conclusions by analyzing such data. Numerical displays have become a minor aspect of statistics and few, if any, professional statisticians make their living today solely by constructing tables and graphs.

1.2 STATISTICS AND EVERYDAY LIFE

Fact finding through the collection and interpretation of data is not confined to professional researchers. It pervades the everyday life of all people who strive, consciously or unconsciously, to understand matters of interest concerning society, living conditions, the environment, and the world at large. In learning about the state of unemployment, pollution from industrial wastes, the performance of competing football teams, the effectiveness of analgesics, and other interests of contemporary life we gather facts and figures and then interpret them or attempt to understand the interpretations that others make. Thus, learning takes place every day through an often implicit analysis of factual information.

Sources of factual information range from individual experience to reports in the news media, government records, and articles in professional journals. Weather forecasts, market reports, cost-of-living indexes, and the results of public opinion polls are some other examples. Statistical methods are employed extensively in the preparation of such reports. Reports that are based on sound statistical reasoning and the careful interpretation of conclusions are truly informative. Frequently, however, the deliberate or inadvertent misuse of statistics leads to erroneous conclusions and distortions of truth. For the general public, the basic consumers of these reports, some idea of statistical reasoning is essential to properly interpret the data and evaluate the conclusions that are drawn. Statistical reasoning provides criteria for determining the conclusions that are actually supported by data and those that are not. In all fields of study where inferences are drawn from analyses of data, the credibility of conclusions also depends greatly on the use of statistical methods in the data collection stage. Statistical methods play an important role in a modern democratic state. For instance, if elected officials can determine the wishes of their constituency by fairly rapid sampling methods, the formulation of public policy can be more responsive to the will of the people.

1.3 ELEMENTS OF SCIENTIFIC INQUIRY

The fundamental importance of statistical methodology is best appreciated when viewed in light of the general process of learning, sometimes called *scientific method.*

Although scientific inquiry is not rigidly structured, it can be described as any process of expending effort to learn about hidden regularities of some aspect of what appears at times to be a chaotic world. Models or theories are tentatively postulated that purport to explain a phenomenon, logical deductions are derived from the postulated model and then weighed against the factual findings, the model is modified, and the search continues for better explanations. The specifics of the learning process are as diverse as the disciplines of study, but some basic steps that form the core of most scientific investigations are listed below. A few scientific inquiries are briefly outlined in Section 1.5.

Specification of objective: Whenever the present state of knowledge concerning something of interest is deemed inadequate, methods of investigation may be considered to improve understanding. This could be further focused on more specific goals such as to substantiate a new theory or to scrutinize an existing theory with respect to the extent that logical deductions drawn from it are verified by factual findings. In some situations, the goal may simply be the creation of a data base of information which accurately reflects the present state of affairs. For instance, the average amounts students spend weekly on leisure activities could be collected to study that component of student spending. At other times, the objective may be far more extensive and not only be to gain an understanding of the factors operating in an environment but also to determine the possibilities for their use in the control or modification of some facet of a phenomenon. Understanding the chemistry of solid waste disposal at a plant and its consequent use for purification of surrounding river water is an objective of this form.

Gathering information: The procurement oı objective information pertaining to the purpose of the study is crucial to any investigation. This process may involve a wide variety of activities, ranging from sophisticated experiments in controlled environments, to field trials, socioeconomic surveys and polls, and even historical records. In the present era of progressive instrumentation and mechanization of record keeping, the increasing quantification of observations is a fact of life. Information is typically gathered in the form of data that numerically measure some characteristics or a record of some qualitative traits possessed by the individuals or elements under study, or both.

Analysis of data: Data collected by an appropriate process of experimentation or observation serve as the basic source for acquiring new knowledge about the matter under study. It is then necessary to examine the data set and extract information pertaining to the issues raised in the specification of objectives. A careful analysis of data is crucial to ascertain the new knowledge gained and to evaluate its strengths and weaknesses.

Statement of findings: The significance of the information provided by the data must then be assessed in the context of what was known at the initial stage of the investigation when the objectives were specified. Data analyses are designed to answer such questions as: "What generalities can be drawn on the phenomenon under study from evidence supplied by the data?" "Is a stated conjecture contradicted by the data?" "Do the data suggest a new theory to explain the phenomenon?" The results of the analysis are then employed to answer these questions and also to weigh the uncertainties involved in the answers obtained. Learning often takes the form of the suggested revision of an existing theory which itself may require further investigation through the collection and analysis of facts.

Thus the basic nature of learning is typically a repetition of this cycle in one form or another. Rarely, if ever, is a truth unraveled in one or even in a few operations of the cycle and changing conditions in many fields demand an indefinite continuation of the process of repetition.

1.4 THE ROLE OF STATISTICS IN SCIENTIFIC INQUIRY

The subject of statistics is comprised of the art and science of the collection, interpretation, and analysis of data and the ability to draw logical generalities that relate to the phenomenon under investigation. In view of the essential stages of scientific method just described, it is apparent that statistics extensively permeates the domain of all scientific investigation.

Specifically, at the stage of information gathering, statistics guides the researcher toward the appropriate ways and means to gather informative data, including a determination of the kind and extent of the data, so that the conclusions drawn from an analysis can be stated with a contemplated degree of precision. In areas of study in which experimentation is costly, the type and amount of data required to provide a desired level of credibility in the conclusions must be carefully determined beforehand. In other areas, too, such decisions are crucial to the ultimate validity and effectiveness of the conclusions drawn from an analysis of the data. The branch of statistics that deals with the planning of experiments and collecting informative data is called *experimental design* or *sampling design*.

After the data have been collected, there is an even greater need for statistical methods. Some of these methods are designed to summarize the information contained in the data to focus attention on the salient features and disregard the nonessential details. A more important group of methods for analyzing the data are intended to draw generalities or inferences about the phenomenon under study. The subject dealing with statistical methods that summarize and describe the prominent features of data is commonly known as *descriptive statistics*. Although historically the primary activity, today descriptive summary is only a narrow part of the realm of activities that fall under the purview of the subject of statistics. A major thrust of the subject is currently the evaluation of information present in data and the assessment of the new learning gained from this information. This is the area of *inferential statistics*, and its associated methods are known as the methods of *statistical inference*. Using these methods provides a basis of reasoning to logically interpret observed facts, to ascertain the extent to which these facts support or contradict a postulated model, and to suggest particular revisions of the existing theory or, perhaps, to plan further investigations.

The different areas of statistics mentioned above are not disjoint entities intended for use one at a time in individual stages of an investigation. Rather, they are integrated into an intertwined system of activities where the methods used in one area may have strong bearing upon those used in others. To decide upon the process and extent of data to be collected, one must have a perception of the inferential procedures contemplated for use and the strength of the inferences desired. On the other hand, methods of analyzing data and drawing conclusions are heavily contingent upon the process by which the data were generated.

1.5 ILLUSTRATIVE SITUATIONS OF THE COLLECTION AND ANALYSIS OF DATA

To clarify the preceding generalities, a few examples are given here. They illustrate some typical situations in which the cognitive process of investigating a phenomenon involves the collection and analysis of data in which statistical methods are consequently indispensable learning aids.

Plant breeding: Experiments involving the cross-fertilization of different genetic types of plant species to produce high-yield hybrids are of considerable interest to agricultural scientists. As a simple example, suppose that the yields of two hybrid varieties are to be compared under specific climate conditions. The only way to learn about the relative performance of these two varieties is to grow them at a number of sites, collect data on their yields, and then analyze the data.

Clinical diagnosis: Early detection is of paramount importance for the successful surgical treatment of many types of cancer. Because frequent in-hospital checkups are expensive and inconvenient, doctors are searching for effective diagnostic processes that patients can administer themselves. To determine the merits of a new process in terms of its rate of success in detecting true cases and avoiding false detections, the process must be field tested on a large number of persons, who must then undergo in-hospital checkups for comparison.

Training programs: Training or teaching programs in many fields, designed for a specific type of clientele (college students, industrial workers, minority groups, physically handicapped people, retarded children, etc.) are continually monitored, evaluated, and modified to improve their usefulness to society. To learn about the comparative effectiveness of different programs, it is essential to collect data on the achievement or growth of skill of subjects at the completion of each program.

Animal migration: Biologists study the migratory habits of birds and animals by tagging them with identification numbers in relevant geographical locations and subsequently tracing them to other locations. The data obtained by such methods not only help us to understand the animal world but they also alert conservationists to situations that require action to protect endangered species.

Socioeconomic surveys: In the interdisciplinary areas of sociology, economics, and political science, studies are undertaken on such aspects as the economic well-being of different ethnic groups, consumer expenditure patterns at different income levels, and attitudes toward pending legislation. Such studies are typically based on data obtained by interviewing or contacting a representative sample of persons selected by a statistical process from the enormous population that forms the domain of study. The data are then analyzed and interpretations of the issue in question are made.

1.6 POPULATION AND SAMPLE

Although the foregoing examples are drawn from widely differing fields and only sketchy descriptions of the scope and objectives of the studies are provided, a few common characteristics are readily discernible.

First, the most apparent feature underlying all these areas of study is the fact that the collection of data by appropriate processes of experimentation or observation is essential to acquire new knowledge. Second, some amount of variability in the outcomes is unavoidable in spite of the fact that the same or closely similar conditions prevailed during repetitions of each experiment or observation. For instance, in the plant breeding example, it is unrealistic to expect each plant of a particular variety to have exactly the same yield, because nature does not follow such a rigid law. Likewise, a training program for individuals with similar backgrounds produces variability in measurements of achievement. The presence of some inherent variation in the outcomes under constant experimental conditions tends to obscure the effect of a change in these conditions. An important ingredient of the statistical analysis of data is the formulation of appropriate models that represent the inherent variability found in nature.

A third notable feature of the examples in Section 1.5 is the fact that it is either physically impossible or practically unfeasible to collect and study an exhaustive set of data pertaining to a specific area of investigation. When data are obtained from laboratory experiments or field trials, no matter how much experimentation has been performed, more can always be conducted. In public opinion or consumer expenditure studies, a complete body of information would emerge only if data were gathered from every individual in the nation. For example, to collect an exhaustive set of data related to the damage sustained by all cars of a particular model and year under collision at a specified speed, every car of that model coming off the production lines would have to be subjected to a collision! The complete set of observations that could be collected by making unlimited repetitions of an experiment or by keeping an exhaustive record of all elements within the purview of the study is so vast that we can at best visualize it in our imaginations. Such a vast set of data may be regarded as the source of complete information, but the limitations of time, resources, and facilities, and sometimes the destructive nature of the testing mean that we must work with incomplete information—the data that are actually collected in the course of an experimental study.

The fundamental ideas emanating from our discussion highlight a distinction between the data set that is actually acquired through the process of observation and the vast collection of all potential observations that can be conceived in a given context. The statistical nomenclature for the former is *sample*; for the latter it is *population*, *statistical population*, or *target population*. A general definition of a population must be postponed until several other concepts have been introduced. To emphasize the distinction between sample and population at this stage, we consider situations in which each measurement (or record of a qualitative trait) in a data set originates from a distinct source called a *sampling unit* or, more simply, a *unit*. These sources can be trees, animals, farms, families, or other elements, depending on the domain of study. The sample data then consist of measurements corresponding to a collection of units that are included in an actual experiment. This collection forms a part of a far larger collection of units about which we wish to make inferences. The set of measurements that would result if all the units in the larger collection could be observed is defined as the population.

> A (statistical) *population* is the complete set of possible measurements or the record of some qualitative trait corresponding to the entire collection of units for which inferences are to be made. The population represents the target of an investigation, and the objective of the process of data collection is to draw conclusions about the population.

> *A sample* from a statistical population is the set of measurements that are actually collected in the course of an investigation.

Some further elaborations should clarify the differences between the concepts of population and sample. It is important to note that in contrast to its ordinary usage, the term "population" in statistics does not imply a collection of living beings. A statistical population is a collection of numbers that represent the totality of measurements of some characteristic for the entire group of units that is the target of an investigation. The characteristic may or may not be associated with a human population. In studying the yield of a particular type of plant under specific climatic conditions, the statistical population of yields is the collection of all yield measurements that we can only imagine would be compiled if the plant were extensively cultivated at all geographical locations with the particular climatic conditions and this process was repeated over years and years. In this context, the statistical population has nothing at all to do with any human population. Moreover, we wish to learn about the abstract concept of the totality of yield measurements. A sample is a part of this infinite population, or the set of yield measurements actually recorded in the course of an experiment that result from growing a number of these plants at a few locations with the given climatic conditions. Obviously, the sample data will vary when this experiment is repeated on different occasions, whereas the population (even though it does not exist in reality) is regarded as a stable body of numbers. In other contexts, a population may consist of a concrete set of numbers, even though the set may be immensely large and unrecorded. The yearly maintenance expenditures on cars for all families in the United States during 1977 is an example of this type of population.

From our perception of a statistical population as the compendium of all potential observations on some facet of nature, the process of experimental investigation can be regarded as an effort to gain an understanding of the population on the basis of incomplete information gathered by sampling. The subject of statistics provides the methodology to make inductive inferences about the population at large from the collection and analysis of sample data. These methods enable one to derive plausible generalizations and then to assess the extent of uncertainty underlying these generalizations. Statistical concepts are also essential during the planning stage of an investigation when decisions as to the mode and extent of the sampling process must be made, so that adequately informative data can be generated within the limitations of the available resources.

> The major objectives of statistics are (*a*) to make *inferences* about a population from an analysis of information contained in sample data, and (*b*) to make assessments of the extent of uncertainty involved in these inferences. A third objective, no less important, is to *design the process and the extent of sampling* so that the observations form a basis for drawing valid and accurate inferences.

The design of the sampling process is often the most important step, especially in controlled experiments in which various factors that influence measurement can be preplanned. A good design for the process of data collection permits a straightforward analysis and efficient inferences to be made, whereas sophisticated methods of data analysis in themselves do not salvage much information from the data produced by a poorly planned experiment.

1.7 STATISTICS INTERACTING WITH OTHER FIELDS

The early use of statistics in the stereotypic compilation and passive presentation of data has been largely superseded by the modern role of providing analytical tools with which data can be efficiently gathered, understood, and interpreted. Statistical concepts and methods permit valid conclusions about the population to be drawn from the sample. Given its extended goal, the subject of statistics has penetrated all fields of human endeavor in which the substantiation of assertions and the ramification of information must be grounded in data-based evidence.

The few brief examples given in Section 1.5 are not intended to demarcate the sphere of statistical application but are presented to illustrate the diversity of statistical application. The use of statistical methods in various areas of the humanities, science, and engineering has produced many interactive subjects, such as biostatistics, psychometry, engineering statistics, business statistics, econometrics, and demography. In many other areas in which compound names have not yet emerged, such as political science, meteorology, forestry, and ecology, the subject of statistics already plays a predominant role.

The basic concepts and the core of statistical methodology are almost identical in all the diverse areas of statistical application. Differences in emphasis do arise because certain techniques are more useful in one area than in another. However, because of strong methodological similarities, examples drawn from a wide range of statistical applications are helpful in developing a basic understanding of various statistical methods, their potential uses, and their vulnerabilities to misuse.

REFERENCES

1. *Careers in Statistics*. American Statistical Association. (A copy may be obtained by writing to: American Statistical Association; 806 15th Street, N.W.; Washington, D.C., 20005.)

2. *Statistics: A Guide to the Unknown*. Tanur, J. (ed.) San Francisco; Holden-Day, Inc., 1972.

CHAPTER 2

Descriptive Study of Data

2.1 INTRODUCTION

The process of data collection may involve such diverse activities as laboratory experiments, field trials, public opinion surveys, and the examination of historical records. A geologist makes several drillings to obtain measurements with which to study the oil reserve in a rock structure. An engineer may investigate sound distortion in mass-produced amplifiers. A demographer interested in the age structure of the population in the midwest may consult census records related to a survey of several thousand people. A researcher in child nutrition may provide an experimental diet for 50 undernourished children over a period of time and record their weight gains. Whatever the collection process may be, the resulting data set usually consists of numerical measurements that can range in complexity from a few figures to hundreds or even thousands of numbers. Chapter 2 deals with the methods of data summarization and the study of its basic features by means of graphical presentation as well as the calculation of numerical measures.

The measurements recorded in a data set are the basic pieces of information nature conveys to the investigator. Confronted with the complexities of a large number of measurements, however, the mind cannot readily grasp the overall content of the information recorded in the data set. By simply looking at a tabulated set of 200 measurements, for instance, we are unable to form mental images of the value that the measurements tend to cluster around, the pattern of clustering, or the extent of variability. A summary highlighting such salient features would help us to understand a data set and to avoid the chaotic effect created by presenting the complete mass of 200 numbers.

The summarization and exposition of the important aspects of a data set is commonly called *descriptive statistics*. This area includes the condensation of data in the form of tables, their graphical presentation, and computations of numerical indicators of center and variability. These methods are versatile and can be applied to both situations in which the data set is obtained by sampling a small fraction of the population and situations in which the data set almost completely enumerates the population—(in a census, for example). In the former case, a summary description is usually followed by closer examination and further analysis of the data, so that inferences on the population at large can be made. Powerful techniques for making statistical inferences and discussion of associated concepts form the core of the remaining chapters in this book. With a census, the factual findings are often presented in reports for dissemination to the public and also may become part of the data base used in assessing the effect of proposed government policies. Typical questions might concern the amount of timber available for making lumber or the number of new people to be covered by some federal insurance if the age limits are changed.

As in the case of summarization and commentary on a long, wordy document, it is difficult to present concrete steps for summary description that work well for all types of data. However, a few important aspects that deserve special attention are outlined below to provide general guidelines for this process. Each data set encountered should be examined in light of these points.

Main aspects of describing a data set

(*a*) Summarization and description of the overall pattern of the data by:

- (*i*) Presentation of tables and graphs.
- (*ii*) Examination of the overall shape of the graphed data for important features, including symmetry or departures from it.
- (*iii*) Scanning the graphed data for any unusual observations which seem to stick far out from the major mass of the data.

(*b*) Computation of numerical measures for:

- (*i*) A typical or representative value that indicates the center of the data.
- (*ii*) The amount of spread or variation present in the data.

2.2 DESCRIPTION OF DATA BY GRAPHS AND TABLES

The two main graphical methods used to represent a data set are the *dot diagram* and the *histogram*. Dot diagrams are employed when there are relatively few observations, (say, less than 20 or 25); histograms are used for larger numbers.

2.2.1 Dot Diagram

When the data consist of a small set of numbers, they can be graphically represented by drawing a line with a scale covering the range of values of the measurements and plotting the individual measurements on this line as prominent dots. The resulting diagram is called a *dot diagram*.

Example 2.1 To study the possible effects of a new noise pollution ordinance, 18 power lawn mowers were observed and their noise levels were recorded to the nearest decibel. The following data were obtained:

95	120	117	99	110	107	125	98	85
127	105	114	103	112	92	101	122	120

These measurements extend from the smallest value of 85 to the largest value of 127. Drawing a line segment with a scale from 80 to 130, we can plot the data as shown in the dot diagram in Figure 2.1. This dot diagram exhibits a fairly even distribution of points with a center between 100 and 110. ■

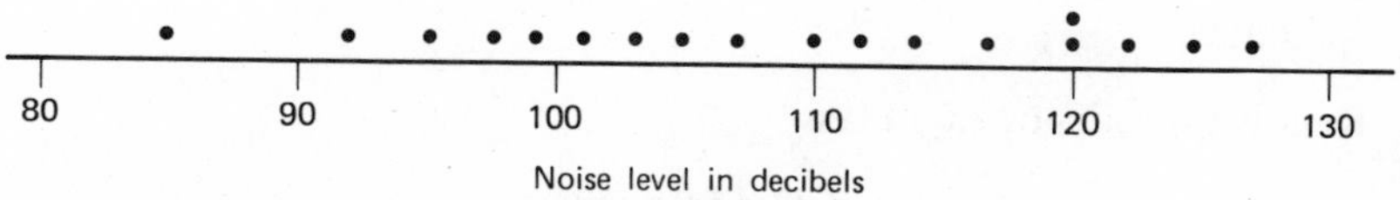

Figure 2.1 Dot diagram of noise-level data given in Example 2.1.

Example 2.2 Waiting to cross a busy street on the way to class one morning, Professor J. noted the following times in seconds between cars travelling in the same direction: 6, 3, 5, 6, 4, 3, 5, 4, 6, 3, 4, 5, 4, 18. The dot diagram in Figure 2.2 reveals that the times are closely grouped around a central value of 4 or 5, except for 18, which substantially deviates from the other times. When Professor J. examined the circumstances that led to this observation, he realized that a stop light a block away had caused the wide deviation of 18 seconds. Unusual observations should always be investigated to see if they can be explained and to determine whether they should be included in any further analyses. ■

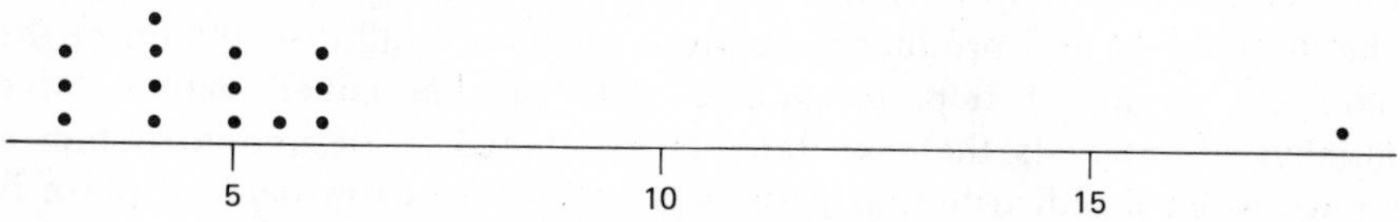

Figure 2.2 Dot diagram of the data given in Example 2.2.

2.2.2 Frequency Distribution

When the data consist of a large number of measurements, plotting a dot diagram can be laborious. Moreover, overcrowding the dots can result in a diagram that obscures details around areas where observations are heavily concentrated. In such cases, it is convenient to condense the data by grouping nearby observations together and constructing a *frequency table* as illustrated in Example 2.3. The main steps in this process are outlined below.

Constructing a frequency distribution

(*a*) Find the minimum and the maximum value in the data set.

(*b*) Choose a number of subintervals or cells of equal length that cover the range between the minimum and the maximum without overlapping. These are called *class intervals*, and their end points are known as *class boundaries*.

(*c*) Count the number of observations in the data that belong to each class interval. The count in each class is the *class frequency* or *cell frequency*.

(*d*) Determine the *relative frequency* of each class by dividing the class frequency by the total number of observations in the data:

Relative frequency of a class

$$= \frac{\text{Class frequency}}{\text{Total number of observations}}$$

Thus the relative frequency of a class is the fraction of the observations belonging to that class.

The choice of the number and position of the class intervals is primarily a matter of trial and error. The number of classes usually ranges from 5 to 15, depending on the number of observations in the data. Grouping observations into cells sacrifices information concerning how the observations are distributed within each cell. If there are too few cells, the loss of information is serious. If there are too many cells and the data set is small, the frequencies from one cell to the next tend to jump up and down in a chaotic manner and produce no pattern for the overall distribution of the data. As an initial step, frequencies may be determined with a large number of intervals that can later be combined as desired to obtain a choice when the distribution pattern is visible. An extra decimal place is often used to define the class boundaries so that no observation can fall exactly on the boundary and cause ambiguity in classification.

Counting frequencies in the cells is often facilitated by placing a tally mark in the appropriate cell for each observation in the data and then counting the number of tally marks in each cell. Alternately, modern computers can readily order the data from smallest to largest so that they can be counted. The construction of a frequency distribution is illustrated in Example 2.3.

Example 2.3 One of the major indicators of air pollution in large cities and industrial belts is the concentration of ozone in the atmosphere. From massive data collected by Los Angeles County authorities, 78 measurements of ozone concentration in the downtown Los Angeles area during the summers of 1966 and 1967 are recorded in Table 2.1. Each measurement is an average of hourly readings taken every fourth day.

TABLE 2.1

78 Measurements of Ozone Concentration (in Parts per Hundred Million) in the Downtown Los Angeles Atmosphere During the Summers of 1966 and 1967 (Courtesy of G. Tiao)

3.5	1.4	6.6	6.0	4.2	4.4	5.3	5.6
6.8	2.5	5.4	4.4	5.4	4.7	3.5	4.0
2.4	3.0	5.6	4.7	6.5	3.0	4.1	3.4
6.8	1.7	5.3	4.7	7.4	6.0	6.7	11.7
5.5	1.1	5.1	5.6	5.5	1.4	3.9	6.6
6.2	7.5	6.2	6.0	5.8	2.8	6.1	4.1
5.7	5.8	3.1	5.8	1.6	2.5	8.1	6.6
9.4	3.4	5.8	7.6	1.4	3.7	2.0	3.7
6.8	3.1	4.7	3.8	5.9	3.3	6.2	7.6
6.6	4.4	5.7	4.5	3.7	9.4		

To construct a frequency distribution, we first scan the data and note that the minimum and the maximum readings are 1.1 and 11.7, respectively. Our initial step is to make an elaborate frequency distribution by choosing class intervals of length .5 (see Table 2.2). Since the data are recorded to the first decimal place, the class boundaries are specified with a second decimal figure to avoid the possibility of any reading falling exactly on the boundary. The relative frequency of each class is then calculated by dividing its frequency by the total frequency. For example, the first class interval of 1.05–1.55 has a frequency of 4, so that its relative frequency is $\frac{4}{78}$ or approximately .051.

With a fine partition, as in Table 2.2, the frequencies seem to fluctuate erratically and no overall pattern of distribution is apparent. By combining classes, we successively obtain the frequency distributions with class intervals of 1.0 and 2.0 and record them in Tables 2.3 and 2.4, respectively. Table 2.4 exhibits a fairly conspicuous pattern of frequencies; therefore the choice of any coarser interval is undesirable.

TABLE 2.2

FREQUENCY DISTRIBUTION OF THE DATA IN TABLE 2.1 WITH CLASS INTERVALS OF .5

CLASS INTERVAL	TALLY	FREQUENCY	RELATIVE FREQUENCY
1.05– 1.55	||||	4	.051
1.55– 2.05	|||	3	.038
2.05– 2.55	|||	3	.038
2.55– 3.05	|||	3	.038
3.05– 3.55	~~||||~~ ||	7	.090
3.55– 4.05	~~||||~~ |	6	.077
4.05– 4.55	~~||||~~ ||	7	.090
4.55– 5.05	||||	4	.051
5.05– 5.55	~~||||~~ ||	7	.090
5.55– 6.05	~~||||~~ ~~||||~~ |||	13	.167
6.05– 6.55	~~||||~~	5	.064
6.55– 7.05	~~||||~~ |||	8	.103
7.05– 7.55	||	2	.026
7.55– 8.05	||	2	.026
8.05– 8.55	|	1	.013
8.55– 9.05		0	
9.05– 9.55	||	2	.026
9.55–10.05		0	
10.05–10.55		0	
10.55–11.05		0	
11.05–11.55		0	
11.55–12.05	|	1	.013
Total		78	1.001

TABLE 2.3

FREQUENCY DISTRIBUTION OF THE DATA IN TABLE 2.1 WITH CLASS INTERVALS OF 1.0

CLASS INTERVAL	FREQUENCY	RELATIVE FREQUENCY
1.05– 2.05	7	.090
2.05– 3.05	6	.077
3.05– 4.05	13	.167
4.05– 5.05	11	.141
5.05– 6.05	20	.256
6.05– 7.05	13	.167
7.05– 8.05	4	.051
8.05– 9.05	1	.013
9.05–10.05	2	.026
10.05–11.05	0	
11.05–12.05	1	.013
Total	78	1.001

TABLE 2.4
FREQUENCY DISTRIBUTION OF THE DATA IN TABLE 2.1 WITH CLASS INTERVALS OF 2.0

CLASS INTERVAL	FREQUENCY	RELATIVE FREQUENCY
0.05– 2.05	7	.090
2.05– 4.05	19	.244
4.05– 6.05	31	.397
6.05– 8.05	17	.218
8.05–10.05	3	.038
10.05–12.05	1	.013
TOTAL	78	1.000

In a frequency distribution, the relative frequencies must always sum to 1. The totals 1.001 in Tables 2.2 and 2.3, respectively, result from rounding off the relative frequencies to three decimals. ■

2.2.3 Relative Frequency Histogram

After summarizing a large data set in the form of a frequency distribution, it can be graphically presented in a *relative frequency histogram*, which is a visual representation of the distribution pattern.

> *Relative frequency histogram*
>
> To draw a relative frequency histogram, the class intervals are marked on the horizontal axis of a graph. On each interval a vertical rectangle is then drawn whose *area* is equal to the relative frequency of that interval.
>
> $$\text{Height of a rectangle} = \frac{\text{Relative frequency of a class}}{\text{Width of the class interval}}$$
>
> This is because the area of a rectangle = height × width.

The area of each rectangle in a histogram represents the proportion of the observations occurring in the class interval on which the rectangle stands. Therefore, the total area of all the rectangles in a histogram is 1. The convention of using the area of the rectangles rather than their heights to represent relative frequencies has a distinct advantage: The eye seems to instinctively compare areas when comparing two parts of a histogram or two different histograms. When two histograms are based on class intervals of different widths, the property of having a total area equal to 1 makes them comparable.

Example 2.4 Referring to the ozone concentration data in Table 2.1, the histograms corresponding to the frequency distributions in Tables 2.2, 2.3, and 2.4 are plotted in Figure 2.3*a*, *b*, and *c*. In each case note that the height of a rectangle is obtained by dividing the relative frequency by the width of the class interval. For instance, the heights of the first rectangles in Figures 2.3*a*, *b*, and *c* are $.051/.5=.102$, $.09/1=.09$, and $.09/2=.045$, respectively. The histogram in Figure 2.3*a* appears ragged because the class intervals are too short. Wider intervals make the histograms in Figure 2.3*b* and *c* more regular and smoother in shape, although more information is lost when wider class intervals are used. ■

The rule requiring equal class intervals is inconvenient when the data are spread over a wide range but are highly concentrated in a small part of the range with relatively few numbers elsewhere. Using smaller intervals where the data are highly concentrated and larger intervals where the data are sparse helps to reduce the loss of information due to grouping. Tabulations of income, age, and other characteristics in official reports are often made

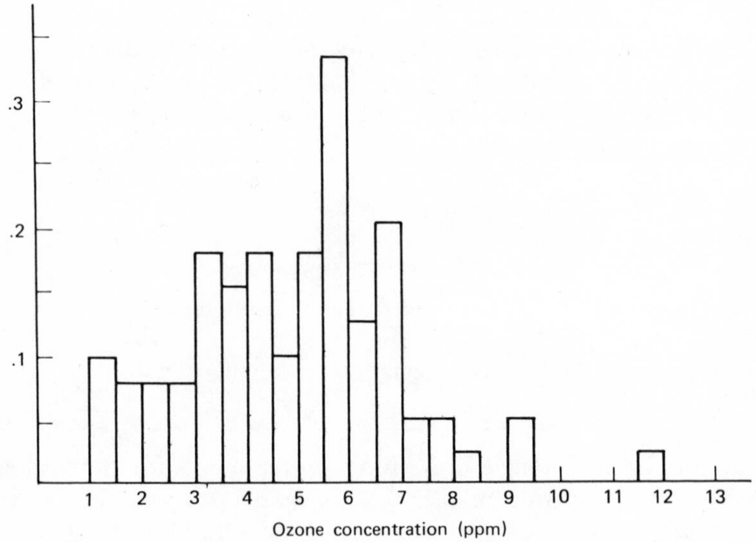

Figure 2.3*a* Relative frequency histogram for the data given in Table 2.2

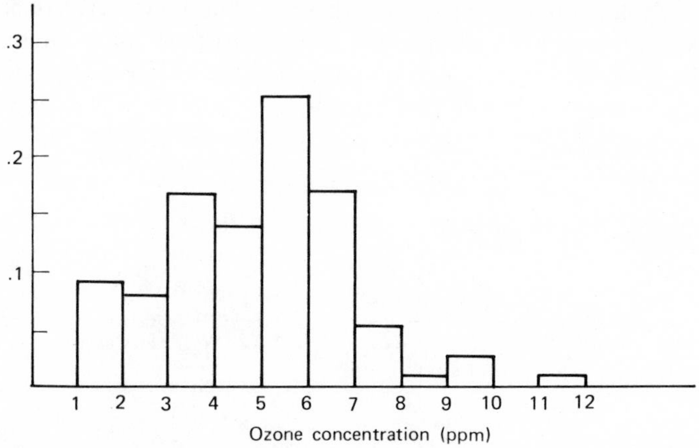

Figure 2.3*b* Relative frequency histogram for the data given in Table 2.3.

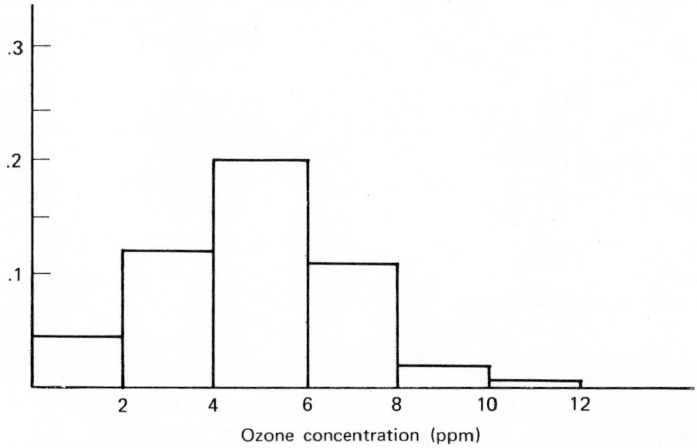

Figure 2.3*c* Relative frequency histogram for the data given in Table 2.4.

with unequal class intervals. When the intervals are not all equal, the histogram must be plotted according to the convention of using the area of the rectangles to represent the relative frequencies to convey the correct pattern of distribution.

Example 2.5 A survey conducted by the U.S. Census Bureau of the 1970 family incomes of 2,643,226 white families in rural farm areas provided the frequency distribution shown in Table 2.5. This frequency table not only has unequal class intervals but the last interval is open-ended. Because the relative frequency in this class is quite small, when plotting the histogram we approximate this interval by terminating it to the same length as the one before it. The histogram is presented in Figure 2.4. ■

TABLE 2.5
FREQUENCY DISTRIBUTION OF INCOME OF 2,643,226 WHITE FAMILIES IN RURAL FARM AREAS IN 1970

FAMILY INCOME (IN THOUSAND DOLLARS)	RELATIVE FREQUENCY (%)
0– 1	4.0
1– 2	5.4
2– 3	6.8
3– 4	7.3
4– 5	7.2
5– 6	8.0
6– 7	7.4
7– 8	7.2
8– 9	7.0
9–10	5.9
10–15	19.9
15–25	10.3
over 25	3.5
Total	99.9

SOURCE: U.S. Bureau of Census Report

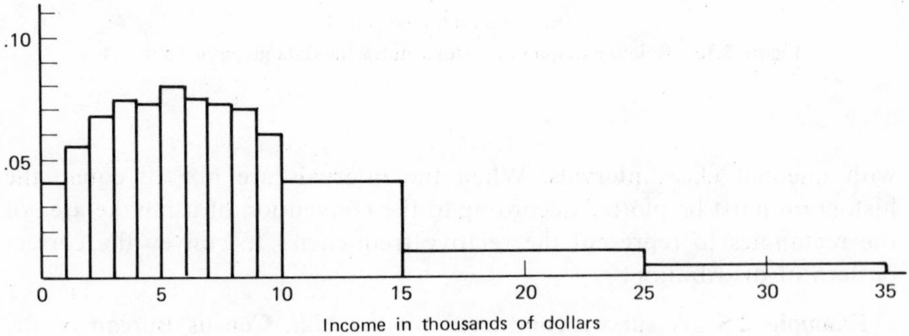

Figure 2.4 Relative frequency histogram for the data given in Table 2.5

2.2.4 Relative Frequency Line Diagram

Sometimes the data consist of counts, such as the number of children in families or the number of traffic fatalities per day, rather than measurements on a continuous scale. If the number of distinct values in such a data set is not too large, a frequency distribution is constructed using the individual values as the classes instead of using class intervals. The data are then presented in the form of a relative frequency line diagram.

Relative frequency line diagram

The distinct values are located on the horizontal axis. Vertical lines with heights equal to the relative frequencies are then drawn at these values.

Lines replace the rectangles to emphasize that the frequencies are not really spread over intervals. A histogram could also be constructed by drawing rectangles centered on the distinct values in the data, provided the relative frequencies are considered to be those of the midpoints. Although both line and rectangular diagrams are used for count data, the line diagram must never be drawn for measurements on a continuous scale.

Example 2.6 A survey of the number of children in 25 households provided the frequency distribution shown in Table 2.6. The line diagram of this frequency distribution appears in Figure 2.5, where the lengths of the vertical lines represent the relative frequencies. ■

TABLE 2.6
FREQUENCY TABLE FOR THE NUMBER OF CHILDREN LIVING IN 25 HOUSEHOLDS

NUMBER OF CHILDREN	FREQUENCY	RELATIVE FREQUENCY
0	1	1/25 = .04
1	4	4/25 = .16
2	10	10/25 = .40
3	6	6/25 = .24
4	2	2/25 = .08
5	2	2/25 = .08
TOTAL	25	1.00

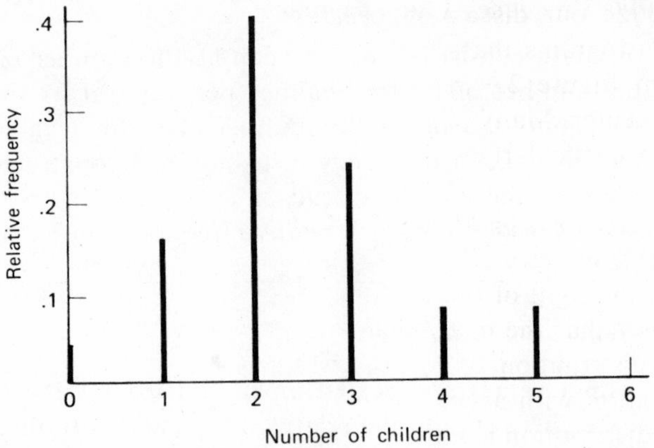

Figure 2.5 Relative frequency line diagram for the data given in Table 2.6

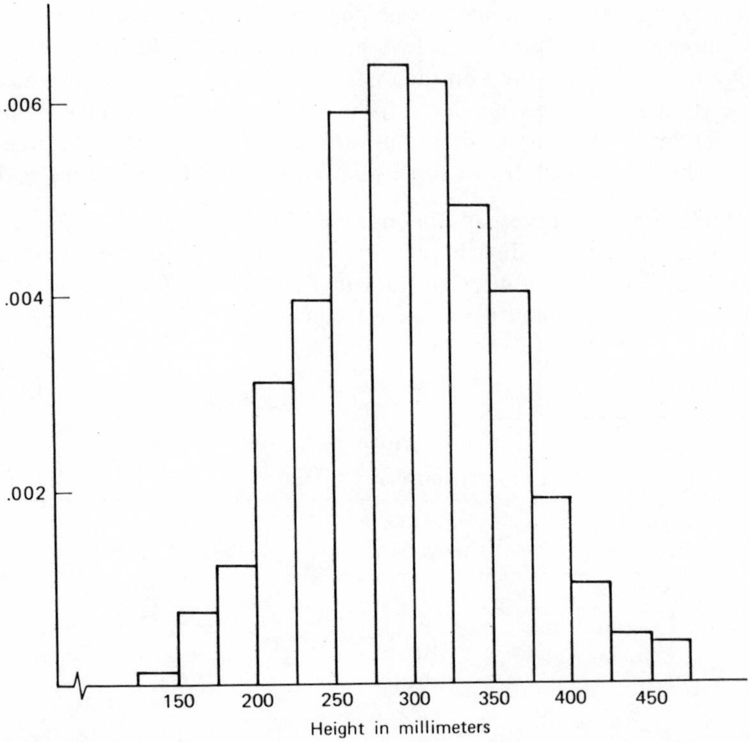

Figure 2.6 Relative frequency histogram of 1465 measurements of 3-year-old red pine (courtesy of A. Ek).

We conclude our discussion of graphic techniques with three relative frequency histograms that illustrate different shapes of distributions. The histogram in Figure 2.6 exhibits a fairly symmetric shape with a single peak at the center. Many standard statistical procedures developed in later chapters are particularly suited to these single-peak symmetric distributions.

Figure 2.7 shows a histogram with a single peak that does not have the property of symmetry. The heights of the rectangles taper off more gradually to the right of the peak than they do to the left. To describe this feature we say that the distribution is "asymmetric with a long right tail," or that the distribution is "*skewed* to the right." It is easy to visualize a parallel situation with asymmetry present in the form of a long left-hand tail. Such a distribution is said to be "skewed to the left."

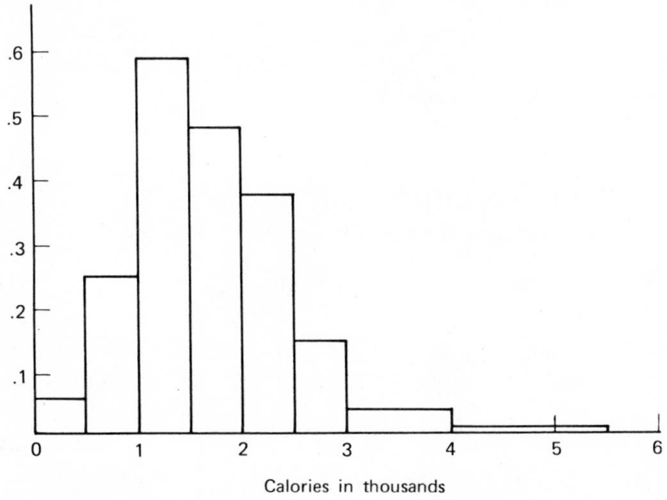

Figure 2.7 Relative frequency histogram of the caloric intake of 204 15 and 16-year-old females in states with low-income ratios (SOURCE: Ten-State Nutrition Survey, 1968–70).

Figure 2.8 is much more complex in shape, because its distribution is not only skewed to the right but it also has two peaks. For a descriptive study, we must identify the locations of the peaks and consider possible explanations for the presence of multiple peaks. Later we will find that greater care must be taken in statistically analyzing such data. Fortunately, distributions with multiple peaks are seldom encountered in statistical practice.

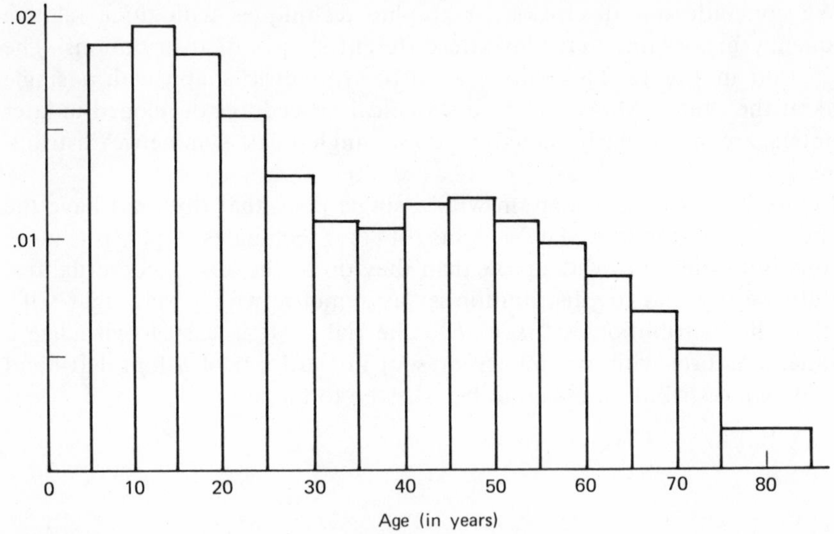

Figure 2.8 Relative frequency histogram of the age distribution of whites in the United States in 1970 (SOURCE: U.S. Bureau of the Census).

2.3 SYMBOLS FOR DATA SETS AND THE OPERATION OF ADDITION

The graphic procedures described in Section 2.2 help us to visualize the pattern of a data set. To obtain a more objective summary description and a comparison of data sets, we must go one step further and formulate quantitative measures for important aspects such as the location of center of the data and the amount of variability present in the data. Numerical measures of these aspects are introduced in the following sections. Because data are normally obtained by sampling from a large population, our discussion of numerical measures is restricted to data arising in this context. Moreover, when the population is finite and is completely sampled, the same arithmetic operations can be carried out to obtain numerical measures for the population.

To effectively present the ideas and associated formulas, it is convenient to represent a data set by symbols to prevent the discussion from becoming anchored to a specific set of numbers. A data set consists of a number of measurements symbolically represented by $x_1, x_2, \ldots, x_n$. The last subscript n denotes the number of measurements in the data, and $x_1, x_2, \ldots$ represent the first observation, the second observation, and so on. For instance, a data set consisting of the five measurements 2.1, 3.2, 4.1, 5.6, and 3.7 is represented in symbols by x_1, x_2, x_3, x_4, x_5, where $x_1 = 2.1$, $x_2 = 3.2$, $x_3 = 4.1$,

$x_4=5.6$, and $x_5=3.7$. Throughout our study of statistics, we will be adding the measurements in a data set or some other numbers computed from these measurements. To avoid a detailed and repeated writing of this operation, the symbol Σ (the Greek capital letter *sigma*) is used as mathematical shorthand for the operation of addition.

Summation notation Σ

The notation $\Sigma_{i=1}^{n} x_i$ represents the sum of n numbers $x_1, x_2, \ldots, x_n$ and is read as *the sum of all* x_i *with i ranging from 1 to* n, or

$$\sum_{i=1}^{n} x_i = x_1 + x_2 + \cdots + x_n$$

The term following the sign Σ indicates the quantities that are being summed, and the notations on the bottom and the top of the Σ specify the range of the terms being added. For instance

$$\sum_{i=1}^{3} x_i = x_1 + x_2 + x_3$$

$$\sum_{i=1}^{4} (x_i - 2) = (x_1 - 2) + (x_2 - 2) + (x_3 - 2) + (x_4 - 2)$$

Example 2.7 Suppose that the four measurements in a data set are given as $x_1=3$, $x_2=5$, $x_3=4$, $x_4=3$. Compute the numerical values of (*a*) $\sum_{i=1}^{4} x_i$, (*b*) $\sum_{i=1}^{4} 3x_i$, (*c*) $\sum_{i=1}^{4} (x_i - 2)$, (*d*) $\sum_{i=1}^{4} x_i^2$, (*e*) $\sum_{i=1}^{4} (x_i - 2)$.

(*a*) $\displaystyle\sum_{i=1}^{4} x_i = x_1 + x_2 + x_3 + x_4 = 3 + 5 + 4 + 3 = 15$

(*b*) $\displaystyle\sum_{i=1}^{4} 3x_i = 3x_1 + 3x_2 + 3x_3 + 3x_4 = 3\left(\sum_{i=1}^{4} x_i\right) = 3 \times 15 = 45$

(*c*) $\displaystyle\sum_{i=1}^{4} (x_i - 2) = (x_1 - 2) + (x_2 - 2) + (x_3 - 2) + (x_4 - 2)$

$$= \sum_{i=1}^{4} x_i - 4(2) = 15 - 8 = 7$$

$$(d)\ \sum_{i=1}^{4} x_i^2 = x_1^2 + x_2^2 + x_3^2 + x_4^2 = 3^2 + 5^2 + 4^2 + 3^2 = 59$$

$$(e)\ \sum_{i=1}^{4} (x_i - 2)^2 = (x_1 - 2)^2 + (x_2 - 2)^2 + (x_3 - 2)^2 + (x_4 - 2)^2$$
$$= (3-2)^2 + (5-2)^2 + (4-2)^2 + (3-2)^2$$
$$= 1 + 9 + 4 + 1 = 15$$

Alternatively, noting that $(x_i - 2)^2 = x_i^2 - 4x_i + 4$, we can write

$$\sum_{i=1}^{4} (x_i - 2)^2 = \sum_{i=1}^{4} (x_i^2 - 4x_i + 4)$$

$$= (x_1^2 - 4x_1 + 4) + (x_2^2 - 4x_2 + 4) + (x_3^2 - 4x_3 + 4) + (x_4^2 - 4x_4 + 4)$$

$$= \sum_{i=1}^{4} x_i^2 - 4\left(\sum_{i=1}^{4} x_i\right) + 4(4)$$

$$= 59 - 4(15) + 16 = 15$$ ∎

Here we list a few basic properties of the summation symbol for later use. These properties should be apparent from the numerical demonstration in Example 2.7.

Some basic properties of summation

If a and b are fixed numbers

$$\sum_{i=1}^{n} bx_i = b \sum_{i=1}^{n} x_i$$

$$\sum_{i=1}^{n} (bx_i + a) = b \sum_{i=1}^{n} x_i + na$$

$$\sum_{i=1}^{n} (x_i - a)^2 = \sum_{i=1}^{n} x_i^2 - 2a \sum_{i=1}^{n} x_i + na^2$$

2.4 MEASURES OF CENTER

Perhaps the most important aspect of studying the distribution of a sample of measurements is the position of a central value, that is, a representative value about which the measurements are distributed. Any numerical measure intended to represent the center of a data set is called a *measure of location* or *central tendency*. The two most commonly used measures of center are the *mean* and the *median*.

The *sample mean* or average of a set of n measurements $x_1, x_2, \ldots, x_n$ is the sum of these measurements divided by n. The mean is denoted by $\bar{x}$, which is expressed operationally

$$\bar{x} = \frac{\sum_{i=1}^{n} x_i}{n}$$

According to the concept of "average," the mean represents a center of a data set. If we picture the dot diagram of a data set as a thin horizontal bar on which balls of equal size are placed at the positions of the data points, then the mean $\bar{x}$ represents the point on which the bar will balance. The computation of the sample mean and its geometrical interpretation are illustrated in Example 2.8.

Example 2.8 The birth weights in pounds of five babies born in a hospital on a certain day are 9.2, 6.4, 10.5, 8.1, 7.8. The mean birth weight for these data is

$$\bar{x} = \frac{9.2 + 6.4 + 10.5 + 8.1 + 7.8}{5} = \frac{42.0}{5} = 8.4 \text{ pounds}$$

The dot diagram of the data appears in Figure 2.9, where the sample mean (marked by the arrow) is the balancing point or center of the picture. ■

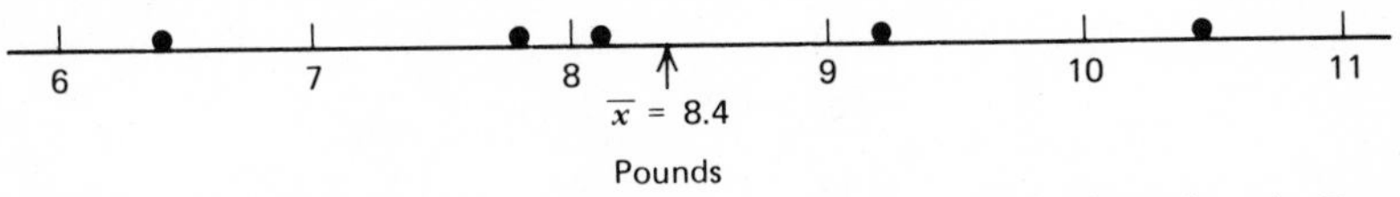

Figure 2.9 Dot diagram and the sample mean for the birth-weight data given in Example 2.8.

> The *sample median* of a set of n measurements $x_1, \ldots, x_n$ is the middle value when the measurements are arranged from smallest to largest. If n is an odd number, there is a unique middle value and it is the median. If n is an even number, there are two middle values and the median is defined as their average.

Roughly speaking, the median is the value that divides the data into two equal halves. In other words, 50% of the data lie below the median and 50% lie above it.

Example 2.9 Find the median of the birth-weight data given in Example 2.8.

The measurements, ordered from smallest to largest, are

$$6.4, \quad 7.8, \quad \boxed{8.1}, \quad 9.2, \quad 10.5$$

The middle value is 8.1, and the median is therefore 8.1 pounds. ■

Example 2.10 The monthly income in dollars for eight members of a real estate firm are 500, 750, 600, 550, 550, 700, 2000, 550. Calculate the mean and the median income.

The sum of the eight monthly incomes is 6200, and the mean is therefore $\bar{x} = 6200/8 = 775$. Thus, the mean monthly income for the group is \$775. Whether this can be considered a "typical" monthly income is questionable, because seven of the eight incomes are below \$775. The single large value of \$2000 drastically inflates the mean.

To find the median, first we order the data. The ordered values are

$$500, \quad 550, \quad 550, \quad \boxed{550, \quad 600}, \quad 700, \quad 750, \quad 2000$$

Note that the two middle values are 550 and 600. Therefore, the median income is

$$\frac{550+600}{2} = \$575$$

Here the median of \$575 appears to be a more sensible measure of the center than the mean. ■

Example 2.10 demonstrates that the median is not affected by a few very small or very large observations, whereas the presence of such extremes will have a significant effect on the mean. For extremely asymmetrical distributions, the median is likely to be a more sensible measure of center than the mean. That is why government reports on income distribution quote the median income as a summary, rather than the mean. When the distribution is not highly asymmetric, the mean is preferred and is more widely used, because the median lacks some theoretical advantages when making inferences of the kind described in later chapters.

If the number of observations is quite large (greater than, say, 25 or 30), it is sometimes useful to extend the notion of the median and to divide the ordered data set into quarters. Just as the point for division into halves is called the median, the points for division into quarters are called *quartiles*. However, rather than confine ourselves to division in quarters, we consider a division into more general fractions and proceed to define the *percentiles*. Note that a fraction p is equivalently expressed as $100p\%$. For instance, $\frac{1}{4}$ is synonymous with 25%.

> The *sample 100pth percentile* is a value such that after the data are ordered from smallest to largest, at least $100p\%$ of the observations are at or to the left of (below) this value *and* at least $100(1-p)\%$ are at or to the right of (above) this value.

The quartiles are simply the 25th, and 75th percentiles.

> *Sample quartiles*
>
> Lower (first) quartile $Q_1 = 25$th percentile
> Second quartile (or median) $Q_2 = 50$th percentile
> Upper (third) quartile $Q_3 = 75$th percentile

We adopt the convention of taking an observed value for the sample percentile except when two adjacent values satisfy the definition, in which case their average is taken as the percentile. This coincides with the way the median is defined when the sample size is even. When all values in an interval satisfy the definition of a percentile, the particular convention used to locate a point in the interval does not appreciably alter the result in large data sets, except perhaps for the determination of extreme percentiles (those before the 5th or after the 95th percentile).

First we illustrate the method of counting for percentiles with an example involving only a few observations. Then we cite an example with a moderately large data set in which the quartiles can be more useful in summarizing data.

Example 2.11 Nine industrial workers were tested for gripping strength and the measurements were 128.3, 106.5, 93.9, 116.6, 152.4, 125.0, 132.1, 105.8, 136.7. Obtain the quartiles for this data.

First we order the data from smallest to largest. The ordered values are

93.9 105.8 106.5 116.6 125.0 128.3 132.1 136.7 152.4

The first quartile is the 25th percentile. From the definition

The number of observations at or below Q_1 is at least $.25 \times 9 = 2.25$

The number of observations at or above Q_1 is at least $.75 \times 9 = 6.75$

In other words, at least three observations must be at or below the first quartile and at least seven observations must be at or above it. The observation 106.5 meets these requirements. A check of the two adjacent points shows that they do not qualify. Counting down three observations from the largest value, we find that 132.1 is the third quartile. The median or 50th percentile is the center value 125.0.

The first quartile 106.5 tells us that any work requiring a gripping strength greater than 106.5 could not be performed by at least 25% of the workers tested. ■

Normally the quartiles are not calculated unless there are at least 25 observations. The next example illustrates this situation.

Example 2.12 The data from 50 measurements of the traffic noise level at an intersection are already ordered from smallest to largest in Table 2.7. Locate the quartiles and also compute the 10th percentile.

TABLE 2.7

MEASUREMENTS OF TRAFFIC NOISE LEVEL IN DECIBELS

52.0	55.9	56.7	59.4	60.2	61.0	62.1	63.8	65.7	67.9
54.4	55.9	56.8	59.4	60.3	61.4	62.6	64.0	66.2	68.2
54.5	56.2	57.2	59.5	60.5	61.7	62.7	64.6	66.8	68.9
55.7	56.4	57.6	59.8	60.6	61.8	63.1	64.8	67.0	69.4
55.8	56.4	58.9	60.0	60.8	62.0	63.6	64.9	67.1	77.1

Courtesy of J. Bollinger.

To determine the first quartile, we must count at least $.25 \times 50 = 12.5$ observations from the smallest measurement and at least $.75 \times 50 = 37.5$ from the largest. We can see that the 13th ordered observation 57.2 has 13 values at or to its left and 38 at or to its right. Consequently, 57.2 is the first quartile. Counting down 13 observations from the largest measurement, we find that 64.6 is the third quartile. The median is $(60.8+61.0)/2 = 60.9$. For the 10th percentile, at least $.10\times 50=5$ points must lie at or to the left and $.90\times 50=45$ points must lie at or to the right. Both the 5th and 6th smallest observations satisfy this condition, so we take their average $(55.8+55.9)/2=55.85$ as the 10th percentile. This means that only 10% of the 50 measurements of noise level were quieter than 55.85 decibels. ■

Before concluding this discussion, we introduce two more descriptive measures of center, the *trimmed mean* and the *Winsorized mean*, which have gained importance in recent years. Recall that the sample mean is sensitive to the existence of a few unusually large or small observations, whereas the median ignores all but one or two center values after the data are ordered. The trimmed and Winsorized means may be viewed as compromises between the sample mean and the median.

Trimmed sample mean

(*a*) *Remove* all observations below the first quartile. *Remove* all observations above the third quartile.
(*b*) Calculate the mean of the remaining observations.

Winsorized sample mean

(*a*) *Replace* each observation below the first quartile with the value of the first quartile. *Replace* each observation above the third quartile with the value of the third quartile. All other observations remain unchanged.
(*b*) Calculate the mean of all the observations thus modified.

Both the trimmed and Winsorized means are undisturbed by the presence of a small fraction of unusual or erroneous observations that are extremely small or large. The first measure "trims off" the observations that lie outside the center half, the second moves these observations over to the corresponding quartile before averaging. Statisticians find that these two measures are almost as good as the mean in symmetric distributions when there are no unusual observations, but both these measures are usually better in the presence of wild values. When a distribution has a single long tail, these provide alternative indexes of a central value to those provided by the mean and the median.

Example 2.13 Calculate the trimmed and Winsorized sample means for the strength data given in Example 2.11.

The ordered data and the quartiles were found to be

93.9	105.8	106.5	116.6	125.0	128.3	132.1	136.7	152.4
		↑ First quartile				↑ Third quartile		

To determine the trimmed mean, we average the five observations between and including the first and third quartiles. Here the trimmed sample mean

$$= \frac{106.5 + 116.6 + 125.0 + 128.3 + 132.1}{5}$$

$$= \frac{608.5}{5} = 121.7$$

To determine the Winsorized mean, we replace the two observations below the first quartile with 106.5 and the two observations above the third quartile with 132.1. We then compute the mean of the nine modified observations. Here the Winsorized sample mean

$$= \frac{106.5 + 106.5 + 106.5 + 116.6 + 125.0 + 128.3 + 132.1 + 132.1 + 132.1}{9}$$

$$= \frac{1085.7}{9} = 120.6$$ ■

2.5 MEASURES OF VARIATION

In addition to locating the center of the data, an important aspect of a descriptive study of data is numerically measuring the extent of variation around the center. Two data sets may exhibit similar positions of center but may be remarkably different with respect to variability. Figure 2.10 shows the dot diagrams of two data sets. Note that the dots in 2.10*b* are more scattered than the dots in 2.10*a*. In this section, some numerical measures are presented that facilitate an objective comparison of the degree of variation in different data sets.

Since the sample mean $\bar{x}$ is a measure of center, the variation of the individual data points about the center is reflected in the differences $(x_1 - \bar{x}), (x_2 - \bar{x}), \ldots, (x_n - \bar{x})$, which we call *deviations from the mean*, or simply *deviations*. To find an indicator of the overall variation, we may consider taking the average of these deviations. However, noting that

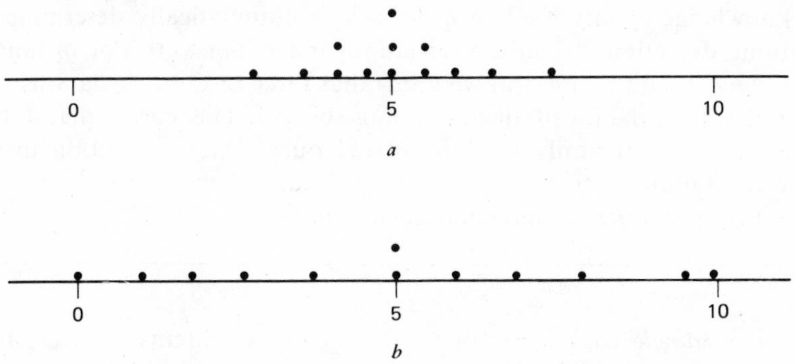

Figure 2.10 Dot diagrams with similar center values but different variations.

$\sum_{i=1}^{n} x_i = n\bar{x}$, we obtain

$$\sum_{i=1}^{n} (x_i - \bar{x}) = \sum_{i=1}^{n} x_i - n\bar{x} = n\bar{x} - n\bar{x} = 0$$

Thus the average deviation is *always* 0. This property results from the fact that some of the deviations are positive, others are negative, and the total of the positive deviations exactly cancels the total of the negative deviations. This canceling property is illustrated in Figure 2.11.

Thus to formulate a numerical measure of variation, we must eliminate the signs of the deviations before averaging them. Squaring a number removes the sign and a measure called the *sample variance* is based on averaging the squared deviations $(x_1 - \bar{x})^2, (x_2 - \bar{x})^2, \ldots, (x_n - \bar{x})^2$. We note

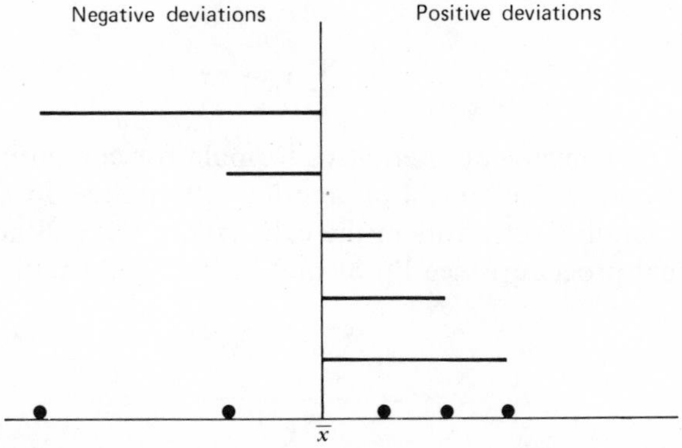

Figure 2.11 The positive deviations $(x - \bar{x})$ cancel the negative deviations.

that knowledge of any $n-1$ of n deviations automatically determines the remaining deviation, because n deviations must sum to 0. For instance, if we have $n=4$ data points and we know that three of these deviations are 5, -4, and 1, then the fourth deviation must be -2. This means that out of n deviations there are only $n-1$ unrelated ones. One convention used to define the sample variance is to divide the sum of the squared deviations by $n-1$, the number of unrelated deviations.

The *sample variance* s^2 of a set of n measurements $x_1, \ldots, x_n$ is defined as

$$s^2 = \frac{\sum_{i=1}^{n} (x_i - \bar{x})^2}{n-1}$$

Squaring each deviation to obtain $(x_i - \bar{x})^2$, summing and then dividing by $n-1$ is the recommended procedure for calculating s^2. Using the properties of the summation sign discussed in Section 2.3, we have

$$\begin{aligned}\sum_{i=1}^{n} (x_i - \bar{x})^2 &= \sum_{i=1}^{n} (x_i^2 - 2\bar{x}x_i + \bar{x}^2)\\ &= \sum_{i=1}^{n} x_i^2 - 2\bar{x}\sum_{i=1}^{n} x_i + n\bar{x}^2\\ &= \sum_{i=1}^{n} x_i^2 - 2n\bar{x}^2 + n\bar{x}^2\\ &= \sum_{i=1}^{n} x_i^2 - n\bar{x}^2\end{aligned}$$

This gives us a convenient alternative formula for computing the variance on an electronic calculator. For another alternative formula that may reduce the rounding-off errors in the calculation, but still not to the extent of the original procedure, see Problem 7 in the Mathematical Exercises.

$$s^2 = \frac{\sum_{i=1}^{n} x_i^2 - n\bar{x}^2}{n-1}$$

Because the variance involves a sum of squares, its unit is the square of the unit in which the measurements are expressed. For example, if the data pertain to measurements of weights in pounds, the variance is expressed in (pound)^2. To obtain a measure of variability in the same unit as the data, we take the square root of the variance, called the *standard deviation*. The standard deviation rather than the variance serves as a basic measure of variability.

Sample standard deviation

$$s = \sqrt{\text{variance}} = \sqrt{\frac{\sum_{i=1}^{n}(x_i - \bar{x})^2}{n-1}}$$

Example 2.14 In a psychological experiment, a stimulating signal of fixed intensity was used on six experimental subjects. Their reaction times, recorded in seconds, were 4, 2, 3, 3, 6, 3. Calculate the mean, variance, and standard deviation for the data.

These calculations can be conveniently carried out in schematic form:

							Total
x	4	2	3	3	6	3	$21 = \Sigma x_i$, $\bar{x} = \frac{21}{6} = 3.5$
$(x - \bar{x})$	.5	−1.5	−.5	−.5	2.5	−.5	0
$(x - \bar{x})^2$	.25	2.25	.25	.25	6.25	.25	$9.50 = \Sigma(x_i - \bar{x})^2$

Sample mean $\bar{x} = 3.5$ seconds

Sample variance $s^2 = \dfrac{\Sigma(x_i - \bar{x})^2}{n-1} = \dfrac{9.50}{5} = 1.9$ (seconds)^2

Sample standard deviation $s = \sqrt{1.9} = 1.4$ seconds

To use the alternative formula for variance, we must compute Σx_i and Σx_i^2 as follows:

							Total
x	4	2	3	3	6	3	$21 = \sum x_i$, $\bar{x} = \dfrac{21}{6} = 3.5$
x^2	16	4	9	9	36	9	$83 = \sum x_i^2$

$$s^2 = \frac{\sum x_i^2 - n\bar{x}^2}{n-1} = \frac{83 - 6(3.5)^2}{5} = \frac{83 - 73.5}{5} = \frac{9.5}{5} = 1.9$$

$$s = \sqrt{1.9} = 1.4$$

■

Unlike the simple interpretation of $\bar{x}$ as the balancing point for the distribution of the set of measurements, a physical interpretation of the standard deviation s is not quite so transparent. In comparing two data sets, a higher value of s in one reflects the presence of greater variation in that data set than in the other. However, in the context of a single data set, the meaning of the numerical value of s in relation to the scatter of points is not so clear. A result due to the Russian mathematician P. Chebyshev provides a connection between the value of s and fractions of the data points located in intervals surrounding the mean. The result is stated below without proof.

Chebyshev's rule

For all data sets:

(*a*) The interval $\bar{x}-2s$ to $\bar{x}+2s$ contains *at least* $\left(1-\frac{1}{2^2}\right)=\frac{3}{4}$ of the data.

(*b*) The inverval $\bar{x}-3s$ to $\bar{x}+3s$ contains *at least* $\left(1-\frac{1}{3^2}\right)=\frac{8}{9}$ of the data.

(*c*) In general, for any multiplier $k>1$ the interval $\bar{x}-ks$ to $\bar{x}+ks$ contains *at least* $\left(1-\frac{1}{k^2}\right)$ fraction of the data.

Chebyshev's rule guarantees the inclusion of a minimum fraction of the data in an interval that is centered at $\bar{x}$ and extends to a specified multiple of s in both directions. In individual cases, of course, the actual fraction of the data included may be considerably larger. The implication of Chebyshev's rule is illustrated in Example 2.15.

Example 2.15 Examine Chebyshev's rule in the context of the ozone-concentration data given in Table 2.1.

Using a computer or a desk calculator to compute $\bar{x}$ and s for such a large data set saves the tedium of performing the mathematical operation outlined in Example 2.14. The computed values are

$$\bar{x}=4.97$$

$$s=2.00$$

To illustrate, we consider three intervals $\bar{x}\pm 1.5s$, $\bar{x}\pm 2s$, and $\bar{x}\pm 3s$.

Chebyshev's rule guarantees that at least

$$1-\frac{1}{(1.5)^2}=.56$$

$$1-\frac{1}{2^2}=\frac{3}{4}=.75$$

$$1-\frac{1}{3^2}=\frac{8}{9}=.89$$

fractions of the set of 78 measurements must belong to these three intervals, respectively. The actual numbers included in these intervals are counted by scanning the data set. For example, the observation 3.5 is contained in $\bar{x} \pm 1.5s$, or the interval 1.97 to 7.97. The results are given in Table 2.8. Note that the observed fractions are considerably larger than the minimums guaranteed by Chebyshev's rule. ■

TABLE 2.8
USING THE OZONE-CONCENTRATION DATA TO ILLUSTRATE CHEBYSHEV'S RULE

	INTERVAL		FRACTION OF THE 78 OBSERVATIONS INCLUDED	
k	$\bar{x}-ks$ TO $\bar{x}+ks$		CHEBYSHEV'S RULE	ACTUALLY OBSERVED
1.5	1.97	7.97	at least .56	$\frac{68}{78}=.87$
2	.97	8.97	at least .75	$\frac{75}{78}=.96$
3	−1.03	10.97	at least .89	$\frac{77}{78}=.99$

Another measure of variation that is sometimes employed is

> *Sample range* = Largest observation − Smallest observation

The range gives the length of the interval spanned by the observations.

Example 2.16 The ozone-concentration data given in Table 2.1 contained

$$\text{Smallest observation} = 1.1$$
$$\text{Largest observation} = 11.7$$

Therefore, the length of the interval covered by these observations is

$$\text{Sample range} = 11.7 - 1.1 = 10.6$$ ■

As a measure of spread, the range has two attractive features: it is extremely simple to compute and to interpret. However, it suffers from the serious disadvantage that it is much too sensitive to the existence of a very large or very small observation in the data set. Also it ignores the information present in the scatter of the intermediate points. The range is most often used when the sample size is smaller than eight.

To circumvent the problem of using a measure that may be thrown far off the mark by one or two wild or unusual observations, a compromise is made by measuring the interval between the first and third quartiles:

Sample interquartile range = Third quartile − First quartile

The sample interquartile range represents the length of the interval covered by the center half of the observations. This measure of the amount of variation is not disturbed if a small fraction of the observations are very large or very small. The sample interquartile range is now sometimes quoted in government reports on income and other distributions that have long tails in one direction, in preference to variance as the measure of spread.

Example 2.17 Calculate the range and the interquartile range for the noise-level data given in Example 2.12.

$$\begin{aligned}\text{Sample range} &= \text{Largest observation} - \text{Smallest observation}\\ &= 77.1 - 52.0\\ &= 25.1 \text{ decibels}\end{aligned}$$

In Example 2.12, the quartiles were found to be $Q_1 = 57.2$ and $Q_3 = 64.6$. Therefore the sample interquartile range

$$\begin{aligned}&= Q_3 - Q_1\\ &= 64.6 - 57.2\\ &= 7.4 \text{ decibels}\end{aligned}$$ ■

2.6 COMMENTS ON THE CHOICE OF NUMERICAL MEASURES

Several numerical measures have been introduced for use in the descriptive summary of data. The alternative measures of center (or variation) place varying emphasis on particular features of the shape of the distribution of observations. Typically, these measures yield somewhat different numerical values. The natural question at this stage is what measures should be chosen to describe the center and the variation of a given data set? A few points that should prove helpful in choosing the appropriate measures are:

(*a*) Purpose to which the descriptive summary is intended.
(*b*) Ease of interpretation.
(*c*) Degree of protection from wild observations.
(*d*) Potential for use in the process of statistical inference.

Simplicity of calculation can be considered an additional criterion, but in the computer age numerical work is no longer a major consideration except in cases of extremely large data sets.

We discuss these points primarily in the context of measures of central tendency, where the various measures proposed are the mean, median, trimmed mean, and Winsorized mean. We have omitted a number of less important measures for the sake of brevity. The income data given in Example 2.10 serve to illustrate that the intended purpose is a motivational factor in the choice of an appropriate measure. If the "typical" income of the group is considered to be an amount that half of the group earns less than and half of the group earns more than, then the median is the natural measure. However, if the total income of the group is the point of interest, this information is contained in the mean rather than the median. This is also the case in reporting the "per-capita income" of a class of workers.

Data often contain unusual observations, and Example 2.10 also illustrates that the mean is much more sensitive to unusually small or large observations than the other measures. However, the purpose of the study again determines the manner in which extreme observations should be handled in the descriptive report. Generally, the suspicious observations should be identified and checked for erroneous recording, if possible.

When the measurements are distributed so that the histogram has a peak in the center and tapers off symmetrically on both sides, all four measures of center yield numerical values that are fairly close to one another. In such cases, the use of the mean is usually recommended, because of its excellent theoretical properties (to be encountered later when we make statistical inferences about the population based on sample data). For extremely asymmetric distributions, or those with long tails, the mean should be used with caution and care must be exercised in its interpretation.

Much of the preceding discussion applies to the choice of a measure of variation, with the standard deviation sharing the merits and the drawbacks of the mean. The interquartile range is much less sensitive to unusual observations, and the range, although simple to calculate, is rarely used except for very small samples.

Complicated distributions like that of the U.S. age distribution shown in Figure 2.8 require a descriptive summary that includes the location of multiple peaks and that also gives some measure of clustering about these values, possibly in terms of suitably selected percentiles. Most importantly, neither the mean nor the median should be blindly used to report data of this form, because they fail to summarize an important aspect of the distribution. Moreover, they should not be used unless they have meaning in terms of the purpose of the study. The overall shape of the distribution must always be considered when a measure is employed for descriptive summary. We can state some general conclusions based on the preceding discussion.

(*a*) A graphic presentation depicting the overall shape of the data is an essential ingredient in descriptive summary.
(*b*) When several measures are compatible with the purpose of the summary description, it is worthwhile to report all of them rather than to arbitrarily select one.
(*c*) The overall shape of the distribution must always be considered when interpreting descriptive measures.
(*d*) The individual values of unusual observations should be reported in a footnote.

2.7 DESCRIPTIVE MEASURES FOR GROUPED DATA

When the data consist of a large number of measurements, a frequency distribution provides a condensation of the information present in the data. It is also convenient to use this summary table to calculate approximate values of the descriptive measures.

We must remember that by grouping data into class intervals we incur some loss of information. Calculations based on grouped data are mathematically convenient but less precise than calculations based on the complete data set. With the advent of high-speed computers that can handle large data sets easily, it is strongly recommended that descriptive measures be calculated on ungrouped data.

Although the computational advantage of grouped data has been lost, these techniques are still useful, because the original data are often given in the form of a frequency table with specified class intervals. This is true of survey data sets in which people are asked to report their income or age by checking questionnaire boxes and each box represents a specified interval of income or age.

To illustrate the computational methods used to determine the mean and the standard deviation, consider the data given in Table 2.9 for the frequency distribution of age at first marriage of 200 males under 40.

TABLE 2.9
FREQUENCY DISTRIBUTION OF AGE AT FIRST MARRIAGE OF 200 MALES

CLASS INTERVAL (AGE IN YEARS)	CLASS MIDPOINT m_i	FREQUENCY f_i
14.5–19.5	17	18
19.5–24.5	22	74
24.5–29.5	27	62
29.5–34.5	32	26
34.5–39.5	37	20
	Total	200

Table 2.9 shows that 18 observations out of 200 are located in the 14.5–19.5 interval, 74 observations are located in the 19.5–24.5 interval, and so on. Without knowing the exact positions of the observations in each interval, we approximate that all the observations are located at the midpoint. The class midpoints are given in the second column of the table. Thus according to this approximation, we act as if we have a data set of 200 measurements, where the measurement 17 is repeated 18 times, 22 is repeated 74 times, and so on. The sum of these observations can then be computed as

$$17\times 18+22\times 74+27\times 62+32\times 26+37\times 20$$

We divide this total by $n=200$ to obtain the mean $\bar{x}$. The calculations are shown in Table 2.10.

Motivated by the above argument, we can now state the following general formula for the grouped mean:

Grouped sample mean

If the frequency distribution has k class intervals with midpoints $m_1, m_2, \ldots, m_k$ and corresponding frequencies $f_1, f_2, \ldots, f_k$, then

$$\text{Grouped mean} \quad \bar{x} = \frac{\sum_{i=1}^{k} m_i f_i}{n} = \frac{\sum_{\text{Cells}} (\text{Midpoint} \times \text{Frequency})}{\text{Total frequency}}$$

Although the same notation $\bar{x}$ is used to denote the grouped mean and the ungrouped mean, it should be noted that the numerical value of $\bar{x}$ is likely to be slightly different in the two cases.

Regarding the calculation of variance, we again imagine that all the observations in a class interval are located at the midpoint of the interval. The squared deviations from the mean are then $(m_1-\bar{x})^2, (m_2-\bar{x})^2, \ldots, (m_k-\bar{x})^2$, which are repeated with frequencies $f_1, f_2, \ldots, f_k$, respectively. Thus the sum of the squared deviations is

$$\sum_{i=1}^{k} (m_i-\bar{x})^2 f_i$$

which when divided by $n-1$ yields the sample variance s^2. Because the total frequency n is usually large for grouped data, dividing the sum of squared deviations by $n-1$ is virtually equivalent to dividing it by n.

Grouped sample variance

$$s^2 = \frac{\sum_{i=1}^{k} (m_i-\bar{x})^2 f_i}{n}$$

As we did with the sample variance for ungrouped data, we can obtain an alternative formula for s^2 by expanding $(m_i-\bar{x})^2$ in the preceding expression and then summing term by term:

Alternative formula for grouped sample variance

$$s^2 = \frac{\sum_{i=1}^{k} m_i^2 f_i}{n} - \bar{x}^2 \qquad \text{(for hand calculation)}$$

Example 2.18 Calculate the mean and the standard deviation of age at first marriage from the frequency distribution given in Table 2.9.

The calculations appear in Table 2.10, where the products $m_i f_i$ and $m_i^2 f_i$ are listed in the fourth and fifth columns, respectively.

$$\bar{x} = \frac{5180}{200} = 25.9 \quad \text{years}$$

$$s^2 = \frac{\sum m_i^2 f_i}{n} - \bar{x}^2 = \frac{140220}{200} - (25.9)^2 = 30.29$$

$$s = \sqrt{30.29} = 5.50 \quad \text{years}$$

TABLE 2.10

CALCULATION OF $\bar{x}$ AND s FOR THE FREQUENCY DISTRIBUTION IN TABLE 2.9

CLASS INTERVAL (AGE IN YEARS)	CLASS MIDPOINT m_i	FREQUENCY f_i	$m_i f_i$	$m_i^2 f_i$
14.5–19.5	17	18	306	5,202
19.5–24.5	22	74	1,628	35,816
24.5–29.5	27	62	1,674	45,198
29.5–34.5	32	26	832	26,624
34.5–39.5	37	20	740	27,380
	Total	200	5,180	140,220

■

The formulas for the grouped mean and the variance illustrated in Example 2.18 are based on the approximation that the observations within each class interval are all located at the center of the interval. A more realistic approximation is achieved if the observations are considered to be spread evenly over the intervals. Although the latter convention does not affect the numerical value of $\bar{x}$, it does make the calculation of s^2 somewhat more involved and we will not pursue this calculation further.

Now we are ready to illustrate the calculation of median and other percentiles for the grouped data given in Table 2.9. The frequency distribution of Table 2.9 is reproduced in Table 2.11, where the Cumulative Frequency column is computed by successively adding the entries in the Frequency column.

TABLE 2.11

FREQUENCY DISTRIBUTION OF AGE AT FIRST MARRIAGE

CLASS INTERVAL (AGE IN YEARS)	FREQUENCY	CUMULATIVE FREQUENCY
14.5–19.5	18	18
19.5–24.5	74	92
24.5–29.5	62	154
29.5–34.5	26	180
34.5–39.5	20	200
Total	200	

The Cumulative Frequency column in Table 2.11 shows the number of observations that are below the upper end point of each interval. For example, the number of observations below 24.5 is $18+74=92$, which is the second entry in this column.

Suppose we want to calculate the first quartile Q_1, which is the percentile with $p=.25$. There are $n=200$ observations, so the first quartile can be located by counting $np=200\times.25=50$ observations, beginning with the smallest one. From the Cumulative Frequency column, it is clear that the first quartile is in the interval 19.5–24.5. Thus among the 74 observations in this interval, we are trying to determine the position of the $50-18=32$nd ordered observation. Assuming that the 74 observations are evenly spread over this interval, the position of the 32nd observation can be determined by taking the fraction 32/74 of the interval from the left end point. The length of the class interval is $24.5-19.5=5$, so the distance of the quartile from the lower end point is $(32/74)\times5=2.16$. Therefore,

$$Q_1=19.5+\frac{32}{74}\times5$$

$$=19.5+2.16=21.66 \text{ yrs}$$

This process of determining the fraction of an interval is called *linear interpolation*. Returning to the above example, we are now in a position to formulate the main steps for the computation of a percentile from grouped data. These steps apply even when the frequency distribution does not have equal class intervals.

Steps to compute the 100pth percentile from grouped data

(*a*) Compute the Cumulative Frequency column.

(*b*) Identify the interval in which the cumulative frequency attains or just exceeds the value np, where n is the total number of observations. Call this interval the *100p percentile class* and label its lower end point L.

(*c*) Compute the fraction $\frac{np-a}{f}$, where a is the cumulative frequency of the preceding interval and f is the frequency of the $100p$ percentile class.

(*d*) Compute

$$100p\text{th percentile}=L+\frac{(np-a)}{f}h$$

where $h=$ length of the $100p$th percentile class.

Example 2.19 Compute the median and the interquartile range for the frequency distribution reproduced in Table 2.11.

To compute the median, we take $p=.5$ and obtain $np=200(.5)=100$. Scanning the Cumulative Frequency column in Table 2.11, we see that the median class is 24.5–29.5, which has frequency $f=62$ and lower end point $L=24.5$. Following the linear interpolation scheme, we then obtain

$$\text{Median}=24.5+\frac{(100-92)}{62}\times 5=25.15 \text{ yrs}$$

We have already computed the first quartile and have obtained the value $Q_1=21.66$ yrs. To compute Q_3, we take $p=3/4$, so that $np=200(3/4)=150$. The Cumulative Frequency column shows that Q_3 is located in the interval 24.5–29.5. Again following the linear interpolation scheme of computation, we obtain

$$Q_3=24.5+\frac{(150-92)}{62}\times 5=29.18 \text{ yrs}$$

Finally, the interquartile range is

$$Q_3-Q_1=29.18-21.66=7.52 \text{ yrs}$$

■

*2.8 CODING THE DATA

The coding of a data set refers to the operation of subtracting (or adding) a constant to each observation and then dividing (multiplying) by another constant. This operation is intended to simplify the calculation of descriptive measures, but this purpose has been largely superseded by the current widespread use of electronic computers. Usually no time is saved by employing the coding procedure outlined below. However, the results are still useful indicators of the effects that changing the unit of measurement can have on the descriptive measures.

If a typical observation in the original data is denoted by x_i, the corresponding coded value is in the form $u_i=(x_i-b)/a$ for some suitably selected numbers a and b. The values x_i and u_i are related by $x_i=au_i+b$. Thus we obtain $\Sigma x_i=\Sigma au_i+\Sigma b=a\Sigma u_i+\Sigma b$. Dividing both sides of this equation by n gives us $\bar{x}=a\bar{u}+b$. Further

$$(x_i-\bar{x})^2=(au_i+b-a\bar{u}-b)^2=a^2(u_i-\bar{u})^2$$

so that

$$s_x^2=\sum(x_i-\bar{x})^2/(n-1)=a^2\sum(u_i-\bar{u})^2/(n-1)=a^2s_u^2$$

In summary

If $x_i = au_i + b$	or	$u_i = \frac{1}{a}(x_i - b)$
Then		
$\bar{x} = a\bar{u} + b$		$\bar{u} = \frac{1}{a}(\bar{x} - b)$
$s_x^2 = a^2 s_u^2$		$s_u^2 = \frac{1}{a^2} s_x^2$
$s_x = \lvert a \rvert s_u$		$s_u = \frac{1}{\lvert a \rvert} s_x$

***Example 2.20** The measurement of the length (in feet) of a section of new highway was repeated five times and the following values were obtained:

$$100.17, \quad 100.19, \quad 100.18, \quad 100.17, \quad 100.18$$

Code the data appropriately and find the mean and the standard deviation.

By eye, select a number near the center of the observations like $b = 100.18$. Because the observations differ from this amount by units of .01, we set $a = .01$. That is,

$$u_i = \frac{1}{.01}(x_i - 100.18)$$

so the coded values are

$$-1, \quad 1, \quad 0, \quad -1, \quad 0$$

From these values, we can easily compute the mean

$$\bar{u} = \frac{(-1+1+0-1+0)}{5} = -.2$$

and the variance

$$s_u^2 = \frac{(-1+.2)^2 + (1+.2)^2 + (.2)^2 + (-1+.2)^2 + (.2)^2}{5-1} = .70$$

$$s_u = \sqrt{s_u^2} = .837$$

Returning to the original x values, we now have $x_i = (.01)u_i + 100.18$

$$\bar{x} = (.01)\bar{u} + 100.18 = 100.178$$

and

$$s_x^2 = (.01)^2 s_u^2 = (.01)^2(.70) = .00007$$

$$s_x = (.01)s_u = (.01)(.837) = .00837$$

■

The method of coding can also be applied to grouped data. These calculations are illustrated in Example 2.21.

***Example 2.21** Use the method of coding to calculate the mean and the standard deviation of the frequency distribution presented in Table 2.9.

Setting the class midpoint $27 = b$ and the class width $5 = a$, we compute the coded midpoints $u_i = (m_i - b)/a = (m_i - 27)/5$ in the second column of Table 2.12. Further calculations follow the outline of Table 2.10, except that the coded value u_i is used in place of m_i.

TABLE 2.12

CALCULATION OF $\bar{x}$ AND s BY CODING

CLASS MIDPOINT m_i	CODED MIDPOINT $u_i = (m_i - 27)/5$	FREQUENCY f_i	$u_i f_i$	$u_i^2 f_i$
17	−2	18	−36	72
22	−1	74	−74	74
27	0	62	0	0
32	1	26	26	26
37	2	20	40	80
	Total	200	−44	252

$$\bar{u} = \frac{-44}{200} = -.22 \qquad \bar{x} = 27 - .22 \times 5 = 25.9$$

$$s_u^2 = \frac{252}{200} - (-.22)^2 = 1.26 - .0484$$

$$= 1.2116 \qquad s_x^2 = 5^2 s_u^2 = 25(1.2116) = 30.29$$

$$s_u = \sqrt{1.2116} = 1.10 \qquad s_x = \sqrt{s_x^2} = 5.50$$

■

EXERCISES

1. The following measurements of the diameters (in feet) of Indian mounds in southern Wisconsin were gathered by examining reports in

the *Wisconsin Archeologist* (courtesy of J. Williams): 22, 24, 24, 30, 22, 20, 28, 30, 24, 34, 36, 15, 37. Plot a dot diagram.

2. In a genetic study, a regular food was placed in each of 20 vials and the number of flies of a particular genotype feeding on each vial was recorded. The counts of flies were also recorded for another set of 20 vials that contained grape juice. The following data sets were obtained (courtesy of C. Denniston and J. Mitchell).
   (*a*) Plot separate dot diagrams for the two data sets.
   (*b*) Make a visual comparison of the two distributions with respect to their relative locations and spreads.

NO. OF FLIES (REGULAR FOOD)

15 20 31 16 22 22 23 33 38 28 25 20 21 23 29 26 40 20 19 31

NO. OF FLIES (GRAPE JUICE)

6 19 0 2 11 12 13 12 5 16 2 7 13 20 18 19 19 9 9 9

3. The following measurements of weight (in grams) have been recorded for a common strain of 70 31-day-old rats (courtesy of J. Holtzman).
   (*a*) Choose appropriate class intervals and group the data into a frequency distribution.
   (*b*) Calculate the relative frequency of each class interval.
   (*c*) Plot the relative frequency histogram.

120	116	94	120	112	112	106	102	118	112
116	98	116	114	120	124	112	122	110	84
106	122	124	112	118	128	108	120	110	106
106	102	140	102	122	112	110	130	112	114
108	110	116	118	118	108	102	110	104	112
122	112	116	110	112	118	98	104	120	106
108	110	102	110	120	126	114	98	116	100

4. The data below were obtained from a detailed record of purchases over several years (courtesy of A. Banerjee). The usage times (in weeks) per ounce of toothpaste for a household taken from a consumer panel were:

.74	.45	.80	.95	.84	.82	.78	.82	.89	.75	.76	.81
.85	.75	.89	.76	.89	.99	.71	.77	.55	.85	.77	.87

   (*a*) Plot a dot diagram of the data.
   (*b*) Find the relative frequency of the usage times that do not exceed .80.

(*c*) Group the data into a frequency distribution.
(*d*) Plot the relative frequency histogram.

5. At a university computing center, the daily numbers of computer stoppages due to machine error were recorded for a period of 70 days and the following data were obtained.
   (*a*) Construct a frequency distribution of the number of stoppages per day.
   (*b*) Calculate the relative frequencies.
   (*c*) Plot a line diagram of the data.
   (*d*) In what proportion of the days did more than three stoppages occur?

0	0	2	0	0	0	3	3	0	0
1	8	5	0	0	4	3	0	6	2
0	3	1	1	0	1	0	1	1	0
2	2	0	0	0	1	2	1	2	0
0	1	6	4	3	3	1	2	4	0
0	3	1	2	0	0	0	0	0	1
1	0	2	0	2	4	4	0	2	2

6. Professor J. counted the number of peas per pod in 60 pea pods randomly taken from a day's pick from the garden and obtained the following frequency distribution:

NO. OF PEAS PER POD	FREQUENCY
1	2
2	4
3	21
4	18
5	10
6	4
7	1
	60

   (*a*) Plot a line diagram of the data.
   (*b*) Find the relative frequency of the pods that contain at least four peas.

7. The frequency distribution of the number of lives lost annually in major tornadoes in the United States between 1900 and 1973 appears below (SOURCE: *U.S. Environmental Data Service*).
   (*a*) Plot the relative frequency histogram.

(*b*) Comment on the shape of the distribution.

NO. OF DEATHS	FREQUENCY
$\leqslant 24$	8
25–49	16
50–74	16
75–99	11
100–149	6
150–199	2
200–249	4
250 and over	1
	64

8. The frequency distribution of the average weekly earnings of 171 draftsmen in Milwaukee in May 1972 follows (SOURCE: Bureau of Labor Statistics).
   (*a*) Plot the relative frequency histogram.
   (*b*) Comment on the shape of the distribution.

EARNINGS (IN DOLLARS)	FREQUENCY
100–110	9
110–120	18
120–130	23
130–140	23
140–150	26
150–160	22
160–170	18
170–180	15
180–190	5
190–200	8
200–210	2
210–220	0
220–230	0
230–240	2
	171

9. Five measurements in a data set are $x_1=7$, $x_2=5$, $x_3=6$, $x_4=8$, $x_5=6$. Compute the numerical values of

   (*a*) $\sum_{i=1}^{5} x_i$ (*b*) $\sum_{i=2}^{4} x_i$ (*c*) $\sum_{i=1}^{5} 2x_i$

(*d*) $\sum_{i=1}^{5}(x_i-6)$ (*e*) $\sum_{i=3}^{5}(x_i-6.4)$ (*f*) $\sum_{i=1}^{5}3$

10. Demonstrate your familiarity with the summation notation by evaluating the following expressions when $x_1=1$, $x_2=-2$, $x_3=4$, and $x_4=5$:

(*a*) $\sum_{i=1}^{4}x_i$ (*b*) $\sum_{i=1}^{4}4x_i$ (*c*) $\sum_{i=1}^{4}(x_i-3)$

(*d*) $\sum_{i=2}^{4}(x_i-4)$ (*e*) $\sum_{i=1}^{4}(x_i-2)$ (*f*) $\sum_{i=1}^{3}(x_i-4)^2$

(*g*) $\sum_{i=1}^{4}x_i^2$ (*h*) $\sum_{i=1}^{4}(x_i-2)^2$ (*i*) $\sum_{i=1}^{4}(x_i^2-4x_i+4)$

11. Calculate the mean, median, variance, standard deviation, and range for each of the following sets:
(*a*) 3, 6, 2, 5, 4
(*b*) 12, 14, 18, 15, 12
(*c*) 1, 3, 1, 1, 5
(*d*) 5, 15, 10, 15
(*e*) −2, 1, −1, 0, 3, −2, −1

12. For each set calculate s^2 according to both formulas by following the outline given in Example 2.14.
(*a*) 2, 5, 4, 3, 3
(*b*) −1, 2, 0, −2, 1, −1
(*c*) 12, 11, 11, 12, 11, 11, 12
(*d*) 101, 102, 101, 101, 100

13. For the data given for the diameters of the Indian mounds in Exercise 1 find the (*a*) mean, (*b*) median, (*c*) quartiles, (*d*) range, (*e*) interquartile range, (*f*) variance and standard deviation.

14. In an epidemiological study, the total organochlorines and PCB's present in milk samples were recorded from 40 donors in Colorado (SOURCE: *Pesticides Monitoring Journal*, June 1973). The measurements were ordered from lowest to highest. For the data set, find the (*a*) median and quartiles, (*b*) 20th percentile and 70th percentile, (*c*) range, (*d*) interquartile range.

27	43	52	53	53	53	61	63	63	65
68	70	72	75	83	95	96	97	101	105
110	115	115	115	115	126	127	134	145	152
153	182	190	197	197	282	322	322	342	521

15. Referring to the data given in Exercise 14:
    (*a*) Compute the mean, trimmed mean, and Winsorized mean.
    (*b*) Compute the standard deviation.
    (*c*) Find the relative frequency of the measurements lying within an interval of $\bar{x} \pm 2s$ and compare it with the bound obtained by Chebyshev's rule.
16. Referring to the data on usage times per ounce of toothpaste given in Exercise 4:
    (*a*) Compute the mean and standard deviation from the ungrouped data.
    (*b*) Compute the mean and standard deviation from the frequency distribution that you obtained in Exercise 4(*c*).
    (*c*) Obtain the median of the ungrouped data.
    (*d*) Compute the median from the grouped data.
17. Referring to the data on the average weekly earnings of draftsmen given in Exercise 8, compute the (*a*) mean, (*b*) standard deviation, (*c*) median, (*d*) quartiles, (*e*) interquartile range.
18. *Obtaining percentiles of grouped data by graphic method*: Refer to the frequency distribution of age at first marriage given in Table 2.11, where the last column indicates the cumulative frequencies. The cumulative relative frequency, which is defined as the ratio of the cumulative frequency to the sample size, gives the proportion of observations less than or equal to the upper boundary of the corresponding class interval. These cumulative relative frequencies and the upper boundaries for the data in Table 2.11 are computed below and are plotted in the following figure, in which consecutive points are joined by straight lines.

UPPER BOUNDARY OF CLASS INTERVAL	CUMULATIVE RELATIVE FREQUENCY
19.5	18/200 = .09
24.5	92/200 = .46
29.5	154/200 = .77
34.5	180/200 = .90
39.5	200/200 = 1.00

This figure is called a plot of the *sample cumulative distribution function*, and it is used to determine percentiles graphically:

To find the 100*p*th percentile, locate *p* on the vertical axis and draw a horizontal line through this point that intersects the sample cumulative distribution curve. The abscissa of this point of intersection is the required percentile.

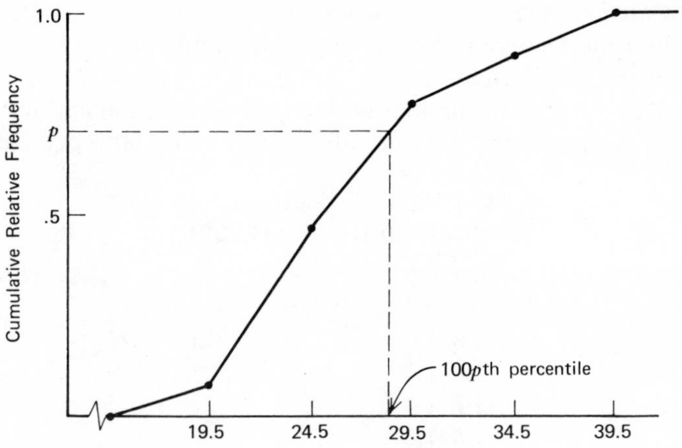

Determine the quartiles from this plot of the sample cumulative distribution function.

19. Again referring to the data on the average weekly earnings of draftsmen in Exercise 8:
    (*a*) Compute the cumulative relative frequencies.
    (*b*) Plot the sample cumulative distribution function.
    (*c*) Graphically determine the median and the quartiles, and then compare them with the results of Exercise 17 (*c*) and (*d*).
20. The following data represent the scores that 50 students earned on a college qualification test (courtesy of R. Johnson).
    (*a*) Group the data into a frequency distribution.
    (*b*) Calculate the mean and the standard deviation from the grouped data.
    (*c*) Calculate the median and the interquartile range from the grouped data.

162	171	138	145	144	126	145	162	174	178
167	98	161	152	182	136	165	137	133	143
184	166	115	115	95	190	119	144	176	135
194	147	160	158	178	162	131	106	157	154
118	146	172	165	139	94	135	162	168	114

21. The table below shows the frequency distribution of the modulus of elasticity measurements in million pounds per square inch, of a particular grade and size of southern pine lumber (SOURCE: U.S. Forest Service Research Paper, FPL-64, July 1966).
    (*a*) Calculate the mean and the standard deviation.

(*b*) Calculate the median and the quartiles.

(*c*) Plot the relative frequency histogram and comment on the shape of the distribution.

(*d*) Plot the sample cumulative distribution function and use this plot to determine the quartiles and the 20th and 60th percentiles.

ELASTICITY (IN MILLION psi)	FREQUENCY
0.4–0.6	1
0.6–0.8	5
0.8–1.0	15
1.0–1.2	28
1.2–1.4	22
1.4–1.6	7
1.6–1.8	1
	79

22. Blood cholesterol levels were recorded for 43 persons sampled in a medical study group and the following data were obtained (courtesy of G. Metter).

(*a*) Group the data into a frequency distribution.

(*b*) Plot the relative frequency histogram and comment on the shape of the distribution.

(*c*) Compute the mean and the standard deviation from the ungrouped data.

(*d*) Compute the mean and the standard deviation from the grouped data.

(*e*) Plot the sample cumulative distribution function and use the graph to determine the median and the quartiles.

239	212	249	227	218	310	281	330	226	233
223	161	195	233	249	284	245	174	154	256
196	299	210	301	199	258	205	195	227	244
355	234	195	179	357	282	265	286	286	176
195	163	297							

23. The winning times of the men's 400-meter freestyle swimming in the Olympics (1908–1972) appear below.

(*a*) Draw a dot diagram and label the points according to time order.

(*b*) Explain why it is not reasonable to group the data into a frequency distribution.

WINNING TIMES IN MINUTES AND SECONDS

YEAR	TIME	YEAR	TIME
1908	5:35.8	1948	4:41.0
1912	5:24.4	1952	4:30.7
1920	5:26.8	1956	4:27.3
1924	5:04.2	1960	4:18.3
1928	5:01.6	1964	4:12.2
1932	4:48.4	1968	4:09.0
1936	4:44.5	1972	4:00.27

24. *Transformations may reduce long tails*: Referring to the data given in Exercise 14:
    (*a*) Plot the relative frequency histogram of the original data.
    (*b*) Taking the logarithm of each observation with base 10 [log(27) = 1.43, etc.], construct a histogram of the transformed data.
    (*c*) Compare the shapes of the histograms in (*a*) and (*b*) and comment.
25. The *mode* of a collection of observations is defined as the observed value with largest relative frequency. The mode is sometimes used as a center value. There can be more than one mode in a data set. Find the mode for the data given in Exercise 6.

*26. Use the method of coding to calculate the mean, variance, and standard deviation for each of the following sets:
    (*a*) 1001, 1003, 1001, 1002, 1003
    (*b*) .005, .002, .004, .005, .007, .003
    (*c*) −27.1, −27.6, −27.4, −27.3, −27.4

*27. Use the method of coding to evaluate the sample mean and the standard deviation for the grouped data given in Exercise 21. (*Hint*: Subtract 1.1 and divide by 0.2.)

28. *Pie Charts*. When the basic categories are not quantifiable as with energy sources classified into coal, natural gas, crude petroleum, etc., graphic displays are still more dramatic and effective than tables. The distribution of a total amount over categories is often presented in the form of a pie chart like the figure below displaying the data on 1930 per capita energy consumption. A circle is divided into sectors like the slices of a pie according to the proportion in individual categories. For the data of 1930, the sector for natural gas is determined by

$$\text{angle} = (\text{proportion in the category}) \times 360°$$

$$= \frac{16}{181} \times 360 = 31.8°$$

PER CAPITA ENERGY CONSUMPTION IN MILLION B.T.U.'S

YEAR	NATURAL GAS	COAL	CRUDE PETROLEUM	OTHER	TOTAL
1930	16	111	46	8	181
1974	105	62	158	21	346

(SOURCE: 1975 Statistical Abstract of the U.S.)

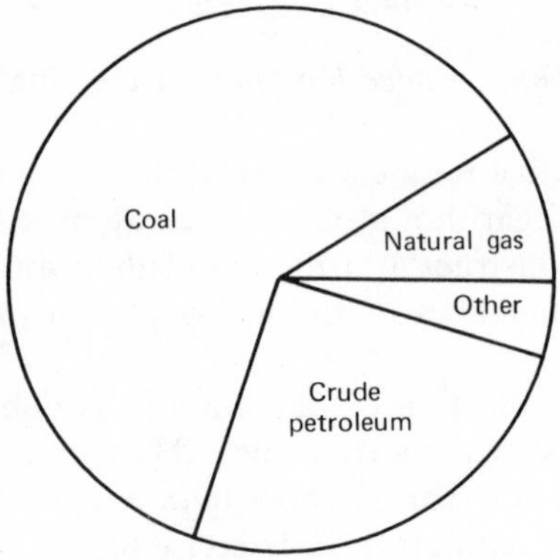

Pie chart for 1930 per capita energy consumption

(*a*) Construct the pie chart for 1974.
(*b*) What changes do you note from 1930 to 1974 in the relative importance of the various sources?

29. *Bar Charts*. Another form of graphically depicting the data that are classified in a number of categories is the bar chart. Equal width rectangular bars are constructed over each category with height equal to the frequency or other measurement associated with the category. The bar chart for the data of 1930 per capita energy consumption given in Exercise 28 is presented below.

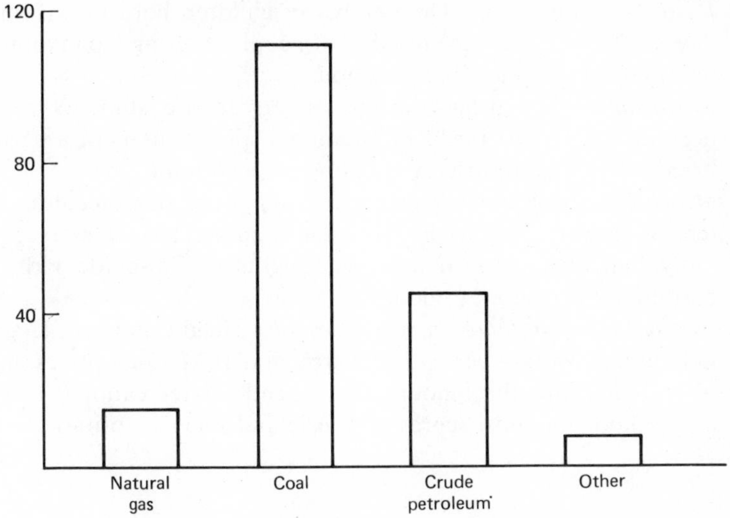

Bar chart for 1930 per capita energy consumption

(*a*) Construct a bar chart for the 1974 data given in Exercise 28.
(*b*) What comparisons can you draw between the two years 1930 and 1974? You may wish to draw the two bars for the same source side by side in different colors.

(Many creative variants of the basic bar chart are found in newspapers and magazines. Find one.)

CLASS PROJECTS

To develop a feeling of the variability and uncertainty in nature, the laboratory, society, commerce, and industry, you should attempt to gather your own data on a topic that is of interest to you. You should use the methods described in this chapter to conduct a descriptive study of your data. Then save the data for use in future discussions and applications of statistical inference.

The procedures you use to select a sample from a population will limit or enhance the quality of any later inferences drawn by statistical techniques. The specifics of sampling methods are discussed later. At this point, you should use your common sense to select a "representative" sample. You may use the examples given or choose your own data sets. Collect at least one data set of each of the following types.

(*a*) *Data involving counts*: The number of children born in a town each day, traffic fatalities, the number of people entering a bank line each minute during a busy banking hour.

(*b*) *Proportions*: The proportion of undergraduate students living in dormitories, female births of an animal species, users of a particular brand of shampoo, drivers violating a speed limit.

(*c*) *Measurements on a continuous scale*: Length of pine needles, diameters of flowers, lengths of television commercials, sales receipts of individual customers at a store, weights of 5th-grade girls, daily barometric pressure at noon.

(*d*) *Measurements of more than one variable*: Height and arm lengths of individuals, weekly *changes* in Xerox and IBM stock prices, income of students and the amount they spend on recreational activities, high-school students' scores on music and science aptitude tests.

MATHEMATICAL EXERCISES

1. Use the definition of Σ to verify that $\sum_{i=1}^{n} bx_i = b\sum_{i=1}^{n} x_i$.

2. Verify that $\sum_{i=1}^{n} (bx_i + a) = b\sum_{i=1}^{n} x_i + na$.

3. Show generally that $\sum_{i=1}^{n} (x_i - a)^2 = \sum_{i=1}^{n} x_i^2 - 2a\sum_{i=1}^{n} x_i + na^2$.

4. Show $\sum_{i=1}^{n} (x_i - b)^2 = \sum_{i=1}^{n} (x_i - \bar{x})^2 + n(\bar{x} - b)^2$. In other words, show the sum of squares about $b = \bar{x}$ is smallest.

Hint: Write $x_i - b = (x_i - \bar{x}) + (\bar{x} - b)$. Then

$$(x_i - b)^2 = (x_i - \bar{x})^2 + 2(x_i - \bar{x})(\bar{x} - b) + (\bar{x} - b)^2$$

and

$$\sum_{i=1}^{n} 2(\bar{x} - b)(x_i - \bar{x}) = 2(\bar{x} - b)\sum_{i=1}^{n} (x_i - \bar{x}) = 0$$

5. Establish Chebyshev's rule: The interval $\bar{x} - ks$ to $\bar{x} + ks$ contains at least the fraction $1 - (1/k^2)$ of the data.

Hint: For any positive a

$$a\cdot(\text{no. of obs. outside } \bar{x}-\sqrt{a} \quad \text{to} \quad \bar{x}+\sqrt{a}\,) \leqslant \sum_{\substack{\text{all obs.}\\ \text{outside}}} (x_i-\bar{x})^2$$

$$\leqslant \sum_{i=1}^{n} (x_i-\bar{x})^2$$

$$=(n-1)s^2$$

Next let $\sqrt{a} = ks$, and divide by $n\cdot a$ to obtain

$$\frac{1}{n}\left[\text{no. of obs. outside } \bar{x}-ks \quad \text{to} \quad \bar{x}+ks\right] \leqslant \frac{(n-1)s^2}{nk^2s^2} \leqslant \frac{1}{k^2}$$

6. Use the properties of summation to verify the equivalence of the two formulas for determining variance from grouped data.
7. Show that the sample variance s^2 can also be expressed by the alternative formula

$$s^2 = \frac{\displaystyle\sum_{i=1}^{n} x_i^2 - \frac{\left(\displaystyle\sum_{i=1}^{n} x_i\right)^2}{n}}{n-1}$$

CHAPTER 3

Elements of Probability

3.1 INTRODUCTION

To study a phenomenon in nature that is not governed by a precisely known physical law, we perform experiments, observe social or economic development, or otherwise accumulate information in the form of sample data. Methods of condensing and summarizing data were discussed in Chapter 2 to help us appreciate the main characteristics of a data set. However, a major goal of our study is to progress beyond the stage of the mere summarization of data to an analysis of the information contained in the sample data, so that we can reach conclusions about the larger population from which the sample is drawn. The logical foundation of the process of drawing statistical inferences about a population by analyzing sample data rests on the subject of probability.

From an intuitive viewpoint, the probability of an uncertain event is a numerical measure of how likely it is that the event will occur. For instance, a low probability indicates that it is extremely unlikely that the event will occur. When the weather forecaster predicts less than a 10% chance of rain, we feel that rain is unlikely and we do not carry our umbrellas. The full usefulness of the idea of assigning a measure to the uncertainty of an event to draw statistical inferences about a population can be appreciated only after the concept has been pursued to a reasonable extent. However, before beginning the study of probability, we digress briefly to consider one example that indicates the role probability plays in statistical reasoning:

> A public health researcher studied extensive reports published in the last decade and found that about 30% of the female population in the age group 20–30 years were overweight. The researcher conjectures that this percentage has not changed in the current period. To verify or refute her conjecture, she investigates a sample of 100 females and finds that at least 45 of them are overweight. What evidence does this sample information provide to validate the conjecture?

We must realize that the occurrence of at least 45 overweight persons in a sample of 100 is possible whether or not the population percentage is equal to 30%. Therefore, the sample information cannot lead to an absolute *proof* of the conjecture or its contradiction. However, a conclusion as to the validity of the conjecture can be arrived at by the following line of statistical reasoning. Tentatively accepting the conjecture that the population percentage of overweight females is 30%, we can employ the theory of probability to compute the chance of observing at least 45 overweight females in a sample of 100. Suppose that the numerical value of this probability turns out to be quite low (say, .0005). If the conjecture is true, then the outcome that the researcher observed would be extremely unlikely.

This calculation and reasoning cast serious doubt on the validity of the conjecture. Therefore, our statistical conclusion is to declare that the sample evidence is grossly contradictory to the hypothesis presented by the researcher.

This kind of reasoning is called *testing a statistical hypothesis*, and it is to be explored in greater detail in later chapters. For now, we are concerned with determining the numerical values of the probabilities of events such as [45 or more overweight] in the example here. This determination always depends on an assumption about the nature of the population and the method of sampling. With an assumed structure for the population, probability theory assigns numerical values to the probabilities of various events. Throughout this chapter, we therefore unrealistically assume that the population structure is known. Statistical inference, the central topic of later chapters, goes in the reverse direction. Namely, after observing an event by sampling, calculations of probability are employed to determine which population structures might provide reasonable explanations for the sample outcome and which structures would make the event nearly impossible.

3.2 SAMPLE SPACE AND EVENTS

The concept of probability is relevant to experiments that have somewhat uncertain outcomes. These are the situations in which, despite every effort to maintain fixed conditions, some variation of the result in repeated trials of the experiment is unavoidable. As used in this text, the term "experiment" is not restricted to laboratory experiments but includes any activity that results in the collection of data pertaining to phenomena that exhibit variation. The domain of probability encompasses all phenomena for which outcomes cannot be exactly predicted in advance.

An *experiment* is the process of collecting data relevant to a phenomenon that exhibits variation in its outcomes.

We begin our study of probability by introducing the concepts of *sample space* and *events*. To facilitate the development of these concepts, first we give a few examples of experiments with uncertain outcomes.

Experiment (a) Note the sex of the first two newborns in town tomorrow.

Experiment (b) Let each of 10 persons taste a cup of instant coffee and a cup of percolated coffee. Record how many people prefer the instant coffee.

Experiment (c) Administer an antibiotic to patients suffering from a viral infection until one has an adverse reaction.

Experiment (d) Measure the concentration of dust in the air near a construction site.

Experiment (e) Give a child a specific dose of a multivitamin in addition to a normal diet. Observe the child's height and weight gains after 12 weeks.

In each of these examples, the experiment is described in terms of what is to be done and what aspect of the result is to be recorded. Although each experimental outcome is unpredictable, we can describe the collection of all possible outcomes.

The collection of all possible distinct outcomes of an experiment is called the *sample space* of outcomes, and each distinct outcome is called a *simple event*, an *elementary outcome,* or an *element of the sample space.* The sample space is denoted by $\mathcal{S}$.

In a given situation, the sample space is presented either by listing all possible results of the experiment, using convenient symbols to identify the results, or by making a descriptive statement characterizing the set of possible results. The sample spaces for the preceding five experiments can be presented as follows:

Experiment (a) $\mathcal{S} = \{BB, BG, GB, GG\}$, where, for example, BG denotes the birth of a boy first and a girl second.

Experiment (b) $\mathcal{S} = \{0, 1, 2, \ldots, 10\}$

Experiment (c) If we write N to denote "no adverse reaction" and R to denote "adverse reaction," the experimental results can be described by the sequences of N that are terminated by a single R; that is, $\mathcal{S} = \{R, NR, NNR, NNNR, \ldots\}$.

Experiment (d) If the concentration of dust is measured as the proportion of dust per unit volume of air, the measurement can be any number between 0 and 1. Here the sample space can be described as $\mathcal{S} = \{t : 0 \leqslant t \leqslant 1\}$, which represents the set of all numbers between 0 and 1. This is read "the set of numbers t such that t is between 0 and 1, including both 0 and 1."

Experiment (e) Here the experimental result consists of the measurements of two characteristics, height and weight, both of which are measured on continuous scales. Denoting the measurements of gain in height and weight x and y, respectively, the sample space can be described as

$$\mathcal{S} = \{(x,y) : x \text{ nonnegative}, y \text{ positive, 0 or negative}\}.$$

The elements of the sample space constitute the ultimate breakdown of the results into distinct possibilities. Each time the experiment is performed, one and only one elementary outcome can occur. Several elementary outcomes may often exhibit some common descriptive feature, and taken together, they constitute a *composite event* having the stated feature. According to this definition, an elementary outcome itself may be a composite event when it is the only outcome that satisfies the descriptive property. Therefore, *elementary outcomes* as well as *composite events* will be referred to as *events* in this text.

> A collection of elementary outcomes characterized by some descriptive feature is called an *event*. An event is a subset of the sample space $\mathcal{S}$. The first few capital letters in the alphabet $A, B, C, \ldots$ are commonly used symbols for events.
>
> We say that an event A occurs when any one of the elementary outcomes in A occurs.

To illustrate an event, consider the sample space of Experiment (a) in which the set of two newborns is recorded. Let A be the event of exactly one girl. This descriptive feature identifies the event as the collection

$A = \{BG, GB\}$. Referring to Experiment (c) (trying an antibiotic until an adverse reaction is observed), let C be the event that at most five patients receive the antibiotic. Then this event represents the collection $C = \{R, NR, NNR, NNNR, NNNNR\}$. In the case of Experiment (d) (measuring the concentration of dust in the air), suppose a federal standard requires the dust concentration to be below .1. Then event D, described as the event of a violation of the federal standard, is characterized by $D = \{t : .1 \leqslant t \leqslant 1\}$.

Sample spaces vary in complexity, depending on the nature of the experiment. In both Experiments (a) and (b), $\mathbb{S}$ consists of a finite number of elements. These are examples of *finite* sample spaces. In Experiment (c), the number of elementary outcomes is not finite because the list never terminates, but the elements can be arranged one after another in a sequence. This is called a *countably infinite sample space*. The remaining two experiments have substantially different sample spaces from the others. Experiment (d) is conducted using a continuous measurement scale, and the sample space consists of all real numbers between 0 and 1. This is an example of a *continuous sample space*, because it not only represents an infinite collection but the numbers cannot be arranged in a sequence. Finally, Experiment (e) is a *two-dimensional continuous sample space*, because the outcomes are measurements of two variables on continuous scales.

> A sample space consisting of either a finite or a countably infinite number of elements is called a *discrete sample space*. When the sample space includes all the numbers in some interval of the real line, it is called a *continuous sample space*.

For ease of exposition, first we develop the basic principles of probability in the context of sample spaces having a finite number of elementary outcomes, which are usually represented by the general notation $e_1, e_2, e_3, \ldots$. After we become familiar with the concepts of probability in these spaces, we then give a general definition of probability in Section 3.8 and consider some of its important properties.

3.3 THE PROBABILITY OF AN EVENT

Intuitively, we first perceive the probability assigned to an event as a numerical gauge with which to measure how likely it is that the event will take place when the experiment is performed. To quantify the phrase "how likely," it is natural to take the fraction of times the event would occur in repeated trials of the experiment.

> Our *intuitive* concept of a numerical measure for the probability of an event will be in terms of the proportion of times the event is expected to occur when the experiment is repeated under identical conditions.

> The symbol $P(A)$ is used to denote the probability of an event A.

The appropriate process to use in determining probabilities for events depends on the nature of the experiment and the associated sample space. In some situations, the proportion of times each elementary outcome is expected to occur can be determined by logical derivation without actually performing the experiment. In other cases, it is necessary to repeat the experiment a large number of times to obtain information about the frequency of occurrence of the different outcomes. These two types of situations, elaborated below, help to guide us toward a formal definition of probability.

3.3.1 Equally Likely Elementary Outcomes

When some symmetry inherent in the experiment ensures that each outcome is as likely to occur as any other, the sample space is said to have *equally likely elementary outcomes*. For example, consider the experiment of tossing a balanced die and recording the number on the face that shows up. Each of the 6 faces on this symmetric cube is as likely to appear as any other. Without performing the experiment, we can logically conclude that the proportion of times each particular face is expected to occur is $\frac{1}{6}$. Any other assignment of probability would not treat each face equally. If the sample space is described as

$$\mathcal{S} = \{e_1, e_2, e_3, e_4, e_5, e_6\}$$

the probabilities of the individual elements are then given by

$$P(e_1) = P(e_2) = \ldots = P(e_6) = \frac{1}{6}$$

Now suppose that A is the event of rolling an even number; that is,

$A=\{e_2,e_4,e_6\}$. Since the proportion of times A occurs is the sum of the proportions of times e_2, e_4, and e_6 occur, we must have

$$P(A)=P(e_2)+P(e_4)+P(e_6)=\frac{3}{6}=\frac{1}{2}$$

We can express the concept behind this illustration in the following general statement:

> If a sample space consists of k elementary outcomes $\{e_1,\ldots,e_k\}$ that are all equally likely to occur, the probability of each elementary outcome is $1/k$. If an event A consists of m of these k elements, then
>
> $$P(A)=\frac{m}{k}$$
>
> A sample space possessing the characteristic of equally likely elements is said to have a *uniform probability model.*

Uniform probability models often arise in the context of sampling operations where the experiment consists of a blindfolded selection of members, from a group of individuals or objects, and then measurement of some specific characteristic. This concept of sampling, which is formally introduced in Section 3.5 as *random sampling*, forms a basis for the application of uniform probability models.

Example 3.1 A food inspector intends to visit one of the 20 fast-food establishments selling hamburgers to record the bacterial count in the hamburger. To randomly choose the place to visit, the inspector writes the addresses of the 20 places on 20 cards, shuffles the cards, and then draws a card from the pile without looking at the addresses. Suppose that unknown to the inspector, the hamburger in 5 of the 20 establishments has a high bacterial level. What is the probability that the inspector will observe a high bacterial count in the hamburger at the place he visits?

We can denote the 5 places that have a high bacterial level by $H_1,H_2,\ldots,H_5$ and the 15 places that have a low bacterial level by $L_1,L_2,\ldots,L_{15}$. Since any of the 20 places may be selected for the visit, the sample space consists of 20 elementary outcomes, or

$$\mathcal{S}=\{H_1,\ldots,H_5,L_1,\ldots,L_{15}\}$$

It is logical to assign equal probability to each of these outcomes, since according to the sampling process just described, any place is as likely to be selected as any other. Thus we have a uniform probability model in which each elementary outcome has a probability of 1/20. If we denote the event of observing a high bacterial level by A, we have $A = \{H_1, H_2, \ldots, H_5\}$ and therefore $P(A) = 5/20 = 1/4$. ■

With regard to Example 3.1, note that the experimental outcome that is to be recorded is simply whether the sampled place has a high (H) or a low (L) bacterial level in the hamburger. We could also have considered a sample space consisting of only two elementary outcomes $\mathcal{S} = \{H, L\}$. However, in this formulation of the sample space, the two elementary outcomes are not equally likely because H and L do not occur in equal numbers among the 20 places. Consequently, in calculating the probability of an event, a specification of the sample space may be chosen that is not the most natural or the simplest construction but that conforms to the uniform probability model.

3.3.2 Stability of Relative Frequency

In many other experiments, it is simply impossible to find a mode of specifying the sample space in which the elementary outcomes can be regarded as equally likely. When tossing a lopsided die, the faces cannot be regarded as equally likely. In a study of infant mortality, we may wish to ascertain the probability of an infant's death in the first month of life. Suppose we choose to designate the occurrence of death in the first month, second month, and so on, of life as the elementary outcomes $e_1, e_2, \ldots$. Extensive mortality data confirm that these events are far from equally likely. In such situations, the probability of an event cannot be ascertained by logical derivation using symmetry, as it can in the context of a uniform probability model. For the concept of probability, we must then consider how frequently an event occurs when the experiment is repeatedly performed under identical conditions. When an experiment is repeated N times, we define:

Relative frequency of an event A in N trials

$$r_N(A) = \frac{\text{Number of times } A \text{ occurs in } N \text{ trials}}{N}$$

where $r_N(A)$ is simply the proportion of times the event A actually occurs in a given set of N repetitions of the experiment. This value changes for different sets of N repetitions. The fraction $r_N(A)$ also typically fluctuates as the number N of repetitions changes. However, a common experience drawn from experiments in many fields is that as long as experimental conditions do not change appreciably, the fraction $r_N(A)$ tends to stabilize at a single numerical value when the number of repetitions N is increased. This behavior is illustrated in Figure 3.1, where values of N are shown in steps of 50 with a break in the scale to indicate larger numbers of repetitions.

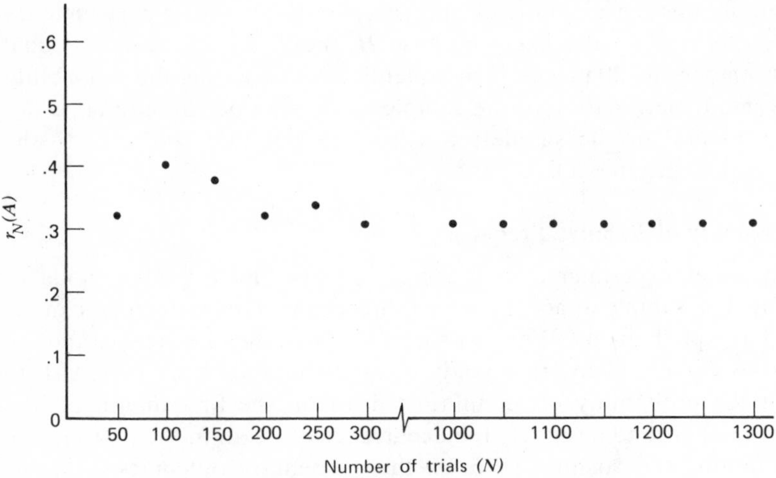

Figure 3.1 Stabilization of relative frequency.

The limiting value where $r_N(A)$ tends to stabilize as N increases indefinitely could be taken as the definition of $P(A)$, the probability of event A. However, our main purpose in considering this behavior of relative frequency is to motivate the assignment of a numerical value $P(A)$ to the probability of event A.

> Empirical evidence for assigning a numerical value $P(A)$ to the probability of an event A is derived from the observed stabilization of the relative frequency of A after many repeated trials.

The property of the long-range stabilization of relative frequencies is based on the findings of experimenters in many fields who have undertaken the strain of studying the behavior of $r_N(A)$ under prolonged repetitions of their experiments. French gamblers, who provided much of the early impetus for the study of probability, performed experiments tossing dice and coins, drawing cards, and playing other games of chance thousands and thousands of times. They observed the stabilization property of relative frequency and applied this knowledge to achieve an understanding of the uncertainty involved in these games. Demographers have compiled and studied volumes of mortality data to examine the relative frequency of the occurrence of such events as death in particular age groups. In each context, the relative frequencies were found to stabilize at specific numerical values as the number of cases studied increased. Life and accident insurance companies actually thrive on this stability property of relative frequencies.

After being introduced to the logical basis of the uniform probability model and to the empirical evidence regarding the stability property, we are now ready to formally define probability for discrete sample spaces. Any general definition of probability should preserve the important features of these two concepts that are themselves in harmony. In particular, the relative frequency viewpoint suggests that the probability of an event A, being the proportion of times A occurs in repeated trials, should be a number between 0 and 1. Moreover, because some elementary outcome must occur at every repetition, the probability assigned to the whole sample space is 1. Finally, because the relative frequency of occurrence for an event A is the sum of the relative frequencies of all the elementary outcomes in A, the probability of A is considered to be the sum of the probabilities of the elementary outcomes that comprise A. We can now summarize these conditions for any model of probability.

Conditions for a model of probability for discrete sample spaces

Probability is a function, defined on events, that satisfies the following conditions:

(i) For all events A, $0 \leq P(A) \leq 1$

(ii) $P(A)$ is the sum of the probabilities of the elementary outcomes belonging to A, or

$$P(A) = \sum_{\substack{\text{all } e_i \\ \text{in } A}} P(e_i)$$

(iii) $P(\mathcal{S}) = \sum_{\substack{\text{all } e_i \\ \text{in } \mathcal{S}}} P(e_i) = 1$

To implement the assignment of probability to events in a discrete sample space $\mathbb{S}$, we need to assign probability only to each elementary outcome in $\mathbb{S}$. Any assignment of numbers $P(e_i)$ to the elementary outcomes will satisfy the three conditions for probability whenever these numbers are nonnegative and the sum $\Sigma P(e_i)$, for all e_i in $\mathbb{S}$, is 1. For any specified event A, each of these assignments yields a numerical value of $P(A)$ according to the second condition of probability, and this $P(A)$ is a number between 0 and 1. From a practical standpoint, however, probabilities must be assigned so that the numerical value obtained for the probability of an event agrees with the long-range relative frequency of that event. Otherwise, an arbitrary assignment of numbers such as $P(e_i)$ satisfying the above conditions of probability is only a matter of academic interest.

Example 3.2 The experiment of observing the sex of the older and the younger children in two-children families has four possible outcomes: (boy, boy) $= e_1$, (boy, girl) $= e_2$, (girl, boy) $= e_3$, and (girl, girl) $= e_4$. Do each of the following assignments of probability satisfy the conditions of a probability model?

(*a*) $P(e_1)=P(e_2)=P(e_3)=P(e_4)=\frac{1}{4}$
(*b*) $P(e_1)=\frac{3}{16}, P(e_2)=\frac{3}{8}, P(e_3)=\frac{1}{4}, P(e_4)=\frac{3}{16}$
(*c*) $P(e_1)=\frac{1}{2}, P(e_2)=\frac{1}{4}, P(e_3)=\frac{1}{4}, P(e_4)=\frac{1}{2}$

We observe that in (*a*) and (*b*) the probabilities assigned to the elementary outcomes are nonnegative numbers that sum to 1 but that in (*c*) the numbers sum to 1.5. Thus (*a*) and (*b*) both satisfy the conditions of probability, whereas (*c*) does not.

Many assignments other than (*a*) and (*b*) would also satisfy these conditions. However, to be of any practical usefulness, the assignment of probabilities should closely resemble the relative frequencies when a large number of two-children families are observed. In fact, one aspect of statistical inference is to verify that a postulated probability model is in agreement with the observations made from the sample data. ■

3.4 OPERATIONS WITH EVENTS AND BASIC LAWS OF PROBABILITY

According to the conditions for a model of probability, the probability of any event A can be computed by

$$P(A)=\sum_{\substack{\text{all } e_i \\ \text{in } A}} P(e_i)$$

In other words, $P(A)$ is the sum of the probabilities of all elementary outcomes that are in A. When an event A is complex in nature, computing $P(A)$ by adding the probabilities of the elementary outcomes may be tedious. On the other hand, it may be possible to express A in terms of some other simpler events with probabilities that can be more readily determined. As these events are combined in various ways, we will want to know how probability behaves. With this purpose in mind, first we describe some operations with events and then we present a few basic properties of probability.

It is convenient to think of the sample space $\mathcal{S}$ as a set of points in a diagram where each point corresponds to a specific elementary outcome of the experiment. The geometrical pattern of the points is irrelevant, but the points must be carefully identified with the specific elementary outcomes. An event is then presented in the diagram as the set of points that satisfies the descriptive feature of the event. Such a presentation is called a *Venn diagram*.

Example 3.3 When an experimental stimulus is given to an animal, it either responds or fails to respond. In other words, there are only two possible outcomes when a stimulus is applied to an animal: Either the animal responds (R) or it does not (N). The experiment consists of administering the stimulus to three animals in succession and recording R or N for each animal. Construct a Venn diagram of the sample space and present the following events:

A: Only one animal responds.

B: There is a response in the first trial.

C: Both the first and second animals fail to respond.

There are two possible results R and N for the first animal, and each of these results may be followed by either of two results for the second animal. Each of these $2\times2=4$ combined results may be followed by an R or an N result for the third animal. Therefore, for three animals the number of elementary outcomes will be $2\times2\times2=8$. These outcomes are listed below and are identified by the symbols $e_1, e_2, \ldots, e_8$:

$$RRR\,(e_1) \quad RRN\,(e_2) \quad RNR\,(e_3) \quad NRR\,(e_4)$$

$$RNN\,(e_5) \quad NRN\,(e_6) \quad NNR\,(e_7) \quad NNN\,(e_8)$$

The compositions of the events A, B, C are

$$A = \{e_5, e_6, e_7\}$$

$$B = \{e_1, e_2, e_3, e_5\}$$

$$C = \{e_7, e_8\}$$

The Venn diagram appears in Figure 3.2, where the rectangle represents the sample space. ■

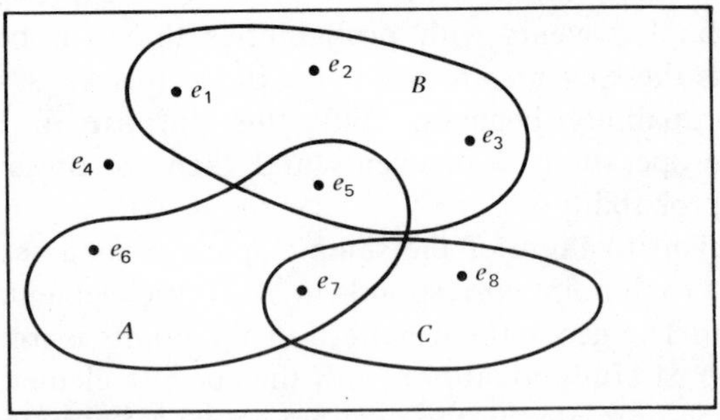

Figure 3.2 Venn diagram of the events in Example 3.3

We can now define three basic event operations, called *union*, *intersection*, and *complementation*:

> The *union* of two events A and B is the set of all elementary outcomes that are in A, in B, or in both A and B. The union of A and B is denoted by $A \cup B$, which is the same as $B \cup A$. The occurrence of $A \cup B$ means that at least one of A and B occur.

> The *intersection* of two events A and B is the set of all elementary outcomes that belong simultaneously to both A and B. The intersection of A and B is denoted by AB, which is the same as BA. The occurrence of AB means that both A and B occur.

> The *complement* of an event A is the set of all elementary outcomes that are not in A. The complement of A is denoted by A^c. The statement that A does not occur is synonymous with the statement that A^c occurs.

These three operations are illustrated in Figure 3.3. Note that $A \cup B$ is a larger set *containing* sets A and B and that AB is a set *contained in* both A and B.

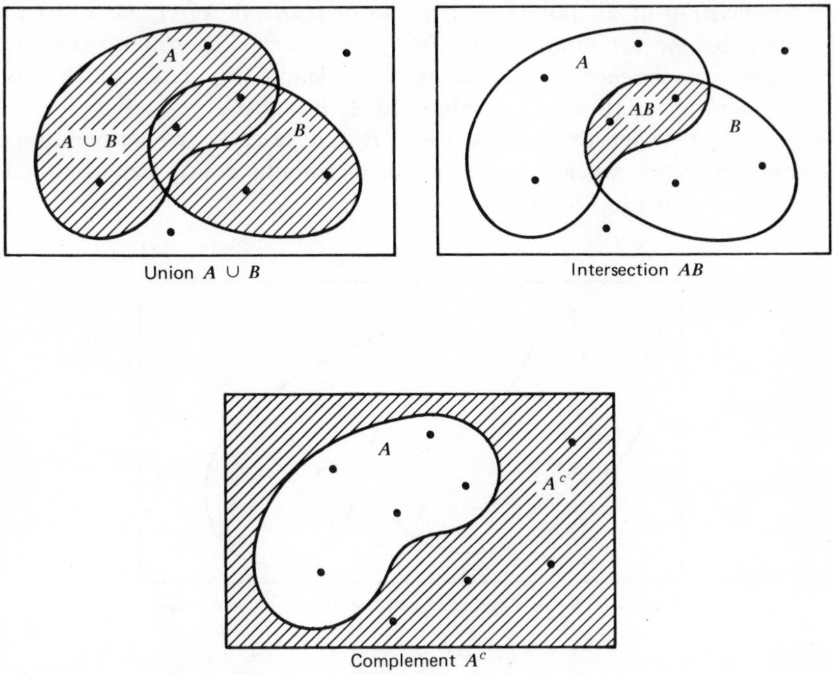

Figure 3.3 Venn diagrams showing the basic operations.

Example 3.4 Let the events A, B, and C be those described in Example 3.3. Give the compositions of the following events: $A \cup B$, AC, BC, B^c, $(A \cup B)^c$, $A^c \cup B^c$.

Here the sample space has 8 elementary outcomes, and the events A, B, and C are given as

$$A = \{e_5, e_6, e_7\} \quad B = \{e_1, e_2, e_3, e_5\} \quad C = \{e_7, e_8\}$$

Employing the definitions of the basic operations, we obtain

$A \cup B = \{e_1, e_2, e_3, e_5, e_6, e_7\}$

$AC = \{e_7\}$

BC is empty.

$B^c = \{e_4, e_6, e_7, e_8\}$

$(A \cup B)^c = \{e_4, e_8\}$

B^c is given above, and $A^c = \{e_1, e_2, e_3, e_4, e_8\}$. Thus

$A^c \cup B^c = \{e_1, e_2, e_3, e_4, e_6, e_7, e_8\}$ ■

The definitions of *union* and *intersection* suggest that these operations can be performed on any number of events, not just two. For example, the event consisting of all points that are in *at least one* of A, B, and C, is identified by the set notation $A \cup B \cup C$. Similarly, the statement of the *simultaneous occurrence* of all three events is denoted by ABC, which is the event consisting of all points common to A, B, and C.

A relation between two events A and B that deserves special mention is the condition when *no* elementary outcomes simultaneously belong to both A and B, as shown in Figure 3.4.

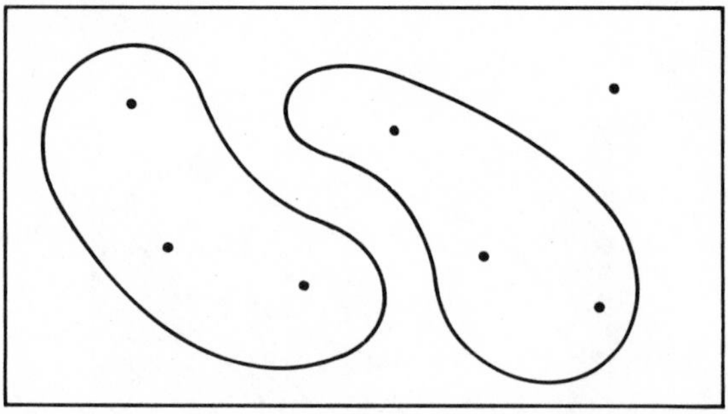

Figure 3.4 Venn diagram of two mutually exclusive events.

> Two events A and B are called *disjoint* or *mutually exclusive* events if they do not have any elementary outcome in common; that is, if their intersection AB is empty. Disjoint events cannot occur simultaneously.

For general guidance, we list some commonly used statements that signify event operations and their corresponding symbolic representations:

Verbal statement	*Event operation*
At least one of A or B	$A \cup B$
Both A and B	AB
Not A	A^c

We now turn to some formulas that allow us to express the probability of an event, which is related to other events by some operation, in terms of the probabilities of the components.

Law of complementation

$$P(A^c) = 1 - P(A)$$

To understand this relation, we recall that sum of $P(e)$ over all e in $\mathbb{S}$ is 1. Since A and A^c are two disjoint sets that together comprise the entire $\mathbb{S}$, we have

$$P(A) + P(A^c) = \sum_A P(e) + \sum_{A^c} P(e)$$
$$= \sum_{\mathbb{S}} P(e) = 1$$

This relationship can be used to compute $P(A)$ when A^c is simple in nature and its probability $P(A^c)$ is easy to compute directly.

Addition law for the union of disjoint events

If two events A and B are disjoint, then

$$P(A \cup B) = P(A) + P(B)$$

General addition law

For any two events A and B

$$P(A \cup B) = P(A) + P(B) - P(AB)$$

To derive these two formulas, we note that $P(A \cup B)$ is the sum of $P(e)$ for all e in the set $(A \cup B)$. Each such e is to be counted only once, but

$$P(A) + P(B) = \sum_A P(e) + \sum_B P(e)$$

Since the intersection AB is the part common to both A and B, this sum has double counted $P(e)$ for those outcomes e that are in AB. To avoid this double counting, we must subtract $P(AB)$ from this sum to obtain $P(A \cup B)$. This gives us the general addition law of probability. When the events are disjoint, AB is empty and its probability is 0. Thus the first addition law is a special case of the second.

The addition law expresses the probability of a larger event $A \cup B$ in terms of the probabilities of the smaller events A, B, and AB. Some applications of the addition and complementation laws are given in the following examples.

Example 3.5 In an examination, a student is asked to check "true" or "false" for each of 6 statements. Suppose that instead of bothering to read and think about the statements before answering, the student decides to check "true" or "false" at random. (*a*) What is the probability that the student will have at least one correct answer? (*b*) What is the probability that the student will have 5 or 6 correct answers?

The operation of marking at random means that the answer to each statement is equally likely to be right (R) or wrong (W). There are two possible outcomes R and W for each statement. Using the method of counting explained in Example 3.3, the number of possible outcomes for the set of 6 statements is $2\times2\times2\times2\times2\times2=64$. The process of random checking implies that all 64 elementary outcomes in the sample space are equally likely; that is, it is a uniform probability model with 64 elementary outcomes.

(*a*) Let the event of having at least one right answer be denoted by A, which consists of many elementary outcomes. On the other hand, A^c is the event of having all answers wrong, which has the single outcome $\{(WWWWWW)\}$. We have $P(A^c)=1/64$ and by the law of complementation

$$P(A)=1-\frac{1}{64}=\frac{63}{64}$$

(*b*) Let A_5 and A_6 denote the events of having exactly 5 and exactly 6 right answers, respectively. Then A_5 consists of all the sequences having 5 R's and 1 W, for which there are the 6 outcomes

$$(WRRRRR), (RWRRRR), (RRWRRR),$$

$$(RRRWRR), (RRRRWR), (RRRRRW)$$

Therefore $P(A_5)=6/64$. Since A_6 consists of the single outcome $(RRRRRR), P(A_6)=1/64$. Finally, A_5 and A_6 are disjoint and therefore

$$\begin{aligned} P(5 \text{ or } 6 \text{ correct answers}) &= P(A_5 \cup A_6) \\ &= P(A_5)+P(A_6) \\ &= \frac{6}{64}+\frac{1}{64}=\frac{7}{64} \end{aligned}$$ ■

Example 3.6 Table 3.1 gives the proportions of a group of high-school students in different categories of performance in mathematics and physical education. If one student is selected at random from this group, what is the probability that the student will be (*a*) Good in at least one subject? (*b*) Not poor in mathematics?

TABLE 3.1

PROPORTIONS OF STUDENTS IN DIFFERENT CATEGORIES OF PERFORMANCE

		PHYSICAL EDUCATION		
		Good	Average	Poor
MATHEMATICS	Good	.06	.11	.18
	Average	.12	.18	.10
	Poor	.16	.05	.04

The elementary outcomes in the sample space of this experiment are the 9 performance categories shown in Table 3.1. Using the operational notion of random selection described in Example 3.1, the probability that the selected student will be in any specific category is precisely the proportion of the population belonging to this category. For instance, the probability that he or she is good in both the subjects is .06, which is the proportion in the upper left-hand corner of Table 3.1.

(*a*) Let A be the event that the student is good in mathematics and let B be the event that the student is good in physical education. The event of being good in at least one subject is then denoted by $A \cup B$, and we obtain

$$P(A) = .06 + .11 + .18 = .35$$

$$P(B) = .06 + .12 + .16 = .34$$

$$P(AB) = .06$$

By the addition law of probability

$$\begin{aligned} P(A \cup B) &= P(A) + P(B) - P(AB) \\ &= .35 + .34 - .06 \\ &= .63 \end{aligned}$$

This can also be seen directly from the fact that the total fraction having the property of being good in at least one subject is

$$.06 + .11 + .18 + .12 + .16 = .63$$

(*b*) Letting D be the event that the student is poor in mathematics, we want to compute $P(D^c)$. Noting that

$$P(D) = .16 + .05 + .04 = .25$$

and using the rule of complementation, we obtain

$$P(D^c) = 1 - .25 = .75$$

■

3.5 RULES OF COUNTING AND THEIR USE IN UNIFORM PROBABILITY MODELS

As we noted earlier, uniform probability models are particularly useful when the sampling scheme is governed by a chance mechanism that makes all possible selections equally likely. In the uniform probability model, all elementary outcomes in the sample space are equally likely and the probability of an event A is given by the ratio

$$P(A) = \frac{\text{Number of elements in } A}{\text{Number of elements in } \mathcal{S}}$$

Thus the calculation of the probability of an event is essentially reduced to counting the number of elements belonging to the event and the number of elements in the entire space $\mathcal{S}$. When convenient methods of counting are available to calculate probability, we can forego listing all the elements of $\mathcal{S}$. Some basic rules for counting that can aid us in this direct calculation of probability are explored in this section.

Product rule

When an experiment consists of two parts such that the first part can have k distinct results, and if with each result of the first part there can be l distinct results of the second part, then the total number of possible results of the experiment is $k \times l$.

The product rule can be readily extended to experiments involving more than two parts. This extension and other implications of the product rule are illustrated in the following examples.

Example 3.7 When preparing a lawsuit on a patent infringement, a company has a choice of three lawyers and also a choice of one of two courts in which to initially try the case. How many decisions can be reached on lawyer and court selections?

There are $k=3$ choices for the lawyer, and for each of these, there are $l=2$ choices for the court. The number of different decisions possible is therefore $k \times l = 3 \times 2 = 6$. ■

Example 3.8 Four fungicides a, b, c, and d are to be tested by applying each to a different tree, and the four trees are to be selected from a row of 10 trees. How many assignments of the four fungicides are possible?

Considering the assignment of a, b, c, and d as the four parts of an experiment, any of the 10 trees can be chosen for a. For every choice of a tree for a, there are 9 trees left, any of which can be chosen for b. Thus as far as the assignment of a and b are concerned, the number of possible choices is 10×9 according to the product rule. Extending this agrument to the remaining two fungicides c and d, the total number of possible assignments of all four fungicides is $10 \times 9 \times 8 \times 7 = 5040$. ■

Rule of permutations or arrangements

The number of different orderings or arrangements that can be formed with r objects selected from a group of n distinct objects is denoted by P_r^n, which is read "the number of permutations of r out of n."

We can now proceed to derive a computational formula for the count P_r^n. Note that Example 3.8 involved precisely the same kind of count described here. The four fungicides a, b, c, and d can be viewed as $r=4$ positions or slots to which four trees are to be assigned by choosing from $n=10$ trees.

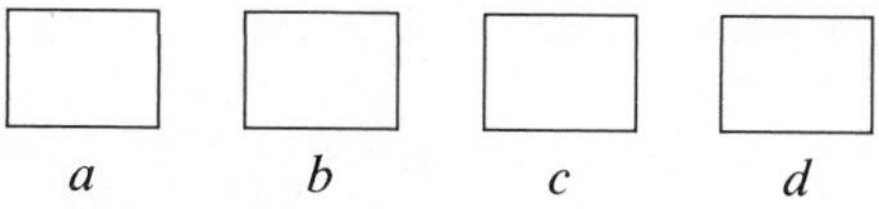

The first slot can be filled with any of the 10 trees. After filling the first slot, there are 9 trees left, any of which can be assigned to the second slot. Proceeding to fill the third and fourth slots in this way, the total number of possible assignments is $10 \times 9 \times 8 \times 7$. Therefore, the computational formula for P_4^{10} is $10 \times 9 \times 8 \times 7 = 5040$.

Motivated by this example, the computational formula for the general symbol P_r^n can now be presented as

> The *number of possible arrangements* of r objects selected from n distinct objects is
>
> $$P_r^n = \underbrace{n(n-1)(n-2)\cdots}_{r \text{ factors}}$$
>
> $$= n(n-1)(n-2)\cdots(n-r+1)$$

Example 3.9 There are 15 people competing in a bicycle race. In how many ways can the first, second, and third prizes be awarded to persons participating in the race?

Identifying the first, second, and third prizes as the three slots, we want to find the number of possible arrangements of 3 persons that can be made in these slots when the persons are chosen from the 15 racers. By the rule of permutations, the answer is $P_3^{15} = 15 \times 14 \times 13 = 2730$. ■

Example 3.10 An urn contains 12 articles, 8 of which are good (marked $G_1, G_2, \ldots, G_8$) and 4 of which are defective (marked D_1, D_2, D_3, D_4). Suppose that 3 articles are to be drawn from the urn one after another without replacement, and the results of the individual draws are to be recorded. (*a*) Keeping track of the order of selections, how many ordered samples (arrangements) are possible? (*b*) How many ordered samples are possible under the condition that the first two draws produce defective articles and the third yields a good piece? (*c*) If the articles are drawn at random (that is, in a manner such that at each draw all the articles remaining in the urn are equally likely to be selected), what is the probability of obtaining defectives in the first two draws and a good piece in the third draw?

(*a*) The 12 articles are distinct objects in the urn from which 3 articles are to be selected one after another. Thus the number of arrangements or possible ordered samples is $P_3^{12} = 12 \times 11 \times 10 = 1320$.

(*b*) The number of ways the first two draws can produce defective articles is the number of arrangements of two objects taken from the group $\{D_1, D_2, D_3, D_4\}$; this number is P_2^4. The number of ways the third draw can produce a good piece when the first two draws have yielded defective articles is P_1^8 by the same reasoning. Finally, by the product rule, the number of ways the first two draws can produce defectives and the third can yield a good piece is

$$P_2^4 \times P_1^8 = (4 \times 3) \times 8 = 96$$

(*c*) Randomly drawing 3 articles ensures that all P_3^{12} ordered samples are equally likely. The event A of obtaining defectives in the first two draws and a good piece in the third consists of $P_2^4 \times P_1^8$ ordered samples. Therefore

$$P(A) = \frac{P_2^4 \times P_1^8}{P_3^{12}} = \frac{96}{1320}$$ ■

> The number of possible arrangements of n distinct objects taken all at a time is
>
> $$P_n^n = n(n-1)(n-2) \times \cdots \times 2 \times 1$$
> $$= 1 \times 2 \times \cdots \times n$$
>
> This is also denoted by $n!$ and is read "*n factorial*."

The previous rule dealt with enumerating all possible arrangements when choosing r objects out of n distinct objects. But in many situations, we are interested only in the number of possible choices of a group of r objects out of n objects, and the various arrangements among the chosen r objects are of no concern to us. We now want to derive a formula that provides a count of the number of combinations of r out of n objects with no consideration for their order of arrangement. An example will help to clarify the distinction between these two situations.

Example 3.11 Enumerate all the possible (*a*) combinations and (*b*) arrangements of three letters chosen from the four letters A, B, C, and D:

COMBINATIONS	ARRANGEMENTS
$\{A, B, C\}$	$ABC, ACB, BAC, BCA, CAB, CBA$
$\{A, B, D\}$	$ABD, ADB, DAB, DBA, BAD, BDA$
$\{A, C, D\}$	$ACD, ADC, CAD, CDA, DAC, DCA$
$\{B, C, D\}$	$BCD, BDC, CBD, CDB, DBC, DCB$

All the arrangements in the same row correspond to the same collection or combination of letters. Alternatively, the three letters in each combination, when arranged among themselves, provide $3! = 3 \times 2 \times 1 = 6$ arrangements. The counting rule for the number of arrangements of 3 letters out of 4 is given by $P_3^4 = 24$. Since each set of $3! = 6$ arrangements identifies the same collection, the number of possible collections can be determined as

$$\frac{P_3^4}{3!} = \frac{24}{6} = 4$$

■

> *Rule of combinations*
>
> The number of possible collections of r objects chosen from a group of n distinct objects is denoted by the symbol $\binom{n}{r}$, which is read "the number of combinations of r out of n." The computational formula is given by
>
> $$\binom{n}{r} = \frac{P_r^n}{r!}$$
>
> $$= \frac{n(n-1)\cdots(n-r+1)}{r!}$$

The general formula is obtained from the following correspondence and an application of the product rule:

$$\begin{pmatrix}\text{Arrange } r \text{ objects}\\ \text{selected from } n\end{pmatrix} \begin{matrix}\text{is the}\\ \text{same as}\end{matrix} \begin{pmatrix}\text{First select } r\\ \text{objects from } n\end{pmatrix} \begin{matrix}\text{and}\\ \text{then}\end{matrix} \begin{pmatrix}\text{arrange the } r\\ \text{selected objects}\end{pmatrix}$$

$$P_r^n \qquad = \qquad \binom{n}{r} \qquad \times \qquad r!$$

Example 3.12 Compute the values of $\binom{5}{3}$, $\binom{7}{2}$, $\binom{7}{5}$, and $\binom{10}{3}$.

$$\binom{5}{3} = \frac{5 \times 4 \times 3}{1 \times 2 \times 3} = 10 \qquad \binom{7}{2} = \frac{7 \times 6}{1 \times 2} = 21$$

$$\binom{7}{5} = \frac{7 \times 6 \times 5 \times 4 \times 3}{1 \times 2 \times 3 \times 4 \times 5} = 21 \qquad \binom{10}{3} = \frac{10 \times 9 \times 8}{1 \times 2 \times 3} = 120$$

■

We now list a few properties of the formula of combinations that are useful in applications.

Properties of the combination formula

$$\binom{n}{r} = \binom{n}{n-r} = \frac{n!}{r!(n-r)!}$$

$$\binom{n}{n} = 1, \quad \binom{n}{1} = n$$

To verify the first relation, we write our original formula for $\binom{n}{r}$, or

$$\binom{n}{r} = \frac{n(n-1)\cdots(n-r+1)}{r!}$$

If we then multiply the numerator and the denominator by the common factors $(n-r)(n-r-1)\cdots(2)(1)$, the ratio remains unchanged and we have

$$\binom{n}{r} = \frac{n(n-1)\cdots(n-r+1)\{(n-r)\cdots 2\cdot 1\}}{r!\qquad\{(n-r)\cdots 2\cdot 1\}}$$

$$= \frac{n!}{r!(n-r)!}$$

Also, replacing r with $(n-r)$ merely interchanges the positions of the two factorials in the denominator, and this gives us the identity $\binom{n}{r} = \binom{n}{n-r}$. Intuitively, this equality can also be visualized by considering the fact that for every collection of r drawn from n, there are $(n-r)$ items that remain to form a collection of size $(n-r)$. Thus the number of collections of r items out of n must be the same as the number of collections of $(n-r)$ items out of n, which is exactly what the first relation says.

This identity is often convenient in simplifying computations. For example, to compute $\binom{20}{18}$ we could calculate

$$\binom{20}{18} = \frac{20\times 19\times 18\times\cdots\times 3}{1\times 2\times 3\times\cdots\times 18}$$

and then cancel the common factors in the numerator and the denominator. However, much of this work can be avoided if we use the relation $\binom{20}{18} = \binom{20}{2}$ and then evaluate $\binom{20}{2}$ as $(20\times 19)/(1\times 2) = 190$.

Note that $\binom{n}{n}=1$ because all n objects taken together form a single collection. On the other hand, the formula

$$\binom{n}{r}=\frac{n!}{r!(n-r)!}$$

gives

$$\binom{n}{n}=\frac{n!}{n!0!}=\frac{1}{0!}$$

where 0! is not yet defined. For this formula to remain valid when $r=n$ we must then define $0!=1$.

> Definition: $0!=1$
>
> Consequently, $\binom{n}{n}=1=\binom{n}{0}$

Example 3.13 Consider in Example 3.10, the urn containing 12 articles 8 of which are good (marked $(G_1, G_2, \ldots, G_8)$ and 4 of which are defective (marked D_1, D_2, D_3, D_4). Now suppose that 3 articles will be drawn from the urn simultaneously and that the collection of objects drawn will be recorded. (*a*) How many distinct results are possible? (*b*) How many results are possible under the condition that two of the items drawn are defective and one is a good piece? (*c*) If three articles are drawn at random (that is, in a manner such that each collection of size 3 has an equal chance of being selected), what is the probability of obtaining two defective pieces and one good piece?

In contrast to Example 3.10, here we are not concerned with arranging the sampled articles according to the order of selection. This example requires counting the number of possible collections; thus the formula of combinations should be employed.

(*a*) The number of possible unordered samples of 3 out of 12 articles is $\binom{12}{3}=(12\times 11\times 10)/(1\times 2\times 3)=220$.

(*b*) The number of combinations of 2 defectives out of the group of 4 defectives is $\binom{4}{2}$, and the number of combinations of 1 good piece out of the group of 8 good pieces is $\binom{8}{1}$. By the product rule, the number of possible results meeting the specified condition is

$$\binom{4}{2}\binom{8}{1}=\frac{4\times 3}{1\times 2}\cdot\frac{8}{1}=48$$

(*c*) According to the process of sampling, the $\binom{12}{3}$ possible samples of 3 articles out of 12 are all equally likely. The number of possible samples having exactly 2 defectives and 1 good piece is $\binom{4}{2}\binom{8}{1}$. Using the uniform probability model, we then obtain

$$P(2 \text{ defectives and } 1 \text{ good}) = \frac{\binom{4}{2}\binom{8}{1}}{\binom{12}{3}} = \frac{48}{220} \qquad \blacksquare$$

Beginners often confuse situations when the rule of permutations should be used with situations when the rule of combinations should be employed.

> As a general guideline, it is helpful to remember that the rule of permutations should be used when an event is described in terms of a specific ordering of the results. When the order in which the results occur is irrelevant, the rule of combinations is the appropriate counting rule to use.

Example 3.14 An advisory committee on prison reform consists of 15 members, of whom 9 favor, 4 oppose, and 2 are indifferent about a specific program. A reporter wishes to randomly select 3 persons from this committee and record their views on a television newscast. (*a*) What is the probability that at least two of the selected persons will be in favor of the program? (*b*) What is the probability that the first two persons selected will favor the program and the third will oppose it?

(*a*) Here the event of interest does not involve any ordering of the persons in the sampling process. Three persons can be selected out of 15 in $\binom{15}{3} = 455$ ways, which are all equally likely to occur according to the concept of random selection. Let A_2 and A_3 be the events of having exactly 2 and 3 persons, respectively, in the sample, who favor the program. The problem requires us to calculate $P(A_2 \cup A_3)$. These events are disjoint, and we have

$$P(A_2 \cup A_3) = P(A_2) + P(A_3)$$

To calculate $P(A_2)$, we note that the number of ways of selecting 3 persons from 15 such that 2 of them come from the group of 9 who favor the program and 1 comes from the remaining 6 is given by

$$\binom{9}{2}\binom{6}{1}=\frac{9\times 8}{1\times 2}\times 6=216$$

Thus, employing the uniform probability model, we obtain

$$P(A_2)=\frac{\binom{9}{2}\binom{6}{1}}{\binom{15}{3}}=\frac{216}{455}$$

Similarly

$$P(A_3)=\frac{\binom{9}{3}\binom{6}{0}}{\binom{15}{3}}=\frac{84}{455}$$

The required probability is then

$$P(\text{at least two in favor})=\frac{216}{455}+\frac{84}{455}=\frac{300}{455}$$

(*b*) This problem involves recognizing the order of selection of the persons, and therefore we must employ the rule of permutations. There are $P_3^{15}=15\times 14\times 13$ ordered selections of 3 persons from a group of 15 that are all equally likely due to the process of random selection. The number of selections in which the first two people are in favor and the third person is opposed is

$$P_2^9\times P_1^4=(9\times 8)(4)=288$$

and thus we obtain

$$P(\text{first two in favor, third opposed})=\frac{9\times 8\times 4}{15\times 14\times 13}=\frac{288}{2730} \qquad \blacksquare$$

Until now we have used the term "random sampling" with the intuitive understanding that the selection process makes the occurrence of any particular sample as likely as any other. In addition to illustrating the uniform probability model, this concept of sampling is the backbone for the role of probability in making statistical inferences about a population from information contained in a sample. Because its role is extremely important in the logical basis of statistical inference, here we formally define a random sample when the order of selection is not important.

> A sample of size n selected from a population of N distinct objects is said to be a *random sample* if each collection of size n has the same probability $1/\binom{N}{n}$ of being selected.

Note that this is a conceptual rather than an operational definition of a random sample. On the surface it might seem that a haphazard selection by the experimenter would result in a random sample. Unfortunately, because of an inherent susceptibility of the human mind to inclinations and aversions, one person's haphazard selection may very well reveal a pattern to another experimenter. Therefore, this job must be entrusted to some device that cannot think; in other words, some sort of mechanization of the selection process is needed to make it truly haphazard! We already mentioned the simple device of first identifying the N population units with N cards and then drawing cards after the lot is shuffled (See Example 3.1). This method is easy to understand but rather awkward to apply to large-size populations. More streamlined mechanical devices are available for convenient applications, and these are discussed in Chapter 16.

3.6 CONDITIONAL PROBABILITY

The probability of an event A must often be modified after information is obtained as to whether or not a related event B has taken place. Information about some aspect of the experimental results may therefore necessitate a revision of the probability of an event concerning some other aspect of the results. The revised probability of A when it is known that B has occurred is called the *conditional probability of A given B* and is denoted by $P(A|B)$. To illustrate how such a modification is made, we consider an example that will lead us to the formula for conditional probability.

Example 3.15 A group of executives are classified according to the status of body weight and incidence of hypertension. The proportions in the various categories appear in Table 3.2. (*a*) What is the probability that a person selected at random from this group will have hypertension? (*b*) What is the probability that a person, selected at random from this group, who is found to be overweight will also have hypertension?

Let A denote the event that a person is hypertensive, and let B denote the event that a person is overweight.

(*a*) Since 20% of the group is hypertensive and the individual is selected at random from this group, we conclude that $P(A) = .2$. This is the unconditional probability of A.

TABLE 3.2
BODY WEIGHT AND HYPERTENSION

	OVERWEIGHT	NORMAL WEIGHT	UNDERWEIGHT	TOTAL
Hypertensive	.10	.08	.02	.20
Not hypertensive	.15	.45	.20	.80
Total	.25	.53	.22	1.00

(*b*) When we are given the information that the selected person is overweight, the categories in the second and third columns of Table 3.2 are not relevant to this person. The first column shows that among the subgroup of overweight persons, the proportion having hypertension is .10/.25. Therefore, given the information that the person is in this subgroup the probability that he or she is hypertensive is

$$P(A|B)=\frac{.10}{.25}=.4$$

Noting that $P(AB)=.10$ and $P(B)=.25$, we have derived $P(A|B)$ by taking the ratio $P(AB)/P(B)$. In other words, $P(A|B)$ is the proportion of the population having the characteristic A among all those having the characteristic B. ■

The *conditional probability of A given B* is denoted by $P(A|B)$ and is defined by the formula

$$P(A|B)=\frac{P(AB)}{P(B)}$$

whenever $P(B)>0$. Equivalently, this formula can be written

$$P(AB)=P(B)P(A|B)$$

This version is called the *multiplication law of probability*.

Similarly, the conditional probability of B given A can be expressed

$$P(B|A)=\frac{P(AB)}{P(A)}$$

which gives the relation $P(AB)=P(A)P(B|A)$. Thus the multiplication law of probability states that the conditional probability of an event multiplied by the probability of the conditioning event gives the probability of the intersection.

The multiplication law can be used in one of two ways, depending on convenience. When it is easy to compute $P(A)$ and $P(AB)$ directly, these values can be used to compute $P(A|B)$, as in Example 3.15. On the other hand, if it is easy to calculate $P(B)$ and $P(A|B)$ directly, these values can be used to compute $P(AB)$.

The multiplication law can be readily extended to more than two events. For example, with three events A, B, and C, the formula is

$$P(ABC)=P(A)P(B|A)P(C|AB)$$

Example 3.16 A farmer has a box containing 30 eggs, 5 of which have blood spots. He checks 3 eggs by taking them at random one after another from the box. What is the probability that the first two eggs will have spots and the third will be clear?

Let S_1, S_2, C_3 denote the events of choosing spotted eggs in the first two draws and a clear egg in the third draw. We want to find the probability of the intersection $S_1S_2C_3$ of these three events. Using the multiplication law, we write

$$P(S_1S_2C_3)=P(S_1)P(S_2|S_1)P(C_3|S_1S_2)$$

Since the drawings are made at random, we have $P(S_1)=5/30$. To find $P(S_2|S_1)$, note that when S_1 has occurred there are 29 eggs left in the box, 4 of which are spotted. Thus the conditional probability of picking an S in the second draw is directly evaluated as $P(S_2|S_1)=4/29$. Using the same reasoning, $P(C_2|S_1S_2)=25/28$. Placing this in the preceding multiplication formula, we obtain

$$P(S_1S_2C_3)=\frac{5}{30}\cdot\frac{4}{29}\cdot\frac{25}{28}=\frac{25}{1218}$$

The problem can also be calculated using the appropriate rules of counting. Since the order of the draws is relevant, we should use the rule of permutations. There are P_3^{30} possible ordered samples when selecting 3 eggs out of 30, and these are equally likely. The number of ordered samples in which the first two eggs are spotted and the third is clear is $P_2^5\times P_1^{25}$. Thus

$$P(S_1S_2C_3)=\frac{P_2^5\times P_1^{25}}{P_3^{30}}=\frac{5\times4\times25}{30\times29\times28}=\frac{25}{1218}$$ ■

A situation that merits special attention occurs when the conditional probability $P(A|B)$ turns out to be the same as the unconditional probability $P(A)$. Information about the occurrence of B then has no bearing on the assessment of the probability of A. Therefore, when we have the equality $P(A|B)=P(A)$, we say that events A and B are independent.

> Two events A and B are *independent* if
>
> $$P(A|B)=P(A)$$
>
> Equivalent conditions are
>
> $$P(B|A)=P(B)$$
>
> or
>
> $$P(AB)=P(A)P(B)$$

The last form follows by recalling that $P(A|B)=P(AB)/P(B)$, so that the condition $P(A|B)=P(A)$ is equivalent to

$$P(AB)=P(A)P(B)$$

which may be used as an alternative definition of independence. The other equivalent form is obtained by

$$P(B|A)=P(AB)/P(A)=P(A)P(B)/P(A)=P(B)$$

The form $P(AB)=P(A)P(B)$ shows that the definition of independence is symmetric in A and B.

We introduced the condition of independence in the context of checking a given assignment of probability to see if $P(A|B)=P(A)$. A second use of this condition is in the assignment of probability when the experiment consists of two physically unrelated parts. When events A and B refer to separate parts of an experiment, AB is assigned the probability $P(AB)=P(A)P(B)$.

Example 3.17 Consider the experiment of tossing a fair coin two times. The symmetry of the experiment indicates that the outcomes $\{HH, HT, TH, TT\}$ are equally likely. Let three events be defined:

A: Head in the first toss.

B: Head in the second toss.

C: Both tosses are heads or both are tails.

Which pairs of events are independent?

Since events A and B refer to two physically unrelated parts of the experiment, we can intuitively say that they should be independent. For the other two pairs, the situation is not obvious. To check the condition of independence for all three pairs, we present the events in a Venn diagram in Figure 3.5.

Since each elementary outcome in $\mathcal{S}$ has a probability of 1/4, we have

$$P(A)=\frac{1}{2} \quad P(AB)=\frac{1}{4}$$
$$P(B)=\frac{1}{2} \quad P(AC)=\frac{1}{4}$$
$$P(C)=\frac{1}{2} \quad P(BC)=\frac{1}{4}$$

These probabilities yield

$$P(A)P(B)=\frac{1}{4}=P(AB)$$

$$P(A)P(C)=\frac{1}{4}=P(AC)$$

$$P(B)P(C)=\frac{1}{4}=P(BC)$$

Therefore, according to the multiplication formula for independence, A and B are independent and so are the other two pairs. ■

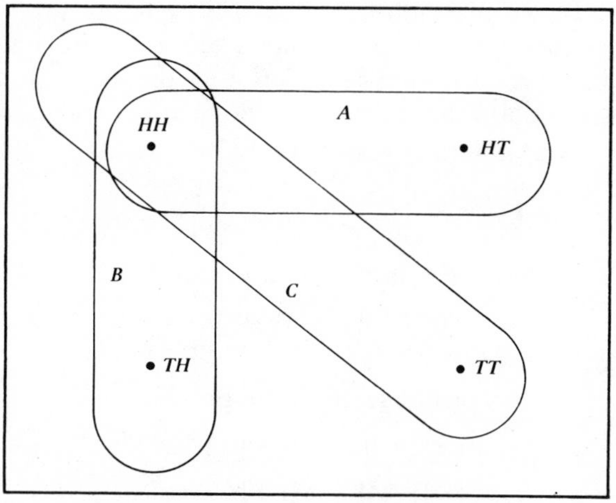

Figure 3.5 Venn diagram of two coin tosses.

Example 3.18 Are the two events $A = \{\text{hypertensive}\}$ and $B = \{\text{overweight}\}$ independent for the population in Example 3.15?

Referring to that example

$$P(A) = .2$$

$$P(A|B) = \frac{P(AB)}{P(B)} = \frac{.10}{.25} = .4$$

so that the two events A and B are not independent. ■

Warning: Students often confuse the terms "mutually exclusive events" and "independent events." These two properties are quite different; in fact, one implies that the other cannot be true. Consider events A and B whose probabilities are not zero. When they are mutually exclusive, the intersection AB is empty and $P(AB)=0$. If these events were also independent, they would have to meet the requirement $P(A)P(B)=P(AB)$, which cannot be true because the product of two nonzero numbers cannot be zero. As an extreme example, the events A and A^c are mutually exclusive, but intuitively they are quite dependent in the sense that as soon as we are told that A has occurred we are certain that A^c did not occur. We can verify this by calculating $P(A^c|A) = P(A^cA)/P(A) = 0/P(A) = 0$.

The following example illustrates an interesting application of the concept of conditional probability.

Example 3.19 An imperfect clinical test: A blood disease is present in 2% of a population in a serious form, in 10% in a light form, and is not present at all in the remaining 88%. An imperfect clinical test is 90% successful in detecting the disease in serious cases. In other words, if a person has a serious case of the disease, the probability is .9 that the clinical test will be positive and the probability is .1 that it will be negative. Moreover, the success rate of the clinical test is .6 among light cases; among unaffected persons, the probability that the test will be positive is .1. A person selected at random from the population is given the test, and the result is positive. What is the probability that this person has a serious case of the disease?

First we introduce some convenient symbols to denote the various events:

Serious case:	S	Test positive: $+$
Light case:	L	Test negative: $-$
No disease:	N	

The following probabilities are given in the problem:

$$P(S) = .02 \qquad P(+|S) = .9$$

$$P(L) = .10 \qquad P(+|L) = .6$$

$$P(N) = .88 \qquad P(+|N) = .1$$

The problem is to determine the conditional probability $P(S|+)$. We proceed by using the formula for conditional probability

$$P(S|+)=\frac{P(S+)}{P(+)}$$

From the given probabilities and the multiplication law, we can compute

$$P(S+)=P(S)P(+|S)=.02\times.9=.018$$

$$P(L+)=P(L)P(+|L)=.10\times.6=.060$$

$$P(N+)=P(N)P(+|N)=.88\times.1=.088$$

Now the events $S+$, $L+$, and $N+$ are three disjoint events whose union is the event $+$. Applying the addition rule of probability for the union of disjoint events, we obtain

$$\begin{aligned}P(+)&=P(S+)+P(L+)+P(N+)\\&=.018+.060+.088\\&=.166\end{aligned}$$

Finally

$$P(S|+)=\frac{P(S+)}{P(+)}=\frac{.018}{.166}=.11$$

*3.7 BAYES' THEOREM

Example 3.19 illustrates a situation in which an event of interest called A occurs in conjunction with one and only one of the events $B_1, B_2, \dots, B_k$, which form a partition of the sample space into disjoint sets. This is illustrated in Figure 3.6 for the case $k=3$.

Moreover, the probabilities of $B_1, B_2, \dots, B_k$, as well as the conditional probabilities $P(A|B_1), \dots, P(A|B_k)$ are known. We are then interested in finding the conditional probability $P(B_1|A)$ from the given information. That Example 3.19 is precisely in this form can be recognized by identifying A with the event $+$ and B_1, B_2, B_3 with the events S, L, and N, respectively.

To derive a formula for $P(B_1|A)$ in terms of the given information, we first write

$$P(B_1|A)=\frac{P(AB_1)}{P(A)}$$

Noting that A is the union of disjoint events $AB_1, AB_2, \dots, AB_k$, we have

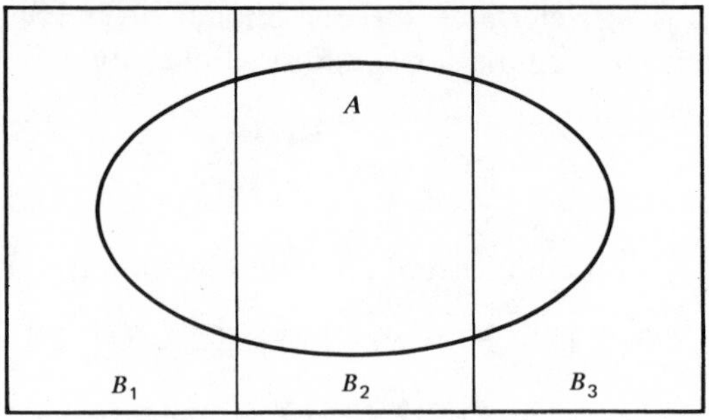

Figure 3.6 Structure for Bayes' Theorem.

$P(A)=P(AB_1)+P(AB_2)+\cdots+P(AB_k)$. Each term in this sum can be obtained by writing

$$P(AB_j)=P(B_j)P(A|B_j)$$

and these combine to give the result

> *Bayes' Theorem*
>
> $$P(B_1|A)=\frac{P(B_1)P(A|B_1)}{\sum_{j=1}^{k} P(B_j)P(A|B_j)}$$

This formula was brought into focus by Reverend Thomas Bayes (1702–1761) and is therefore known as *Bayes' Theorem*. Although a straightforward consequence of the concept of conditional probability, Bayes' Theorem has far-reaching implications in the development of a certain philosophy of statistical inference designated *Bayesian inference*. This is due to the interpretation given to this result. The B_j are thought of as the possible states of nature to which the experimenter assigns subjective probabilities. These probabilities, which may be based on personal feelings rather than on data (indeed data may be totally lacking), are then combined with the experimental evidence A.

Initially, the experimenter is aware of several possible states of nature $B_1, B_2, \ldots, B_k$ but does not know precisely which one of them truly prevails. For example, to a pharmacologist, two unknown states of nature may be that one ingredient is more effective or not more effective than another; to an advertizing agency, they may be the effectiveness of various color combinations in a display. Based on the present state of knowledge about the situation or on experimental evidence gathered from similar situations, the experimenter may make some assessment of the probabilities $P(B_1), P(B_2), \ldots, P(B_k)$ that reflects his or her feelings as to how likely the various alternative states of nature are to prevail. Such probabilities are called *prior* or *pre-experimental probabilities* for the states of nature. Now, the experimenter proceeds with the observation or experiment and collects data, which we denote by A. The experimenter may then determine the probability of the experimental evidence A given that it has been observed under a specific state of nature B_j. These are the conditional probabilities $P(A|B_j)$, $j=1,\ldots,k$. Bayes' Theorem then allows the experimenter to calculate the conditional probabilities $P(B_j|A), j=1,\ldots,k$, which are nothing more than an expression of his or her revised belief concerning the different states of nature after observing the experimental evidence. These revised probabilities are called *posterior* or *post-experimental probabilities*; they form the basis of whatever inferences the experimenter wishes to make about the unknown state of nature.

Although this line of reasoning has been criticized by some schools of thought because prior probabilities can be affected by the experimenter's own biased viewpoint, it is considerably appreciated by researchers in many fields, including business, economics, and education. However, due to the limited scope of this book, we do not explore the philosophy and the techniques of Bayesian inference.

*3.8 GENERAL RULES OF PROBABILITY

Our discussion of the properties of probability in Sections 3.3–3.6 has been confined to discrete sample spaces where the elementary outcomes are either finite or can be arranged in a sequence. In other experiments, measurements are taken on a continuous scale of such characteristics as height, weight, and temperature. The resulting sample space consists of all real numbers in an interval and is called a *continuous sample space*. Much of our development including the interpretation of the probability of an event as the long-range relative frequency and most of the properties of probability remain valid for these spaces. However, a notable exception is that the relation

$$P(A) = \sum_{\text{all } e \text{ in } A} P(e)$$

is meaningless when the elementary outcomes e in A are not only infinite but also cannot be arranged in a sequence. In mathematics, the summation operation is not defined if the terms to be summed cannot be arranged in a sequence. This is the case, for instance, when A is an interval and the elementary outcomes are the points in this interval. Because of this difficulty, one cannot assign a probability $P(e)$ to the elementary outcomes e of $\mathbb{S}$ first and then obtain the probability of an event A by adding $P(e)$ for all e in A. We must handle the probabilistic structure of a continuous sample space by considering the various events (typically, the subintervals of $\mathbb{S}$) and then assigning numerical values to the probabilities of these events.

We present the general definition of probability by stating conditions that any assignment must satisfy if it is to be considered a probability. The conditions are motivated by the behavior of relative frequencies, and they coincide with the properties of probability in discrete spaces.

> *Probability* P is a numerical valued function that is defined on the events in a sample space $\mathbb{S}$ and that satisfies the following conditions:
>
> (*i*) $0 \leqslant P(A) \leqslant 1$ for all events A
>
> (*ii*) $P(\mathbb{S}) = 1$
>
> (*iii*) For *disjoint* events $A_1, A_2, \ldots$
>
> $$P(A_1 \cup A_2 \cup \ldots) = P(A_1) + P(A_2) + \ldots$$

The rules of probability that we previously demonstrated in the context of discrete sample spaces can now be derived as consequences of these three conditions. To obtain the rule of complementation, we note that A and A^c are two disjoint events and $A \cup A^c = \mathbb{S}$. By conditions (*iii*) and (*ii*) we have

$$P(A \cup A^c) = P(A) + P(A^c)$$

$$P(\mathbb{S}) = 1$$

from which we obtain $P(A^c) = 1 - P(A)$. The derivation of the addition law for the union of two events is left as an exercise for the reader. The definition of conditional probability remains the same as the definition for discrete sample spaces, when $P(B) > 0$.

> *Three important laws of probability for a general sample space*
>
> $$P(A) = 1 - P(A^c)$$
>
> $$P(A \cup B) = P(A) + P(B) - P(AB)$$
>
> $$P(AB) = P(B)P(A|B)$$

***Example 3.20** To study fish movement in lakes, experimenters often release tagged fish at specific points in the lake and use an electronic device to measure the direction of their positions after a certain period of time. The model of no preferred orientation can be identified as that of spinning a pointer on a disk, as described below.

A disk of unit circumference has a pointer pivoted at the center that rotates freely. After spinning, the pointer is as likely to stop in some specified arc of the circle as in any other arc of equal length.

The experiment consists of spinning the pointer and recording its stopping position (see Figure 3.7).

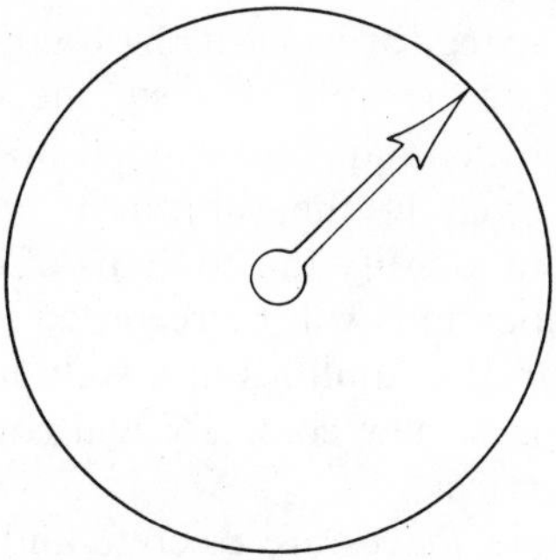

Figure 3.7 Spinner model.

Let A be the event that the pointer stops in the upper semicircle of the disk. According to the symmetry in the experiment derived from the model of no preferred orientation, it is reasonable to assign $P(A)=\frac{1}{2}$. More generally, for any specified arc

$$P(\text{stopping position lies in the arc}) = \frac{\text{Length of arc}}{\text{Circumference}}$$

This is an analog of the uniform probability model for a continuous sample space. The elementary outcomes are the possible stopping positions, and the sample space is the interval of values on the circumference of the circle. Moreover, each elementary outcome is an arc of length 0 and therefore has a probability of 0, according to the preceding assignment. ■

EXERCISES

1. Describe the sample space for each of the following experiments:
   (*a*) A space probe will measure the temperature 50 miles above a planet at 2:00 P.M. next Monday.

(*b*) The number of students out of 20 who will pass beginning swimming and graduate to the intermediate class.
(*c*) In an unemployment survey, 1000 persons will be asked to answer "yes" or "no" to the question "Are you employed?" Only the number answering "no" will be recorded.
(*d*) A geophysicist wants to determine the natural gas reserve in a particular area. The volume will be given in cubic feet.
(*e*) The order in which the birthdate of a student, registered with the Selective Service Board, is drawn in next years's draft lottery will be recorded.

2. For the experiments in Exercise 1, which sample spaces are discrete and which are continuous?
3. Construct a sample space for each of the following experiments:
   (*a*) Someone claims to be able to taste the difference between the same brand of bottled, tap, and canned draft beer. A glass of each is poured and given to the subject in an unknown order. The subject is asked to identify the contents of each glass. The number of correct identifications will be recorded.
   (*b*) The number of traffic fatalities in a state next year.
   (*c*) The length of time a new color TV will continue to work satisfactorily without service.

   Which of these sample spaces are discrete and which are continuous?
4. Identify these events in Exercise 1:
   (*a*) {Temperature less than 40°}
   (*b*) {At least half the students pass}
   (*c*) {Less than or equal to 5.5% unemployment}
   (*d*) {Between 1 and 2 million cubic feet}
   (*e*) {Called first or last}
5. Identify these events in Exercise 3:
   (*a*) {Not more than one correct identification}
   (*b*) {Less accidents than last year}
   NOTE: If you don't know last year's value, use 345.
   (*c*) {Longer than the 90-day warranty but less than 1000.4 days}
6. A driver is stopped for erratic driving, and the alcohol content of his blood is checked. Specify the sample space and the event

$$A = \{\text{Level exceeds legal limit}\} \text{ if the legal limit is .10.}$$

7. Between 9:00 and 9:30 A.M. one day, a garage mechanic will check the headlight alignment of two cars. For each car, the result will be recorded as follows:
   O : Both lights are in alignment.
   L : Only the left light is out of alignment.
   R : Only the right light is out of alignment.

LR: Both lights are out of alignment.

(*a*) Using these symbols, list the sample space.

(*b*) Give the compositions of the following events:

A = The first car has both lights out of alignment.

B = The left light is out of alignment in both cars.

C = Exactly one of the cars has both lights in alignment.

(*c*) Give the compositions of the events $A \cup B$, AB, and $(A \cup B)^c$.

8. Children joining a kindergarten class will be checked one after another to see if they have been inoculated for polio (I) or not (N). Suppose that the checking is to be continued until one noninoculated child is found or four children have been checked, whichever occurs first. List the sample space for this experiment.

9. *Life table*: The following table lists the number of survivors in each age group for a cohort of 1000 white male births in the United States. (For instance, 798 persons survived at least to their 10th birthday; 772 at least to their 20th birthday; etc.)

(*a*) Compute the relative frequencies of the events "death in age 0–9," "death in age 10–19," and so on. For the remainder of this exercise, use the relative frequencies as approximations for the probabilities.

(*b*) For a new birth, what is the probability of survival beyond age 40?

(*c*) For a new birth, what is the probability of death between the ages of 40 and 60? Before age 10?

AGE	NUMBER OF SURVIVORS
0–9	1000
10–19	798
20–29	772
30–39	743
40–49	712
50–59	665
60–69	573
70 and over	407

SOURCE: Condensed from Vital and Health Statistics, Series 3, Number 16, Department of Health, Education and Welfare Publication No. (HSM) 73-1400.

10. The sample space for the response of a single person's attitude toward a political issue consists of the three elementary outcomes $e_1 = \{\text{Unfavorable}\}$, $e_2 = \{\text{Favorable}\}$ and $e_3 = \{\text{Undecided}\}$. Are the following assignments of probability permissible?

(*a*) $P(e_1) = .4,\quad P(e_2) = .5,\quad P(e_3) = .1$

(*b*) $P(e_1) = .4,\quad P(e_2) = .4,\quad P(e_3) = .4$

(*c*) $P(e_1) = .5,\quad P(e_2) = .5,\quad P(e_3) = 0$

11. The medical records of the male diabetic patients reporting to a clinic during one year provide the following percentages:

AGE OF PATIENT	LIGHT CASE		SERIOUS CASE	
	Diabetes in parents		Diabetes in parents	
	Yes	No	Yes	No
Below 40	15	10	8	2
Above 40	15	20	20	10

Suppose a patient is chosen at random from this group, and the events A, B, and C are defined

A: He has a serious case.

B: He is below 40.

C: His parents are diabetic.

(*a*) Find the probabilities $P(A), P(B), P(BC), P(ABC)$.

(*b*) Describe the following events verbally and find their probabilities: (*i*) A^cB^c, (*ii*) $A^c \cup C^c$, (*iii*) A^cBC^c.

12. A sample space has 8 elementary outcomes $e_1, \ldots, e_8$. Suppose the events A, B, and C have the following compositions: $A = \{e_1, e_2, e_3, e_4\}$, $B = \{e_2, e_4, e_5, e_6\}$, $C = \{e_3, e_4, e_6, e_7\}$. Given $P(A) = .4$, $P(B) = .5, P(C) = .5, P(AB) = .2, P(AC) = .2, P(BC) = .2, P(ABC) = .1$, draw a Venn diagram to determine the probabilities of the elementary outcomes.

13. Referring to Exercise 12, give the compositions of the following events and find their probabilities: $A \cup B$, $(A \cup B \cup C)^c$, AB^c.

14. An experimenter wants to study the effect of temperature and baking time on quality of bread by testing 3 temperatures and 5 time periods. How many trials are required for the experimenter to try each temperature with each time period?

15. A biologist studying the internal temperature mechanism of fish wants to test 3 warm temperatures in one half of a divided tank and 2 cold temperatures in the other half. How many pairs of temperature settings are necessary?

16. An agronomist wants to study the yield of 3 different varieties in 4 different row spacings. How many plots are required if each variety-spacing combination is tested on a single plot? How many plots are required if each combination is tested repeatedly on four plots?

17. To determine the optimal operating conditions for a sewage plant, 2 recycle times, 2 blower speeds, and 2 amounts of sludge to be removed are studied. How many sets of conditions (runs) are necessary? If the complete experiment is to be repeated a second time, what is the total number of runs necessary?

18. Referring to Exercise 17, suppose that instead of 3 variables, the plant manager feels that there are 9 variables that may influence the quality of output. If each variable is tested at two values, how many runs will be required if each set of conditions is tested once?
19. Evaluate (*a*) P_3^{10}, (*b*) 5!, (*c*) $\binom{13}{3}, \binom{13}{10}$.
20. In how many ways can the first 4 teams finish in a 9-team league? Ignore ties in your answer.
21. As stimuli, a psychologist can use a bell, a light, or an electric shock. In how many orders can these stimuli be presented if each is used once? If only two are presented?
22. Suppose 20 ingredients are available for making pizza.
    (*a*) How many different pizzas can be made with:
        (*i*) Exactly one ingredient?
        (*ii*) Exactly two ingredients?
        (*iii*) Exactly three ingredients?
    (*b*) Answer (*ii*) if the two halves of a pizza can be different.
23. It is believed that a gas additive will increase mileage. On a single car, a different amount is to be tried in each of 9 tanks of gas. How many orders are required to run the experiment?
24. Suppose 9 mice are available for a study of a possible carcinogen and 4 of them will form a control group that is not to receive the substance.
    (*a*) In how many ways can the control group be selected?
    (*b*) If all selections are equally likely, what is the probability that a particular one, Mike Mouse will be included in the control group?
25. Out of 12 people applying for an assembly job, 3 cannot do the work. Suppose two persons will be hired.
    (*a*) How many distinct pairs are possible?
    (*b*) In how many of the pairs will 0 or 1 people not be able to do the work?
    (*c*) If two persons are chosen in a random manner, what is the probability that neither will be able to do the job?
26. After a preliminary screening, the list of qualified jurors consists of 10 males and 7 females. The 5 jurors the judge selects from this list are all males. Did the selection process seem to discriminate against females? Answer this by computing the probability of having no female members in the jury if the selection is random.
27. A preferred customer at a store is offered the choice of a mystery gift package from a group of 12 packages. Unknown to the customer, 3 of the packages contain wallets, 5 contain pocketbooks, and 4 contain toys. If the customer chooses a package at random, what is the probability that it will contain a wallet?

28. Referring to Exercise 27, suppose the customer prefers to have a wallet and is allowed to pick two packages at random, open both, and choose one of the gifts. What is the probability that the customer will get a wallet?
29. The Jones and Smith families each have 2 girls and 1 boy. One day, Mrs. Smith announces she will hold a lottery and award a special treat to 2 of the 6 children. She will write the children's names on 6 cards and draw 2 cards at random to decide the winners.
   (*a*) What is the probability that both winners will be from the same family?
   (*b*) What is the probability that both winners will be girls?
30. A forest contains 25 elks of a rare species, 5 of which are captured, tagged, and released. After some time, the forester will capture 6 elks and count the number of tagged animals. What is the probability that 2 of the 6 elks captured on the second occasion will be tagged? Assume that no elks migrated into or out of the forest between the first and second captures. (NOTE: Normally, the number of animals in a forest is unknown. The type of capture–recapture experiment described here is used to estimate the size of an animal population.)
31. The following data relate to the proportions in a population of drivers:

$$A = \{\text{Defensive driver training last year}\}$$

$$B = \{\text{Accident in current year}\}$$

The probabilities are given in the following Venn diagram. Find $P(B|A)$. Are A and B independent?

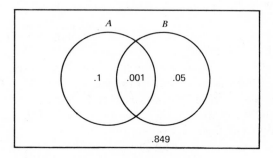

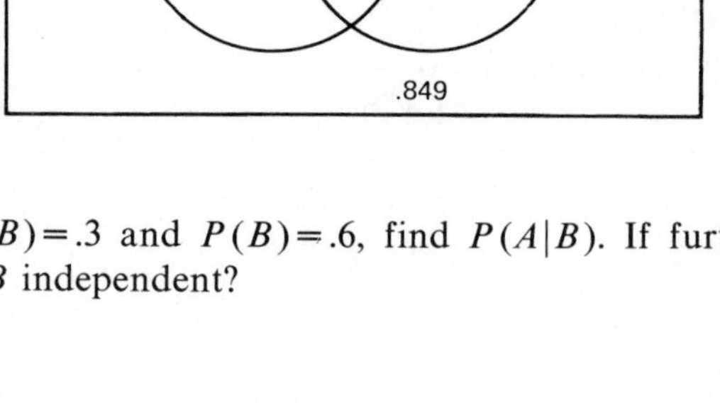

32. Given $P(AB) = .3$ and $P(B) = .6$, find $P(A|B)$. If further $P(A) = .5$, are A and B independent?

33. Referring to Exercise 11:
    (*a*) Suppose a patient will be chosen at random from the group of patients who are below age 40. What is the probability that this patient will have a serious case of the disease? Explain how this can be interpreted as a conditional probability.
    (*b*) Calculate the following conditional probabilities and interpret them in light of your answer to (*a*): $P(A^c|B)$, $P(C|A)$.
34. The following probabilities are given for two events A and B:

$$P(A)=\frac{1}{2}, \quad P(B)=\frac{1}{4}, \quad P(A|B)=\frac{1}{3}$$

    (*a*) Using the appropriate laws of probability, compute $P(A^c)$, $P(AB)$, and $P(A\cup B)$.
    (*b*) Draw a Venn diagram to determine $P(AB^c)$ and $P(A^c \cup B^c)$.
35. Of three events A, B, and C, suppose events A and B are independent and events B and C are mutually exclusive. Their probabilities are $P(A)=.5$, $P(B)=.3$, and $P(C)=.1$. Express the following events in set notations and calculate their probabilities:
    (*a*) Both B and C occur.
    (*b*) At least one of A and B occurs.
    (*c*) B does not occur.
    (*d*) All three events occur.
36. Mr. Hope, a character apprehended by Sherlock Holmes, was driven by revenge to commit two murders. He presented two seemingly identical pills, one containing a deadly poison, to an adversary who selected one while Mr. Hope took the other. The entire procedure was then to be repeated with the second victim. Mr. Hope felt that Providence would protect him, but what is the probability of the success of his endeavor?
37. An orbiting satellite has 3 panels of solar cells, all of which must be active to provide an adequate power output. The panels function independently of one another. The chance that a single panel will fail during the mission is .02. What is the probability that there will be adequate power output during the entire mission time? (This probability is called the *reliability* of the system.)
38. Referring to Exercise 7, suppose the probabilities of observing O, L, R, and LR for each car are, respectively, .1, .2, .2, and .5. The results for the two cars can be regarded as independent.
    (*a*) Assign probabilities to all the elementary outcomes in the sample space.
    (*b*) Compute $P(A)$, $P(B)$, and $P(C)$.

(*c*) Are A and B independent?

(*d*) Are B and C mutually exclusive?

39. An engineering system has two components that function independently. Suppose

$$P[\text{Component 1 fails}] = .1$$

$$P[\text{Component 2 fails}] = .2$$

Find the probability that the system does not fail in the following situations:

(*a*) The components are connected in series; that is, the system works only when both components function.

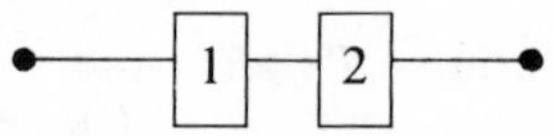

(Series system)

(*b*) The components are in parallel; that is, the system works if either component functions.

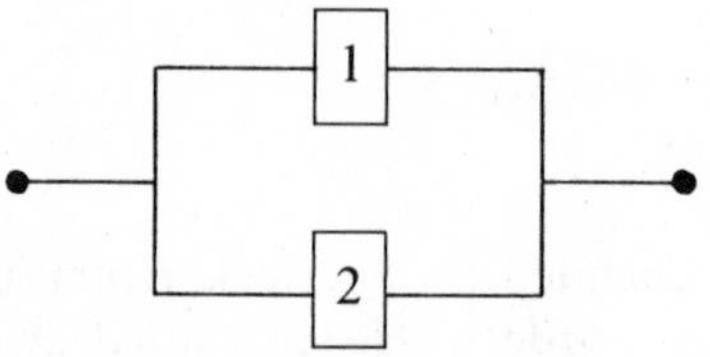

(Parallel system)

40. Suppose independent components having $P[\text{Function}] = .9$ must be placed in parallel to have the probability that the system functions greater than or equal to .99. How many components would be needed?
41. Suppose 60% of a particular breed of mice exhibit aggressive behavior when injected with a given dose of a stimulant. An experimenter will apply the stimulant to 3 mice one after another and will observe the presence or absence of aggressive behavior in each case.
    (*a*) List the sample space for the experiment. (Use A to denote aggressive and N to denote nonaggressive.)
    (*b*) Assuming that the behaviors of different mice are independent, determine the probability of each elementary outcome.
42. Referring to Exercise 41, find the probability that:
    (*a*) Two or more mice will be aggressive.
    (*b*) The first two mice will be aggressive, and the third mouse will be nonaggressive.
    (*c*) Exactly two mice will be aggressive.

43. The chance of a mechanical failure in a system used to prevent a radiation leak in a nuclear power plant is .002. An additional sensing device is designed to detect any failure in the mechanical system and activate a safety shutdown system to stop the radiation leak. The probability that this sensing device will fail to operate when needed is .03. Evaluate the hazard of a radiation leak from the plant; that is, find the probability that both devices will fail.

44. It is suspected that pollution from a chemical plant may be retarding the growth of trees in a nearby forest. Prior to investigating, it was felt that $P(\text{Presence of pollutant})=.6$. It has also been determined that $P(\text{Retarded growth}|\text{pollution})=.8$ and that $P(\text{Retarded growth}|\text{absense of pollution})=.2$. If an investigation reveals retarded growth, find the revised probability that it is caused by pollution.

45. A factory has three production lines 1, 2, and 3 contributing 20%, 30%, and 50%, respectively, to its total output. The percentages of substandard items produced by lines 1, 2, and 3 are, respectively, 15, 10, and 2.
    (*a*) If an item is chosen at random from the total output, what is the probability that it will be substandard?
    (*b*) Suppose that an item chosen at random from the total output is found to be substandard. What is the probability that the item is from line 1?

*46. Martha, concerned about passing a rare recessive gene a onto any future sons, told a genetic counselor that her grandmother was a carrier of this gene having the pair Aa. This sex-linked gene is such that any male has a single gene. If it is a, he will have symptoms of abnormality; hemophilia is one example. Martha's grandfather, father and husband are all normal so they are A. Genetic theory tells us that each son takes one gene in the pair from the mother, and that the selection is with probability $\frac{1}{2}$. A daughter takes one from the mother's pair with the same probabilities and the one from the father. Moreover, the results for different offspring are independent. Consequently, Martha's mother is either Aa or AA, each with probability $\frac{1}{2}$.
    (*a*) Show that Martha herself, is then Aa with probability $\frac{1}{4}$ and could pass a onto her children.
    (*b*) Suppose, however, the counselor is also given the additional information that Martha's mother also had four normal sons with A. Show the counselor how to obtain the modified probability that Martha's mother is Aa.
    (*c*) Find the modified probability that Martha is Aa. {*hint*: $P[\text{Four } A \text{ sons}|\text{Mother } Aa]=(\frac{1}{2})^4$ and $P[\text{Four } A \text{ sons}|\text{Mother } AA]=1$}.

CLASS PROJECTS

1. Take a sample of 50 telephone numbers from your local directory and record the last digit of each number.
   (*a*) Obtain the frequencies of $0, 1, \ldots, 9$. Find the relative frequency of event $A = \{\text{Last digit is an odd number}\}$.
   (*b*) Aggregate the data from other students in the class in groups of 4. Draw a graph of the relative frequency of event A for $n = 50, 200, 400, \ldots$. Examine the stabilization of the relative frequencies.
2. Toss a pair of dice 50 times, recording the sum of the numbers that face up each time.
   (*a*) Find the relative frequency of event $A = \{\text{Sum is 8 or more}\}$.
   (*b*) Aggregate the data from other students in the class in groups of 4. Then draw a graph of the relative frequencies of A and examine the stabilization property. (NOTE: If the dice are unbiased, the probability of A is $15/36$.)
3. Choose a book of facts and figures that contains a large collection of different kinds of numbers (the *World Almanac*; yearbooks on trade, agriculture, shipping; census reports; tables of physical constants; planetary distances, etc., are a few examples). Different students should choose different sources if possible.
   (*a*) Record the first digits of 100 numbers chosen at random from throughout the book. Do not observe too many of the same kinds of numbers; be diverse in your choice. For a decimal number, record the first significant digit (for example, 3 for the number .00358).
   (*b*) Construct a frequency distribution of the digits 1 through 9.
   (*c*) Construct a pooled frequency distribution by collecting the observations of all students in the class. Calculate the relative frequencies.
   (*d*) A curious result called the "Law of Anomalous Numbers" or sometimes "Benford's Law" suggests that the first digits of numbers from a wide diversity of sources may have the probabilities given in the following table. Are the relative frequencies close to these probabilities?

First digit	1	2	3	4	5	6	7	8	9
Probability	.301	.176	.125	.097	.079	.067	.058	.051	.046

MATHEMATICAL EXERCISES

1. Show that $(A \cup B)^c = A^c B^c$

$$(AB)^c = A^c \cup B^c$$

In other words, show that under complementation a union changes to an intersection, and vice versa.

2. If B is a subset of A, show that $P[AB^c] = P(A) - P(B)$.
3. For any events A, B, C, show that

$$P(ABC) = P(A)P(B|A)P(C|AB)$$

4. An example that shows that three events A, B, and C are pairwise independent but all three taken together are not necessarily independent: Suppose

$$\mathbb{S} = \{e_1, e_2, e_3, e_4\} \text{ with } P(e_1) = P(e_2) = P(e_3) = P(e_4)$$

$$A = \{e_1, e_2\}, \quad B = \{e_2, e_3\}, \quad C = \{e_3, e_1\}$$

Find AB, AC, BC, and ABC. Show that $\frac{1}{4} = P(AB) = P(AC) = P(BC)$, so that the events in each pair are independent. Also verify that $0 = P(ABC) \neq P(A)P(B)P(C) = \frac{1}{2} \times \frac{1}{2} \times \frac{1}{2} = \frac{1}{8}$.

5. If A and B are independent, show that A and B^c are independent. [*Hint*: $P(A) = P(AB^c) + P(AB)$ or $P(AB^c) = P(A) - P(AB)$, and $P(B^c) = 1 - P(B)$]
6. Assuming the general conditions for probability given in Section 3.8, show that $P(A \cup B) = P(A) + (B) - P(AB)$.
7. Let A, B, C be any events. Draw a Venn diagram to show that

$$P(A \cup B \cup C) = P(A) + P(A^c B) + P(A^c B^c C)$$

and

$$P(A \cup B \cup C) = P(A) + P(B) + P(C) - P(AB) - P(AC) - P(BC) + P(ABC)$$

CHAPTER 4

Random Variables and Probability Distributions

4.1 RANDOM VARIABLES

We recall that the probability model for an experiment is specified by its sample space, which is the collection of all basic outcomes or simple events, and by the probabilities associated with the simple events. However, these simple events are not necessarily numerical. For example, in flipping a single coin, the sample space is $\{H, T\}$, where the two simple events are described by their attributes rather than by numbers. When investigating the incidence of a disease among 100 industrial workers, the simple events for each worker may be "serious attack," "mild attack," and "healthy." Although simple events may describe qualitative characteristics, we are most often interested in studying some numerical aspect of a phenomenon. In an examination of the 100 industrial workers, the information of particular interest may be the numbers in each of the three categories. Focusing our attention on quantitative measures, we introduce the concept of a random variable.

> A *random variable* X is a numerical valued function defined on a sample space. In other words, a number $X(e)$, providing a measure of the characteristic of interest, is assigned to each simple event e in the sample space.

The diagram in Figure 4.1 illustrates this notion with a sample space having the four outcomes e_1, e_2, e_3, e_4.

The question may arise as to why the word "random" is used when X is just a real valued function defined on the sample space. The qualification "random" is made because we do not know beforehand which particular

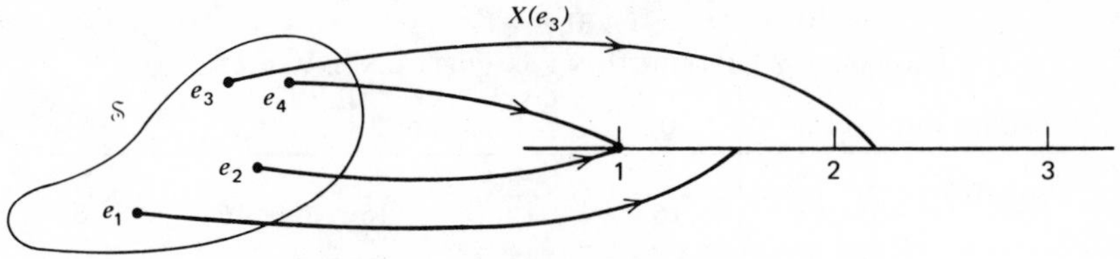

Figure 4.1 A random variable on a sample space.

simple event will occur and what value X will assume. One picturesque interpretation is that "Mother Nature" selects the event and because the choice is uncontrolled by us, we say that it is random.

Another point to note is that X is called a variable although in essence it is a function defined on the sample space. Later we will concern ourselves with some functions $g(X)$ of X, where the values of X play the role of the variable. To avoid calling these "functions of functions," we say that X is a variable and $g(X)$ is a function of X.

Example 4.1 Suppose that two products A and B are judged by four consumers who then express a preference for A or B. The outcome when the first and third consumers prefer A and the other two consumers prefer B is denoted by $ABAB$. The number of outcomes is $2\cdot2\cdot2\cdot2=2^4=16$. These are

AAAA	*AAAB*	*AABB*	*ABBB*	*BBBB*
	AABA	*ABAB*	*BABB*	
	ABAA	*ABBA*	*BBAB*	
	BAAA	*BAAB*	*BBBA*	
		BBAA		
		BABA		

where each simple event e consists of a sequence of four symbols A or B.

Now suppose that the products are alike in quality and that the consumers express their preferences independently. Then the 16 simple events in the sample space are equally likely, and each has a probability of $\frac{1}{16}$. Let a random variable X be defined as $X=$ number of persons preferring A to B. We have grouped the simple events so that all events in a column have the same value for X. For instance, for the events in the third column, X has the value 2. Adding the probabilities of the simple events that are assigned the same value of X, we obtain the summary shown in Table 4.1. This table is called the *probability distribution* of the random variable $X=$ number of persons preferring A, when two products A and B are equally popular. It shows how the total probability 1 is distributed over the different values of the random variable.

TABLE 4.1

Probability Distribution for Preference Variable

Distinct values of X	0	1	2	3	4
Probability	$\frac{1}{16}$	$\frac{4}{16}$	$\frac{6}{16}$	$\frac{4}{16}$	$\frac{1}{16}$

From the distribution of probability summarized in Table 4.1, we can obtain the probability of the events that depend on X. For instance

$$P[X \geqslant 2] = \frac{6}{16} + \frac{4}{16} + \frac{1}{16} = \frac{11}{16}$$

$$P[1 \leqslant X \leqslant 3] = \frac{4}{16} + \frac{6}{16} + \frac{4}{16} = \frac{14}{16}$$ ■

In this chapter, we are concerned with random variables that can assume only a finite number of values or possibly an infinite number of values that can be arranged in a sequence (see Example 4.3). These are called *discrete random variables*.

Many other random variables assume values on a continuous scale, such as the expansion of a section of a bridge support or the duration of a symptom in a patient undergoing medical treatment. The concept of a probability distribution for these random variables will be considered in Chapter 7.

The usefulness of the summary of the distribution of probability given in Example 4.1 motivates us to define the notion of a probability distribution.

> The *probability distribution* or simply, *the distribution of a discrete random variable* is a list of the *distinct* values x_i of X together with their associated probabilities $f(x_i) = P[X = x_i]$. Often a formula can be used in place of a detailed list.

Whenever there is only a small number of distinct values as there is in Example 4.1, it is convenient to present the probability distribution in a table as shown in Table 4.2.

TABLE 4.2

Table of the Probability Distribution

Distinct values of X	x_1	x_2	· · ·	x_k
Probability	$f(x_1)$	$f(x_2)$	· · ·	$f(x_k)$

It is customary to use the capital letter X to denote a random variable and the lower case letter x to denote a value of the random variable. Subscripts are used with x to refer to individual values; $f(x_i)$ denotes the probability of the value x_i of X. Thus the numbers represented by $f(x_i)$ are all between 0 and 1 and $\Sigma_{i=1}^{k} f(x_i) = 1$.

At this point, let us digress briefly to indicate the prime role the probability distribution plays in statistical inference. Suppose that in our preference example we do not know whether A and B are going to be equally popular among the population of consumers. From a sample of 4 persons, the data will be X = number of persons who prefer A to B. With the initial hypothesis that A and B are equally preferred, probability theory provides a table of the probability distribution. What if the data then show that $X=4$? Under the hypothesis of "equal preference," this outcome has probability $\frac{1}{16} = .0625$. The observed value $X=4$ then tells us that either (a) the hypothesis is false, or (b) an unlikely event has occurred. The smaller the probability in (b), the more heavily we lean toward making statement (a). This reasoning, which assesses the probability of the observed or more extreme outcomes, lies at the heart of the statistical testing of hypotheses, which we explore in Chapter 6.

Example 4.2 Sampling from a lot of defectives and nondefectives: A large store places its last 15 clock radios in a clearance sale. Unknown to anyone, 5 of these radios are defective. If a customer tests 3 different clock radios selected at random, what is the distribution of X = number of defective radios in the sample?

Because we are interested only in the total number of defectives in the sample, the order in which the defective and the good radios occur is irrelevant. We can then directly calculate the probabilities of the X values by using formulas for combinations:

$$f(0) = P[X=0] = \frac{\binom{10}{3}}{\binom{15}{3}} = \frac{120}{455} = \frac{24}{91}$$

$$f(1) = P[X=1] = \frac{\binom{10}{2}\binom{5}{1}}{\binom{15}{3}} = \frac{45 \times 5}{455} = \frac{45}{91}$$

$$f(2) = P[X=2] = \frac{\binom{10}{1}\binom{5}{2}}{\binom{15}{3}} = \frac{10 \times 10}{455} = \frac{20}{91}$$

$$f(3) = P[X=3] = \frac{\binom{5}{3}}{\binom{15}{3}} = \frac{10}{455} = \frac{2}{91}$$

These probabilities appear in Table 4.3.

TABLE 4.3
PROBABILITY DISTRIBUTION OF THE NUMBER OF DEFECTIVES

Value x	0	1	2	3
Probability $f(x)$	$\frac{24}{91}$	$\frac{45}{91}$	$\frac{20}{91}$	$\frac{2}{91}$

■

Example 4.3 The chance that monkeys will have a positive (successful) reaction during a trial of an experiment is $\frac{1}{3}$. Trials are performed until there is a success, and the random variable X is defined as the number of trials performed.

The sample space is composed of simple events, which consist of a number of F's (failures) followed by a single S (success), or

$$\mathcal{S} = \{S, FS, FFS, FFFS, \ldots\}$$

Assuming that the trials are independent, we obtain the probability distribution of X shown in Table 4.4.

TABLE 4.4
PROBABILITY DISTRIBUTION OF THE NUMBER OF TRIALS REQUIRED TO OBTAIN A POSITIVE REACTION

Value x	1	2	3	4	...
Probability $f(x)$	$\frac{1}{3}$	$\left(\frac{2}{3}\right)\frac{1}{3}$	$\left(\frac{2}{3}\right)^2\frac{1}{3}$	$\left(\frac{2}{3}\right)^3\frac{1}{3}$	...

It is impossible to make a complete list of all the values and their probabilities. However, we can construct a formula that gives the probability $f(x)$ for any specified value x. This is

$$f(x) = P[X = x] = \left(\frac{2}{3}\right)^{x-1} \cdot \frac{1}{3} \qquad x = 1, 2, \ldots$$

From the formula for $f(x)$, we can compute the probability for any event expressed in terms of X. For example

$$P[X \geqslant 3] = 1 - P[X \leqslant 2] = 1 - \frac{1}{3} - \frac{2}{3} \cdot \frac{1}{3} = \frac{4}{9}$$ ■

Example 4.3 shows that a probability distribution is often conveniently presented by a formula for the probability function $f(x)$ and the range of possible values of X. When such a formula is available, we do not need to construct a table showing all the values and their probabilities. The probability distribution given in Example 4.3 is called a *Geometric distribution* because the successive probabilities $f(1)$, $f(2), \ldots$ form a geometric progression. The formula for the sum of a geometric series can be used to verify that the total probability is 1.

A probability distribution should not be confused with the frequency distribution introduced in Chapter 2. The probability distribution is a theoretical model that assigns probabilities to the values of a random variable; the model is generated by some plausible hypothesis about the underlying chance mechanism. On the other hand, a frequency distribution can be constructed only after a data set has been obtained by repeating an experiment or observation several times. On different occasions, experiments will typically yield different frequency distributions, but a probability distribution is determined by the postulated hypothesis concerning the chance mechanism. As an example, suppose we postulate the hypothesis that a die is fair. The probability distribution of X = number of points is shown in Table 4.5. After actually tossing the die 200 times, one may obtain the frequency distribution shown in Table 4.6. When the sample size is large, the relative frequencies are close to the theoretical probabilities. If the frequency distribution shown in Table 4.6 actually occurred, we would naturally suspect the model of "fair die" that generated the theoretical probability distribution.

TABLE 4.5
PROBABILITY DISTRIBUTION FOR A FAIR DIE

x	1	2	3	4	5	6
$f(x)$	$\frac{1}{6}$	$\frac{1}{6}$	$\frac{1}{6}$	$\frac{1}{6}$	$\frac{1}{6}$	$\frac{1}{6}$

TABLE 4.6
FREQUENCY DISTRIBUTION OF 200 TOSSES OF A DIE

x	1	2	3	4	5	6
Frequency	60	24	20	16	20	60
Relative Frequency	.30	.12	.10	.08	.10	.30

4.2 GRAPHIC PRESENTATION OF A PROBABILITY DISTRIBUTION

It is usually true that we cannot appreciate the salient features of a probability distribution by looking at the numbers in a table the way we can by studying a graphic display. The two main ways to graph a discrete probability distribution are the line diagram and the probability histogram. These diagrams are illustrated for the distribution shown in Table 4.7.

TABLE 4.7

Value x	1	2	3	4
Probability $f(x)$	$\frac{1}{8}$	$\frac{1}{4}$	$\frac{1}{2}$	$\frac{1}{8}$

4.2.1 Line Diagram

The distinct values of the random variable X are located on the horizontal axis. At each value x, a vertical line is drawn whose height is equal to its probability $f(x)$, as shown in Figure 4.2.

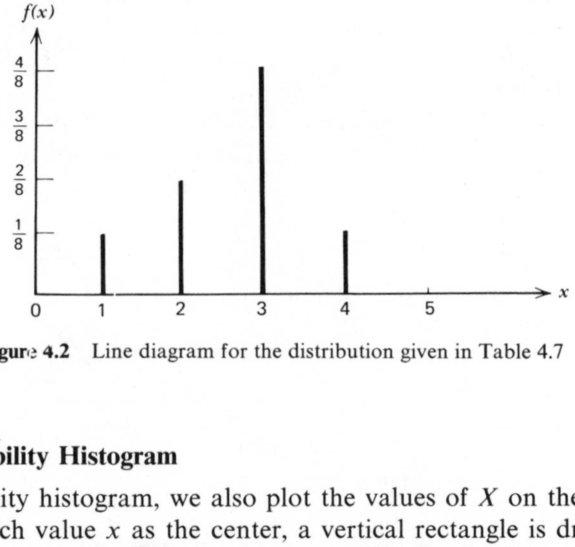

Figure 4.2 Line diagram for the distribution given in Table 4.7

4.2.2 Probability Histogram

In a probability histogram, we also plot the values of X on the horizontal axis. With each value x as the center, a vertical rectangle is drawn whose *area* is equal to the probability $f(x)$, as shown in Figure 4.3. Note that in plotting a probability histogram, the area of a rectangle must be equal to the probability of the value x at the center. If the width of a rectangle is 1, as it is in Figure 4.3, the height and the area have the same value.

We now illustrate plotting a histogram when the x values do not increase in steps of one, using the distribution shown in Table 4.8.

In Figure 4.4, the values of X proceed by steps of .5, so that the width of each rectangle is .5. At $x=1.0$, the probability is .3. The height of the corresponding rectangle must be .6, so that area$=$height$\times$width$=.6\times.5$

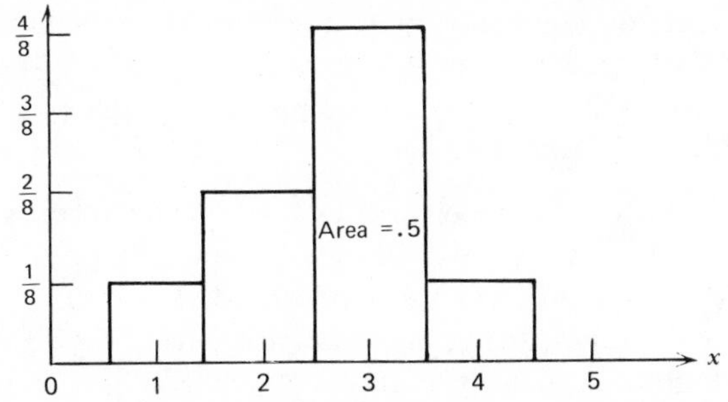

Figure 4.3 Probability histogram for the distribution given in Table 4.7

TABLE 4.8
PROBABILITY DISTRIBUTION

Value x	0	.5	1.0	1.5	2.0
Probability $f(x)$	.1	.2	.3	.25	.15

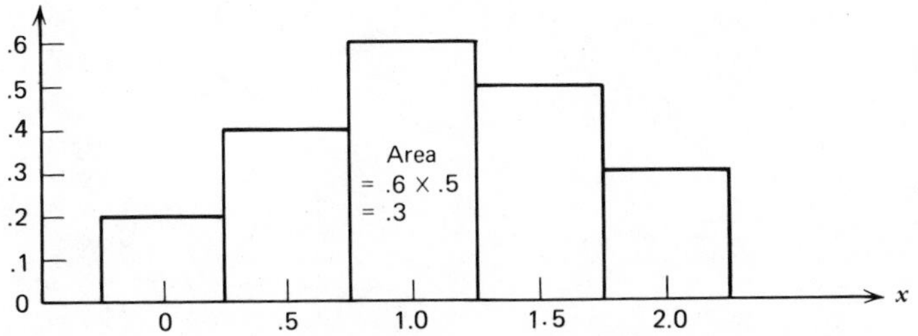

Figure 4.4 Probability histogram for the distribution given in Table 4.8

= .3. Thus in plotting a probability histogram, the heights of the rectangles are determined by dividing the probabilities by the widths of the rectangles.

The probability histogram is recommended for distributions with equal spaced x values. When the spacing of the x values is unequal, the line diagram should be used.

We recall that a probability distribution tells us how the total probability 1 is distributed over the different values of the random variable. Probability distributions of different random variables can exhibit different forms. We are often interested in comparing two or more probability distributions to determine the nature and the extent of their similarities and dissimilarities. For example, we may want to compare the distribution of the number of traffic accidents on clear days with the number on cloudy days or the distribution of daily sales of one brand with daily sales of another brand. A visual comparison is naturally accomplished by the probability histograms, as illustrated in Example 4.4.

Example 4.4 A store manager uses extensive sales records to assess the probability distributions for two brands of hi-fi units. The probability distributions for X = number of units of brand A and for Y = number of units of brand B sold during a week are given in Table 4.9; their probability histograms appear in Figure 4.5.

TABLE 4.9
PROBABILITY DISTRIBUTIONS OF THE SALES OF TWO BRANDS OF HI-FI

Brand A	Value x	0	1	2	3	4	5
	Probability $f(x)$	.1	.1	.2	.3	.2	.1
Brand B	Value y	0	1	2			
	Probability $f(y)$	.23	.48	.29			

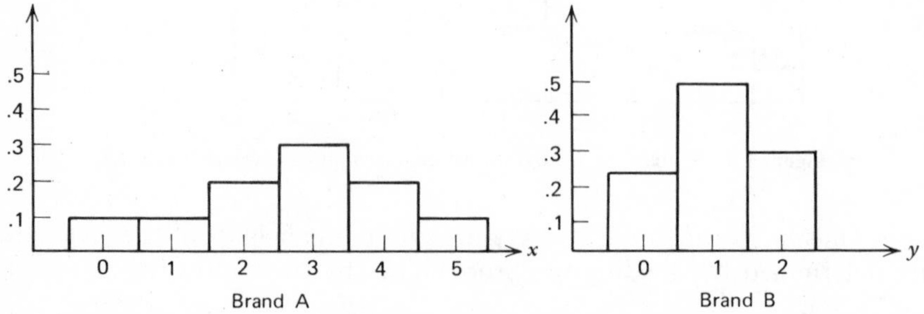

Figure 4.5 Probability histograms for the two distributions given in Table 4.9.

It is clear from the probability histograms that these two distributions differ in two important respects:

(*a*) The first distribution is centered a little below 3, whereas the second distribution is centered at about 1.

(*b*) The first distribution is more spread out or dispersed than the second. ■

A visual comparison of probability histograms, while illuminating, is subjective in the sense that different people may draw different conclusions about the extent of dissimilarity. It is therefore desirable to assign concrete numerical measures to specific aspects of the distributions, so that a comparison is not based solely on subjective assessment. In the next section, we introduce numerical measures for the *center* and the *spread* of a probability distribution. The commonly used measure of the center is called the *mean*, *expectation*, or *expected value* and that for the spread is called the *variance*.

4.3 EXPECTATION AND ITS PROPERTIES

The data sets we encountered in Chapter 2 were first explored by graphic procedures and their main features, such as mean and variance, were then assigned numerical values. Because probability distributions are theoretical models in which the probabilities are viewed as long-range relative frequencies, similar measures of center and spread can be defined for them.

We recall that the mean of a set of numbers 0, 2, 2, 1, 2, 3, 0, 1, 2, 1 can be computed as

$$0\left(\frac{2}{10}\right)+1\left(\frac{3}{10}\right)+2\left(\frac{4}{10}\right)+3\left(\frac{1}{10}\right)=\sum \text{(Value} \times \text{ Relative Frequency)}$$

If we continue to record values of a random variable X in a large series of repetitions of the experiment, the relative frequencies approach the probabilities as the number of repetitions increases. It is then natural to define the mean of a random variable X or of its probability distribution

$$\text{Mean of } X = \sum \text{(Value} \times \text{ Probability)}$$

The mean of X is also called the *expected value* or *expectation* of X and is denoted by $E(X)$. When X has possible values $x_1, x_2, x_3, \ldots$ and when $f(x_i) = P[X = x_i]$:

> *Expected value or Expectation of X*
>
> $$E(X) = \sum_{\substack{\text{all distinct} \\ \text{values of } X}} x_i f(x_i)$$

To compute $E(X)$, we multiply each possible value of the random variable by its probability and add these products.

Example 4.5 Calculation of expected value: We now calculate the means of the two probability distributions in Table 4.9 whose diagrams in Figure 4.5 displayed a difference in center. The numerical measure also shows a difference in center for the two distributions.

TABLE 4.10

CALCULATION OF EXPECTED VALUE FOR THE TWO DISTRIBUTIONS IN TABLE 4.9

	x	0	1	2	3	4	5	Total	
Brand A	$f(x)$	.1	.1	.2	.3	.2	.1	1.0	
	$xf(x)$	0	.1	.4	.9	.8	.5	2.7	$E(X)=2.7$

	y	0	1	2	Total	
Brand B	$f(y)$	.23	.48	.29	1.00	
	$yf(y)$	0	.48	.58	1.06	$E(Y)=1.06$

■

For a probability distribution, the value of $E(X)$ has an important physical interpretation as the center of gravity of a mass distribution. If a metal block is cut in the shape of the probability histogram, then $E(X)$ represents the point on the base at which the block will balance (see Figure 4.6). This further justifies its use as a measure of the center of a distribution. Motivated by this notion of center, $E(X)$ is also called the *population mean* and is usually denoted by μ.

> The *population mean* for the variable X is denoted by μ or μ_X.
>
> $$\mu = E(X)$$

It should not be confused with the sample mean $\bar{x}$, which was introduced in Chapter 2. As its name suggests, the sample mean $\bar{x}$ is based on the

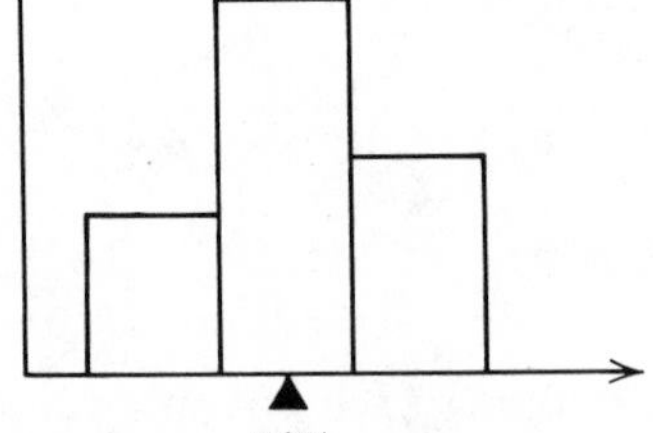

Figure 4.6 $E(X)$ as the center of gravity.

observations in a sample taken from the population, whereas $E(X)$ is computed from the probability model for the population.

Example 4.6 Gambling example–explanation of the term "expectation": A casino allows a gambler to toss a fair die and to receive as many dollars as the number of dots that appear on the die. The gambler has to pay a fee of c dollars to play each game. Disregarding this fee for the moment, the gambler's income per game is a random variable with probability distribution given in Table 4.11.

TABLE 4.11
PROBABILITY DISTRIBUTION OF INCOME PER GAME

Income x	1	2	3	4	5	6	
Probability	$\frac{1}{6}$	$\frac{1}{6}$	$\frac{1}{6}$	$\frac{1}{6}$	$\frac{1}{6}$	$\frac{1}{6}$	$E(X)=\$3.5$

The gamble is "fair" if $c=\$3.5$, because then the long-run average winnings are 0. If the fee is \$4 a game, the expected net gain of the casino is \$.50 per game. ■

In the gambling example, the income in any single trial cannot be \$3.5, so $E(X)$ is not necessarily the value we expect in a single trial. What is meant by $E(X)=\$3.5$ is that if the game is played a large number of times, the ratio (total income)/(number of games played) will be approximately \$3.5.

Let us return to the "preference" distribution in Table 4.1 and suppose that for some reason we are interested in the new random variable $g(X)=(X-2)^2$, which is a function of X. The expected value of this new random variable can be computed in two ways:

(*a*) By computing the probability distribution of $(X-2)^2$ from that of X to obtain the first two lines of Table 4.12. The expected value can then be computed in the normal way, using the products of the entries.
(*b*) Directly from the probability distribution of X. The probability distribution of X is given in the first two rows of Table 4.13. For each value x of X, we evaluate $(x-2)^2$ and multiply by the probability that $X=x$. The sum of the products is then the expected value.

TABLE 4.12
CALCULATION OF $E[(X-2)^2]$ FROM THE PROBABILITY DISTRIBUTION OF $(X-2)^2$

Distinct values of $(X-2)^2$	0	1	4	Total	
Probability	$\frac{3}{8}$	$\frac{4}{8}$	$\frac{1}{8}$	1	
Value × probability	0	$\frac{4}{8}$	$\frac{4}{8}$	1	$E[(X-2)^2]=1$

TABLE 4.13

CALCULATION OF $E[(X-2)^2]$ FROM THE PROBABILITY DISTRIBUTION OF X

Distinct values of X	0	1	2	3	4	Total	
$f(x)$	$\frac{1}{16}$	$\frac{4}{16}$	$\frac{6}{16}$	$\frac{4}{16}$	$\frac{1}{16}$	1	
$(x-2)^2$	4	1	0	1	1		
$(x-2)^2 f(x)$	$\frac{4}{16}$	$\frac{4}{16}$	0	$\frac{4}{16}$	$\frac{4}{16}$	1	$E[(X-2)^2]=1$

The two methods of calculation lead to identical results (why?) The second method motivates the definition of the expectation of a function $g(X)$ of a random variable:

$$E[g(X)] = \sum_{\substack{\text{all distinct} \\ \text{values of } X}} g(x_i) f(x_i)$$

To obtain the expected value of $g(X)$, we evaluate $g(x_i)$ at each value x_i of X, multiply by the probability that X equals x_i, and then add these products.

The operation of determining expectation has certain properties that we record here for future use. Both a and b are constants, and X is the random variable.

Properties of Expectation

(i) $E(a)=a$
(ii) $E(bX)=bE(X)$
(iii) $E(X+a)=E(X)+a$
(iv) $E(a+bX)=a+bE(X)$
(v) $E(a+bX+cX^2)=a+bE(X)+cE(X^2)$

These properties follow from the definition of the expectation of a function $g(X)$ of X. For example, to derive property (iv), we take $g(X)=a+bX$ in the expression for $E[g(X)]$ and obtain

$$E(a+bX) = \sum (a+bx_i) f(x_i) = a\sum f(x_i) + b\sum x_i f(x_i)$$

$$= a + bE(X)$$

More generally, the expectation of any sum of terms is the sum of the expectations of the individual terms. Properties (i)–(v) are some of the important special cases of this result.

Example 4.7 Let X be the number of brand A hi-fi units sold in a week whose probability distribution is given in Table 4.10. Suppose that a profit of \$50 is realized on each unit sold and that the weekly fixed cost is \$20. What is the expected net profit?

For X units sold in a week, the profit is \$$50X$. Subtracting the fixed cost, the net profit is $50X-20$. Then by property (iv)

$$\begin{aligned} E(50X-20) &= 50E(X)-20 \\ &= 50\times 2.7-20 \\ &= \$115 \end{aligned}$$

■

4.4 VARIANCE: A MEASURE OF SPREAD

Because $\mu=E(X)$ provides a measure of center for a distribution, it is natural to consider the deviations of X from μ to determine a measure of spread of the distribution. In other words

Deviation $=(X-\mu)$	has values	$(x_1-\mu)$,	$(x_2-\mu)$,	...,	$(x_k-\mu)$
with corresponding probabilities		$f(x_1)$,	$f(x_2)$,	...,	$f(x_k)$

However, we find that

$$E(\text{deviation})=E(X-\mu)=\sum(x_i-\mu)f(x_i)=\sum x_i f(x_i)-\mu\sum f(x_i)=0$$

This is so because positive deviations are being counterbalanced by negative ones. To determine a measure of spread, we need to consider only the magnitude of the deviations and not their signs. One way to remove the sign from $(x_i-\mu)$ is to consider the squared deviation $(x_i-\mu)^2$. This gives us a measure of spread of a distribution, which is called *variance*.

> *Variance of X*
>
> $$\text{Var}(X)=E[(X-\mu)^2]=E(X^2)-\mu^2$$
>
> The variance of X is also denoted by σ^2 or σ_X^2.

The second version of the formula can be obtained from

$$(X-\mu)^2 = X^2 - 2\mu X + \mu^2$$

Expectation property (v) gives

$$E(X-\mu)^2 = E(X^2) - 2\mu E(X) + \mu^2 = E(X^2) - 2\mu^2 + \mu^2 = E(X^2) - \mu^2$$

There is one difficulty in using variance as our basic measure of variation. If X is measured in some unit (for instance, dollars in our gambling example), then $\mu = E(X)$ is measured in the same unit but $\mathrm{Var}(X)$ is measured in $(\text{dollars})^2$. To express our measure of spread in the same unit in which X is expressed, we must take the square root of the variance. This is called the *standard deviation.*

Standard deviation of X

$$\mathrm{sd}(X) = \sqrt{\mathrm{Var}(X)}$$

The standard deviation of X is also denoted by σ or σ_X.

Example 4.8 Computation of Variance: To illustrate, we again consider the distribution of the weekly sales of brand A hi-fi, which are reproduced in the first two rows of Table 4.14. The necessary calculations for variance and standard deviation appear in the last two rows of the table. It should also be noted that $E(X^2) = \Sigma x_i^2 f(x_i)$ is *not* the same as $[E(X)]^2$.

TABLE 4.14

CALCULATION OF VARIANCE FOR THE WEEKLY SALES OF BRAND A

x	0	1	2	3	4	5	Total
$f(x)$	.1	.1	.2	.3	.2	.1	1
$xf(x)$	0	.1	.4	.9	.8	.5	$2.7 = E(X)$
$x^2f(x)$	0	.1	.8	2.7	3.2	2.5	$9.3 = E(X^2)$

$$\mathrm{Var}(X) = 9.3 - (2.7)^2 = 2.01$$

$$\mathrm{sd}(X) = \sqrt{2.01} = 1.42$$

■

The standard deviation does not have the same intuitive appeal as a measure of spread that expectation has as a measure of center. If the standard deviations for two distributions are .6 and 2.5, we can only make the comparative statement that the second distribution is more dispersed than the first.

Because variance is defined in terms of expectation, variance inherits certain properties that we list here for future reference. Both a and b denote constants.

Properties of

	Variance	*Standard Deviation*
(i)	$\mathrm{Var}(X)$ cannot be negative.	$\mathrm{sd}(X)$ cannot be negative.
(ii)	$\mathrm{Var}(X+a)=\mathrm{Var}(X)$	$\mathrm{sd}(X+a)=\mathrm{sd}(X)$
(iii)	$\mathrm{Var}(bX)=b^2\,\mathrm{Var}(X)$	$\mathrm{sd}(bX)=\lvert b\rvert\,\mathrm{sd}(X)$
(iv)	$\mathrm{Var}(a+bX)=b^2\,\mathrm{Var}(X)$	$\mathrm{sd}(a+bX)=\lvert b\rvert\,\mathrm{sd}(X)$

The properties of standard deviation follow by taking the square root of the results for variance. Property (i) is clear from the definition of variance. Property (ii) states that adding a constant to the values of a random variable does not change the variance. In fact if a constant a is added to each value of X, the probability distribution is merely shifted by a units; its shape is otherwise unchanged. If we write $Y=X+a$ then $E(Y)=E(X)+a$, so that $[Y-E(Y)]=[X-E(X)]$ and hence $\mathrm{Var}(Y)=\mathrm{Var}(X)$. The proof of property (iii) is left as an exercise, and property (iv) is a combination of properties (ii) and (iii).

One important application of these properties is in the transformation of a random variable X with mean $=\mu_X$ and sd $=\sigma_X$ to one with mean $=0$ and sd $=1$. A random variable Z with mean $=0$ and sd $=1$ is called a *standardized random variable*. If X has mean $E(X)=\mu_X$ and variance $\mathrm{Var}(X)=\sigma_X^2$, then

Standardized random variable

$$Z=\frac{X-\mu_X}{\sigma_X} \quad \text{has } E(Z)=0 \quad \text{and } \mathrm{Var}(Z)=1$$

Example 4.9 To illustrate the use of the properties of variance and standard deviation, we return to Examples 4.7 and 4.8, where $X=$number of hi-fi units sold per week and a profit of \$50 is realized for each unit sold. Subtracting the fixed cost of \$20, the weekly net profit from the sales of this hi-fi model is a random variable $Y = 50X - 20$. Then

$$E(Y) = 50E(X) - 20 = \$115$$

$$\text{Var}(Y) = 50^2\,\text{Var}(X) = 50^2(2.01) = 5025$$

$$\text{sd}(Y) = |50|\text{sd}(X) = 50(1.44) = \$72$$

■

4.5 JOINT DISTRIBUTION OF TWO RANDOM VARIABLES

When an experiment is conducted, two or more random variables are often observed simultaneously not only to study their individual probabilistic behaviors but also to determine the degree of relationship between the variables. To describe the model, we recall that the probability distribution of a single random variable X is a list of the distinct values of X and their corresponding probabilities. Extending this idea to the joint probability distribution of two random variables X and Y would require a list of all the distinct pairs of values of X and Y and the associated probabilities. More specifically, suppose that X and Y can have k and l numbers of distinct values, or

$$\text{Values of } X\text{:} \quad x_1, x_2, \ldots, x_k$$

$$\text{Values of } Y\text{:} \quad y_1, y_2, \ldots, y_l$$

There are $k \times l$ distinct pairs of values $(x_i, y_j), i = 1, \ldots, k$ and $j = 1, \ldots, l$ for (X, Y). We let $f(x_i, y_j)$ denote the probability that X and Y simultaneously assume the values x_i and y_j, respectively, or

$$f(x_i, y_j) = P\left[X = x_i \quad \text{and} \quad Y = y_j\right]$$

The joint probability distribution consisting of a list of all pairs of values (x_i, y_j) and the corresponding probabilities $f(x_i, y_j)$ can then be conveniently presented in the form of a two-way table such as Table 4.15.

There the values of X appear in the left-hand margin, the values of Y appear in the upper margin, and the probabilities $f(x_i,y_j)$ are recorded in the cells.

TABLE 4.15
JOINT PROBABILITY DISTRIBUTION OF X AND Y

		Values of Y			
		y_1	y_2	$\cdots$	y_l
	x_1	$f(x_1,y_1)$	$f(x_1,y_2)$	$\cdots$	$f(x_1,y_l)$
	x_2	$f(x_2,y_1)$	$f(x_2,y_2)$	$\cdots$	$f(x_2,y_l)$
	.	.	.	$\cdots$	.
Values of X	.	.	.	$\cdots$	.
	.	.	.	$\cdots$	.
	x_k	$f(x_k,y_1)$	$f(x_k,y_2)$	$\cdots$	$f(x_k,y_l)$

The *joint probability distribution* of two discrete random variables X and Y is specified by a two-way table showing the distinct values of X and Y in the two margins and having as cell entries the probabilities corresponding to the pairs of values. The cell probabilities are often presented as a formula instead of a two-way table.

Example 4.10 There are 10 shiny cars in a lot, of which 5 are in good condition (G), 2 have defective transmissions (DT), and the other 3 have defective steering mechanisms (DS). Suppose that 2 cars are chosen at random, and let the random variables of interest be X = number of cars with defective transmissions and Y = number of cars with defective steering mechanisms among the sampled cars. We want to obtain the joint probability distribution of X and Y.

The sampling scheme is illustrated below:

2	DT	Sample 2 $\longrightarrow$	X = number of DT in sample
3	DS		Y = number of DS in sample
5	G		

The possible values of X are 0, 1, and 2; those for Y are also 0, 1, and 2. These values are listed in the two margins of Table 4.16.

TABLE 4.16
JOINT DISTRIBUTION FOR EXAMPLE 4.10

		Values of Y		
		0	1	2
Values of X	0	$\frac{10}{45}$	$\frac{15}{45}$	$\frac{3}{45}$
	1	$\frac{10}{45}$	$\frac{6}{45}$	0
	2	$\frac{1}{45}$	0	0

The entries in the nine cells of Table 4.16 are computed by using the results for combinations to obtain the probability of each pair of values of X and Y. For example

$$f(0,0)=P[X=0,Y=0]=P[\text{both are } G]=\frac{\binom{5}{2}}{\binom{10}{2}}=\frac{10}{45}$$

$$f(1,0)=P[X=1,Y=0]=P[\text{one is } DT \text{ and one is } G]=\frac{\binom{2}{1}\binom{5}{1}}{\binom{10}{2}}=\frac{10}{45}$$

$$f(1,2)=P[X=1,Y=2]=0$$

Other entries are computed by using the product of combinations in the same way. Because all possible pairs of values are accounted for in the table, the sum of the nine cell entries has a total probability of 1. ■

The joint distribution given in Example 4.10 is determined from the specified sampling scheme by applying the appropriate combinatorial formulas. In other situations, an assessment of the joint distribution may have to be based on the principle of the stability of relative frequencies after a long series of observations have been made on the values of the random variables. This is illustrated in Example 4.11.

Example 4.11 A small factory operates on two shifts daily. In a study concerning the pattern of worker absenteeism, the random variables of interest are: X = number of absentees from the morning shift and Y = number of absentees from the evening shift of the same day. Based on a long series of past attendance records, the personnel manager provides the assessment of the joint probability distribution of X and Y shown in Table 4.17.

TABLE 4.17
JOINT DISTRIBUTION OF DAILY ABSENTEES FROM TWO SHIFTS

x \ y	0	1	2	3	Row sum
0	.05	.05	.10	0	.20
1	.05	.10	.25	.10	.50
2	0	.15	.10	.05	.30
Column sum	.10	.30	.45	.15	1

Here the cell probabilities are ascertained from the relative frequency of occurrence of the individual pairs of values in a long series of records. For instance, the entry .25 in the cell $(x=1, y=2)$ means that on 25% of the days one worker is absent from the morning shift and two workers are absent from the evening shift. The first row sum of .2 indicates that on 20% of the days no one is absent from the morning shift. Similarly, the third column sum of .45 indicates that on 45% of the days two workers are absent from the evening shift. ■

In the remainder of this section, we discuss the various kinds of information that can be extracted from a joint distribution, and we illustrate the points with reference to the joint distribution given in Table 4.17.

(*a*) *The probability of any event concerning X and Y*: For example, to compute $P[X+Y=3]$ from Table 4.17, we note that $x+y=3$ for the cells (0,3), (1,2), and (2,1). Adding these probabilities, we obtain $P[X+Y=3] = 0 + .25 + .15 = .40$. Similarly

$$P[X=2] = .15 + .10 + .05 = .30$$

$$P[X>Y] = .05 + 0 + .15 = .20$$

(*b*) *The probability distribution of any function of the two random variables X and Y*: In Example 4.11, let $Z = X + Y$ be the total number of absentees in the two shifts combined. The possible values of Z are $0, 1, 2, \ldots, 5$ and the probability for each case is computed as indicated in (*a*) for $X+Y=3$. The resulting distribution is given in Table 4.18.

TABLE 4.18
DISTRIBUTION OF $Z = X + Y$ IN EXAMPLE 4.11

z	0	1	2	3	4	5	Total
$f_Z(z)$	.05	.10	.20	.40	.20	.05	1.00

(*c*) *The probability distribution of X alone*: This can be obtained immediately from the joint distribution of X and Y. Referring to Table 4.17, the distinct values of X are already listed in the left margin. The probability of each value is the corresponding row sum. Therefore, the values of X listed in the left margin of Table 4.17 together with the row sums written in the right margin consistitute the probability distribution of X. Because the probabilities obtained are marginal totals from the two-way table, the distribution of X is also called the *marginal distribution of* X. The word "marginal" is actually redundant. It only reflects the fact that some other random variable in addition to X was originally considered in the problem and that the X distribution was obtained from the two-way table. In the same manner, we obtain the probability distribution of Y by summing each column and recording the entries in the bottom margin as shown in Table 4.17.

(*d*) *The mean and standard deviation of X or Y*: Any numerical measure for the X distribution such as the expectation μ_X or the variance σ_X^2 is computed from the marginal distribution of X alone without referring back to the joint distribution. The same procedure applies to the Y distribution. These computations are illustrated in Table 4.19 with the marginal distributions of X and Y obtained from Table 4.17.

TABLE 4.19

THE MEANS AND STANDARD DEVIATIONS OF X AND Y FOR THE DISTRIBUTION GIVEN IN TABLE 4.17

x	0	1	2	Total
$f_X(x)$	.2	.5	.3	1.0
$xf_X(x)$	0	.5	.6	1.1
$x^2f_X(x)$	0	.5	1.2	1.7

y	0	1	2	3	Total
$f_Y(y)$	.10	.30	.45	.15	1.00
$yf_Y(y)$	0	.30	.90	.45	1.65
$y^2f_Y(y)$	0	.30	1.80	1.35	3.45

$\mu_X = 1.1$ $\qquad$ $\mu_Y = 1.65$

$\sigma_X^2 = 1.7 - (1.1)^2 = .49$ $\qquad$ $\sigma_Y^2 = 3.45 - (1.65)^2 = .7275$

$\sigma_X = .70$ $\qquad$ $\sigma_Y = .85$

(*e*) *Additivity of expectation*: Expectation possesses the property of additivity. That is, for any two random variables X and Y having some joint distribution, the expected values add according to $E(X+Y) = E(X) + E(Y)$. Instead of providing a formal proof of this result here, we demonstrate the property for the joint distribution given in Table 4.17. Computing the expected value from the distribution of $Z = X + Y$ in Table 4.18, we obtain $E(Z) = 2.75$. On the other hand, from the computation using the marginal distributions, we obtain $\mu_X + \mu_Y = 1.1 + 1.65 = 2.75$, which shows that $E(X+Y) = E(X) + E(Y)$.

4.6 COVARIANCE AND CORRELATION

Covariance between X and Y is a numerical measure of the joint variation of the two random variables and is defined as the expected value of the product $(X-\mu_X)(Y-\mu_Y)$. Intuitively, we can say that X and Y vary in the same direction if the probability is high that large values of X are associated with large values of Y and small values of X are associated with small values of Y. In such a case, both the values of the deviations $(X-\mu_X)$ and $(Y-\mu_Y)$ are positive or negative with a high probability, so that the product $(X-\mu_X)(Y-\mu_Y)$ is predominantly positive. Consequently, the expected value of the product is positive and large. On the other hand, if X and Y tend to vary in opposite directions, the positive values of $(X-\mu_X)$ are more often associated with the negative values of $(Y-\mu_Y)$ and vice versa. The product is then predominantly negative, and the expected value is negative. In this sense, the sign and the magnitude of $E[(X-\mu_X)(Y-\mu_Y)]$ is likely to reflect the direction and the strength of the relationship between X and Y.

Covariance between X and Y

$$\text{Cov}(X,Y) = E[(X-\mu_X)(Y-\mu_Y)]$$
$$= E(XY) - \mu_X\mu_Y$$

To obtain the second form of the formula from the first, we write

$$(X-\mu_X)(Y-\mu_Y) = XY - \mu_X Y - \mu_Y X + \mu_X\mu_Y$$

Taking expectation on both sides and invoking the additivity property of expectation, we obtain

$$\text{Cov}(X,Y) = E(XY) - \mu_X\mu_Y - \mu_Y\mu_X + \mu_X\mu_Y$$
$$= E(XY) - \mu_X\mu_Y$$

The second form is easier to compute. Because μ_X and μ_Y are obtained from the marginal distributions, it remains for us to show how to calculate $E(XY)$. From the general concept of expectation as the sum of (value)$\times$(probability), we have

$$E(XY) = \sum_{\text{all cells}} \begin{bmatrix} \text{value of } XY \\ \text{for a cell} \end{bmatrix} \times [\text{Cell probability}]$$

The computation of $E(XY)$ is illustrated in Table 4.20.

The value of Cov(X, Y) depends on the units of measurement associated with X and Y. It is desirable to have a measure of relationship for two variables that does not depend on the units of measurement. This is accomplished by dividing the covariance by the standard deviations for X and Y. The resulting measure is called the *correlation coefficient* between X and Y.

Correlation coefficient between X and Y

$$\text{Corr}(X, Y) = \frac{\text{Cov}(X, Y)}{\sigma_X \sigma_Y}$$

Example 4.12 Computation of covariance and correlation: We now illustrate the computation of the covariance and the correlation coefficient with reference to the joint distribution given in Table 4.17. The means and the standard deviations already appear in Table 4.19, so we only need to obtain $E(XY)$. The necessary computations are made in Table 4.20, where the numbers in parentheses are the values of (XY) for the individual cells. These values are then multiplied by the probabilities and the products are added.

TABLE 4.20
COMPUTATION OF $E(XY)$ FOR TABLE 4.17

x \ y	0	1	2	3
0	.05 (0)	.05 (0)	.10 (0)	0 (0)
1	.05 (0)	.10 (1)	.25 (2)	.10 (3)
2	0 (0)	.15 (2)	.10 (4)	.05 (6)

$$\begin{array}{c} .10 \times 1 = .10 \\ .25 \times 2 = .50 \\ .10 \times 3 = .30 \\ .15 \times 2 = .30 \\ .10 \times 4 = .40 \\ .05 \times 6 = .30 \\ \hline E(XY) = 1.90 \end{array}$$

$$\text{Cov}(X, Y) = 1.90 - 1.65 \times 1.1 = .085$$

$$\text{Corr}(X, Y) = \frac{\text{Cov}(X, Y)}{\sigma_X \sigma_Y} = \frac{.085}{.70 \times .85} = .14$$

■

Properties of Correlation

(*a*) Corr(X, Y) is always a number between -1 and $+1$. The two extreme values $+1$ and -1 are attained when X and Y are related by a straight line with a positive slope or a negative slope, respectively.

(*b*) Corr(X, Y) remains unchanged if constants are added to the variables or if the variables are multiplied by constants having the same sign. For example, if $U=5X+2$ and $V=2Y+3$, then Corr(U, V) = Corr(X, Y)

4.7 INDEPENDENCE OF TWO RANDOM VARIABLES

As we noted in Chapter 3, two events A and B are independent if $P(AB)=P(A)P(B)$. Using the same reasoning, we can say that two random variables X and Y are independent if the event that X assumes a specific value x_i is independent of the event that Y assumes a specific value y_j no matter what specific values are selected. By the definition of independent events, we then have

$$P[X=x_i, Y=y_j]=P[X=x_i]P[Y=y_j]$$

This results in the definition:

Random variables X and Y are *independent* if $f(x_i,y_j)=f_X(x_i)f_Y(y_j)$ for *all* pairs of values (x_i,y_j) in the joint distribution. In other words, each cell probability is the product of the two corresponding marginal totals.

Example 4.13 A coin is to be tossed three times. Let X = number of heads in the first 2 tosses and Y = number of heads in the third toss. The joint probability distribution of X and Y appears in Table 4.21. Are X and Y independent? Because X and Y refer to physically unrelated parts of an experiment, we would expect them to be independent. This is confirmed by the joint distribution, in which it can be seen that each cell entry is the product of its marginal totals.

TABLE 4.21

x \ y	0	1	Row sum
0	$\frac{1}{8}$	$\frac{1}{8}$	$\frac{1}{4}$
1	$\frac{2}{8}$	$\frac{2}{8}$	$\frac{1}{2}$
2	$\frac{1}{8}$	$\frac{1}{8}$	$\frac{1}{4}$
Column sum	$\frac{1}{2}$	$\frac{1}{2}$	1

■

Referring to the joint distribution given in Table 4.16, we note that X and Y are not independent, because $f(0,0)=10/45$ but $f_X(0)f_Y(0)=(28/45)\times(21/45)\neq 10/45$. The random variables for the joint distribution given in Table 4.17 are also not independent.

If X and Y are independent, then $E(XY)=E(X)E(Y)$. This is so because $f(x_i,y_j)$ factors as $f_X(x_i)f_Y(y_j)$, by the definition of independence, and we obtain

$$E(XY)=\sum_i \sum_j x_i y_j f_X(x_i) f_Y(y_j)$$

$$=\sum_i x_i f_X(x_i) \sum_j y_j f_Y(y_j)$$

$$=E(X)E(Y)$$

Therefore $\text{Cov}(X,Y)=0$.

> Independence of X and Y implies that $\text{Cov}(X,Y)=0$ and that $\text{Corr}(X,Y)=0$.

The converse of the result is not necessarily true. Two random variables may be uncorrelated and still be dependent (see Exercise 25).

We conclude by stating some important rules about the expected value and the variance of sums of random variables in general and about the special case when they are independent. We have already noted that expectation exhibits the property of additivity.

Additivity of expectation:

$$E(X+Y)=E(X)+E(Y)$$

This holds quite generally whether or not X and Y are independent. However, the same is not true for variance. Writing $Z=X+Y$, we always have $\mu_Z=\mu_X+\mu_Y$, and

$$\begin{aligned}\mathrm{Var}(Z)&=E(Z-\mu_Z)^2=E\left[(X-\mu_X)+(Y-\mu_Y)\right]^2\\&=E\left[(X-\mu_X)^2+(Y-\mu_Y)^2+2(X-\mu_X)(Y-\mu_Y)\right]\\&=\mathrm{Var}(X)+\mathrm{Var}(Y)+2\,\mathrm{Cov}(X,Y)\end{aligned}$$

We can obtain a formula for $\mathrm{Var}(X-Y)$ in the same way. Summarizing, the general formulas for the variance of a sum and a difference are

$$\mathrm{Var}(X+Y)=\mathrm{Var}(X)+\mathrm{Var}(Y)+2\,\mathrm{Cov}(X,Y)$$

$$\mathrm{Var}(X-Y)=\mathrm{Var}(X)+\mathrm{Var}(Y)-2\,\mathrm{Cov}(X,Y)$$

If X and Y are independent, then $\mathrm{Cov}(X,Y)=0$, in which case both $\mathrm{Var}(X+Y)$ and $\mathrm{Var}(X-Y)$ reduce to $\mathrm{Var}(X)+\mathrm{Var}(Y)$.

If X and Y are independent

$$\mathrm{Var}(X+Y)=\mathrm{Var}(X)+\mathrm{Var}(Y)$$

$$\mathrm{Var}(X-Y)=\mathrm{Var}(X)+\mathrm{Var}(Y)$$

EXERCISES

1. Suppose that a factory supervisor records whether the day or the night shift has a higher production rate for each of the next 3 days. List the possible outcomes and, for each, record the number of days X that the night shift has a higher production rate. (Assume there are no ties.)

2. Each week a grocery shopper buys either canned (C) or bottled (B) soft drinks. The type of soft drink purchased in 4 consecutive weeks is to be recorded.
   (*a*) List the sample space.
   (*b*) If a different type of soft drink is purchased than in the previous week, we say that there is a switch. Let X denote the number of switches. Determine the value of X for each elementary outcome. (Example: for $BBBB$, $x=0$; for $BCBC$, $x=3$).
   (*c*) Suppose that for each purchase $P(B)=\frac{1}{2}$ and the decisions in different weeks are independent. Assign probabilities to the elementary outcomes and obtain the distribution of X.
3. An animal either dies (D) or survives (S) in the course of a surgical experiment. The experiment is to be performed first with two animals. If both survive, no further trials are to be made. If exactly one animal survives, one more animal is to undergo the experiment. Finally, if both animals die, two additional animals are to be tried.
   (*a*) List the sample space.
   (*b*) Assume that the trials are independent and that the probability of survival in each trial is $\frac{2}{3}$. Assign probabilities to the elementary outcomes.
   (*c*) Let X denote the number of survivals. Obtain the probability distribution of X.
4. Of 6 candidates seeking 3 positions at a counseling center, 2 have degrees in social science and 4 do not. If 3 candidates are selected at random, find the probability distribution of X, the number having social science degrees among the selected persons.
5. For the probability distribution

x	2	3	4	5	6
$f(x)$	.1	.3	.3	.2	.1

   (*a*) Draw a line diagram and a probability histogram.
   (*b*) Calculate $E(X)$.
   (*c*) Find $P[X \geqslant 4]$ and $P[2 < X \leqslant 4]$.
6. A student buys a lottery ticket for \$1. For every 1000 tickets sold, 2 bicycles are to be given away in a drawing.
   (*a*) What is the probability that the student will win a bicycle?
   (*b*) If each bicycle is worth \$160, determine the student's expected gain.
7. An insurance company sells to 1500 individuals policies that protect their canoes against theft for a period of 2 years. Based on a cost of \$300 to replace a stolen canoe, the company determines that it will break even if the probability of the theft of an individual's canoe

during the policy period is .15. Assume that the probability of more than one theft per individual is 0.

(*a*) What should be the selling price of the insurance policy?

(*b*) If the probability of theft is actually .1, what is the company's expected gain per policy given the selling price determined in (*a*)?

8. For the probability distribution

x	0	1	2	3
$f(x)$	.3	.4	.2	.1

find:

(*a*) $P[X \geqslant 2]$

(*b*) $P[0 < X \leqslant 2]$

(*c*) $E(X)$

(*d*) $\text{Var}(X)$; $\text{sd}(X)$

9. Referring to Exercise 5, let $Y = (2X - 8)^2$.

(*a*) Calculate $E(2X-8)^2$ from the distribution of X.

(*b*) Obtain the distribution of Y and calculate $E(Y)$.

10. Referring to Exercise 3, calculate the mean and the standard deviation of X.

11. *Definition*: The *median* of a distribution is the value m_o of the random variable such that $P[X \leqslant m_o] \geqslant .5$ and $P[X \geqslant m_o] \geqslant .5$. In other words, the probability at or below m_o is at least .5, and the probability at or above m_o is at least .5.

(*a*) Find the median of the distribution given in Exercise 5.

(*b*) Find the median of the distribution given in Exercise 8.

12. For the distribution $P[X = x] = \binom{3}{x}\left(\frac{1}{3}\right)^x\left(\frac{2}{3}\right)^{3-x}$, $x = 0, 1, 2, 3$, determine:

(*a*) $E(X)$

(*b*) $\text{Var}(X)$ and $\text{sd}(X)$

(*c*) the standardized form of X

(*d*) the distribution of the standardized variable Z

13. For scores X on a nationally administered aptitude test, the mean is $E(X) = 120$ and $\text{Var}(X) = 100$. Find the mean and the variance of:

(*a*) $\dfrac{X - 120}{10}$

(*b*) $\dfrac{X - 100}{20}$

14. Standardize the following random variables:

(*a*) X has mean 12 and standard deviation 5

(*b*) Y has mean 20 and variance 16.

15. A manufacturer of electrical appliances can buy a component in lots of 50 from either of two wholesalers. The price quotations are \$500 per

lot from Supplier A and \$510 per lot from Supplier B. If any piece is defective, the cost of reworking it is \$7. Based on the quality inspection of some of the lots, suppose that the following probability distributions are assessed for the number of defectives per lot.

Supplier A

Number of defectives per lot	0	1	2	3	4
Probability	.1	.2	.3	.3	.1

Supplier B

Number of defectives per lot	0	1	2
Probability	.4	.4	.2

From which supplier should the manufacturer buy the components to minimize the expected cost? (NOTE: The cost of a lot is $\$(c+7X)$, where c = purchase price and X = number of defectives.)

16. Referring to Exercise 2, let Y denote the number of purchases of bottled soft drinks made in 4 weeks and let X represent the number of switches, as defined in (*b*).
    (*a*) Make a list of the elementary outcomes, their probabilities, and the corresponding x and y values, and then obtain the joint probability distribution of X and Y.
    (*b*) Find the marginal distributions of X and Y.
17. From the following joint probability distribution of X and Y, find:
    (*a*) $P[X=Y]$, $P[X>Y]$
    (*b*) the distribution of $X+Y$
    (*c*) $E(X)$, $E(Y)$, $\text{Var}(X)$, $\text{Var}(Y)$
    (*d*) $\text{Cov}(X, Y)$, $\text{Corr}(X, Y)$.

x \ y	0	1	2
0	.1	.3	.05
1	.2	.25	.1

18. In a study of the coexistence of two types of insects, let X and Y denote the number of type A and type B insects, respectively, that reside on the same plant. From observations of a large number of plants, suppose that the following joint probability distribution is assessed for the insect counts per plant.
    (*a*) Find the probability that there are more type B insects than type A insects on a plant.
    (*b*) Compute μ_X, μ_Y, σ_X, σ_Y and $\text{Cov}(X, Y)$.
    (*c*) Find the correlation coefficient between X and Y. Interpret the result.

y \ x	1	2	3	4
0	0	.05	.05	.10
1	.08	.15	.10	.10
2	.20	.12	.05	0

19. Referring to Exercise 18, the total number of insects of both types living on a plant is the random variable $T = X + Y$.
   (*a*) Obtain the probability distribution of T and use it to calculate $E(T)$ and $\text{Var}(T)$.
   (*b*) Using your results from Exercise 18 *b*, verify that:

$$E(T) = E(X) + E(Y)$$

$$\text{Var}(T) = \text{Var}(X) + \text{Var}(Y) + 2\text{Cov}(X, Y)$$

20. For X let $E(X) = 0$ and $\text{sd}(X) = 2$, and for Y let $E(Y) = -1$ and $\text{sd}(Y) = 4$. Find:
   (*a*) $E(X - Y)$ and $E(X + Y)$.
   (*b*) $\text{Var}(X - Y)$ and $\text{Var}(X + Y)$ if X and Y are independent.
   (*c*) $E(\frac{1}{2}X + \frac{1}{2}Y)$ and $\text{Var}(\frac{1}{2}X + \frac{1}{2}Y)$ if X and Y are independent.
   (*d*) Repeat (*b*) if, instead of independence, $\text{Cov}(X, Y) = 1$.
21. Let $\text{Var}(X) = 4$ and $\text{Var}(Y) = 9$. If $\text{Cov}(X, Y) = 2$, is $\text{Var}(X - Y)$ larger or smaller than it is when X and Y are independent? What is your answer if $\text{Cov}(X, Y) = -2$?
22. If $\text{Corr}(X, Y) = .3$, are X and Y independent? If not, do X and Y tend to vary together or as opposites?
23. If $\text{Var}(X) = 16$, $\text{Var}(Y) = 25$, and $\text{Cov}(X, Y) = -10$, find:
   (*a*) $\text{Corr}(X, Y)$
   (*b*) $\text{Var}(3X + 2)$
   (*c*) $\text{Corr}(3X + 2, 2Y - 1)$
24. (*a*) Are the random variables X and Y in Exercise 16 independent?
   (*b*) Are the random variables X and Y in Exercise 18 independent?
25. Consider the joint distribution

x \ y	−1	0	1
0	0	$\frac{1}{3}$	0
1	$\frac{1}{3}$	0	$\frac{1}{3}$

   (*a*) Show that X and Y are *not* independent.
   (*b*) Show that $\text{Corr}(X, Y) = 0$.

MATHEMATICAL EXERCISES

1. If a and b are constants, prove that $\mathrm{Var}(a+bX)=b^2\mathrm{Var}(X)$.
2. Show that for any constants a, b, c, and d:
   (*a*) $\mathrm{Cov}(a+bX, c+dY)=bd\,\mathrm{Cov}(X,Y)$
   (*b*) $\mathrm{Corr}(a+bX, c+dY)=\mathrm{Corr}(X,Y)$ when b and d are both positive. How are these two correlations related when b and d are both negative? When b and d have opposite signs?
   (*c*) Use (*b*) to prove that $\mathrm{Corr}\left(\dfrac{X-\mu_X}{\sigma_X}, \dfrac{Y-\mu_Y}{\sigma_Y}\right)=\mathrm{Corr}(X,Y)$.
3. Let X_1, X_2, X_3 be independent. Assuming that all have a mean of μ and a variance of σ^2, show that:
   (*a*) $E\left(\dfrac{X_1+X_2}{2}\right)=\mu$, $\quad \mathrm{sd}\left(\dfrac{X_1+X_2}{2}\right)=\dfrac{\sigma}{\sqrt{2}}$
   (*b*) $E\left(\dfrac{X_1+X_2+X_3}{3}\right)=\mu$, $\quad \mathrm{sd}\left(\dfrac{X_1+X_2+X_3}{3}\right)=\dfrac{\sigma}{\sqrt{3}}$
   (*c*) Generalize these results for the average of n independent random variables

$$\bar{X}=\frac{X_1+X_2+\ldots+X_n}{n}$$

   where $X_1,\ldots,X_n$ all have a mean of μ and a variance of σ^2.
4. Considering the joint distribution $f(x_i,y_j)$, $i=1,\ldots,k$; $j=1,\ldots,l$ and its marginals $f_X(x_i)$, $f_Y(y_j)$, prove that

$$E(X+Y)=E(X)+E(Y)$$

$$\left[\textit{Hint:}\quad E(X+Y)=\sum_i\sum_j (x_i+y_j)f(x_i,y_j)\right.$$

$$\left.=\sum_i\sum_j x_i f(x_i,y_j)+\sum_i\sum_j y_j f(x_i,y_j)\right]$$

5. Prove that the correlation coefficient between two random variables must be a number between -1 and $+1$.

   [*Hint*: Denote $\mathrm{Corr}(X,Y)=\rho$, $Z_1=\dfrac{X-\mu_X}{\sigma_X}$, $Z_2=\dfrac{Y-\mu_Y}{\sigma_Y}$. Then $\mathrm{Cov}(Z_1,Z_2)=\rho$. Because the variance of a random variable is nonnegative

$$\mathrm{Var}(Z_1-Z_2)=\mathrm{Var}(Z_1)+\mathrm{Var}(Z_2)-2\mathrm{Cov}(Z_1,Z_2)\geqslant 0$$

From this inequality, establish that $\rho \leqslant 1$. Similarly, use the fact $\operatorname{Var}(Z_1 + Z_2) \geqslant 0$ to establish $\rho \geqslant -1$.]

6. *Chebyshev inequality*:
   (*a*) Let X be a random variable with a mean of μ. Show that for every constant d

$$P[|X-\mu|>d] \leqslant \frac{\operatorname{Var}(X)}{d^2}$$

Whatever the distribution, we obtain a bound on the probability that X is more than d units from μ.

$$\left[\textit{Hint}: \operatorname{Var}(X) = \sum_{\text{All values}} (x_i-\mu)^2 f(x_i) \right.$$

$$\left. \geqslant \sum_{\substack{\text{All values} \\ \text{with } |x_i-\mu|>d}} (x_i-\mu)^2 f(x_i) \geqslant d^2 \sum_{\substack{\text{all values} \\ \text{with } |x_i-\mu|>d}} f(x_i) \right]$$

   (*b*) For $\bar{X} = \sum_{i=1}^{n} X_i/n$ where $X_1, \ldots, X_n$ are independent and all have a mean of μ and a variance of σ^2, show that $P[|\bar{X}-\mu| \leqslant d] \geqslant 1 - \sigma^2/nd^2$. Note: $\operatorname{Var}(\bar{X}) = \sigma^2/n$.

CHAPTER 5

Distributions for Counts

5.1 IDEA OF A PROBABILITY MODEL

We discussed the probability distributions of discrete random variables and their basic properties in Chapter 4. Examples 4.1, 4.2, and 4.3 illustrate some specific discrete distributions that arise when a sample is taken from a population that consists of only two types of elements and the random variable of interest is the number of elements of one type occurring in the sample. In Example 4.1, which concerns consumer preference between two brands A and B, the probability distribution of the number of persons preferring brand A in a sample of 4 persons is obtained under the assumption that both brands A and B are equally popular. In other words, in the population of all consumers, the fraction preferring brand A is assumed to be $\frac{1}{2}$. In a practical situation, however, the fraction of consumers preferring brand A in the population may be unknown. If this is true, it is no longer possible to compute the numerical probabilities of the various values of X. Instead, we attempt to build a model for the probability distribution of X that has the unknown population fraction as a *parameter*, and the probability for any individual value of X is specified in terms of this unknown parameter. In the sampling process, observations on the random variable X are collected, and the objective becomes to ascertain what values of the unknown parameter best explain the data at hand. This aspect constitutes the domain of statistical inference.

In studying any phenomenon that exhibits variation, we usually represent some population characteristics as parameters, whose values are unknown to us. Here we intend to describe the chance variation of the observations in terms of a probability distribution involving these quantities.

Of course, rarely if ever is an experimenter absolutely certain that the correct probability model is chosen. In the face of the complexities of the phenomenon under study, the experimenter's approach is often to tentatively make postulates about the random mechanism and then to logically derive a model by employing these postulates in combination with the laws of probability.

A *probability model* for the random variable X is a specified form of probability distribution that is assumed to reflect the behavior of X. Probabilities are expressed in terms of unknown parameters that are related to characteristics of the population and the method of sampling.

Here we list a few desirable criteria to provide some general guidelines for the formulation of useful probability models. The full force of these criteria will become more apparent when we introduce a greater variety of probability models in later chapters.

Desirable properties of a model.

Adequacy: The model must be based on postulates that adequately portray the chance mechanism that causes variation in the observations.

Simplicity: Whenever possible, simplifying approximations should be attempted so that the model lends itself to statistical analysis without sacrificing the quality of adequacy.

Parsimony in parameters: Incorporating an undue number of parameters tends to lower the quality of the statistical analysis. When choosing between two models that are both realistic approximations of the underlying chance behavior, the model that has fewer unknown parameters is preferred.

Chapter 5 is devoted to developing a few basic models that are useful in representing the probability distribution of counts pertaining to the number of occurrences of an event in repeated trials of an experiment. Emphasis is first placed on probability distributions that are associated with sampling from dichotomous populations, so named because each element of the population is classified in one of two distinct categories. Selecting a single element of the population and determining its category is regarded as a trial of the experiment, so that each trial can have only one of two possible outcomes. Starting with the sample space consisting of these two outcomes, we embark on a study of distributions for the number of occurrences in repeated experimental trials.

5.2 BERNOULLI TRIALS : SUCCESS – FAILURE

Here we consider successive repetitions of an experiment or observation in which each repetition is called a *trial*. Furthermore, we assume that there are only two possible outcomes for each individual trial. Instead of designating these outcomes e_1 and e_2, the conventional practice is to call them *success* (S) and *failure* (F) to emphasize the point that they are the only possible results. The usage of these terms is prompted by convenience alone and they do not have the same connotation as success or failure in real life. Customarily, the outcome of primary interest to the experimenter is labeled a success even if it is a disastrous event. In a study of traffic fatalities, for example, the occurrence of an accident may be statistically designated a success!

Situations that involve only two possible results in an experimental trial abound in diverse fields. In particular, they include sampling from dichotomous populations. A few examples are:

(*a*) Observe the next chicken egg that hatches and determine if the chick is a *male* or a *female*.
(*b*) Inspect an item from a production line and observe if it is *defective* or *nondefective*.
(*c*) A contractor makes a certain bid for a contract; the outcome is *success* or *failure* in obtaining the contract.
(*d*) Test an antibotic on a mouse and record if the mouse's reaction is *positive* or *negative*.

The simple nature of the results in each trial of an experiment provides a convenient starting point from which to develop probability models for random variables that are defined in terms of repetitions of the trials. The repeated trials are performed under a set of conditions that we state as *postulates*. These postulates not only provide conditions that closely approximate many experimental conditions, but they also produce simple and useful probability models. Repeated trials that obey these postulates are called *Bernoulli trials*, after the French mathematician Jacob Bernoulli.

Bernoulli trials

(*a*) Each trial yields one of two outcomes technically called *success* (S) and *failure* (F).

(*b*) For each trial, the probability of success $P(S)$ is the same and is denoted by $p=P(S)$. The probability of failure is then $P(F)=1-p$ for each trial and is denoted by q, so that $p+q=1$.

(*c*) Trials are independent. The probability of success in a trial does not change given any amount of information about the outcomes of other trials.

Perhaps the simplest example of Bernoulli trials is the prototype model of tossing a fair coin, where the occurrences *head* and *tail* can be labeled S and F, respectively. Thus we have $p=q=\frac{1}{2}$.

Example 5.1 Sampling from a dichotomous population: Consider a lot (population) of items in which each item can be classified as either defective or nondefective.

(*a*) *Sampling with replacement*: Suppose that a lot consists of 15 items of which 5 are defective and 10 are nondefective. An item is drawn *at random*, (i.e., in a manner that all items in the lot are equally likely to be selected). The quality of the item is recorded and it is returned to the lot before the next drawing. The conditions for the Bernoulli trials are satisfied. If the occurrence of a defective is labeled S, we have $P(S)=p=\frac{5}{15}$.

(*b*) *Sampling without replacement*: In situation (*a*), suppose that 3 items are drawn one at a time but without replacement. Then the condition concerning the independence of trials is violated. For the first drawing, $P(S)=\frac{5}{15}$. If the first draw produces an S, the lot then consists of 14 items, 4 of which are defective. Given this information about the result of the first draw, the conditional probability of obtaining an S on the second draw is then $\frac{4}{14}\neq\frac{5}{15}$, which establishes the lack of independence.

As a secondary point, it is interesting to note that condition (*b*) for Bernoulli trials still holds in this sampling: That is, the (unconditional) probability of success is $\frac{5}{15}$ for each trial. For instance, consider the probability of obtaining an S in the second draw. We denote this event by S_2, where the suffix corresponds to the particular drawing. Considering the possible results of the first draw, the event S_2 is the union of two mutually exclusive events S_1S_2 and F_1S_2. By the multiplication law of probability,

$$P(S_1S_2)=P(S_1)P(S_2|S_1)=\frac{5}{15}\cdot\frac{4}{14}$$

and similarly

$$P(F_1S_2)=\frac{10}{15}\cdot\frac{5}{14}$$

Then $P(S_2)=P(S_1S_2)+P(F_1S_2)=\frac{5}{15}$. In the same manner, we can verify that $P(S_3)$ is also $\frac{5}{15}$.

(*c*) *Sampling without replacement from a large population*: Consider sampling 3 items without replacement from a lot of 1500 items, 500 of which are defective. With the notations we used in (*b*) we have

$$P(S_1)=\frac{500}{1500}=\frac{5}{15}$$

and

$$P(S_2|S_1) = \frac{499}{1499}$$

For most practical purposes, the latter fraction can be approximated by $\frac{5}{15}$. Thus, strictly speaking, there has been a violation of the independence of trials, but it is to such a negligible extent that the conditions of the Bernoulli trials can be assumed to be satisfied approximately. The purpose of considering this approximation is to incorporate simplicity into the model. We can conclude that in sampling without replacement from a large population where the sample size is a small fraction of the population size (usually less than 10%), the conditions of the Bernoulli trials are satisfied approximately. ■

The models we described in terms of sampling from a lot consisting of defectives and nondefectives can obviously be adapted to sampling from any population whose elements can be placed in only two categories. In problems such as sampling from a population of voters and observing whether a voter favors a certain candidate or not, or sampling from a population of trees and observing if a tree is infected with a specific disease or not, we easily visualize the similarities between each of these situations and sampling from a lot consisting of defective and nondefective items.

Example 5.1 illustrates several important results:

> If random samples are drawn from a dichotomous population with replacement, the conditions of the Bernoulli trials are satisfied. When the sampling is made without replacement, the condition of the independence of trials is violated. However, if the population size is quite large in comparison to the sample size, the effect of this violation is negligible and the model of the Bernoulli trials can be taken as an approximation.

Example 5.2 further illustrates the kinds of approximations that are sometimes employed when using the model of the Bernoulli trials.

Example 5.2 Testing a new antibiotic: Suppose that a newly developed antibiotic is to be tried on 10 patients who have a certain disease and that the possible outcomes in each case are cure (S) or no cure (F). Each patient has a distinct physical condition and genetic constitution that cannot be perfectly matched by any other patient. Therefore, strictly speaking, it may not be possible to regard the trials made on 10 different patients as 10 repetitions of an experiment under identical conditions, as the postulates of Bernoulli trials demand. We must remember that the postulates of a probability model are abstractions that help to realistically simplify the complex mechanism governing the outcomes of an experiment. Identification with the Bernoulli trials in such situations is to be viewed as an approximation of the real world, and its merit rests on how successfully the model explains chance variations in the outcomes. ■

5.3 THE BINOMIAL DISTRIBUTION

When a *fixed number n* of repeated Bernoulli trials is conducted with success probability p in each trial, we consider the random variable X, which represents the count of the number of successes in n trials. The probability distribution of X is called a *binomial distribution with n trials and success probability p*. The possible values of X are the integers $0, 1, \ldots, n$, and we proceed to derive a formula for the probability $P[X=x]$, where x can be any integer from 0 to n. Before presenting the formula for this function in the case of n trials, it is helpful to consider a special case.

In $n=4$ trials, each of which may result in S or F, there are $2\times2\times2\times2 = 16$ possible outcomes. These are listed in the columns below according to the number of successes. The entry *SFFS*, for instance, denotes the occurrence of S in the first and fourth trials and F in the second and third trials.

	FFFF	SFFF FSFF FFSF FFFS	SSFF SFSF SFFS FSSF FSFS FFSS	SSSF SSFS SFSS FSSS	SSSS
Value of X	0	1	2	3	4
Probability of each sequence	q^4	pq^3	p^2q^2	p^3q	p^4
Number of sequences	$1=\binom{4}{0}$	$4=\binom{4}{1}$	$6=\binom{4}{2}$	$4=\binom{4}{3}$	$1=\binom{4}{4}$

To derive an expression for $P[X=2]$, we consider all the sequences listed in the third column that have 2 S's and 2 F's. Because the trials are independent and in each trial $P(S)=p$ and $P(F)=q$, we obtain from the definition of independent events

$$P(SSFF) = P(S)P(S)P(F)P(F) = p^2q^2$$

In the same way, the probability of every individual sequence in this column is p^2q^2. There are 6 sequences, so we obtain $P[X=2]=6p^2q^2$. The factor 6 is the number of sequences with 2 S's and 2 F's. Even without making a complete list of the sequences, we can obtain this count by noting that the 2 places where S occurs can be selected from the total of 4 places in $\binom{4}{2}=6$ ways, each of the remaining two places always being filled with an F. Continuing in the same line of reasoning, the value $X=1$ occurs with $\binom{4}{1}=4$ sequences, each of which has a probability of pq^3, so that $P[X=1]=\binom{4}{1}pq^3$. After working out the remaining terms, the binomial distribution with $n=4$ trials can be tabulated (see Table 5.1).

TABLE 5.1
BINOMIAL DISTRIBUTION WITH $n=4$ TRIALS

x	0	1	2	3	4
$P[X=x]$	$\binom{4}{0}p^0q^4$	$\binom{4}{1}p^1q^3$	$\binom{4}{2}p^2q^2$	$\binom{4}{3}p^3q^1$	$\binom{4}{4}p^4q^0$

Extending this reasoning to the case of a general number n of Bernoulli trials, we observe that there are $\binom{n}{x}$ sequences that have exactly x successes and $(n-x)$ failures and that the probability of every such sequence is p^xq^{n-x}. Therefore

$$P[X=x]=\binom{n}{x}p^xq^{n-x} \quad \text{for} \quad x=0,1,\ldots,n$$

is the formula for the binomial probability distribution with n trials. We denote this probability by $b(x;n,p)$, which is read *the probability of $X=x$ in a binomial distribution having n trials and success probability p*. Sometimes when n and p are specified and do not vary in the course of a problem, we use the simpler notation $b(x)$ instead of $b(x;n,p)$.

Binomial distribution:

n trials; $p=P(S)$

$$P[X=x]=b(x;n,p)=\binom{n}{x}p^x(1-p)^{n-x}$$

for the possible values $0,1,2,\ldots,n$.

The term *binomial distribution* is derived from an important fact in algebra, called the *binomial expansion theorem*, which deals with an expansion formula for $(a+b)^n$. The following expansions for $n=1$, 2, and 3 should be somewhat familiar.

$$(a+b)^1 = a+b$$

$$(a+b)^2 = a^2+2ab+b^2$$

$$(a+b)^3 = a^3+3a^2b+3ab^2+b^3$$

The binomial expansion, for any positive integer power n is given by

$$(a+b)^n = a^n + \binom{n}{1} b a^{n-1} + \binom{n}{2} b^2 a^{n-2} + \ldots + \binom{n}{x} b^x a^{n-x} + \ldots + b^n$$

and this can be established by mathematical induction. Considering, in particular, $a=q$ and $b=p$, this formula yields

$$(q+p)^n = q^n + \binom{n}{1} p q^{n-1} + \binom{n}{2} p^2 q^{n-2} + \ldots + \binom{n}{x} p^x q^{n-x} + \ldots + p^n$$

Successive terms on the right-hand side of this formula are precisely the binomial probabilities $b(0)$, $b(1), \ldots, b(n)$. Incidentally, from the fact that $q+p=1$, this expansion also confirms that the probabilities in the binomial distribution sum to 1, as they should for any distribution.

To illustrate the manner in which the values of p influence the shape of the distribution, the probability histograms for three binomial distributions with $n=6$ and $p=.5$, .3, and .7, respectively, are presented in Figure 5.1. When $p=.5$, the binomial distribution is symmetric with the highest probability occurring at the center (see Figure 5.1*a*). For values of p smaller than .5, more of the probability mass is shifted toward the smaller values of x and the distribution is skewed with a longer tail to the right-hand side of the histogram. Figure 5.1*b*, where the binomial histogram for $p=.3$ is plotted, illustrates this tendency. On the other hand, Figure 5.1*c* with $p=.7$ illustrates the opposite tendency: the value of p is higher than .5, more probability mass is shifted toward higher values of x, and the distribution is skewed with a longer tail to the left. Considering the histograms in Figure 5.1*b* and *c*, we note that the value of p in one histogram is the same as the value of q in the other. The probabilities in one histogram are exactly the same as those in the other, but their order is reversed. This illustrates a general property of the binomial distribution: when p and q are interchanged, the distribution of probabilities is reversed.

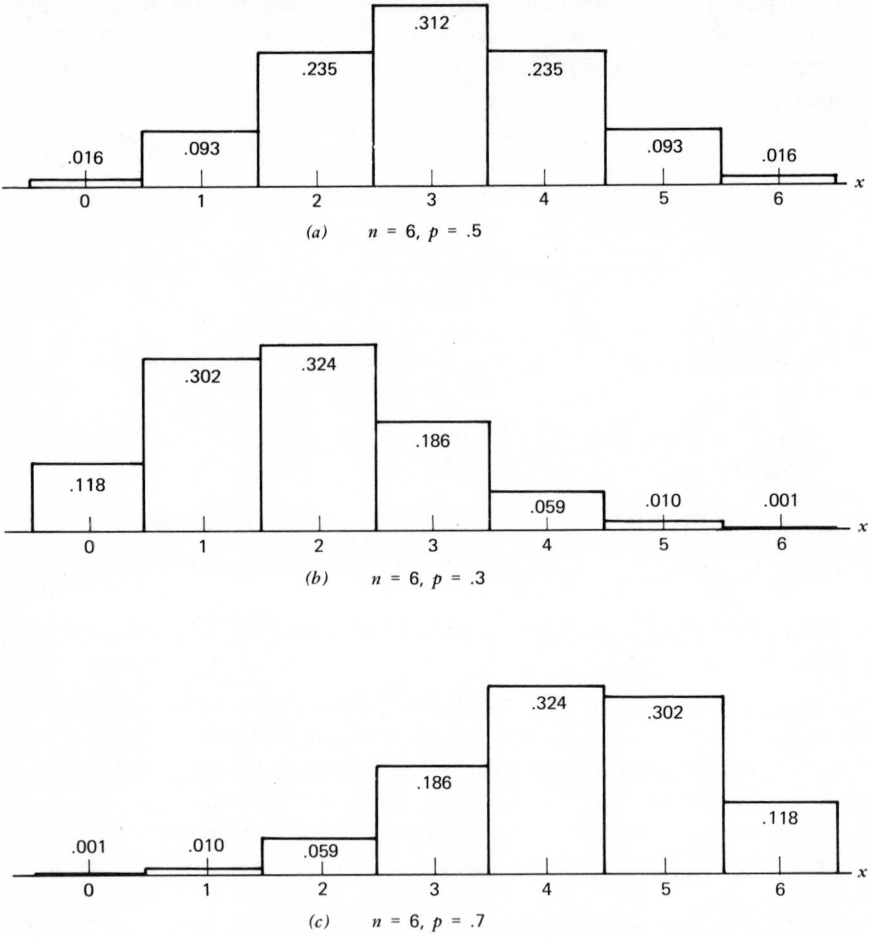

Figure 5.1 Binomial distributions for $n = 6$.

Example 5.3 According to the Mendelian theory of inherited characteristics, a cross fertilization of related species of red and white flowered plants produces offspring of which 25% are red flowered plants. Suppose that a horticulturist wishes to cross 5 pairs of red and white flowered plants. What is the probability that of the resulting 5 offspring:

(*a*) There will be no red flowered plants?
(*b*) There will be 4 or more red flowered plants?

It is natural to assume that the five trials are independent (because they are conducted on different parent plants) and that the experiment conforms to the postulates for Bernoulli trials. Let the random variable X denote the number of red flowered plants among the 5 offspring. If we identify the occurrence of a red as a success S, the Mendelian theory specifies that $P(S)=p=\frac{1}{4}$, and hence X has a binomial distribution with $n=5$ and $p=\frac{1}{4}$. The required probabilities are therefore:

$$(a)\ P[X=0]=b(0)=\left(\frac{3}{4}\right)^5=\frac{243}{1024}=.237$$

$$(b)\ P[X \geqslant 4]=b(4)+b(5)$$
$$=\binom{5}{4}\left(\frac{1}{4}\right)^4\left(\frac{3}{4}\right)^1+\binom{5}{5}\left(\frac{1}{4}\right)^5\left(\frac{3}{4}\right)^0$$
$$=\frac{16}{1024}=.016$$

■

Example 5.4 (A sample size problem): In a biological experiment for which it is rather expensive to procure subjects, the experimenter has a successful outcome with 40% of the subjects. How many subjects should be procured if the experimenter wishes to be 95% confident that at least one trial will be successful?

The conditions for the Bernoulli trials are assumed to hold. Setting X = number of successes, the experimenter wishes to determine n so that

$$P[X \geqslant 1] \geqslant .95 \quad \text{or} \quad 1-.95 \geqslant P[X=0]=q^n$$

where $q=1-p=1-.4=.6$. Considering succesive powers of .6 we obtain

n	2	3	4	5	6
$(.6)^n$	.36	.216	.1296	.07776	.046656

so that $n=6$ is the smallest n such that the probability of observing at least one success is greater than .95.

This calculation can serve as a useful guide in selecting a sample size (see Table 5.2 for various values of p).

TABLE 5.2
SMALLEST SAMPLE SIZE n REQUIRED TO OBTAIN $P[X \geqslant 1] \geqslant .95$

p	.05	.10	.20	.30	.40	.50	.60	.70
Required number of trials n	59	29	14	9	6	5	4	3

Note how rapidly n must be increased for quite small values of p. ■

5.4 HOW TO USE THE BINOMIAL TABLES

When n is even moderately large, the computation of the binomial probabilities becomes tedious because they involve powers of p and q. The binomial model plays an important role in many areas of sampling, and extensive tables have been prepared for the probabilities. The cumulative probabilities $P[X \leqslant c] = \Sigma_{x=0}^{c} b(x; n, p)$ are tabulated for some values of n, p, and c in Appendix Table 2. The probabilities of other events involving X can be obtained by expressing those events in terms of events of the form $[X \leqslant c]$ and applying the appropriate laws of probability. Some examples are:

$$P[X = x] = P[X \leqslant x] - P[X \leqslant (x-1)]$$

$$P[a \leqslant X \leqslant b] = P[X \leqslant b] - P[X \leqslant (a-1)]$$

$$P[X > x] = 1 - P[X \leqslant x]$$

Example 5.5 Suppose it is known that the germination rate of a particular variety of seeds is 80% if they are planted this season. If 15 seeds are planted, determine the probability that:

(*a*) At most 8 will germinate.
(*b*) Ten or more will germinate.
(*c*) The number germinating will be no fewer than 8 and no more than 12.

Designating the germination of a seed S and assuming that the results for individual seeds are independent, we note that the binomial distribution with $n = 15$ and $p = .8$ is appropriate for X = number of seeds that will germinate. To compute the required probabilities, we consult the binomial table for $n = 15$ and $p = .8$.

(*a*) $P[X \leqslant 8] = .018$, which is directly obtained by reading from the row $c = 8$.
(*b*) $P[X \geqslant 10] = 1 - P[X \leqslant 9] = 1 - .061 = .939$
(*c*) $P[8 \leqslant X \leqslant 12] = P[X \leqslant 12] - P[X \leqslant 7] = .602 - .004 = .598$ ■

*Other tables

For a more detailed or extensive table than Appendix Table 2, especially for larger values of n, consult the binomial tables given in [1] or [2]. These include values of p that range up to .5. When the value of p is greater than .5, a symmetry property of the binomial distribution can be invoked and the problem can be recast in the form where p is less than .5. This is achieved by interchanging the terms S and F and considering the original failure probability q to be the new success probability p^*. To be concrete, suppose that we want to compute $b(2;7,.8)$ = probability of 2 successes in 7 trials, where $P(S)=.8$. Having 2 successes is equivalent to having $7-2=5$ failures, and the $P(F)$ in each trial is $1-.8=.2$. Interchanging S and F, we must then find the probability of 5 successes in 7 trials when the success probability in each trial is .2. To do this, we consult the binomial table for $b(5;7,.2)$. We can now state the general relation

$$b(x;n,p)=b(n-x;n,p^*)$$

where $p^*=1-p$. The equality can also be verified formally by writing out the formulas for these two probabilities:

$$b(x;n,p)=\binom{n}{x}p^x(1-p)^{n-x}$$

$$b(n-x;n,p^*)=\binom{n}{n-x}p^{*n-x}(1-p^*)^x$$

These two expressions are the same, because $\binom{n}{x}=\binom{n}{n-x}$ and $p^*=(1-p)$.

5.5 THE MEAN AND VARIANCE OF THE BINOMIAL DISTRIBUTION

Although we already have a general formula which gives the binomial probabilities for any n and p, in later chapters we will need to know the mean and the standard deviation of the binomial distribution. Referring again to the probability histograms of the binomial distributions in Figure 5.1a, we can see that the balance point or the mean is $3=np$ for $n=6$ and $p=.5$. This result coincides with our previous knowledge about the number of heads that would be expected to occur in 6 flips of a fair coin. The other two probability histograms in Figure 5.1 show how the mean increases as p increases, but the exact relationship is not as apparent. The general formulas for the mean and the variance of a binomial distribution are:

For a binomial distribution with n trials; $p=P(S)$

$$\text{Mean} = np$$
$$\text{Variance} = np(1-p)=npq$$
$$\text{sd} = \sqrt{np(1-p)} = \sqrt{npq}$$

To justify these three formulas, we first consider the simple case $n=1$. The probability distribution of Y = number of successes in one Bernoulli trial is given by

y	0	1
Probability	q	p

and we obtain

$$E(Y)=0\cdot q+1\cdot p=p$$

$$E(Y^2)=0\cdot q+1\cdot p=p$$

$$\mathrm{Var}(Y)=E(Y^2)-\left[E(Y)\right]^2=p-p^2=p(1-p)=pq$$

Thus the formulas are seen to hold for $n=1$. We now proceed to use these results to obtain the general expression for the case of n Bernoulli trials. The number of successes X in n trials can be expressed

$$X=X_1+X_2+\ldots+X_n$$

where X_1 is the number of successes in the first trial ($X_1=0$ or 1), X_2 is the number of successes in the second trial ($X_2=0$ or 1), and so on. Because the trials are independent, the counts $X_1,X_2,\ldots,X_n$ are independent random variables, each having the same distribution as that just given for Y above. Using the properties of expectation and variance of a sum of random variables discussed in Section 4.7, we obtain

$$E(X)=E(X_1)+\ldots+E(X_n)=p+\ldots+p=np$$

$$\mathrm{Var}(X)=\mathrm{Var}(X_1)+\ldots+\mathrm{Var}(X_n)=pq+\ldots+pq=npq$$

An alternative but tedious way of obtaining these results is to use the definition of expectation and variance directly with the expression for the binomial probabilities and then to simplify by algebraic manipulation.

5.6 THE HYPERGEOMETRIC DISTRIBUTION

We noted in Example 5.1 that random sampling *without* replacement from a finite population violates the independence condition of the Bernoulli trials and that this is serious when the sample size is more than a small fraction of the population size. To be specific, suppose that we are sampling from a population of N elements that can be divided into two groups according to those who possess a specific characteristic and those who do not. The groups could be male–female, employed–unemployed, defective–nondefective, and so on. Adopting the names *defective* and *nondefective* to technically describe the two categories, we denote the number of defectives in the population by D. Thus the number of nondefectives is $N-D$. We then let X denote the number of defectives in a random sample of n elements. This sampling scheme is illustrated in Figure 5.2.

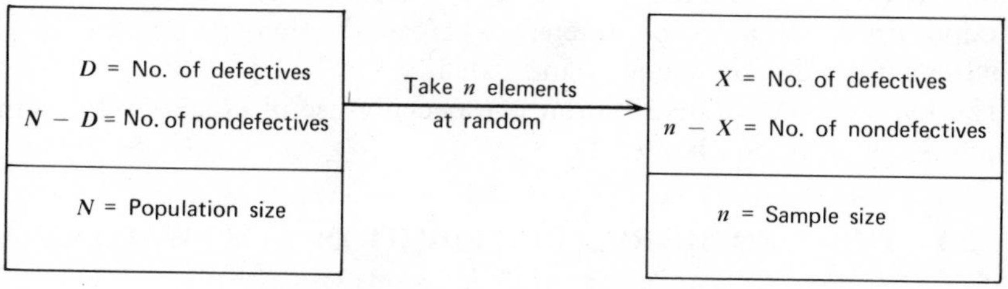

Figure 5.2 Sampling without replacement.

Since the X defectives in the sample must come from the subgroup of D in the population and the $(n-X)$ nondefectives must come from the subgroup of $(N-D)$, the probability distribution of X can be obtained by using the rule of combination. This distribution is called the *hypergeometric distribution*.

Hypergeometric distribution:

N = population size $\quad D$ = number of defectives in population

n = sample size $\quad X$ = number of defectives in sample

$$P[X=x]=\frac{\binom{D}{x}\binom{N-D}{n-x}}{\binom{N}{n}} \quad \text{for } x=0,1,\ldots,n$$

when $n \leqslant D$ and $n \leqslant N-D$.

Mean $= np$, where $p=\frac{D}{N}$ (the population proportion of defectives).

Variance $= npq\left(\frac{N-n}{N-1}\right)$, where $q=1-p$.

Note that the mean of this distribution is the same as it is in the case of the binomial distribution, which is appropriate when the sampling is done one by one and with replacement. However, the variance is multiplied by the factor $(N-n)/(N-1)$, which is called the *finite-population correction factor for variance*. When n/N is quite small, this factor is close to 1 (as it should be), because the binomial distribution approximates the distribution, as Example 5.1*c* illustrates.

Comparing these two distributions, we can see that the binomial distribution has the merit of simplicity in the probability formula. It also has the fraction $p = D/N$ as a parameter, whereas the hypergeometric model requires knowledge of D and N individually.

The hypergeometric distribution is especially useful in survey sampling, which is the topic of Chapter 16.

5.7 THE GEOMETRIC DISTRIBUTION FOR WAITING TIME

The geometric distribution is another discrete distribution arising in the context of the Bernoulli trials. When we perform a fixed number n of these trials, the number of successes is a random variable whose distribution is binomial $b(n,p)$. If instead of fixing the number of trials in advance, we wish to perform Bernoulli trials until the first success occurs, then the number of successes becomes fixed at 1 and the number of trials is a random variable. Setting X = the number of trials required to produce one success, we can derive the probability distribution of X. The possible values of X are $1, 2, \ldots$, where the value $X = x$ occurs if and only if we have a chain of $x-1$ failures followed by 1 success. Note that the trials are terminated as soon as a single success is obtained. Using the condition of the independence of trials, we obtain

$$P[X=x] = P[\underbrace{FF\ldots F}_{x-1}S] = q^{x-1}p, \quad x = 1, 2, \ldots$$

This distribution is called the *geometric distribution*. A special case in which $p = \frac{1}{3}$ appears in Example 4.3.

Geometric distribution:

X = number of Bernoulli trials until the first S occurs.

$$p = P(S), \quad q = 1 - p$$

$$P[X=x] = q^{x-1}p, \quad x = 1, 2, \ldots$$

$$\text{Mean} = \frac{1}{p}, \quad \text{Variance} = \frac{q}{p^2}$$

The derivations of the formulas for the mean and the variance stated above are omitted here.

The geometric distribution is sometimes called a *discrete waiting-time distribution*. This stems from the fact that if it takes a unit of time to perform one Bernoulli trial, then the waiting time to obtain a single success is precisely the random variable X that has a geometric distribution.

The geometric distribution is often useful in studies of a rare population characteristic, such as the incidence of a rare blood disease. We may, of course, choose either to examine a specified number n of individuals and count the number affected or to continue examining individuals until an affected person is found. The first choice conforms to the binomial model, but the observed sample may not contain a single affected person, in which case we would obtain little information about the incidence of the disease. In contrast, the latter type of sampling, which is governed by the geometric distribution, guarantees the presence of the characteristic in the sample.

Example 5.6 Expected waiting time to obtain a success: A biologist wishes to investigate the incidence of a particular disease in a certain tribal population. Let p be the fraction of the population affected by this disease. Individual members are to be randomly chosen and examined one at a time until the first affected person is found. How many persons are expected to be observed using this method of sampling?

If the population is quite large, we can assume that the fraction p is practically constant from trial to trial and that the conditions for the Bernoulli trials are approximately met. From the formula for the mean of the geometric distribution, the expected number of trials required to find a single diseased person is

$$E(X) = \frac{1}{p}$$

To show the manner in which the expected waiting time depends on the incidence rate of the disease in the population, we present a few numerical values in Table 5.3.

TABLE 5.3
EXPECTED WAITING TIME TO OBSERVE ONE SUCCESS

p	.01	.03	.05	.10	.20	.30	.40	.50	.60
Expected waiting time	100	33.3	20	10	5	3.33	2.5	2	1.67

Note that the expected waiting time increases rapidly as p decreases.

If we are interested in the expected waiting time required to obtain two successes, the time is the sum of the two times required to obtain a single success. Thus the expected waiting time for $p = .03$ is $2(33.3) = 66.6$. In general, the expected waiting time to observe K successes is K/p. ■

5.8 RARE EVENTS AND THE POISSON DISTRIBUTION

The binomial distribution describes the chance behavior of the number of times an event S occurs in a given number n of Bernoulli trials where $P(S)=p$ in each trial. When n is very large and beyond the range of tables, it is often difficult to obtain the binomial probabilities by computation. We introduce here an important probability distribution, called the *Poisson distribution*, that is not only useful in modeling many chance phenomena but that also provides an approximation of the binomial probabilities when *n is very large and p is very small and the product np is of moderate magnitude*. (Another approximation when p is not too small will be considered later.) Apart from its role as an approximating distribution, the Poisson distribution also serves as a useful probability model for events occurring randomly over time or space when all that is known is the average number of occurrences np per unit time or space. For an event occurring over time, each instant of time can be regarded as a potential trial in which the event may or may not occur. In a unit of time, there are virtually an infinite number of trials, but normally only a few occurrences take place. In such situations, we need a probability model that involves only the average rate of occurrence np and that does not require the knowledge of n and p individually.

Example 5.7 Incidence of color blindness in a large population was 4% in the previous decade. To study the present situation, an investigator tests a random sample of 200 persons taken from this population. Assuming that a 4% rate of color blindness still prevails, what is the probability of having 5 or less color blind people in the sample?

Here the trials are well defined: each person in the sample is a trial with outcome color blind (S) or normal (F). This is the binomial scheme with $n=200$ and $p=.04$. We wish to determine the probability

$$P[X \leqslant 5] = \sum_{x=0}^{5} \binom{200}{x} (.04)^x (.96)^{200-x}$$

The only difficulty in obtaining a numerical answer is that n and p are beyond the range of binomial tables. ■

Example 5.8 A large number of spindles in a textile mill operate simultaneously and without any recognizable cause, the event "breakage of yarn" (S) occurs every so often. When this happens, the spindle remains idle until an attendant joins the broken pieces together again. Each instant of time is a potential trial in which S can occur. In each unit of time (say, an hour), there are virtually an infinite number of trials, but records indicate that yarn breakage occurs only a few times (say, an average of 10 times per hour). If the random variable X denotes the number of breakages in an hour, what would be an appropriate probability model for X? The binomial distribution does not provide an answer. Although the situation seems similar to that of the Bernoulli process, the binomial distribution is not applicable here because n and p cannot be assessed individually. In this case, our knowledge is limited to *the average number of breaks per hour*. ■

Example 5.9 Plankton counts: Events can occur in space just as they do over time and be otherwise similar in nature. Suppose that we are interested in finding a suitable probability model for the count X of "cyclops" (a zoo-plankton) in a liter of lake water. The unit volume is a liter, which may be regarded as consisting of a large number of micro-volumes, each of which may or may not contain a cyclop. Thus there are virtually an infinite number of trials, but the average number of occurrences (i.e., the average cyclop count per liter) is only a moderate number. ■

The events considered in the preceding examples can be called *rare events*, because each event is expected to occur only a few times in a very large number of potential trials. This leads us to consider a suitable probability model for rare events. Some other examples of rare events are: the number of traffic accidents per week at a busy intersection, the number of tornado touchdowns, the number of particles emitted by a radioactive substance, the daily sales of refrigerators in a store, and the number of misprints on a page.

The Poisson distribution is important in many applications other than those in which it approximates the binomial distribution. To understand the kinds of situations in which it is a useful probability model, we state in somewhat heuristic terms the postulates that lead to a mathematical derivation of the Poisson distribution.

5.8.1 Formal Definition of the Poisson Distribution

An event S occurs in time obeying the following postulates:

(*a*) *Independence*: The number of times S occurs in any interval of time is independent of the number of occurrences of S in any other disjoint time interval.

(*b*) *Lack of clustering*: The chance of two or more occurrences happening simultaneously can be assumed to be 0.

(*c*) *Rate*: The average number of occurrences per unit time is a constant, denoted by m, and it does not change with time.

This formulation is similar for events occurring in space.

Let X be the random variable representing the number of times S occurs in a unit time interval. Under these three postulates, the probability distribution of X is given by the formula:

> *Poisson distribution:*
>
> $$P[X=x]=\frac{e^{-m}m^{x}}{x!}, \quad x=0,1,2,\ldots$$

where e is the *exponential number* and its value is $e=2.71828$ (rounded at the fifth decimal). This is called the *Poisson distribution with parameter m*. Note that m represents the rate or the average number of times that S occurs. Although we omit the mathematical derivation of this distribution, the three postulates stated here can help to identify situations in which the Poisson distribution is appropriate. The cumulative Poisson probabilities $\Sigma_{x=0}^{c} e^{-m}m^{x}/x!$ are given in Appendix Table 3. More elaborate Poisson probability tables are provided in [1] and [3].

In a binomial distribution with very large n, very small p, and $np=m$ of moderate magnitude, the binomial probability $\binom{n}{x}p^{x}q^{n-x}$ is approximately equal to the Poisson probability $e^{-m}m^{x}/x!$. The nature of the approximation is illustrated in Table 5.4, where $m=np=2$ is fixed and n successively increased and p decreased.

TABLE 5.4
THE POISSON APPROXIMATION TO THE BINOMIAL DISTRIBUTION

		BINOMIAL PROBABILITY	
n	p	$X=0$	$X=4$
10	.2	.1074	.0881
20	.1	.1216	.0898
50	.04	.1299	.0902
100	.02	.1326	.0902
POISSON PROBABILITY WITH $m=2$		.1353	.0902

Example 5.10 Poisson approximation to binomial: In the color blindness problem posed in Example 5.7, we have $n=200$, $p=.04$, and $m=np=8$. Consequently, the conditions for the Poisson approximation are satisfied. Using the Poisson probability table with $m=8$, we obtain

$$P[X \leqslant 5]=.191$$

■

We can now state the formulas for the mean and the variance of the Poisson distribution and compare them with the corresponding formulas for the binomial distribution (see Table 5.5).

TABLE 5.5
MEAN AND VARIANCE OF POISSON AND BINOMIAL DISTRIBUTIONS

	POISSON (m)	BINOMIAL (n,p)
Mean	m	np
Variance	m	npq

Recall that the Poisson distribution approximates the binomial distribution when $np = m$ is of moderate magnitude and p is quite small. In this case, q is very close to 1; npq is therefore close to np. This is an intuitive justification as to why the variance of a Poisson distribution is the same as its mean.

Example 5.11 Suppose that the average number of cyclops per liter of lake water is 2. What is the probability that 5 or more cyclops will be found in a sample of 3 liters of lake water?

Applying the reasoning that each liter consists of a great many small units, each of which may contain a single cyclop, and assuming that cyclops do not tend to cluster together, we can conclude that the Poisson distribution may apply. Because there is an average of 2 cyclops per liter, in a 3-liter volume there is an average of $3 \times 2 = 6$ cyclops. Consulting the Poisson probability table with $m = 6$, we obtain

$$P[X \geqslant 5] = 1 - P[X \leqslant 4] = 1 - .285 = .715 \qquad \blacksquare$$

REFERENCES

1. Beyer, W. (ed.), *Handbook of Tables for Probability and Statistics*, 2nd Ed., The Chemical Rubber Company, 1968.

2. *Tables of the Cumulative Binomial Probability Distribution*, Harvard University Press, 1955.

3. Molina, E. C., *Poisson's Exponential Binomial Limit*, Van Nostrand, 1949.

EXERCISES

1. Is the model of the Bernoulli trials plausible in each of the following situations? If so, discuss in what manner (if any) a serious violation of the assumptions can occur.

(*a*) Beetles of a common strain are sprayed with a given concentration of an insecticide and the occurrence of death or survival is recorded in each case.
(*b*) A word association test is given to 10 first-grade children and the amount of time each child takes to complete the test is recorded.
(*c*) Items coming off an assembly line are inspected and classified defective or nondefective.
(*d*) Each of the first 10 days of July is observed and recorded as cloudy or clear.
(*e*) From a group of 50 persons, 20 people are chosen at random and each is classified as left-handed or not.

2. A backpacking party carries 3 emergency signal flares, each of which will light with a probability of .99. Assuming that the flares operate independently, find:
(*a*) The probability that at least one flare lights.
(*b*) The probability that exactly two flares light.
3. A stop light on the way to class is red 40% of the time. What is the probability of hitting a red light:
(*a*) 2 days in a row?
(*b*) 3 days in a row?
(*c*) 2 out of 3 days?
4. If the probability of having a male child is .5, find the probability that the third child is the first son.
5. If the probability of getting caught copying someone else's exam is .2, find the probability of not getting caught in 3 attempts. Assume independence.
6. Using the binomial table, find the probability of:
(*a*) 3 successes in 8 trials when $p = .4$.
(*b*) 7 failures in 16 trials when $p = .6$.
(*c*) 3 or fewer successes in 9 trials when $p = .4$.
(*d*) More than 12 successes in 16 trials when $p = .7$.
(*e*) The number of successes between 8 and 13 (both inclusive), in 16 trials when $p = .6$.
7. (*a*) Using the binomial table, find the probabilities $b(x; n, p)$ for $n = 7$ and $p = .5$. Plot the probability histogram.
(*b*) Repeat (*a*) for $p = .8$ and for $p = .2$.
(*c*) Locate the mean and the median on each histogram.
(*Recall*: The median is a value m_0 of the random variable such that $P[X \leqslant m_0] \geqslant .5$ and $P[X \geqslant m_0] \geqslant .5$.)
8. There are 8 outcomes in the sample space of the binomial with $n = 3$. If each outcome is equally likely, the probabilities of $0, 1, 2, 3$ successes are $\frac{1}{8}, \frac{3}{8}, \frac{3}{8}, \frac{1}{8}$.

(*a*) Compute the mean and the variance directly from this probability distribution.
(*b*) Check the mean and the variance in (*a*), using the special formula for the binomial distribution.

9. A sociologist feels that only half of the high-school seniors capable of graduating from college go to college. Of 17 high-school seniors who have the ability to go to college, find the probability that 10 or more will go to college if the sociologist is correct. Assume that the seniors make their decisions independently. Also find the expected number.
10. A new medication gives 5% of the users an undesirable reaction. If a sample of 13 users receive the medication, find the probability of:
(*a*) 0 undesirable reactions.
(*b*) 1 undesirable reaction.
11. A manufacturer of furniture polish claims that its new product gives a more glossy finish than another leading brand. Of 18 randomly selected homeowners who were asked to use both brands and compare glossiness, 13 preferred the new brand. Assuming there is really no difference in quality between the brands, so that the probability of preference for the new brand is .5, find the probability that 13 or more people will prefer the new product. Comment on the strength of the manufacturer's assertion.
12. *Sampling inspection*: A pharmaceutical company is contracted to supply batches of cattle vaccine to a distributor. Occasionally, some of these vaccines happen to be sterile. The distributor wants to be protected against receiving too many sterile vaccines. Testing each individual vial is impractical, because the test renders the vaccine useless. To monitor the quality of the vaccine, the distributor uses the following screening process. A random sample of 10 vials from each batch is tested, and the number of sterile vials X is counted. If $X=0$ the batch is accepted, and if $X \geqslant 1$ it is rejected. This is called a *single sampling plan* with *sample size* $n=10$ and *acceptance number* $c=0$. Assume that the batch size is large so that the distribution of X is approximately binomial, with $n=10$ and $p=$the unknown fraction of sterile vials in each batch.
(*a*) If $p=.2$, what is the probability that a batch will be accepted under the distributor's sampling inspection plan?
(*b*) Calculate the probability of accepting a batch $P(A)$ for $p=.05$, .1, .2, .3, .4. Plot $P(A)$ vs. p on a graph and join the points with a smooth curve. (This curve is called the *operating characteristic curve* for the sampling plan.)
13. Referring to Exercise 12, consider an alternative sampling plan in which $n=15$ and $c=1$. In other words, a batch is accepted if at most 1

out of the 15 vials tested is sterile.

(*a*) Calculate $P(A)$ for various values of p and plot the operating characteristic curve of this plan.

(*b*) With which plan is there a lower risk of accepting a batch that contains 20% sterile vials?

14. Let p denote the unknown proportion of rocks in a riverbed that are sedimentary in type. Suppose that $X=12$ of a sample of $n=20$ rocks collected in random locations are found to be sedimentary in type.

(*a*) Use the binomial table to find the probability of observing this result when $p=.1, .2, .3, .4, .5, .6, .7, .8, .9$. Graphically plot the probability as a function of the parameter p. Join the points with a smooth curve.

(*b*) From your graph, determine the value $\hat{p}$ of p where the curve reaches its maximum.

(NOTE: The unknown parameter p is being estimated from the sample data. The above principle is used to determine the value of the parameter that maximizes the likelihood of the observed data and is called the *method of maximum likelihood*. For the binomial sample, it can be mathematically shown that $\hat{p}=X/n$.)

15. Only 30% of the people in a large city feel that its mass transit system is adequate. If 20 persons are selected at random, find the probability that 5 or less will feel that the system is adequate. Find the probability that exactly 6 will feel that the system is adequate.

16. A school newspaper claims that 80% of the students support its view on a campus issue. A random sample of 20 students is taken, and 12 students agree with the newspaper. Find $P[12$ or less agree$]$ and comment on the plausibility of the claim.

17. An auditor for the Internal Revenue Service starts each day by selecting 5 of 12 returns for a careful audit. If one day 4 of the 12 actually contain illegitimate deductions, find the probability that 2 of them will be selected by the auditor. Also, find the mean and the standard deviation of the number of returns the auditor finds to contain illegitimate deductions.

18. A sorority has 12 members, 6 of whom support open marriage. Let X denote the number of supporters among 4 girls selected at random from the group.

(*a*) State the probability distribution of X.

(*b*) Numerically tabulate the probabilities.

(*c*) Using the formulas given in this chapter, compute $E(X)$ and $\text{Var}(X)$.

19. Suppose that the sampling in Exercise 18 is made with replacement. Then what is the probability distribution of X? Find the probabilities

and compare them with those in Exercise 18. Also compute $E(X)$ and $\mathrm{Var}(X)$.

20. An administrator studying the quality of high-school curriculums randomly selects 6 out of 15 schools in the area. If 2 of the 15 schools have unsatisfactory curriculums, what is the probability that all 6 of the schools in the sample have satisfactory curriculums? What is the expected value of the number of unsatisfactory curriculums?
21. Suppose that it is reasonable to assume that a sales representative's contacts with individual customers constitute Bernoulli trials, where the execution of a sale is termed a success S. Let X denote the number of customer contacts until the first sale is executed.
    (*a*) State the probability distribution of X.
    (*b*) Suppose that the sales representative's success rate is .2. Find $P[X=2]$, $P[X \geqslant 3]$, and $E(X)$.
22. Referring to Exercise 21, find the probability that the sales representative will have to contact at least 10 customers to execute 2 sales. (*Hint*: Find the probability of fewer than 2 sales in 9 contacts.)
23. For a Poisson random variable X, find:
    (*a*) $P[X=3]$ and $P[X \leqslant 5]$ when $m=4$.
    (*b*) $P[X=0]$ and $P[X>1]$ when $m=.2$.
    (*c*) $P[X=2]$ and $P[3 \leqslant X \leqslant 7]$ when $m=3$.
24. Suppose that the number of insurance claims closely approximates a Poisson distribution with $m = .1$. Find the probability of (*a*) no claim, and (*b*) 1 or fewer claims.
25. The number of tornado touchdowns in a central area of a state is Poisson with $m=2.5$. If $X=$number of touchdowns, find $P[X \geqslant 4]$ and $P[X \leqslant 1]$.
26. Only 3% of the students in a major city have an IQ of 130 or above. For a randomly selected sample of size 50 use the Poisson approximation to find $P[X=2]$ and $P[X \geqslant 3]$, where $X=$number of students with an IQ of 130 or above.
27. Use the Poisson approximation to approximate $b(3; 100, .04)$ and $\Sigma_{x=0}^{5} b(x; 100, .04)$.

MATHEMATICAL EXERCISES

1. *Another calculation of E(X) for the binomial*: Verify that

$$E(X) = \sum_{x=0}^{n} x \binom{n}{x} p^x (1-p)^{n-x}$$

$$= \sum_{x=1}^{n} \frac{n(n-1)\dots(n-x+1)}{(x-1)!} pp^{x-1}(1-p)^{n-x}$$

$$= np \sum_{s=0}^{n-1} b(s; n-1, p) = np$$

2. *Poisson mean and variance*: Show that

$$\mu = E(X) = \sum_{x=0}^{\infty} x \frac{m^x e^{-m}}{x!} = \sum_{x=1}^{\infty} \frac{m^x e^{-m}}{(x-1)!} = m \sum_{s=0}^{\infty} \frac{m^s e^{-m}}{s!} = m$$

$$E[X(X-1)] = \sum_{x=0}^{\infty} x(x-1) \frac{m^x e^{-m}}{x!} = \sum_{x=2}^{\infty} m^2 \frac{m^{x-2} e^{-m}}{(x-2)!} = m^2$$

$$\sigma^2 = E[X(X-1)] + E(X) - [E(X)]^2$$

$$= m$$

CHAPTER 6

Basic Concepts of Testing Hypotheses

6.1 INTRODUCTION

We have seen how simplifying postulates about the behavior of the chance mechanism in repeated trials together with the laws of probability led to the specification of the binomial probability model and how another set of postulates produced the Poisson probability model. In this chapter, we introduce a particular type of statistical inference called *testing of statistical hypotheses*, which is intended to illustrate an important role played by probability distributions in drawing statistical inferences from a set of data.

We begin by focussing our attention on the binomial model. This model involves two quantities: the sample size n, which is known to the experimenter, and the population proportion p, which is usually unknown. Actually, the whole purpose of sampling is to make inferences about this unknown proportion p that occurs as a *parameter* in the probability model.

Broadly speaking, the goal of testing statistical hypotheses is to determine if a conjecture about some feature of a population is strongly supported by information obtained from the sample data. Prompted by the intent of the investigation, this conjecture typically involves an assertion about the value of a population parameter. In fact, we call any statement about the population a *statistical hypothesis*.

> A *statistical hypothesis* is a statement about the population. Its plausibility is to be evaluated on the basis of information obtained by sampling from the population.

Because an assertion may be true or false, two complementary hypotheses come to mind:

Hypothesis H: the assertion is true.
Hypothesis H': the assertion is false.

Using information from the sample observations, the decision maker must select one of these two decisions or *inferences*:

EITHER: *Reject* H' and conclude that H is strongly supported by the data.

OR: *Fail to reject* H' and conclude that H is not strongly supported by the data.

The process by which a choice is made between these two actions is called *testing statistical hypothesis*.

The rejection or nonrejection of a statistical hypothesis is somewhat different from disproving or proving a mathematical proposition. A mathematical proposition is either proved or a counterexample is provided to disprove it. In either case, the conclusion is established beyond any doubt. In contrast to this, there is an element of uncertainty in the conclusion reached when testing a statistical hypothesis by analyzing the experimental data. The following two groups of statements reflect this fundamental difference:

A mathematical proposition: The function $h(x) = 3x^2 - 2x + 1$ has a minimum at $x = 2$.

Results of investigation: At $x = 2$ the value of $h(x)$ is $3(2)^2 - 2(2) + 1 = 9$, whereas its value at $x = 0$ is $3(0)^2 - 2(0) + 1 = 1$.

Conclusion: The mathematical proposition is *false*.

A statistical hypothesis: The proportion of consumers preferring brand A to brand B is $p = .4$ (i.e., 40% of the consumers prefer brand A).

Results of investigation: In a random sample of 15 consumers, 12 are found to prefer brand A to brand B.

Typical conclusion: It is highly unlikely that the statistical hypothesis is true. If the true p is .4, the probability of observing 12 or more successes in 15 trials is .002. However, because it is physically possible to have this observation when $p = .4$, we *cannot be absolutely sure that the hypothesis is false*.

6.2 THE NULL AND THE ALTERNATIVE HYPOTHESES

A discussion of the formulation of a statistical hypothesis testing problem and the steps for solving it requires the introduction of a number of definitions and concepts. Before proceeding with this topic in its full generality, we develop its basic ideas in terms of a specific problem in which the chance behavior is governed by the binomial distribution.

> *Problem:* Experience has shown that the cure rate for a given disease using standard medication is 60%. The cure rate of a new drug is anticipated to be better than the standard medication. Suppose that the new drug is to be tried on a sample of 20 patients and that the number cured X in the 20 is to be recorded. How should the experimental data be used to answer the question: "*Is there substantial evidence that the new drug has a higher cure rate than the standard medication*?".

The cure rate of the new drug is a proportion p whose value can be correctly ascertained only if the drug is administered to a vast number of patients. However, our information is limited to the results obtained from the 20 patients. Assuming that the 20 patients in the experiment can be considered independent observations from the population of all present and potential patients, the probability model for X conforms to a binomial distribution with $n=20$ and p an unknown parameter. In light of the question raised in the statement of the problem, the following two hypotheses are relevant:

The new drug is better than the standard medication: $p>.6$.
The new drug is not better than the standard medication: $p \leqslant .6$.

The unknown state of nature is embodied in the parameter p, which is the success rate of the new drug. Of the two complementary statements concerning the unknown state of nature, one is called the *null hypothesis* H_0 and the other is called the *alternative hypothesis* H_1. To determine which hypothesis should be labeled the null hypothesis, the intrinsic difference between the roles and the implications of these two terms should be clearly understood.

> *Choice of H_0 and H_1*
>
> When an investigation is aimed at establishing an assertion with substantive support obtained from the sample, the negation of the assertion is taken to be the null hypothesis H_0 and the assertion itself is taken to be the alternative H_1.

The word "null" in this context can be interpreted to mean that the assertion purported to be established is actually void.

Before claiming that a statement is valid, adequate evidence must be produced to support it. Consequently, the analyst should consider a statement false unless the contrary is strongly supported by the data. In other words, the null hypothesis H_0 should be regarded as true and should be rejected only when the data strongly testify against it. A close analogy can be made to a court trial where the jury clings to the null hypothesis of "not guilty" unless there is convincing evidence of guilt. The intent of the hearings is to establish the assertion that the accused is guilty rather than to prove that he or she is innocent.

This discussion indicates that the roles assumed by H_0 and H_1 are not symmetric. Instead, the decision maker is somewhat conservative in the treatment of H_0 and does not reject it without overwhelming evidence. The term *null hypothesis* originated from comparative experiments in which a new product or a new technique is compared with the standard to determine if its superiority can be corroborated by experimental evidence. In this context, a null hypothesis is the statement that the difference between the new and standard product is null or zero—that the claim is incorrect. Switching to a new product or technique usually requires large initial expenditures, and a decision maker should not do so unless the new product is clearly better than the old. Thus when potential losses due to wrong decisions are substantially different, it should be regarded as more serious to wrongly reject the null hypothesis than to fail to reject it when it is actually false.

> In testing a null hypothesis H_0 against an alternative H_1, our attitude is to uphold H_0 as true unless the data speak strongly against it, in which case, H_0 should be rejected in favor of H_1. Falsely rejecting H_0 is viewed as a more serious error than failing to reject H_0 when H_1 is true.

In view of these guidelines, the specification of the null and alternative hypotheses in our problem should be

$H_0: p \leqslant .6$ (new drug is not better)
$H_1: p > .6$ (new drug is better)

Now the experimental data from the drug trials is in the form of X, the number of cures out of the 20 patients trying the new drug. Before the experiment is actually performed, X is a random variable with possible values of $0, 1, 2, \ldots, 20$. Every value is physically possible under both H_0 and H_1, so that none of these outcomes can absolutely prove that H_0 is true or that H_1 is true. (Again note the difference from the proof of a mathematical proposition.) However, intuition suggests that large values of

X would strongly indicate that H_0 may be false, whereas small values of X would support H_0. As an objective procedure, we can then specify a course of action: for example, to reject H_0 (in favor of H_1) if $X \geqslant 15$ and to retain H_0 if $X \leqslant 14$. Such a rule is called a *test of the null hypothesis*, and X is called a *test statistic*. Obviously, many other rules could be considered.

> A *test of the null hypothesis* is a course of action specifying the set of values of a random variable X for which H_0 is to be rejected. The random variable whose value serves to determine the action is called the *test statistic*, and the set of its values for which H_0 is to be rejected is called the *rejection region* of the test. A test is completely specified by a test statistic and the rejection region.

6.3 THE TWO TYPES OF ERRORS AND THE POWER FUNCTION OF A TEST

To continue our discussion of the problem concerning the cure rate of a new medicine, we introduce some measures of performance of a test. Common sense suggests that H_0 should be rejected for large values of X, although no reason has been established for the choice of $X \geqslant 15$ as the rejection region as yet. Other possible choices might be $X \geqslant 18$ or $X \geqslant 14$. Before we choose the rejection region, we must consider our chances of making a wrong decision with any rule that we use. Considering the unknown state of nature and the possible results from applying a test, one of the following situations will arise:

	Unknown true state of nature	
Test concludes:	H_0 true ($p \leqslant .6$)	H_0 false ($p > .6$)
Do not reject H_0	Correct	Wrong (type II error)
Reject H_0	Wrong (type I error)	Correct

The decision reached by using a test may be wrong in either of the two ways indicated here: (a) H_0 may be true, and the test may conclude that it should be rejected, or (b) the test may fail to reject H_0 when H_1 is true.

These two types of errors are called the *type I error* and the *type II error*, respectively.

> *Type I error*: rejection of H_0 when H_0 is true.
> *Type II error*: failure to reject H_0 when H_1 is true.

The probability of making a type I error typically depends on the exact value of the parameter p that prevails and is denoted by $\alpha(p)$, with p restricted to the range covered by H_0. The probability of making a type II error also depends on the value of p under the range covered by H_1 and is denoted by $\beta(p)$.

> The *probabilities of the two types of error*
>
> $\alpha = P$ [type I error] $= P$ [rejection of H_0 when H_0 is true]
>
> $\beta = P$ [type II error] $= P$ [not rejecting H_0 when H_1 is true]
>
> The probability α depends on the particular value of the parameter in the range covered by H_0, whereas β depends on the value over the range covered by H_1.

We now illustrate these definitions by computing the two kinds of error probabilities for the test that rejects H_0 if $X \geqslant 15$ and fails to reject H_0 if $X \leqslant 14$. For the sake of brevity, we henceforth write "the test $X \geqslant 15$" when we mean that the rejection region of the test consists of the values of X that are equal to or more than 15. Recall that the distribution of X is binomial $b(20,p)$, so that the rejection probability $P[X \geqslant 15]$ can be obtained from the binomial table for $n = 20$ and for each specified value of p. We denote the rejection probability function of a test by $\gamma(p)$, or

$$\gamma(p) = P\left[\text{the test rejects } H_0 \text{ when the true value of the parameter is } p\right]$$

For the test $X \geqslant 15$, we then have $\gamma(p) = P[X \geqslant 15 | p]$. Consulting the binomial tables, we obtain the numerical values of this probability, which are listed in Table 6.1 and plotted in Figure 6.1. For instance, when $p = .5$, the value of $P[X \geqslant 15] = 1 - P[X \leqslant 14] = 1 - .979 = .021$.

TABLE 6.1

THE PROBABILITIES OF REJECTION OF H_0 FOR THE TEST $X \geqslant 15$

p	.3	.4	.5	.6	.7	.8	.9
$\gamma(p)=P[X \geqslant 15 \mid p]$	.000	.002	.021	.126	.416	.804	.989

Under H_0, p is restricted to the range $p \leqslant .6$, which is to the left of the middle vertical line in Figure 6.1. In this part of the graph, the rejection probability $\gamma(p)$ is, by definition, the same as the type I error probability $\alpha(p)$. Under H_1, the range of p is $p > .6$, which is to the right of the middle vertical line. In this range, $1-\gamma(p)=P[\text{retain } H_0]=P[\text{type II error}]=\beta(p)$. Thus the graph of the rejection probability curve $\gamma(p)$ of a test provides a complete picture of the performance of the test for all possible contingencies with regard to the true state of nature. The two types of error probabilities can be obtained from the two parts of this graph:

Type I error probability $\quad \alpha(p)=\gamma(p)$ for $p \leqslant .6$

Type II error probability $\quad \beta(p)=1-\gamma(p)$ for $p > .6$

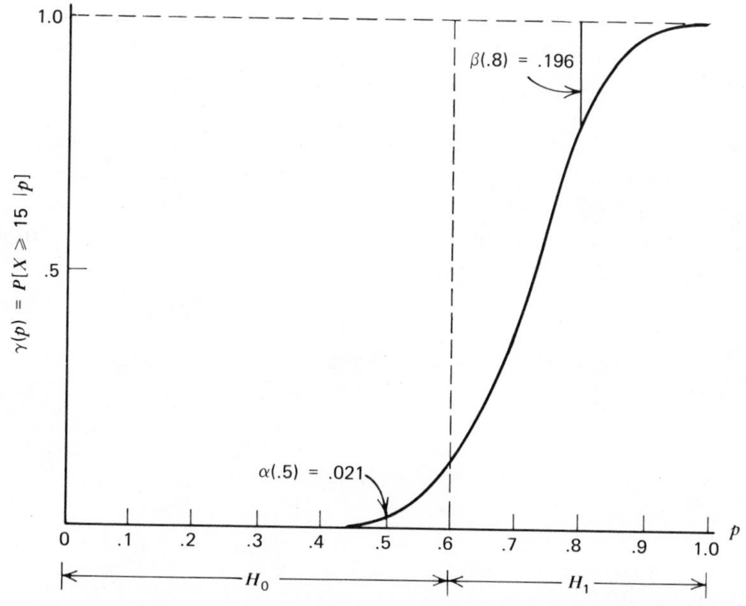

Figure 6.1 Graph of the rejection probability $\gamma(p)$ for the test $X \geqslant 15$.

In the portion $p > .6$ where H_1 is true, the quantity $\gamma(p) = 1 - \beta(p)$ is called the *power of the test at the alternative* p. This is the probability that the test will reject H_0 when H_1 is actually true, and it is in this sense that power indicates the strength or goodness of the test.

> A plot of the rejection probability function γ of a test exhibits the performance of the test by showing the magnitudes of the error probabilities for all possible realizations of the parameter. This curve is called the *power curve* of the test. The ordinate of the curve in the portion in which H_0 is true gives the type I error probability; while, in the other part, subtracting the ordinate from 1 gives the type II error probability.

6.4 SELECTING AMONG SEVERAL TESTS

We now employ the concept of power to compare the performance of the following three tests that might be used to determine whether the new drug is better than the standard medication. The same test statistic X is employed, but different rejection regions are chosen.

Test A: rejection region $X \geqslant 15$
Test B: rejection region $X \geqslant 18$
Test C: rejection region $X \geqslant 14$

Test A has already been studied in Section 6.3, and its power function appears in Table 6.1. The rejection probability functions for the other two tests are computed in the same manner using the binomial table. The results are listed in Table 6.2, and the three power curves are plotted in Figure 6.2.

TABLE 6.2
REJECTION PROBABILITIES FOR TESTS A, B, AND C

p	.3	.4	.5	.6	.7	.8	.9
$(A)\ \gamma(p) = P[X \geqslant 15]$	.000	.002	.021	.126	.416	.804	.989
$(B)\ \gamma(p) = P[X \geqslant 18]$	.000	.000	.000	.004	.035	.206	.677
$(C)\ \gamma(p) = P[X \geqslant 14]$	.000	.006	.058	.250	.608	.913	.998

Comparing the power curves of the three tests, we observe two important features:

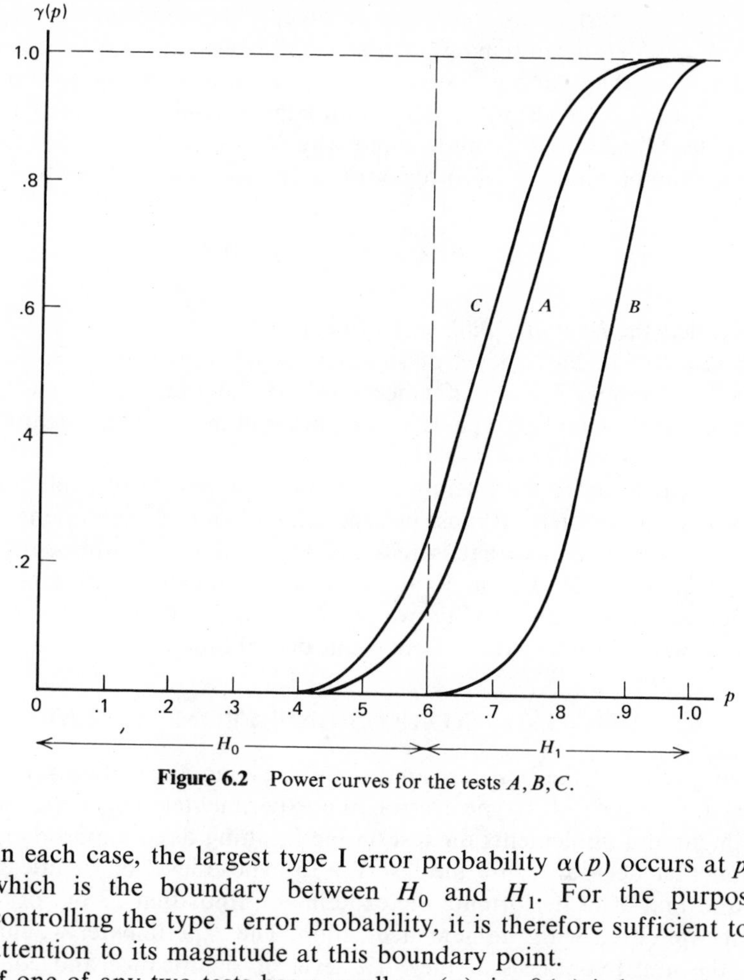

Figure 6.2 Power curves for the tests A, B, C.

(*a*) In each case, the largest type I error probability $\alpha(p)$ occurs at $p=.6$, which is the boundary between H_0 and H_1. For the purpose of controlling the type I error probability, it is therefore sufficient to pay attention to its magnitude at this boundary point.

(*b*) If one of any two tests has a smaller $\alpha(p)$, its $\beta(p)$ is larger than that of the other test. This shows that an error probability of one kind can only be reduced at the cost of increasing the other error probability.

In an ideal test, α and β are as small as possible, but such a solution cannot be obtained in view of observation (*b*), at least as long as the sample size remains fixed. Because the realization of a type I error is deemed more serious than a type II error, the conventional practice is to ensure that α is controlled below a predetermined level of tolerance and then to choose from among these tests the one with the smallest possible β.

If in our study of the new medication we demand that the type I error probability of the tests should not exceed .07, tests A and C will be disqualified but test B (as well as some other tests with rejection regions $X \geqslant 16$, $X \geqslant 17$, etc.) will remain possible candidates. Among the qualifying candidates, we then choose the test $X \geqslant 16$, because it has the lowest type II error probabilities $\beta(p)$. These candidates should be studied in the manner described above to understand why $X \geqslant 16$ is the best test to use.

The maximum value of $\alpha(p)$ for our test is

$$\alpha(.6) = P[X \geqslant 16 | p = .6] = .051$$

and is called the *level of significance of the test* $X \geqslant 16$.

The test $X \geqslant 15$ has a level of significance of .126, which exceeds the specified tolerance. The significance level of the test $X \geqslant 17$ meets the tolerance but has higher type II error porbabilities or, equivalently, less power than that of the test $X \geqslant 16$.

The specification of the tolerance level for the type I error probability is not a statistical problem. It must be ascertained from considerations of the strength of the evidence that is required to reject H_0. Traditionally, low values such as $\alpha = .01$, .05, or .10 are used to perform statistical tests. From the interpretation of probability, we recall that an α of .05 means that H_0 would be wrongly rejected in about 5 out of 100 independent tests.

6.5 DRAWING CONCLUSIONS FROM A TEST

Having selected the best test among those giving a specified protection against the chance of a type I error, the experimenter then performs the experiment and implements the test on the resulting data. Suppose that the experimenter decides to use the test $X \geqslant 16$, whose level of significance is .051, and that after performing the experiment, finds that 18 out of the 20 patients are cured with the new medication. The experimenter would then reject the null hypothesis at the 5.1% level of significance and conclude that there is substantial evidence that the new drug performs better than the standard medicine. If on the other hand, the observed value is $x = 14$, H_0 is not rejected at 5.1% level of significance and the experimenter concludes that the data do not provide convincing evidence that the new drug is better than the standard medicine. Note that in either situation, there is an element of vagueness in the conclusion reached by testing the statistical hypothesis. In particular, the retention of H_0 should be interpreted only as a lack of evidence to reject it. This does not mean that there is enough evidence to support it. Therefore, rather than saying H_0 is accepted, a more appropriate phraseology is to say that H_0 *is not rejected.*

A test conclusion should be stated as

$$H_0 \quad \textit{is rejected}$$

or as

$$H_0 \quad \textit{is not rejected}$$

at level of significance α

This approach to drawing conclusions from a statistical test consists of first establishing a rejection region with a prespecified low α and then determining if the observed value of the test statistic falls in the rejection region. Choosing $\alpha = .051$ in our example, an observed value $x = 16$ leads to the rejection of H_0, as does an observed value of $x = 19$. Apparently, however, $x = 19$ constitutes stronger evidence in support of H_1 than the value $x = 16$ does. To pursue this aspect, we can, as an auxiliary approach, determine how small α can be made subject to H_0 being rejected on the basis of the observed value. For instance, if $x = 19$ is observed, the tests $X \geqslant 19$, $X \geqslant 18$, $X \geqslant 17, \ldots$, will all lead to the rejection of H_0. Consulting the binomial table for $n = 20$ and $p = .6$, the corresponding α's are found to be .001, .004, .016,.... The smallest possible α that would permit the rejection of H_0 on the basis of the observed value is therefore .001. This α value is called the *significance probability* of the observation $x = 19$.

The *significance probability* P^* of an observed value of the test statistic is the smallest α for which this observation leads to the rejection of H_0. In other words, it is the probability under H_0 of the occurrence of the particular observed value or more extreme values. The smaller the magnitude of P^*, the stronger the evidence against H_0.

In our example, the significance probabilities of the observations $x = 16$ and $x = 19$ are .051 and .001, respectively. The use of P^* gauges the strength of evidence against H_0 on a numerical scale, where small values indicate a stronger justification of rejection.

In addition to performing a test of hypothesis with a predetermined α, it is a good statistical practice to record the significance probability as well.

6.6 TESTING WITH A TWO-SIDED ALTERNATIVE

In our cure rate example in the previous sections, the alternative hypothesis is formulated as $H_1: p > .6$ because the investigation is being made to determine if the new drug has a *higher* cure rate than .6. Such an alternative is called a *one-sided alternative* for the obvious reason that the parameter values under H_1 lie to one side of the range of values specified by H_0. Because only large values of X support H_1, the rejection region has the structure $X \geqslant c$ and is called a *one-sided rejection region*.

By contrast, if the intent of the investigation is to find substantive evidence that the new cure rate is *different from* .6, the alternative hypothesis should include values of p in both directions from .6. In other words, the formulation of the hypotheses should then be

$$H_0 : p = .6 \quad \text{vs.} \quad H_1 : p \neq .6$$

The latter alternative is called a *two-sided alternative*. It will be supported if the number of cures X in the sample is either too large *or* too small. The appropriate structure of the rejection region is then

$$\text{Reject } H_0 \quad \text{if either} \quad X \leqslant c_1 \quad \text{or} \quad X \geqslant c_2$$

where the boundaries c_1 and c_2 are to be determined such that the type I error probability is controlled within a specified level.

Example 6.1 A cat-food seller wishes to determine if two flavors A and B appeal differently to cats. Two identical bowls of food, each containing one of the two flavors, are to be presented to 15 cats, and the number of cats X eating from the bowl containing flavor A is to be noted. The data are to be used to infer about the possible presence of a strong preference among cats for either flavor A or flavor B.

If the two flavors appeal equally to cats, the population proportion of cats preferring the type A flavor would be $p = .5$. The inference problem, therefore, concerns testing the null hypothesis $H_0 : p = .5$ vs. the alternative $H_1 : p \neq .5$. Because extremely low as well as high values of X would contradict a lack of preference, the rejection region should be two sided. To illustrate, we consider two choices of the rejection region:

(*a*) Reject H_0 if $X \leqslant 4$ or $X \geqslant 11$.
(*b*) Reject H_0 if $X \leqslant 3$ or $X \geqslant 12$.

Note that under H_0, the test statistic X has a binomial distribution with $n = 15$ and $p = .5$. This distribution is symmetric, and symmetric choices of the rejection region are reasonable and, theoretically, are also the best.

Suppose that we wish to control the type I error probability below .05.

Consulting the binomial table for $n=15$ and $p=.5$, the type I error probability for test (a) is

$$\alpha=P[X\leqslant 4]+P[X\geqslant 11]=.059+.059=.118$$

and for test (b) is

$$\alpha=P[X\leqslant 3]+P[X\geqslant 12]=.018+.018=.036$$

Thus given the specification that α should not exceed .05, it is appropriate to use test (b). The level of significance of this test is $\alpha=.036$.

Having chosen the appropriate test, we now study its ability to detect an alternative by computing the power of the test at various alternative values of p and plotting the power curve. By definition, the power of the test at a given p is

$$\gamma(p)=P[\text{rejection of } H_0|p]$$

$$=P[X\leqslant 3|p]+P[X\geqslant 12|p]$$

Again, the binomial table provides these probabilities. For instance, when $p=.4$, $P[X\leqslant 3]=.091$ and $P[X\geqslant 12]=.002$, so that $\gamma(.4)=.091+.002=.093$. The powers at several alternatives are presented in Table 6.3, and the power curve is plotted in Figure 6.3.

TABLE 6.3
POWERS OF TEST (b)

p	.1	.2	.3	.4	.5	.6	.7	.8	.9
$P[X\leqslant 3]$	.944	.648	.297	.091	.018	.002	.000	.000	.000
+	+	+	+	+	+	+	+	+	+
$P[X\geqslant 12]$	.000	.000	.000	.002	.018	.091	.297	.648	.944
$\gamma(p)$	.944	.648	.297	.093	.036	.093	.297	.648	.944

After conducting the experiment, suppose that the seller finds that 5 out of 15 cats prefer flavor A, or $x=5$. Because this value is not in the rejection region, the seller can conclude that at a level of significance of $\alpha=.036$, H_0 is not rejected and strong evidence of preference is lacking. Alternatively, as an example of the calculation of the significance probability, we note that if $x=5$ were just inside the rejection region, the two-sided region would have to be in the form $X\leqslant 5$ or $X\geqslant 10$, and hence

$$P^*=P[X\leqslant 5]+P[X\geqslant 10]=.151+.151=.302$$

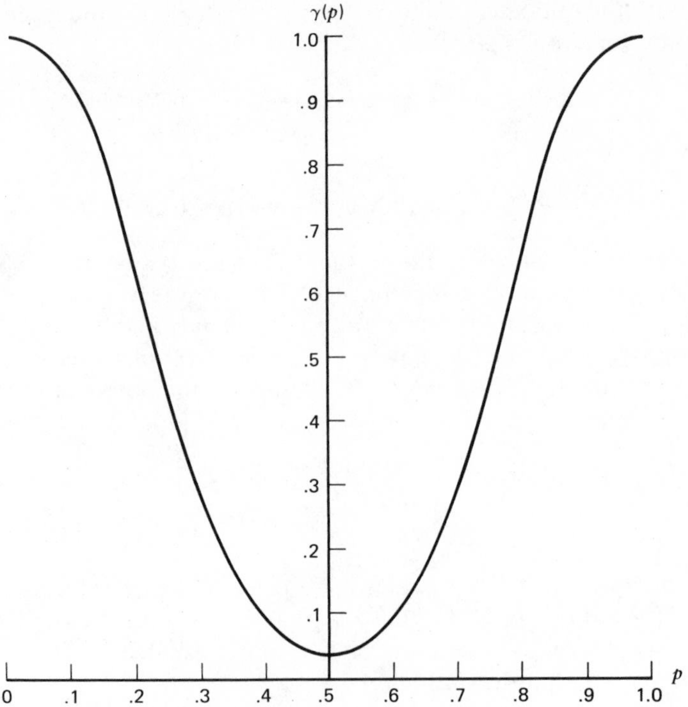

Figure 6.3 Power curve for test (*b*).

This large value of P^* again indicates that strong support for preference is lacking. ■

6.7 GENERAL STEPS IN TESTING HYPOTHESES

Thus far in Chapter 6, we have discussed the basic concepts underlying the formulation of a hypothesis testing problem, treatment of the null and alternative hypotheses, and the determination of a test within the framework of the binomial probability model. The same concepts and guidelines will provide the basis for solving the relevant hypothesis testing problems that we will encounter as other useful probability models are introduced in later chapters. Here we summarize the main steps in testing hypotheses about a general probability model, rather than restricting these guidelines to the binomial model.

(*a*) From the nature of the experimental data and the consideration of the assertions that are to be examined, identify the appropriate probability model and translate each assertion in terms of the range of values of the relevant parameter(s) θ of the model.

(*b*) When evidence is sought to establish a particular assertion, with the support of experimental data, the negation of this assertion is formulated as the null hypothesis H_0 and the assertion itself as the alternative hypothesis H_1. Both H_0 and H_1 are stated in terms of the parameter(s) of the model. Examples are:

$$H_0 : \theta = \theta_0 \quad \text{vs.} \quad H_1 : \theta > \theta_0 \quad \text{or} \quad H_0 : \theta = \theta_0 \quad \text{vs.} \quad H_1 : \theta \neq \theta_0$$

(*c*) Choose a test statistic T whose value is likely to best reflect the plausibility of the hypothesis being tested. The test statistic must be a function of the observable data and must not involve any unknown parameter. Identify the range of possible values of T and determine the values for which H_0 should be rejected in favor of H_1. In other words, formulate the structure of the rejection region by saying that H_0 should be rejected if the observed value of T is too large, too small, or intermediate, as the case may be.

(*d*) Specify the tolerance level of the type I error probability that you wish to assign to the testing process. Use tables that give the probability distributions of T to determine the rejection region, so that the maximum type I error probability does not exceed the specified tolerance level. Typically, it is sufficient to ensure that the type I error probability at the boundary θ_0, between H_0 and H_1, is equal to or just below the specified tolerance. Power should now be studied to ensure that the choice of sample size and α provide reasonable protection against type II errors.

(*e*) After the test or the decision rule has been explicitly formulated and the level of significance has been specified, compute the observed value of T from the experimental data and determine whether it falls in the rejection region. If it does, conclude that the validity of H_1 has been demonstrated by the data at the stated level of significance, but do *not* claim that H_0 has been proved false. If the observed value of T does not fall in the rejection region, conclude that the validity of H_1 has failed to be demonstrated by the data at the stated level of significance. Here again, do *not* claim that the validity of H_0 has been statistically justified, rather state that there is not sufficient experimental evidence to reject H_0.

In addition to establishing a rejection region with prespecified α, as stated in step (*d*), the significance probability can also be considered in drawing conclusions. To do this, record the observed value of T and, referring to the structure of the rejection region defined in step (*c*), determine which values should be considered more extreme than the observed value. Using the distribution of T under H_0, calculate the probability P^* of the occurrence of the observed value or more extreme values. If the test rejects H_0 and the significance probability P^* happens to be very small compared to α, this should be recorded to communicate the strength of support for H_1.

It should be noted here that step (*c*) relies on the judgment and the common sense of the hypothesis tester. Elegant mathematical theory has been developed to derive the test statistic and the rejection region that minimize the type II error probability subject to a tolerance on the type I error probability. For the relatively simple problems within the scope of this text, the tests suggested by intuitive reasoning are the best tests derived by this theory.

Another important aspect that we have not emphasized is the magnitude of the type II error probability. Our strategy has been to control the more serious type I error probability. In addition, we usually want the type II error probability to remain at a tolerably low value, at least for alternatives that are quite distant from the range of values for H_0. With this additional constraint, we must design the experiment beforehand to determine what sample size is required to meet the protection standards for both types of error probabilities.

The general steps in hypotheses testing are illustrated in Example 6.2 in terms of a problem involving the Poisson probability model.

Example 6.2 Discharging warm water from a nuclear power plant into a river is suspected to be detrimental to the habitat of a particular species of zoo-plankton. From extensive records maintained at a regular observation site located one mile upstream, it is known that the average count of this species in mid-July is 1.2 per liter of water. In the immediate vicinity of the nuclear plant, 5 one-liter water samples are collected from the river; the counts of zoo-plankton are found to be 0, 0, 1, 2, 1. (*a*) Formulate a test, with a type I error probability that does not exceed 8%, to substantiate the conjecture that the warm-water discharge reduces the density of this species of zoo-plankton. Using the five counts recorded in mid-July, what will the test conclude? (*b*) If the correct density of this species at the plant location is actually .8 per liter, what is the probability that the test will correctly conclude there has been a reduction in density from that upstream?

To obtain a solution, we follow the line of reasoning in Example 5.11, which suggests that the Poisson distribution is an appropriate probability model for the plankton counts.

(*a*) Let m denote the true average rate of occurrence of plankton per 5-liter volume of water at the plant site. If there is no difference in density from the site upstream, the average rate of occurrence of plankton will be $5\times 1.2=6$ per 5-liter volume. Because we are seeking evidence to substantiate the claim that there has been a decrease in the rate of occurrence at the plant site, the null and the alternative hypotheses should be formulated

$$H_0: m \geqslant 6 \quad \text{and} \quad H_1: m < 6$$

respectively. Letting $X_1, X_2, \ldots, X_5$ denote the counts in the individual one-liter samples (which are random variables before the samples are actually taken), the natural test statistic to use is $T=\Sigma_1^5 X_i$, the total count in the 5-liter sample. The total T has a Poisson distribution with parameter m, and the hypotheses are stated in terms of this parameter.

The possible values of T are $0, 1, 2, \ldots$, and it is evident from the nature of H_1 that small values of T should lead to the rejection of H_0 in favor of H_1. In other words, the form of the rejection region should be $T \leqslant c$, and the cutoff point c should be determined so that the type I error probability does not exceed 8%. It is sufficient to guarantee that the type I error probability is controlled below 8% at $m=6$, the boundary between H_0 and H_1. Consulting the Poisson tables for $m=6$, we find

$$P[T \leqslant 2] = .062$$

$$P[T \leqslant 3] = .151$$

Our test should therefore

$$\begin{array}{lll} \text{reject} & H_0 & \text{if} \quad T \leqslant 2 \\ \text{not reject} & H_0 & \text{if} \quad T \geqslant 3 \end{array}$$

The level of significance of the test is $\alpha = .062$.

After formulating the test, we consult the data and calculate the value of T as $0+0+1+2+1=4$, which is not in the rejection region. Thus we can conclude that at the 6.2% level of significance, the evidence is not convincing that H_0 should be rejected. In other words, the claim that the occurrence rate has been reduced is not substantiated by the data.

(*b*) If the true density or the average rate of occurrence at the plant site is .8 per liter, the rate per 5-liter volume is $5\times .8=4$. Our test will indicate a reduction if $T \leqslant 2$. The probability of the event $T \leqslant 2$, obtained from the Poisson table for $m=4$, is

$$P[T \leqslant 2] = .238$$

so that the power of the test at the alternative $m=4$ is $\gamma(4)=.238$.

At $m=4$, the type II error probability is $\beta(4)=1-\gamma(4)=1-.238=.762$, which is large. If a density of .8 per liter or $m=4$ per 5-liter volume is considered to be a serious deviation that our test should be able to detect, we must enlarge the sample size to reduce the type II error probability. ■

EXERCISES

1. Identify the null and the alternative hypotheses in terms of descriptive statements. The type of answer required is illustrated in (*a*).
   (*a*) A construction engineer wishes to determine if a new cement mix has a better bonding quality than the mix currently in use. The new mix is more expensive, so the engineer would not recommend it unless its better quality is supported by experimental evidence. The bonding quality is to be observed from several cement slabs prepared with the new mix.
   ANSWER:
   Null hypothesis: The new mix is not better.
   Alternative: The new mix is better.
   (*b*) A state labor department wishes to determine if the current rate of unemployment in the state varies significantly from the forecast of 6% made two months ago.
   (*c*) During a flu epidemic, 20% of a city population suffers from flu attacks. A physician theorizes that regular users of vitamin C are less susceptible to the flu. She intends to sample 500 regular users of vitamin C, determine how many of them had the flu, and use the data to document her claim.
   (*d*) An agronomist believes that plants grown from a new strain of seed is likely to be more resistant to a disease than an existing variety. He plans to expose both types of plants to the disease, count the number of incidences of the disease, and use the data to establish his conjecture.
   (*e*) The research and development department of a cigarette company projects that the average tar content of a new blend of tobacco will be less than 5 milligrams per cigarette. This low-tar quality has a good market potential. Data collected from chemical analyses of several cigarettes are to be used to determine if there is strong support for this conjecture.
   (*f*) Referring to (*e*), suppose that the cigarette company is now marketing its new brand with the claim "Average tar content: 5 milligrams per cigarette" printed on the packs. A consumer testing group suspects that the true average tar content of these cigarettes may be higher than the manufacturer's claim. The group intends to analyze several cigarettes and use the data it collects to challenge the company's claim.
2. In testing the null hypothesis $H_0: p = .6$ vs. the alternative $H_1: p < .6$ for a binomial model $b(n,p)$, the rejection region of a test has the structure $X \leqslant c$, where X is the number of successes in n trials. For each of the

following tests, determine the level of significance and the probability of type II error at the alternative $p = .3$.
(*a*) $n = 10, \quad c = 2$
(*b*) $n = 10, \quad c = 3$
(*c*) $n = 19, \quad c = 7$

3. Instead of collecting data, somebody recommends the following procedure for testing a null hypothesis. Flip 3 coins. If all 3 show heads, reject H_0; otherwise, do not reject H_0. Using this procedure:
(*a*) What is the type I error probability?
(*b*) What is the type II error probability?
4. As of last year, only 20% of the employees in an organization used public transportation to commute to and from work. To determine if a recent campaign encouraging the use of public transportation has been effective, a random sample of 25 employees is to be interviewed and the number of employees currently using public transportation X is to be recorded.
(*a*) Formulate the hypotheses in terms of p, the population proportion of employees currently using public transportation.
(*b*) What should the rejection region be if α is to be controlled below .1?
(*c*) Given the chosen rejection region, what is the maximum type I error probability?
5. Referring to the test in Exercise 4:
(*a*) If $p = .4$, what is the probability that the test will fail to reject H_0?
(*b*) Find the power of the test at $p = .3, .4, .5, .6, .7$, and plot the power curve.
(*c*) After interviewing 25 employees, suppose that 10 are currently using public transportation. What conclusion would you draw by using the test?
(*d*) What is the smallest α at which H_0 can be rejected, given the data in (*c*)?
6. Among one-year-old flashlight batteries, only 70% possess a specified strength. Given a proposed new method of storing, it is anticipated that a higher percentage of batteries will possess this specified strength.
(*a*) State the null hypothesis of "no improvement" and the alternative hypothesis of "improvement in strength."
(*b*) Given a sample size of 13, determine the rejection region so that $\alpha \leqslant .07$.
(*c*) What is β for $p = .95$?
(*d*) If only 3 out 13 batteries do not possess the specified strength, what is your conclusion?
7. A psychiatrist believes that more than 50% of the users of sleeping pills

sleep better simply because of the psychological effect of taking the pills. To substantiate this hypothesis with data, he selects a random sample of 20 insomniacs and gives each of them a box of pills to use. These are actually sugar pills but are otherwise identical to a popular brand of sleeping pills currently on the market. Subsequently, 15 of these patients report that the pills have been effective in inducing sleep. Assuming a level of significance close to .05, does this observation support the psychiatrist's conjecture?

8. A study of spiders leads an investigator to conjecture that less than 40% of all spider webs are built on the ground. The data to corroborate this observation consist of a count of the number of webs on the ground X among a sample of 20 webs.
   (*a*) State the null and the alternative hypotheses.
   (*b*) With $\alpha \leqslant .1$, determine the rejection region.
   (*c*) What conclusion would you draw from the observation $x=3$?
   (*d*) Graph the power curve of your test.
9. *Quality control*: When the output of a production process is stable at an acceptable standard, it is said to be "in control." Suppose that a production process has been in control for some time and that the proportion of defectives has been .05. As a means of monitoring the process, the production staff decides to consider the process "out of control" if 2 or more defectives are found in a sample of 15 items.
   (*a*) Find α, the probability of signaling "out of control," when the process is at $p=.05$.
   (*b*) Graph the power curve for this control scheme from the values at $p=.05$, .1, .2, .3, .4.

   (NOTE: In quality control terminology, one usually works with the operating characteristic whose values are equal to $1-$(power), or the probability of accepting H_0.)
10. Referring to Exercise 9:
    (*a*) Graph the power curve for the test with $\alpha \leqslant .05$.
    (*b*) Compare the power curve in (*a*) with the power curve in Exercise 9*b*.
11. Again referring to the situation described in Exercise 9, determine a plan based on $n=25$ with $\alpha \leqslant .04$. Graph the power curve on the same paper you used to graph the power curve of the test in Exercise 9*b*. What has increasing the sample size accomplished?
12. Many psychological experiments are conducted to determine whether animals exhibit a preference between two rewards. In an experiment consisting of 7 trials with different animals, the results are

Trial	1	2	3	4	5	6	7
Reward	1	1	2	1	2	2	2

Assume that the binomial model applies.

(*a*) State the null hypothesis of no preference.

(*b*) State the alternative hypothesis of some preference.

(*c*) If the critical region consists of 0 and 7, determine α and also the value of β at $p=.8$.

(*d*) Use this test to draw conclusions from the above data.

13. In testing the null hypothesis $H_0: p=.6$ vs. the alternative $H_1: p \neq .6$ for a binomial model $b(n,p)$, the rejection region of a test has the structure

$$\text{Reject } H_0 \quad \text{if} \quad X \leqslant c_1 \quad \text{or if} \quad X \geqslant c_2$$

where X is the number of successes in n trials. For each of the following tests, find the level of significance and the probability of type II error at the alternative $p=.3$.

(*a*) $n=10,\quad c_1=1,\quad c_2=9$

(*b*) $n=10,\quad c_1=3,\quad c_2=10$

(*c*) $n=19,\quad c_1=7,\quad c_2=17$

14. A regional marketing manager feels that his company has cornered 40% of the typewriter market in a specific area. Taking $p=.4$ as the null hypothesis, the manager decides to consider this claim reasonable unless a sample of 19 sales shows $X \leqslant 3$ or $X \geqslant 12$, where X denotes the number of typewriters sold by his company.

(*a*) Find α for this test.

(*b*) Find the power of the test for several values of p ranging between .1 and .9, and plot the power curve.

15. Referring to Exercise 14:

(*a*) Determine a test with $\alpha \leqslant .02$.

(*b*) Construct a test based on a sample size of $n=10$ and $\alpha \leqslant .11$.

(*c*) Does the test (b) have a higher power at $p=.3$ than the test in Exercise 14?

16. Suppose that the serious accidents along a certain stretch of highway follow a Poisson distribution with a mean of 1.2 accidents per week. After a reduction in the speed limit along this stretch of highway, it is hoped that this average has decreased. Construct a test for reduction based on a five-week count with $\alpha \leqslant .07$.

17. From an extensive record collected over several years, it is found that only 2% of a certain type of cancer patient are cured by surgery. A chemotherapist claims that her nonsurgical method of treating these patients is more successful than surgery. To obtain experimental evidence to support her claim, she uses her method on 200 patients, 6 of whom are cured. The chemotherapist asserts that the observation of a 3% cure rate in a large sample is sufficient evidence to support her claim.

(*a*) Formulate the hypotheses in terms of the parameter m = expected number of cures out of 200 patients.

(*b*) With α approximately .05, determine the rejection region. Determine if the claim is convincing.

(*c*) If the chemotherapy actually produces a cure rate of 4.5% what is the probability that the test will support the claim that chemotherapy is more successful than surgery?

18. Referring to Exercise 16, graph the power function of your test and find β for an average of .4 accidents per week.

19. A company selling a 50-volume set of encyclopedias mails colorful brochures to potential customers and receives orders by mail. The success rate using this method is found to be 1%. The advertising manager feels that instead of direct-mail advertisements, having sales clerks display a few volumes would produce higher sales. A few sales clerks are hired to test this method. After contacting 200 potential customers, the company receives a total of 7 orders. Does this observation strongly testify to higher sales using the new method (use $\alpha \leqslant .05$)?

20. During the two previous decades, the number of earthquakes in an area average 1.4 per year. In the current 5 year period, the number of earthquakes in this area have been 0, 1, 1, 0, 1.

(*a*) With $\alpha \leqslant .1$, formulate a test of the null hypothesis that the current earthquake rate is the same in the last 5 years as it was before against the alternative that the rate is different.

(*b*) What conclusion can you draw from the above data?

(*c*) Graph the power curve of your test.

CHAPTER 7

The Normal Distribution and Random Samples

7.1 PROBABILITY MODELS FOR CONTINUOUS VARIABLES

We now turn our attention to describing the probability distribution of a random variable that can assume all the values in an interval. Measurements of height, temperature, amount of rainfall, and the waiting time at a checkout counter are all of this type. Although in practice each measurement is recorded only to the nearest integer, tenth, or other limit set by the measuring device, conceptually it is advantageous to work within the framework of an underlying continuous scale.

The probability distribution of a *continuous random variable* can be visualized as a smooth form of the relative frequency histogram based on a large number of observations. To establish some concepts, let us consider the probability distribution of the weights of newborn babies. Initially, we suppose that the birth weights of 100 babies are recorded to the nearest tenth of a pound and that the relative frequency histogram in Figure 7.1*a* is obtained with class intervals of 1 pound. Recall that a relative frequency histogram has the properties:

(*a*) The total area under the histogram is 1.
(*b*) For two points *a* and *b* such that each is a boundary point of some class, the relative frequency of measurements in the interval *a* to *b* is the *area* under the histogram enclosed by this interval.

For example, Figure 7.1*a* shows that a proportion .53 of the 100 measurements lie within the interval 6.5–8.5 pounds. However, this histogram does not show how these measurements are distributed within the class interval. Next, we suppose that the number of measurements is increased to 5000 and that the class interval is decreased to .25 pounds. The resulting relative frequency histogram appears in Figure 7.1*b*. This is a

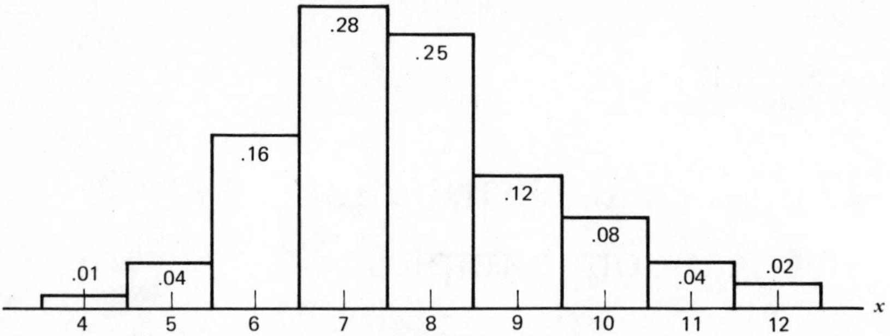

(a) Relative frequency histogram of 100 birth weights with a class interval of 1 pound.

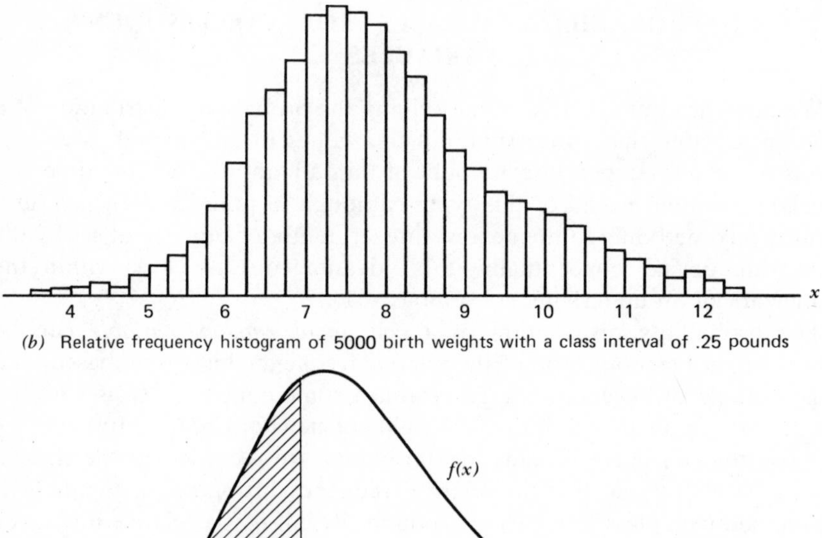

(b) Relative frequency histogram of 5000 birth weights with a class interval of .25 pounds

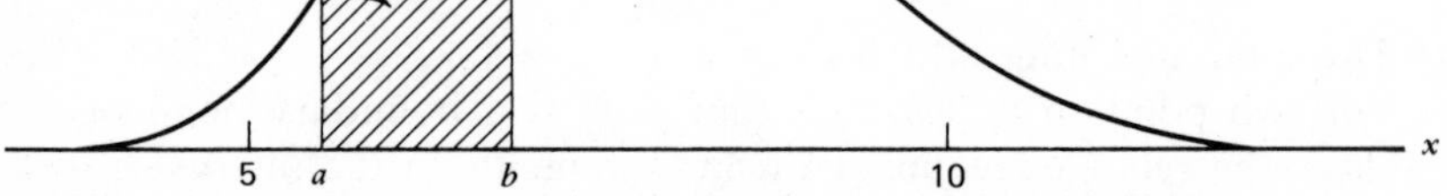

(c) Probability density curve for the continuous random variable X = birth weight.

Figure 7.1 Probability density curve viewed as a limiting form of relative frequency histograms.

refinement of the histogram 7.1*a* in the sense that it is constructed from a larger set of observations and exhibits relative frequencies for finer class intervals. Note that narrowing the class interval without increasing the number of observations would obscure the overall shape of the distribution. The refined histogram 7.1*b* again exhibits both properties (*a*) and (*b*) stated earlier.

Proceeding in this manner, even further refinements of relative frequency histograms can be imagined with larger numbers of observations and smaller class intervals. Accepting this conceptual argument, we ignore the difficulty that accuracy of the measuring device is limited. In the course of refining the histograms, the jumps between consecutive rectangles tend to dampen out, and the top of the histogram approximates the shape of a smooth curve, as illustrated in Figure 7.1*c*. Because probability is interpreted as long-run relative frequency, the curve obtained as the limiting form of the relative frequency histograms represents the manner in which the total probability 1 is distributed over the range of possible values of the random variable X. The mathematical function denoted by $f(x)$ whose graph produces this curve is called the *probability density function* of the continuous random variable X.

Drawing on these ideas, we can now formally state the fundamental properties of a probability density function. These properties are all inherited from the relative frequency histograms from which the concept of the probability density curve is motivated.

> The *probability density function* $f(x)$ describes the distribution of probability for a continuous random variable. It has the properties:
>
> (*a*) The total area under the density curve is 1.
>
> (*b*) $P[a \leqslant X \leqslant b]$ = area under the density curve between a and b.
>
> (*c*) $f(x)$ is positive or 0.

Unlike the description of a discrete probability distribution, the probability density $f(x)$ in the case of the continuous random variable does not represent the probability of $[X = x]$. Instead, a probability density function relates the probability of an interval $[a, b]$ to the *area* under the curve in a strip over this interval. A single point a being an interval with a width of 0, the area above this point is 0 and, therefore, $P[X = a] = 0$. This statement is true for every individual point.

With a continuous random variable, the probability that $X=x$ is *always* 0. It is only meaningful to speak about the probability that X lies in an interval.

The deduction that $P[X=a]=0$ for every single point needs some clarification. In the context of our birth-weight example, the statement $P[X=8.5 \text{ pounds}]=0$ probably seems shocking. Does this statement mean that no child can have a birth weight of 8.5 pounds? To resolve this paradox, we need to recognize that the accuracy of every measuring device is limited, so that here the number 8.5 is actually indistinguishable from all numbers in an interval surrounding it, say, [8.495, 8.505]. Thus the question really concerns the probability of an interval surrounding 8.5, and the area under the curve is no longer 0.

Once the probability density function $f(x)$ of a continuous random variable is specified, the problem of calculating the probability of an interval becomes to compute the area under the curve in a strip supported by the interval. Such calculations involve integral calculus, which is beyond the scope of this text. Fortunately, areas for important distributions have been extensively tabulated, and all we need to do is to consult these tables. When determining the probability of an interval a to b, we need not be concerned if either or both end points are included in the interval. With the probability of $X=a$ and $X=b$ both equal to 0

$$P[a \leqslant X \leqslant b]=P[a<X \leqslant b]=P[a \leqslant X<b]=P[a<X<b]$$

In contrast, these may be different for a discrete distribution.

In most tables, the entire area to the left of each point is tabulated. To obtain the probability of other intervals, we must apply the following rules:

$$P[a<X<b]=(\text{Area to left of } b)-(\text{Area to left of } a)$$

$$P[b<X]=1-(\text{Area to left of } b)$$

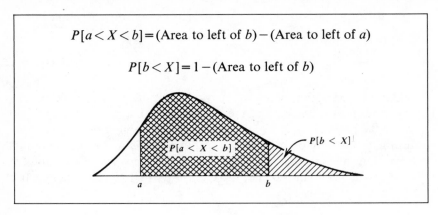

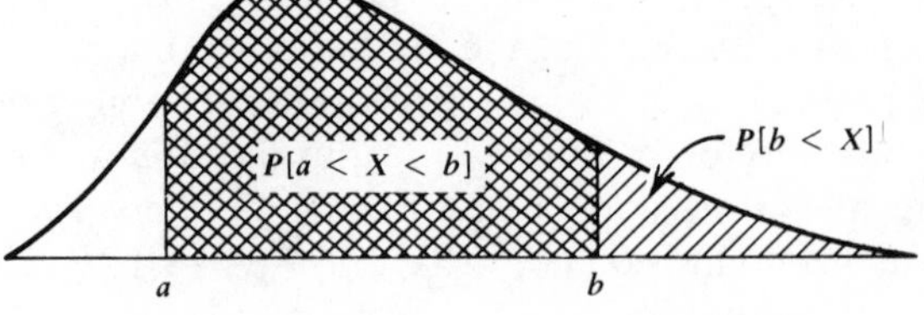

7.1.1 Specification of a Probability Model

The probability model for a continuous random variable is specified by giving the mathematical form of the probability density function. We described the process of the successive refinement of the relative frequency histogram, based on an increasing volume of data, only to establish the concept of a probability density curve. This process should not be regarded as the modus operandi for determining the probability model in a practical situation. However, when a fairly large number of observations of a continuous random variable are available, we may try to approximate the top of the staircase silhouette of the relative frequency histogram by a mathematical curve. In specifying such an approximating curve, the criteria of simplicity of form and parsimony of parameters should be considered, so that the model is amenable to methods of statistical inference.

In the absence of a large data set, we may tentatively assume a reasonable model that may have been suggested by data from a similar source. An alternative approach is to set down some plausible postulates that seem to agree with the real situation and then to derive the distribution mathematically, as we did both with the binomial and the Poisson distributions in the discrete case. Of course, any model obtained in this way must be closely scrutinized to verify that it agrees with any new evidence in the form of further observations.

7.1.2 Features of a Continuous Distribution

As with relative frequency histograms, the probability density curves of continuous random variables possess a wide variety of shapes. A few of these are illustrated in Figure 7.2, where the term *skewed* denotes a long

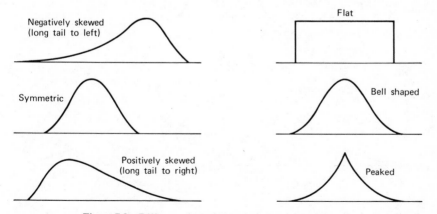

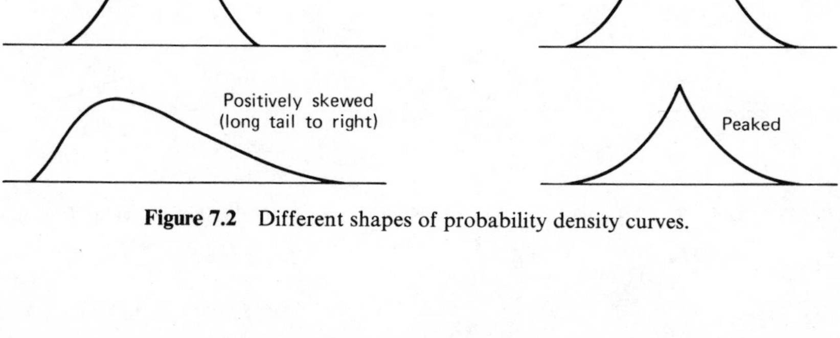

Figure 7.2 Different shapes of probability density curves.

tail in one direction. In statistical literature, a rich collection of functions is available so that an appropriate choice can usually be made for modeling distributions of specific shapes.

A continuous random variable X may also have a mean or an expected value $E(X)$ as well as a variance and a standard deviation. Their interpretations are the same as in the case of discrete random variables, but their formal definitions involve integral calculus and are therefore not pursued here. However, it is instructive to see in Figure 7.3 that the mean $\mu = E(X)$ marks the balance point of the distribution of probability specified by the probability density function. The median, another measure of center, is the value of X that divides the area under the curve into halves.

Percentiles are defined as:

The *population* $100p$-th *percentile* is an x value that has an area p to the left and $1-p$ to the right.

Lower (first) quartile = 25th percentile

Second quartile (or median) = 50th percentile

Upper (third) quartile = 75th percentile

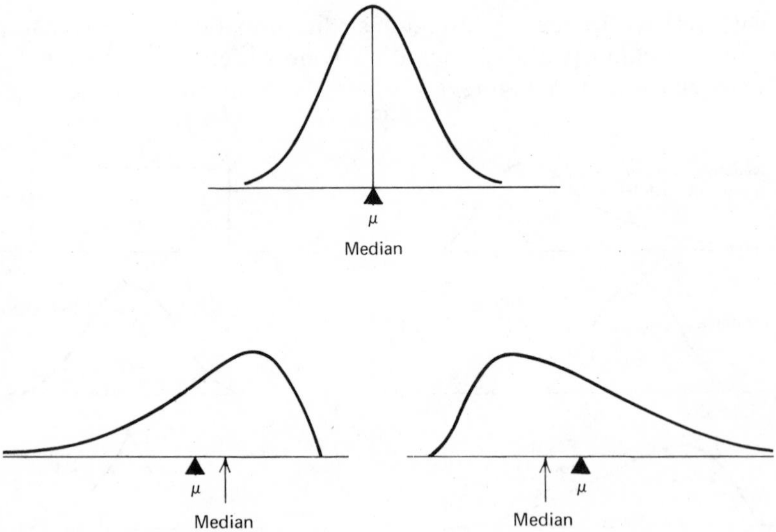

Figure 7.3 Mean as the balance point and median as the point of equal division of the area.

The properties of expectation and variance discussed in Chapter 4 for discrete random variables also apply to continuous random variables and enable us to unify the treatment of the two cases. Here we record some particularly important properties that we have verified for the case of two random variables.

For n random variables $X_1, \ldots, X_n$

$$E(a_1X_1 + \ldots + a_nX_n) = a_1E(X_1) + \ldots + a_nE(X_n)$$

If $X_1, \ldots, X_n$ are independent

$$\text{Var}(a_1X_1 + \ldots + a_nX_n) = a_1^2\text{Var}(X_1) + \ldots + a_n^2\text{Var}(X_n)$$

7.2 THE NORMAL DISTRIBUTION

The normal distribution, which may already be familiar to some readers as the curve with the bell shape, is sometimes associated with the names of Pierre Laplace and Carl Gauss, who figured prominently in its historical development. Gauss derived the normal distribution mathematically as the probability distribution of the error of measurements, which he called the "normal law of errors." Subsequently, astronomers, physicists, and, somewhat later, data collectors in a wide variety of fields found that their histograms exhibited the common feature of first rising gradually in height to a maximum and then decreasing in a symmetric manner. Although the normal curve is not unique in exhibiting this form, it has been found to provide a reasonable approximation in a great many situations. At one time during the early stages of the development of statistics, apparently the feeling was that every real-life data must conform to the bell-shaped normal curve and if it did not the process of data collection should be suspect. It is in this context that the distribution became known as the *normal distribution.* However, careful scrutiny of data has often revealed inadequacies of the normal distribution. In fact, the universality of the normal distribution is only a myth, and examples of quite nonnormal distributions abound in virtually every field of study. Still, the normal distribution plays a central role in statistics, and inference procedures derived from it have wide applicability and form the backbone of current methods of statistical analysis.

Although we are speaking of the importance of the normal distribution, our remarks really apply to a whole class of distributions having bell-shaped densities. Each distribution can be completely specified by giving the value of its mean μ and its standard deviation σ that appear in the formula for the probability density function.

A *normal distribution* has a bell-shaped density

$$\frac{1}{\sqrt{2\pi}\,\sigma} e^{-\frac{(x-\mu)^2}{2\sigma^2}} \quad \text{for } -\infty < x < \infty$$

with mean $= \mu$ and standard deviation $= \sigma$.

The probability of the interval extending

one sd each side of mean: $P[\mu-\sigma<X<\mu+\sigma] \quad =.683$

two sd each side of mean: $P[\mu-2\sigma<X<\mu+2\sigma]=.954$

three sd each side of mean: $P[\mu-3\sigma<X<\mu+3\sigma]=.997$

In the formula for the probability density function, π is the area of a circle having the unit radius, or approximately 3.1416, and e is approximately 2.7183. The specific formula for the normal curve is not of importance to us, but a few details do merit special attention. The curve is symmetric about its mean μ, which locates the peak of the bell (see Figure 7.4). The interval running one standard deviation in each direction from μ has a probability of .683, the interval from $\mu-2\sigma$ to $\mu+2\sigma$ has a probability of .954, and the interval from $\mu-3\sigma$ to $\mu+3\sigma$ has a probability of .997. The curve never reaches 0 for any value of x, but because the tail areas outside

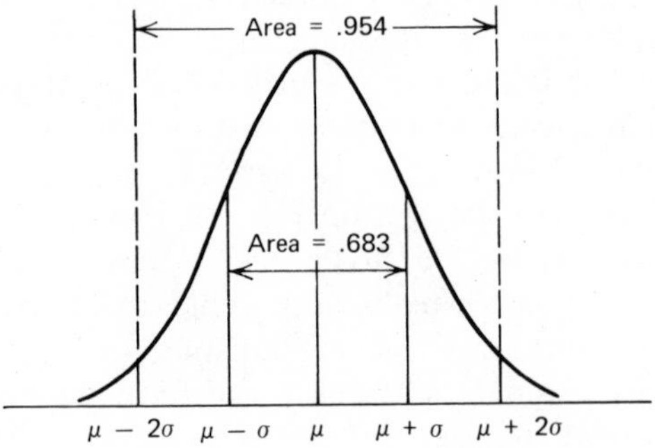

Figure 7.4 Normal distribution.

$(\mu-3\sigma, \mu+3\sigma)$ are very small, we usually terminate the graph at these points.

> *Notation*
>
> The normal distribution with a mean of μ and a standard deviation of σ is denoted by $N(\mu, \sigma)$.

Interpreting the parameters, we can see in Figure 7.5 that a change of mean from μ_1 to a larger value μ_2 merely slides the bell-shaped curve along the axis until a new center is established at μ_2. There is no change in the shape of the curve.

A different value for the standard deviation results in a different maximum height of the curve and changes the amount of the area in any fixed interval about μ (see Figure 7.6). The position of the center does not change if only σ is changed.

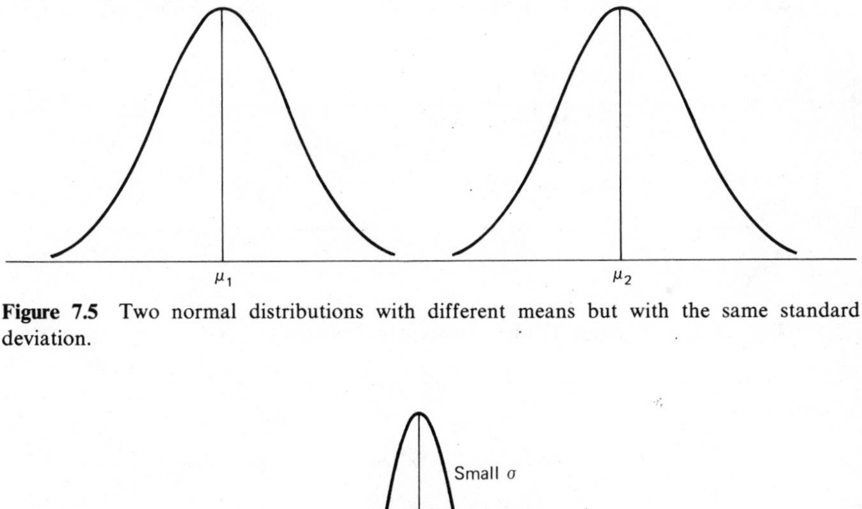

Figure 7.5 Two normal distributions with different means but with the same standard deviation.

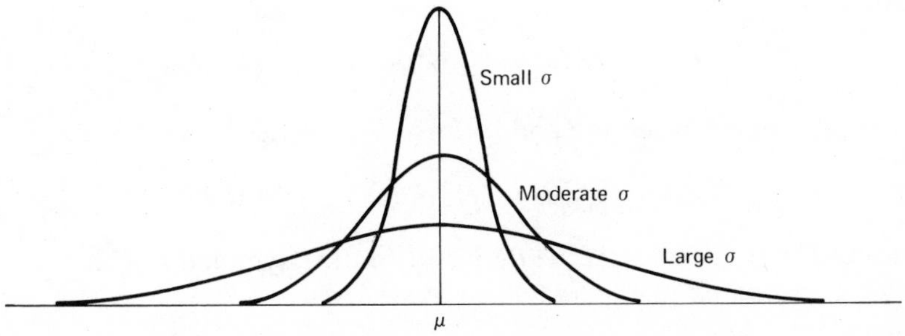

Figure 7.6 Decreasing σ increases the maximum height and the concentration of probability about μ.

The particular normal distribution that has a mean of 0 and a standard deviation of 1 is called the *standard normal distribution*. Its mean and variance coincide with those of the standardized variable defined in Section 4.4. It is customary to denote the standard normal variable by Z. The standard normal curve is illustrated in Figure 7.7.

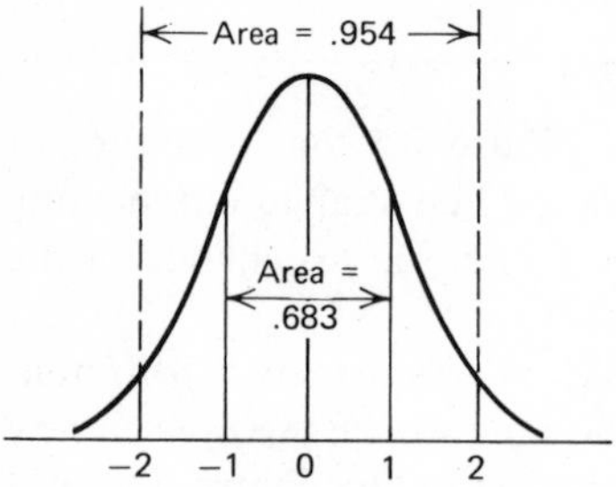

Figure 7.7 The standard normal curve.

> The *standard normal distribution* exhibits a bell-shaped curve with
>
> $$\text{mean} \quad \mu = 0$$
> $$\text{standard deviation} \quad \sigma = 1.$$
>
> The standard normal distribution is denoted by $N(0,1)$.

7.2.1 Use of the Normal Table (Appendix Table 4)

The standard normal table in the Appendix gives the area to the left of a specified value of z:

$$P[Z \leqslant z] = \text{Area under curve to the left of } z$$

For the probability of an interval $[a,b]$

$$P[a \leqslant Z \leqslant b] = [\text{Area to left of } b] - [\text{Area to left of } a]$$

The following properties can be derived from the symmetry of the density about 0 as exhibited in Figure 7.8:

(*a*) $P[Z \leqslant 0] = .5$

(*b*) $P[Z \leqslant -z] = 1 - P[Z \leqslant z]$

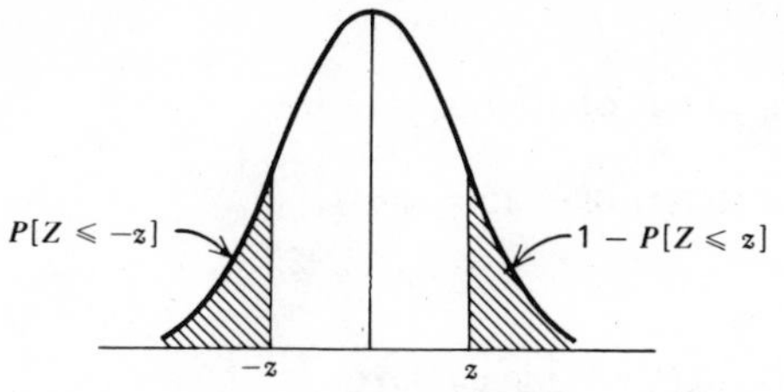

Figure 7.8

(c) If $z>0$

$P[Z \leqslant z]=.5+P[0<Z \leqslant z]$

$P[Z \leqslant -z]=.5-P[0<Z \leqslant z]$

Property (c) is needed for using other normal tables that give only the probabilities $P[0<Z \leqslant z]$.

Example 7.1 Find $P[Z \leqslant 1.52]$ and $P[Z>1.52]$.

From Appendix Table 4, we know that the probability or area to the left of 1.52 is .9357. Consequently, $P[Z \leqslant 1.52]=.9357$. Moreover, because $[Z>1.52]$ is the complement of $[Z \leqslant 1.52]$

$$P[Z>1.52]=1-P[Z \leqslant 1.52]=1-.9357=.0643$$

as we can see in Figure 7.9.

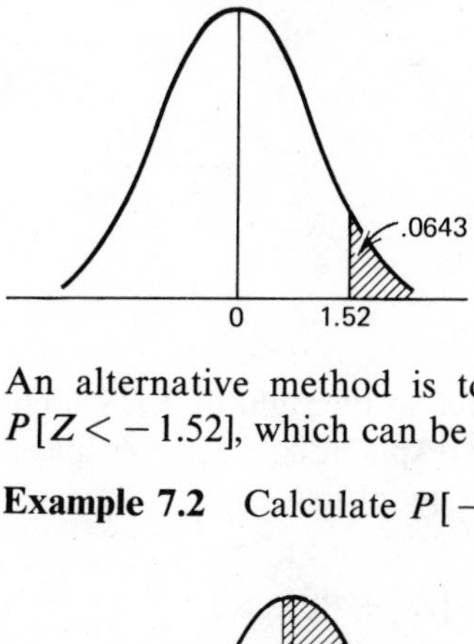

Figure 7.9

An alternative method is to use symmetry to show that $P[Z>1.52]= P[Z<-1.52]$, which can be obtained directly from Appendix Table 4. ■

Example 7.2 Calculate $P[-.15<Z<1.60]$.

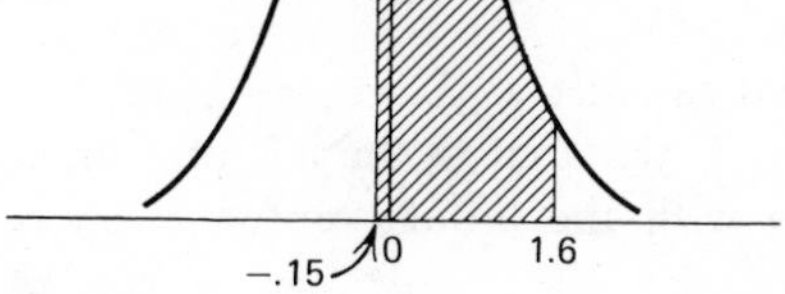

Figure 7.10

From Appendix Table 4, we see that

$$P[Z \leqslant 1.60] = \text{Area to left of } 1.60 = .9452$$

$$P[Z \leqslant -.15] = \text{Area to left of } -.15 = .4404$$

Therefore

$$P[-.15 < Z < 1.60] = .9452 - .4404 = .5048$$

as we can see in Figure 7.10. ■

Example 7.3 Find $P[Z < -1.9 \quad \text{or} \quad Z > 2.1]$.

The two events $[Z < -1.9]$ and $[Z > 2.1]$ are disjoint, so that we add their probabilities

$$P[Z < -1.9 \quad \text{or} \quad Z > 2.1] = P[Z < -1.9] + P[Z > 2.1]$$

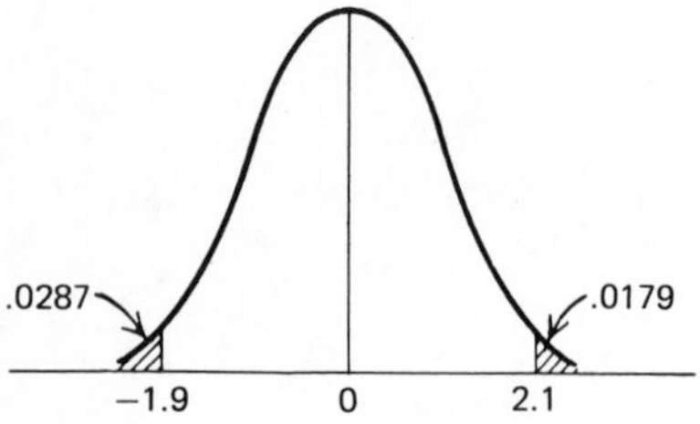

Figure 7.11

As indicated in Figure 7.11, $P[Z > 2.1]$ is the area to the right of 2.1, which is $1 - (\text{Area to left of } 2.1) = 1 - .9821 = .0179$. Appendix Table 4 gives $P[Z < -1.9] = .0287$ directly. Adding these two quantities

$$P[Z < -1.9 \quad \text{or} \quad Z > 2.1] = P[Z < -1.9] + P[Z > 2.1]$$
$$= .0287 + .0179 = .0466$$

■

Example 7.4 Locate the value of z that satisfies $P[Z > z] = .025$.

Using the property that the total area is 1, the area to the left of z must be $1 - .0250 = .9750$. The marginal value with the tabular entry .9750 is $z = 1.96$ (diagrammed in Figure 7.12). ■

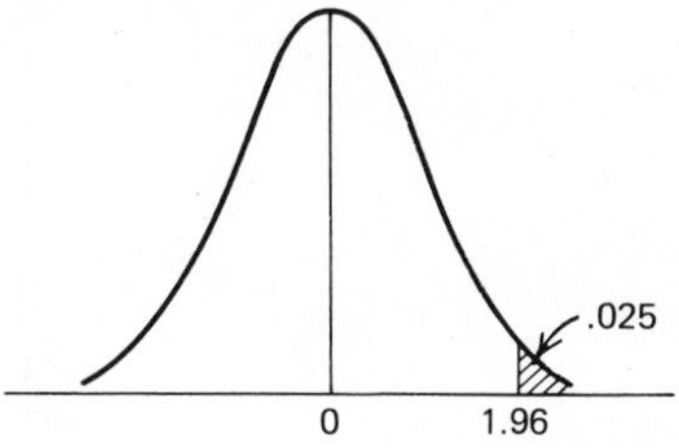

Figure 7.12

Example 7.5 Obtain the value $z > 0$ with $P[-z \leqslant Z \leqslant z] = .90$.
We observe from the symmetry of the curve that

$$P[Z < -z] = P[Z > z] = .05$$

From Appendix Table 4, we see that $z = 1.65$ gives $P[Z \leqslant -1.65] = .0495$ and $z = 1.64$ gives $P[Z < -1.64] = .0505$. Since .05 is half way between these two probabilities, we interpolate between the two z values to obtain $z = 1.645$ (see Figure 7.13). ■

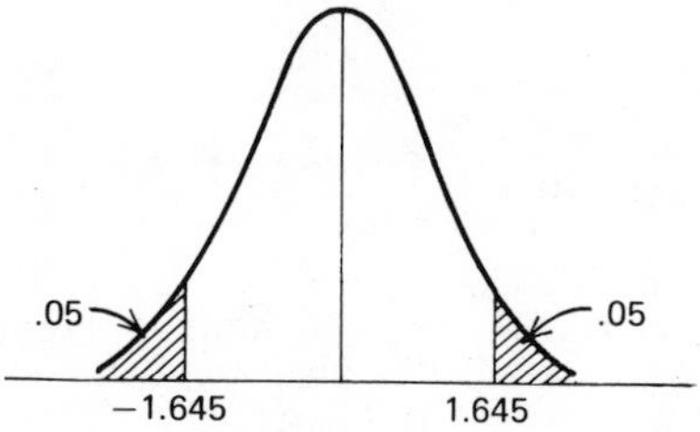

Figure 7.13

The preceding examples illustrate the usefulness of a sketch to depict an area under the standard normal curve. A correct diagram shows how to combine the areas to the left of specified z values given in the normal table.

Fortunately, no new tables are required in calculations of the general normal distribution. The normal distribution has the property that if X is $N(\mu, \sigma)$, then the standardized variable

$$Z = \frac{X - \mu}{\sigma}$$

has the standard normal distribution. The probabilities of intervals can therefore be related to the standard normal distribution by subtracting the mean and then dividing by the standard deviation.

If X is $N(\mu, \sigma)$, then $Z = \dfrac{X-\mu}{\sigma}$ is $N(0,1)$. So

$$P[X \leqslant b] = P\left[\frac{X-\mu}{\sigma} \leqslant \frac{b-\mu}{\sigma}\right] = P\left[Z \leqslant \frac{b-\mu}{\sigma}\right]$$

$$P[a \leqslant X \leqslant b] = P\left[\frac{a-\mu}{\sigma} \leqslant \frac{X-\mu}{\sigma} \leqslant \frac{b-\mu}{\sigma}\right]$$

$$= P\left[\frac{a-\mu}{\sigma} \leqslant Z \leqslant \frac{b-\mu}{\sigma}\right]$$

where the probabilities for Z are obtained from the standard normal table.

The line of reasoning for this relationship follows. The event that X is less than b is the same event as $[X-\mu \leqslant b-\mu]$, and these latter values of X are the same for which $\left[\dfrac{X-\mu}{\sigma} \leqslant \dfrac{b-\mu}{\sigma}\right]$. Now $[a \leqslant X \leqslant b]$ is the intersection of $[X \geqslant a]$ and $[X \leqslant b]$ or, equivalently, $\left[\dfrac{X-\mu}{\sigma} \geqslant \dfrac{a-\mu}{\sigma}\right]$ and $\left[\dfrac{X-\mu}{\sigma} \leqslant \dfrac{b-\mu}{\sigma}\right]$. The intersection is $\left[\dfrac{a-\mu}{\sigma} \leqslant Z \leqslant \dfrac{b-\mu}{\sigma}\right]$. This last event is stated in terms of a standard normal variable $Z = \dfrac{X-\mu}{\sigma}$. The z values are $\dfrac{a-\mu}{\sigma}$ and $\dfrac{b-\mu}{\sigma}$.

Example 7.6 Determine $P[X > -3]$ and $P[-3 < X < 10]$ when X is $N(1,4)$. Here $\mu = 1$ and $\sigma = 4$, so that we subtract 1 and divide by 4:

$$P[X > -3] = P[X - 1 > -3 - 1]$$

$$= P\left[\frac{X-1}{4} > \frac{-3-1}{4}\right] = P[Z > -1] = .8413$$

$$P[-3 < X < 10] = P\left[\frac{-3-1}{4} < \frac{X-1}{4} < \frac{10-1}{4}\right]$$

$$= P[-1 < Z < 2.25] = .8291$$

■

Example 7.7 Scores on a performance test following a training program are approximately $N(14, 2)$. If those who score below 11 must be retrained, what percentage will have to be retrained?

The percentage is 100 times the fraction of scores below 11. This fraction is equal to the probability of a score less than 11, or

$$P[X<11]=P\left[\frac{X-14}{2}<\frac{11-14}{2}\right]$$
$$=P[Z<-1.5]$$
$$=.0668$$

Thus $100\times(.0668)=6.68\%$ will have to be retrained.

The normal distribution in any particular application is only an abstract model, just as a straight line is a model for the side of a building or for a section of highway. The model gives positive probability to negative scores as well as to very large positive values. It is precisely because these probabilities are often quite small that the distribution can provide a realistic model, even for variables that are constrained to a range of positive values. ■

*7.2.2 Further Properties of the Normal Distribution

Two important properties of the normal distribution deserve special attention here. From the properties of expectation, if $Y=a+bX$ then $E(Y)=a+bE(X)$ and $\text{sd}(Y)=|b|\text{sd}(X)$. Moreover, if X is $N(\mu,\sigma)$, then if $b>0$

$$\frac{Y-(a+b\mu)}{|b|\sigma}=\frac{(X-\mu)}{\sigma}$$

has a standard normal distribution. A similar result holds if $b<0$.

> If X is $N(\mu,\sigma)$, then
>
> $$Y=a+bX \quad \text{is} \quad N(a+b\mu, |b|\sigma)$$
>
> Multiplying by a constant b and adding a constant a only changes the mean and the variance of the normal distribution.

The second property of the normal distribution states that the sum of independent normal variables is normal. Calculation of the mean and the variance follows directly from the properties of expectation and variance (see Section 7.1).

> The sum of two independent normals is normal. If X is $N(\mu_1, \sigma_1)$ and Y is $N(\mu_2, \sigma_2)$, then for independent X and Y
>
> $$X + Y \text{ is } N(\mu, \sigma)$$
>
> where
>
> $$\mu = \mu_1 + \mu_2$$
>
> $$\sigma^2 = \sigma_1^2 + \sigma_2^2$$
>
> $$\sigma = \sqrt{\sigma_1^2 + \sigma_2^2}$$

7.3 THE NORMAL APPROXIMATION TO THE BINOMIAL DISTRIBUTION

The binomial distribution $b(n,p)$ was introduced in Section 5.3 as the distribution governing the number of successes X in n independent trials of an experiment that has a success probability of p in each trial. For large n when p is not too close to 0 or 1, the normal distribution serves as a good approximation to the binomial distribution. Bypassing the mathematical proof of this fact, we concentrate on illustrating the manner in which this approximation works.

Recall that a binomial random variable X has a mean of $\mu = np$ and a standard deviation of $\sigma = \sqrt{np(1-p)}$. When n is large but σ^2 is moderate that is, when np or $n(1-p)$ is moderate, a Poisson distribution can be used to approximate the binomial probabilities, as we saw in Section 5.8. The normal approximation is applied in the more typical situation in which n is large and p is not too near 0 or 1. It allows us to treat the binomial X as if it had the normal distribution $N(np, \sqrt{np(1-p)}\,)$ when calculating probabilities. This normal curve closely approximates the top of the binomial probability histogram. To calculate the probability of X assuming values between two integers a and b

$$P[a \leqslant X \leqslant b] = \sum_{x=a}^{b} \binom{n}{x} p^x (1-p)^{n-x}$$

we then proceed as if we are calculating the probability of this interval from the normal distribution. The standardized normal variable is

$$Z = \frac{X - np}{\sqrt{np(1-p)}}$$

so that the preceding binomial probability is approximated by

$$P[a \leqslant X \leqslant b] \approx P\left[\frac{a - np}{\sqrt{np(1-p)}} \leqslant Z \leqslant \frac{b - np}{\sqrt{np(1-p)}}\right]$$

The right-hand side of this approximation can be calculated from the normal probability table. As a rule of thumb, this approximation is usually satisfactory when both np and $n(1-p)$ are greater than 15.

To indicate the closeness of the approximation, the probability histogram of the distribution $b(15,.4)$ appears in Figure 7.14, with the approximating normal curve having $\mu = 15 \times .4 = 6$ and $\sigma = \sqrt{15 \times .4 \times .6} = 1.9$. Although the normal approximation is intended to be used for large values of n, we choose $n = 15$ for the purposes of illustration, so that the exact probabilities can be read from the binomial table, Appendix Table 2. Even for n as small as 15, the approximation appears to be reasonable.

For a more concrete comparison, let us consider the probability $P[7 \leqslant X \leqslant 10]$ for the distribution $b(15,.4)$. The exact value obtained from the

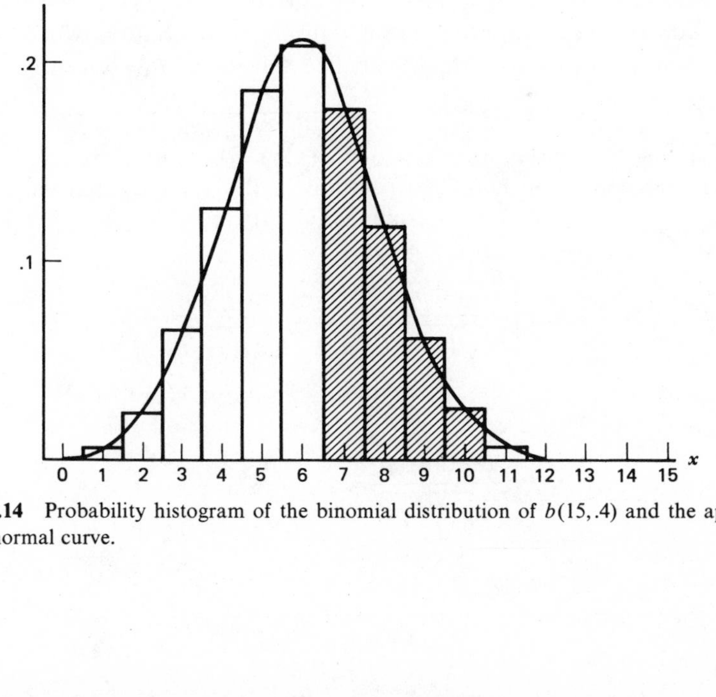

Figure 7.14 Probability histogram of the binomial distribution of $b(15,.4)$ and the approximating normal curve.

binomial table is

$$P[7 \leqslant X \leqslant 10] = .991 - .610 = .381$$

Using the normal approximation with $\mu = 6$ and $\sigma = 1.9$, to three decimal places, gives us

$$\begin{aligned} P[7 \leqslant X \leqslant 10] &\approx P\left[\frac{7-6}{1.9} \leqslant Z \leqslant \frac{10-6}{1.9}\right] \\ &= P[.526 \leqslant Z \leqslant 2.105] \\ &= .982 - .700 \\ &= .282 \end{aligned}$$

This approximation can be improved by making an adjustment in our calculations. Note that the probability $P[7 \leqslant X \leqslant 10] = .381$ is the area of the shaded bars in Figure 7.14. Because these bars extend from 6.5 to 10.5, a better match of the area under the normal curve results if we consider the interval to be $6.5 \leqslant X \leqslant 10.5$. With this adjustment we obtain

$$\begin{aligned} P[7 \leqslant X \leqslant 10] &\approx P\left[\frac{6.5-6}{1.9} \leqslant Z \leqslant \frac{10.5-6}{1.9}\right] \\ &= P[.263 \leqslant Z \leqslant 2.368] \\ &= .991 - .604 \\ &= .387 \end{aligned}$$

This result is considerably closer to the correct value .381, even though n is small. Because we are approximating a discrete distribution with a continuous distribution, this adjustment is called the *continuity correction*.

If X has the binomial distribution $b(n,p)$, where n is large and p is not too near 0 or 1, the distribution of the standardized variable $Z = (X - np)/\sqrt{npq}$ is approximately $N(0, 1)$:

$$P[a \leqslant X \leqslant b] \approx P\left[\frac{a - np}{\sqrt{np(1-p)}} \leqslant Z \leqslant \frac{b - np}{\sqrt{np(1-p)}}\right]$$

(without continuity correction)

$$\approx P\left[\frac{a - .5 - np}{\sqrt{np(1-p)}} \leqslant Z \leqslant \frac{b + .5 - np}{\sqrt{np(1-p)}}\right]$$

(using continuity correction)

Example 7.8 A large-scale survey conducted five years ago revealed that 30% of the adult population were regular users of alcoholic beverages. If this is still the current rate, what is the probability that in a random sample of 1000 adults the number of users of alcoholic beverages will be (*a*) less than 280? (*b*) 316 or more?

Let X denote the number of regular alcoholic beverage users in the random sample of 1000 adults. Under the assumption that the proportion of the population who are users is .30, X has a binomial distribution with $n = 1000$ and $p = .3$. Since

$$np = 300, \quad \sqrt{np(1-p)} = \sqrt{210} = 14.5$$

the distribution of X is approximately $N(300, 14.5)$.

(*a*) Ignoring the continuity correction

$$P[X \leqslant 279] \approx P\left[Z \leqslant \frac{279 - 300}{14.5}\right]$$
$$= P[Z \leqslant -1.448]$$
$$= .074$$

Using the continuity correction, we have

$$P[X \leqslant 279] \approx P\left[Z \leqslant \frac{279.5 - 300}{14.5}\right]$$
$$= P[Z \leqslant -1.414]$$
$$= .079$$

Note that the continuity correction barely alters the result. This is because a change of .5 in the numerator does not change the ratio appreciably when the denominator $\sqrt{np(1-p)}$ is large.

(*b*)
$$P[X \geqslant 316] \approx P\left[Z \geqslant \frac{316 - 300}{14.5}\right]$$
$$= P[Z \geqslant 1.103]$$
$$= 1 - .865$$
$$= .135$$

■

7.4 RANDOM SAMPLING, STATISTIC, AND SAMPLING DISTRIBUTION

We have just demonstrated that the distribution of a binomial count is approximately normal when the number of trials is large. In a number of other important instances, a general result, known as the *central limit theorem*, also justifies the use of the normal distribution as an approximation to the distributions of many random variables arising in the context of drawing inferences from samples. However, before elaborating on this, we must first introduce some cornerstone concepts of statistics. These key concepts are (*a*) random sampling from a probability distribution, and (*b*) sampling distributions. These two concepts underly almost everything that follows in the remainder of the text. An understanding of them is so essential that it is fruitless to proceed beyond this section without it. A few examples will provide some footholds for these important concepts.

Example 7.9 An investigator intends to study the ownership of bicycles per household in a city of 50,000 households. We designate the number of bicycles per family as the random variable X and suppose that, unknown to the investigator, the distribution of X over the population of 50,000 families is as given in Table 7.1.

TABLE 7.1
POPULATION DISTRIBUTION OF X, THE NUMBER OF BICYCLES PER FAMILY

x	0	1	2	3
$f(x)$	.2	.4	.3	.1

Table 7.1 shows that 20% of the families do not own a bicycle, that 40% own one bicycle, and so on. How should the investigator select a sample of households to obtain information on this (unknown) population distribution? What is a logical assessment of the probability distribution for an observation or for a set of observations?

Suppose that the investigator wants to select 60 families and record the number of bicycles owned by them as $X_1, X_2, \ldots, X_{60}$. The families should be selected at random in accordance with the definition in Section 3.5, which requires that each of the $\binom{50{,}000}{60}$ sets of 60 families have the same chance of being selected for the sample. As an operational method, it may be convenient to think of 50,000 cards, each identifying one household, being shuffled and 60 cards being selected from the pile one after another. In this manner, a random sample without replacement is obtained.

Before the sampling is actually made, we know that X_1, the number for the first family selected, is capable of assuming values 0, 1, 2, or 3. Because all 50,000 cards are equally likely to be the first observation, X_1 can have the values 0, 1, 2, or 3 with probabilities coinciding exactly with the proportions of the population having these values. In other words, X_1 is a random variable. Its distribution is the same as the population distribution of X given in Table 7.1. If the first card drawn is restored to the pile, the cards are shuffled, and a second card is drawn, the same distribution applies to X_2 and, moreover, X_2 and X_1 are independent. With this modification, the random sampling is with replacement and the sample observations $X_1, \ldots, X_{60}$ will be independent random variables, each having the probability distribution that characterizes the population. If a card drawn at one stage is not returned to the pile before the next card is drawn, the condition of independence will be violated. However, when the population is of an enormous size compared to the sample size, as is the case in this example, the proportions having each x value remain virtually the same whether or not a few elements are removed. For all practical purposes, we can ignore this minor difference and regard $X_1, \ldots, X_{60}$ as independent random variables, each having the population distribution $f(x)$. This probability structure provides motivation for the definition of a random sample. ■

Example 7.9 also serves to remind us of the fundamental conceptual steps involved in modeling uncertainty.

> An observation, before it is actually taken, is modeled as a random variable X with distribution $f(x)$. The actual numerical value of the observation is considered to be one realization of the random variable.

Example 7.10 A horticulturist wishes to measure the blossoming times for a new hybrid tulip plant under controlled greenhouse conditions. He intends to grow 25 plants and to record the times $X_1, X_2, \ldots, X_{25}$ between germination and the appearance of the first blossom for each plant. What can be said about a probability model for these measurements?

In contrast to the previous example, here the sample is not being selected from a tangible population. We can certainly envision an experiment that involves thousands of plants, but it would not include all the plants that may be grown at other times. In this case, we must take an additional, but short step in our modeling and visualize a continuous distribution for the blossoming time. Although the population is not tangible, a density function still serves to model the variation in the observations.

A plant in a field may affect the growth of others nearby, thus introducing some dependence. If care is taken to avoid such difficulties, the contemplated observations $X_1, \ldots, X_{25}$ for 25 plants can be viewed as outcomes of 25 independent trials of an experiment performed under identical conditions. By analogy with taking a random sample from a tangible population, it is then logical to model $X_1, \ldots, X_{25}$ as independent random variables, each having the same distribution $f(x)$ that characterizes the infinite population. ■

7.4.1 Random Samples

Examples 7.9 and 7.10 illustrate the distributional structure associated with random sampling. This structure applies to both large, tangible populations and the outcomes of independent measurements repeated under identical conditions. The definition of a random sample applies equally when the distribution described by $f(x)$ is discrete or continuous.

> A *random sample* of size n from a population $f(x)$ is a collection of n *independent* random variables $X_1, \ldots, X_n$, *each having the distribution* $f(x)$.

This definition contains the probabilistic condition of independence, which plays a vital role in the inference procedures we will discuss in later chapters. Whether it seems to hold in any specific sampling process must always be checked, like any other tentative assumption, to determine the validity of the inference procedures.

It is important to grasp the sense in which a random sample of size n is viewed as a set of n random variables. Initially, $X_1, \ldots, X_n$ represent the unknown measurements that will arise as the 1st, ..., nth observations in a process of random sampling. Before the sampling is performed, these observable quantities are random variables. Once the measurements are recorded, we have a realization of the n random variables in the form of a data set, which is called a *sample* in the common usage of the term. An observed data set is symbolically denoted by $x_1, \ldots, x_n$. Strictly speaking, $X_1, \ldots, X_n$ should be called the *observables* in a process of random sampling and a set of realized values $x_1, \ldots, x_n$ should be called an *observed random sample*. However, in the interest of brevity, we use the common term *sample observations* for both, with the understanding that before being observed, they are random variables and once observed, they are a set of numbers.

7.4.2 Statistic

In describing the features of sample observations or in drawing inferences about the population, we deal with some relevant measures computed from the sample. Various sample measures, including the sample mean $\overline{X}$, median, and variance s^2, were introduced in Chapter 2 in the context of specific realizations of the sample data. Each of these measures is a specified function of the sample observations. In general terms, a function of the sample observations is called a *statistic*. A statistic serves as a tool to describe some feature of a sample and to make inferences about a feature of the population.

> A *statistic* is a function of the sample observations.

7.4.3 Sampling Distributions

Being a function of the random sample $X_1,\ldots,X_n$, a statistic is also a random variable and has a distribution. Conceptually, this is so fundamental that it should be repeated here. Although in any given situation we are limited to one sample or set of observations and the corresponding single value for the statistic, over different samples the statistic changes value according to a distribution determined from that governing the random sample. The important point is that the behavior of the statistic can be described by some probability distribution.

> Every statistic is, itself, a random variable. Its probability distribution is called the *sampling distribution* of the statistic.

Elegant mathematical devices are available for deriving the distributions of many statistics of common interest. A discussion of these techniques is beyond the scope of this text. Example 7.11 illustrates how the distribution of the sample mean $\bar{X}$ can be determined for a simple situation when the sample size is 2 and the population distribution is discrete.

Example 7.11 The population distribution given in Table 7.2 is repeated from Example 7.9. A random sample (X_1,X_2) of size 2 is to be taken from this population. What is the sampling distribution of the statistic

$$\bar{X}=\frac{X_1+X_2}{2} \qquad ?$$

TABLE 7.2
POPULATION DISTRIBUTION

x	0	1	2	3
$f(x)$	.2	.4	.3	.1

From the definition of a random sample, X_1 and X_2 are independent random variables, each having the distribution given in Table 7.2. Recall that the joint distribution of two independent random variables is obtained by multiplying the marginal probabilities. For instance

$$P[X_1=0, X_2=1] = P[X_1=0]P[X_2=1] = .2 \times .4 = .08$$

The joint distribution obtained in this manner is presented in Table 7.3.

TABLE 7.3
JOINT DISTRIBUTION OF X_1 AND X_2

x_2 \ x_1	0	1	2	3	ROW SUM
0	.04	.08	.06	.02	.2
1	.08	.16	.12	.04	.4
2	.06	.12	.09	.03	.3
3	.02	.04	.03	.01	.1
Column sum	.2	.4	.3	.1	1

To obtain the distribution of $\bar{X}=(X_1+X_2)/2$, we first list the possible values of $\bar{X}$ in Table 7.4. Next for each value of $\bar{X}$, we identify the cells in Table 7.3 whose (x_1, x_2) values yield this value of $\bar{X}$. Then we add these cell probabilities. For example, $\bar{X}=1.5$ when $(X_1, X_2)=(0,3)$, $(1,2)$, $(2,1)$, or $(3,0)$, so that $P[\bar{X}=1.5]=.02+.12+.12+.02=.28$.

TABLE 7.4
SAMPLING DISTRIBUTION OF $\bar{X}=\dfrac{X_1+X_2}{2}$

Value of $\bar{X}$	0	.5	1	1.5	2	2.5	3	Total
Probability	.04	.16	.28	.28	.17	.06	.01	1

■

7.5 DISTRIBUTION OF THE SAMPLE MEAN AND THE CENTRAL LIMIT THEOREM

A statistical inference problem of prime practical consideration involves making an informative statement about the unknown population mean. Not surprisingly, the procedures are based on the sample mean

$$\bar{X}=\frac{X_1+X_2+\dots+X_n}{n}$$

We therefore explore the fundamental properties of the sampling distribution of $\bar{X}$ and explain the role of the normal distribution as a useful approximation.

Some basic measures of the sampling distribution of $\bar{X}$, such as the expectation and the variance, can readily be deduced from the probabilistic structure of a random sample. We denote

Population mean $= \mu$
Population variance $= \sigma^2$

In a random sample, the random variables $X_1, \ldots, X_n$ are independent, and each has the distribution of the population. Consequently

$$E(X_1) = \ldots = E(X_n) = \mu$$

$$\text{Var}(X_1) = \ldots = \text{Var}(X_n) = \sigma^2$$

Since

$$\bar{X} = \frac{1}{n}(X_1 + \ldots + X_n)$$

and n is a constant, the additivity properties of expectation and variance discussed in Section 7.1 can be used to obtain

$$\begin{aligned} E(\bar{X}) &= \frac{1}{n} E(X_1 + \ldots + X_n) \\ &= \frac{1}{n}\left[E(X_1) + \ldots + E(X_n)\right] \\ &= \frac{1}{n}\left[\mu + \ldots + \mu\right] = \frac{n\mu}{n} = \mu \end{aligned}$$

$$\begin{aligned} \text{Var}(\bar{X}) &= \frac{1}{n^2} \text{Var}(X_1 + \ldots + X_n) \\ &= \frac{1}{n^2}\left[\text{Var}(X_1) + \ldots + \text{Var}(X_n)\right] \quad \text{(due to independence)} \\ &= \frac{1}{n^2}\left[\sigma^2 + \ldots + \sigma^2\right] = \frac{n\sigma^2}{n^2} = \frac{\sigma^2}{n} \end{aligned}$$

Mean and standard deviation of $\bar{X}$

The distribution of the sample mean based on a random sample of size n has

$$E(\bar{X}) = \mu \quad (= \text{Population mean})$$

$$\text{Var}(\bar{X}) = \frac{\sigma^2}{n} \quad \left(= \frac{\text{Population variance}}{n} \right)$$

$$\text{sd}(\bar{X}) = \frac{\sigma}{\sqrt{n}} \quad \left(= \frac{\text{Population sd}}{\sqrt{n}} \right)$$

The first result shows that the distribution of $\bar{X}$ is centered at the population mean μ in the sense that expectation serves as a measure of center of a distribution. Second, the standard deviation of $\bar{X}$ is the population standard deviation divided by the square root of the sample size. The averaging produces a statistic that is less variable than an individual observation. With increasing sample size, the standard deviation $\sigma/\sqrt{n}$ decreases and the distribution of $\bar{X}$ tends to become more concentrated around the population mean μ. This decrease in $\sigma/\sqrt{n}$ occurs slowly if n is already large because in order to halve the standard deviation of $\bar{X}$, the sample size must be increased by a factor of 4.

Example 7.12 Compute the mean and the variance for the population distribution given in Table 7.2 and for the distribution of $\bar{X}$ given in Table 7.4 when $n=2$. Verify the relations $E(\bar{X}) = \mu$ and, $\text{Var}(\bar{X}) = \sigma^2/n$.

The computations are performed in the following schematic forms.

POPULATION DISTRIBUTION

x	0	1	2	3	TOTAL
$f(x)$	.2	.4	.3	.1	1
$xf(x)$	0	.4	.6	.3	1.3
$x^2f(x)$	0	.4	1.2	.9	2.5

$$\mu = 1.3$$

$$\begin{aligned} \sigma^2 &= 2.5 - (1.3)^2 \\ &= 2.5 - 1.69 \\ &= .81 \end{aligned}$$

DISTRIBUTION OF $\bar{X} = \dfrac{X_1 + X_2}{2}$

$\bar{x}$	0	.5	1	1.5	2	2.5	3	TOTAL
$f(\bar{x})$	.04	.16	.28	.28	.17	.06	.01	1
$\bar{x}f(\bar{x})$	0	.08	.28	.42	.34	.15	.03	1.3
$\bar{x}^2f(\bar{x})$	0	.04	.28	.63	.68	.375	.09	2.095

$$E(\bar{X}) = 1.3 = \mu$$

$$\begin{aligned}\text{Var}(\bar{X}) &= 2.095 - (1.3)^2\\ &= 2.095 - 1.69\\ &= .405 = \sigma^2/2\end{aligned}$$

■

We can now state two important results concerning the shape of the sampling distribution of $\bar{X}$. The first result gives the exact form of the distribution of $\bar{X}$ when the population distribution is normal:

> *$\bar{X}$ from a normal population is normal*
>
> Based on a random sample of size n from a *normal* population with mean μ and standard deviation σ, the sample mean $\bar{X}$ has the normal distribution with mean μ and standard deviation $\sigma/\sqrt{n}$.

When sampling from a nonnormal population, the distribution of $\bar{X}$ depends on the particular form of the population distribution that prevails. The second more surprising result, known as the *central limit theorem*, states that when the sample size n is large, the distribution of $\bar{X}$ is approximately normal, regardless of the shape of the population distribution. ■

> *Central Limit Theorem*
>
> In random sampling from an arbitrary population with mean μ and standard deviation σ, the distribution of $\bar{X}$ when n is large is approximately normal, with mean μ and standard deviation $\sigma/\sqrt{n}$. In other words
>
> $$Z = \frac{\bar{X} - \mu}{\sigma/\sqrt{n}} \quad \text{is approximately } N(0, 1).$$

Whether the population distribution is continuous, discrete, symmetric, or skewed, the central limit theorem asserts that as long as the population variance is finite, the distribution of the sample mean $\bar{X}$ is nearly normal if the sample size is large. In this sense, the normal distribution plays a central role in the development of statistical procedures. Rather than attempt a mathematical justification here, it may be more instructive to demonstrate how this result operates.

Example 7.13 Demonstration of the Central Limit Theorem: Consider a population having a discrete uniform distribution that places a probability of .1 on each of the integers $0, 1, \ldots, 9$. This may be an appropriate model for the distribution of the last digit in telephone numbers or the first overflow digit in computer calculations. The line diagram of this distribution appears in Figure 7.15.

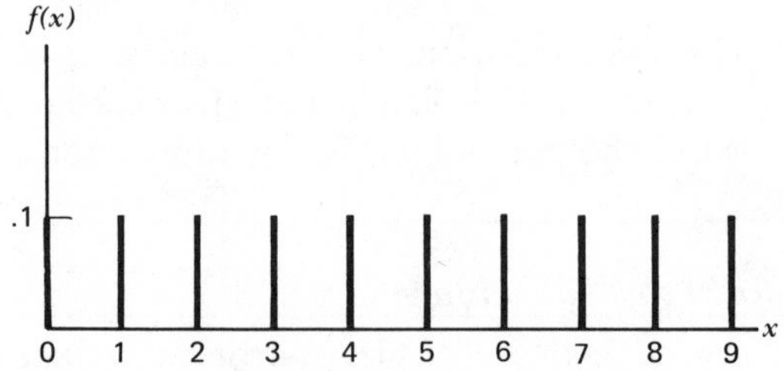

Figure 7.15 Uniform distribution on the integers $0,1,\ldots,9$.

By means of a computer, 100 random samples of size 5 were generated from this distribution and the value of $\bar{X}$ was computed for each sample. The results of this repeated random sampling are presented in Table 7.5. The relative frequency histogram in Figure 7.16 is constructed from the 100 observed values of $\bar{X}$. Consulting Figures 7.15 and 7.16, we can see that although the population distribution is far from normal, the top of the histogram of the $\bar{X}$ values (Figure 7.16) suggests the appearance of a bell-shaped curve, even for the small sample size of 5. For larger sample sizes, the normal distribution would give even a closer approximation. It might be interesting for the reader to collect samples by reading the last digits of numbers from a telephone directory and then to construct a histogram of the $\bar{X}$ values. ■

Another graphic illustration of the central limit theorem appears in Figure 7.17, where the population distribution represented by the solid curve is a continuous asymmetric distribution with $\mu = 2$ and $\sigma = 1.41$. The distributions of the sample mean $\bar{X}$ for sample sizes $n = 3$ and $n = 10$ are plotted as dashed curves on the graph, indicating that with increasing n,

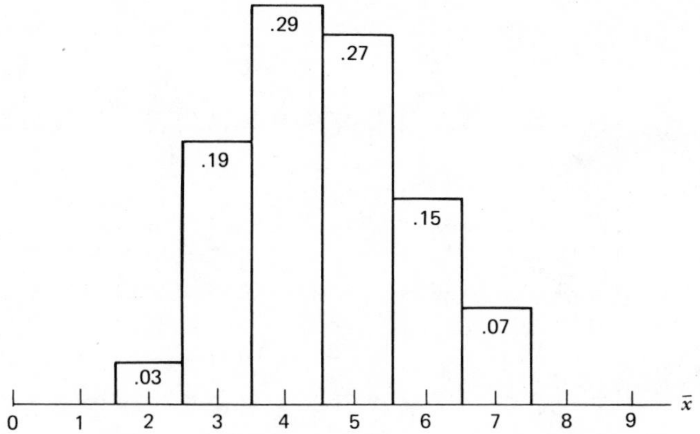

Figure 7.16 Relative frequency histogram of the $\bar{x}$ values recorded in Table 7.5.

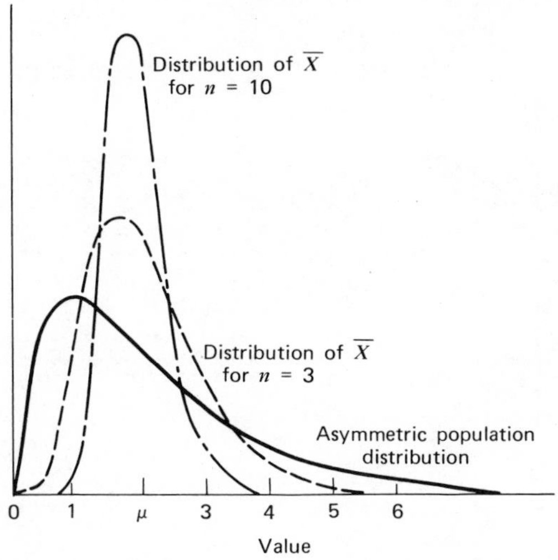

Figure 7.17 Distributions of $\overline{X}$ for $n = 3$ and $n = 10$ in sampling from a skewed population.

the distributions become more concentrated around μ and look more like a normal distribution.

Example 7.14 Suppose that the population distribution of the gripping strengths of industrial workers is known to have a mean of 110 and a standard deviation of 5. For a random sample of 75 workers, what is the probability that the sample mean gripping strength will be between 109 and 111?

TABLE 7.5

SAMPLES OF SIZE 5 FROM A DISCRETE UNIFORM DISTRIBUTION

SAMPLE NUMBER	OBSERVATIONS	SUM	MEAN $\bar{x}$	SAMPLE NUMBER	OBSERVATIONS	SUM	MEAN $\bar{x}$
1	4, 7, 9, 0, 6	26	5.2	51	4, 7, 3, 8, 8	30	6.0
2	7, 3, 7, 7, 4	28	5.6	52	2, 0, 3, 3, 2	10	2.0
3	0, 4, 6, 9, 2	21	4.2	53	4, 4, 2, 6, 3	19	3.8
4	7, 6, 1, 9, 1	24	4.8	54	1, 6, 4, 0, 6	17	3.4
5	9, 0, 2, 9, 4	24	4.8	55	2, 4, 5, 8, 9	28	5.6
6	9, 4, 9, 4, 2	28	5.6	56	1, 5, 5, 4, 0	15	3.0
7	7, 4, 2, 1, 6	20	4.0	57	3, 7, 5, 4, 3	22	4.4
8	4, 4, 7, 7, 9	31	6.2	58	3, 7, 0, 7, 6	23	4.6
9	8, 7, 6, 0, 5	26	5.2	59	4, 8, 9, 5, 9	35	7.0
10	7, 9, 1, 0, 6	23	4.6	60	6, 7, 8, 2, 9	32	6.4
11	1, 3, 6, 5, 7	22	4.4	61	7, 3, 6, 3, 6	25	5.0
12	3, 7, 5, 3, 2	20	4.0	62	7, 4, 6, 0, 1	18	3.6
13	5, 6, 6, 5, 0	22	4.4	63	7, 9, 9, 7, 5	37	7.4
14	9, 9, 6, 4, 1	29	5.8	64	8, 0, 6, 2, 7	23	4.6
15	0, 0, 9, 5, 7	21	4.2	65	6, 5, 3, 6, 2	22	4.4
16	4, 9, 1, 1, 6	21	4.2	66	5, 0, 5, 2, 9	21	4.2
17	9, 4, 1, 1, 4	19	3.8	67	2, 9, 4, 9, 1	25	5.0
18	6, 4, 2, 7, 3	22	4.4	68	9, 5, 2, 2, 6	24	4.8
19	9, 4, 4, 1, 8	26	5.2	69	0, 1, 4, 4, 4	13	2.6
20	8, 4, 6, 8, 3	29	5.8	70	5, 4, 0, 5, 2	16	3.2
21	5, 2, 2, 6, 1	16	3.2	71	1, 1, 4, 2, 0	8	1.6
22	2, 2, 9, 1, 0	14	2.8	72	9, 5, 4, 5, 9	32	6.4
23	1, 4, 5, 8, 8	26	5.2	73	7, 1, 6, 6, 9	29	5.8

24	8, 1, 6, 3, 7	25	5.0	74	3, 5, 0, 0, 5	13	2.6
25	1, 2, 0, 9, 6	18	3.6	75	3, 7, 7, 3, 5	25	5.0
26	8, 5, 3, 0, 0	16	3.2	76	7, 4, 7, 6, 2	26	5.2
27	9, 5, 8, 5, 0	27	5.4	77	8, 1, 0, 9, 1	19	3.8
28	8, 9, 1, 1, 8	27	5.4	78	6, 4, 7, 9, 3	29	5.8
29	8, 0, 7, 4, 0	19	3.8	79	7, 7, 6, 9, 7	36	7.2
30	6, 5, 5, 3, 0	19	3.8	80	9, 4, 2, 9, 9	33	6.6
31	4, 6, 4, 2, 1	17	3.4	81	3, 3, 3, 3, 3	15	3.0
32	7, 8, 3, 6, 5	29	5.8	82	8, 7, 7, 0, 3	25	5.0
33	4, 2, 8, 5, 2	21	4.2	83	5, 3, 2, 1, 1	12	2.4
34	7, 1, 9, 0, 9	26	5.2	84	0, 4, 5, 2, 6	17	3.4
35	5, 8, 4, 1, 4	22	4.4	85	3, 7, 5, 4, 1	20	4.0
36	6, 4, 4, 5, 1	20	4.0	86	7, 4, 5, 9, 8	33	6.6
37	4, 2, 1, 1, 6	14	2.8	87	3, 2, 9, 0, 5	19	3.8
38	4, 7, 5, 5, 7	28	5.6	88	4, 6, 6, 3, 3	22	4.4
39	9, 0, 5, 9, 2	25	5.0	89	1, 0, 9, 3, 7	20	4.0
40	3, 1, 5, 4, 5	18	3.6	90	2, 9, 6, 8, 5	30	6.0
41	9, 8, 6, 3, 2	28	5.6	91	4, 8, 0, 7, 6	25	5.0
42	9, 4, 2, 2, 8	25	5.0	92	5, 6, 7, 6, 3	27	5.4
43	8, 4, 7, 2, 2	23	4.6	93	3, 6, 2, 5, 6	22	4.4
44	0, 7, 3, 4, 9	23	4.6	94	0, 1, 1, 8, 4	14	2.8
45	0, 2, 7, 5, 2	16	3.2	95	3, 6, 6, 4, 5	24	4.8
46	7, 1, 9, 9, 9	35	7.0	96	9, 2, 9, 8, 6	34	6.8
47	4, 0, 5, 9, 4	22	4.4	97	2, 0, 0, 6, 8	16	3.2
48	5, 8, 6, 3, 3	25	5.0	98	0, 4, 5, 0, 5	14	2.8
49	4, 5, 0, 5, 3	17	3.4	99	0, 3, 7, 3, 9	22	4.4
50	7, 7, 2, 0, 1	17	3.4	100	2, 5, 0, 0, 7	14	2.8

Here the population mean and the standard deviation are $\mu = 110$ and $\sigma = 5$, respectively. The sample size $n = 75$ is large, so the central limit theorem ensures that the distribution of $\bar{X}$ is approximately normal with

$$\text{mean of } \bar{X} = 110$$

$$\text{sd of } \bar{X} = \frac{\sigma}{\sqrt{n}} = \frac{5}{\sqrt{75}} = .577$$

$$P[109 < \bar{X} < 111] = P\left[\frac{109-110}{.577} < Z < \frac{111-110}{.577}\right]$$

$$= P[-1.73 < Z < 1.73]$$

$$= .958 - .042 = .916$$

Looking ahead, we might imagine a problem of statistical inference in which one or both of the population parameters μ and σ are unknown. What is available is a sample of strength measurements from which a numerical value of $\bar{X}$ and other statistics can be calculated. Using these observed values, we would then attempt to determine plausible values for the unknown parameters. ■

A natural question that arises is how large should n be for the normal approximation to be used for the distribution of $\bar{X}$? The nature of the approximation depends on the extent to which the population distribution deviates from a normal form. If the population distribution is normal, then $\bar{X}$ is exactly normally distributed for all n, small or large. As the population distribution increasingly departs from normality, larger values of n are required for a good approximation. As a rule of thumb, $n > 30$ is expected to provide a satisfactory approximation for situations in which the population distribution has a single peak and not too heavy a tail.

Before we finish this general discussion, it is important to be aware that the reason a normal distribution provides an approximation to the binomial distribution is also rooted in the central limit theorem. To understand this connection, we attach a counter variable to each Bernoulli trial:

$$X_i = 1 \quad \text{if the } i\text{th trial is a success.}$$

$$X_i = 0 \quad \text{if the } i\text{th trial is a failure.}$$

The random variables $X_1, \ldots, X_n$ are then independent, and each has the distribution

x	0	1
$f(x)$	$(1-p)$	p

with a mean of p and a standard deviation of $\sqrt{p(1-p)}$, as we saw in Section 5.5.

Accordingly, $X_1,\ldots,X_n$ can be regarded as a random sample. As a consequence of the central limit theorem, the distribution of the sample mean $\bar{X}$ is approximately $N(p, \sqrt{p(1-p)}/\sqrt{n})$ when n is large. Now from the definition of the counter variables we note that

$$\text{Sample total } T = \sum_{i=1}^{n} X_i = \text{Number of successes in } n \text{ trials}$$

$$\text{Sample mean } \bar{X} = \frac{T}{n} = \frac{\text{Number of successes in } n \text{ trials}}{n}$$

In our present context, therefore, the sample mean is nothing but the proportion or the relative frequency of successes in n trials. Because the symbol for the population proportion of successes is p, it is customary to denote the sample proportion by $\hat{p}$ instead of the symbol $\bar{X}$.

When n is large, the sample proportion $\hat{p}$ of successes in n Bernoulli trials is approximately distributed

$$N\left(p, \sqrt{\frac{p(1-p)}{n}}\right)$$

and

$$Z = \frac{\hat{p}-p}{\sqrt{p(1-p)/n}} \text{ is approximately } N(0,1).$$

Multiplying the numerator and the denominator of Z by n and noting that $n\hat{p} = T$ (the number of successes in n trials), we can also write

$$Z = \frac{T-np}{\sqrt{np(1-p)}}$$

is approximately $N(0,1)$. T has the binomial distribution $b(n,p)$, and this result shows that with large n, T is approximately $N(np, \sqrt{np(1-p)}\,)$, which is precisely the statement of normal approximation to the binomial.

Example 7.15 Suppose that 60% of a city population favors public funding for a proposed recreational facility. If 150 persons are to be randomly selected and interviewed, what is the probability that the sample proportion favoring this issue will be less than .52?

Here we know $n=150$, $p=.6$, and the problem concerns computing the probability $P[\hat{p}<.52]$. Thus

$$\sqrt{\frac{p(1-p)}{n}}=\sqrt{\frac{.24}{150}}=.040$$

Using the normal approximation

$$P[\hat{p}<.52]\approx P\left[Z<\frac{.52-.60}{.040}\right]=P[Z<-2]=.0228$$

In a statistical inference problem, the population proportion p is unknown and is actually the target of investigation. Instead, a value of $\hat{p}$ is obtained by interviewing a random sample from the city population. Using the observed value, we would like to determine a range of plausible values for p. ■

7.6 CHECKING THE ASSUMPTION OF A NORMAL POPULATION

Here we consider the problem of ascertaining from a random sample whether a population can be adequately described by a normal distribution. One reason for our interest in this question is that the validity of many of the commonly used procedures for statistical inference with small samples depends to some extent on the population being nearly normal. Also in many situations, such as in biological measurements, we may be interested in the complete population distribution and not just in inferences about its mean μ. If a normal distribution is tentatively assumed to be a plausible model, there should be some way to check this assumption once the sample data are obtained. The investigator should not feel that the data were generated from a magical bell-shaped curve. The proper attitude is to be skeptical as to whether it is reasonable to proceed as if normality were approximately true.

It is important not to confuse the present issue with the approximate normal distribution of $\bar{X}$ guaranteed by the central limit theorem. Here we are concerned with the form of the population distribution for a single observation and not with the distribution of $\bar{X}$.

We proceed to describe two simple methods that do involve subjective judgment, but that can be helpful in detecting serious departures from normality.

7.6.1 Examining the Proportions of Observations in Intervals Surrounding the Mean

If the histogram appears to be reasonably symmetric, we can check against thick tails by counting the number of observations in symmetric intervals about the mean. For a normal distribution with mean μ and standard deviation σ, we know that the intervals $(\mu-\sigma, \mu+\sigma)$, $(\mu-2\sigma, \mu+2\sigma)$, and $(\mu-3\sigma, \mu+3\sigma)$ contain the probabilities .683, .954, and .997, respectively. Alternatively, the probabilities outside these intervals are roughly $1/3$, $1/20$, and $1/300$, respectively. Both μ and σ are typically unknown, but with a reasonably large sample size we can expect the sample mean $\bar{X}$ to be close to μ and the sample standard deviation s to be close to σ. From the sample data we can therefore compute $\bar{X}$ and s and then count the numbers of observations outside the intervals $(\bar{X}-s, \bar{X}+s)$, $(\bar{X}-2s, \bar{X}+2s)$, and $(\bar{X}-3s, \bar{X}+3s)$. Dividing these counts by the sample size n, we obtain the relative frequencies, which we compare with the theoretical probabilities $1/3$, $1/20$, and $1/300$. For instance, a large discrepancy between the observed relative frequency $\hat{p}$ and the expected fraction p, say $|\hat{p}-p|/\sqrt{\dfrac{p(1-p)}{n}} > 3$, would indicate lack of normality.

7.6.2 Normal Probability Paper

Another way to check a postulated normal model is to graph the cumulative relative frequencies on a special paper called *probability paper*. More properly called *normal probability paper*, the scale of its vertical axis for the cumulative probability is stretched at low and high probabilities, so that the graph of $P[X \leqslant x]$ is transformed into a straight line, as shown in Figure 7.18.

With a random sample, we would expect the cumulative relative frequencies to mimic the behavior of cumulative probabilities. It has been suggested that at least 15 or 20 observations are required to provide a meaningful pattern.

The steps in constructing a normal probability plot are:

(*a*) Order the n observations from smallest to largest.

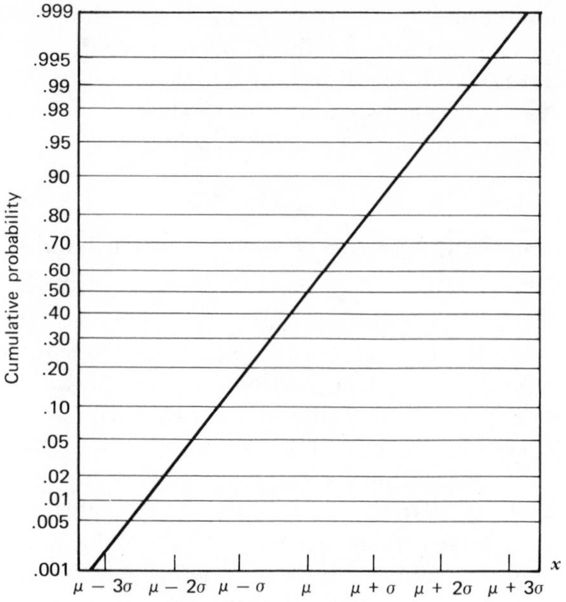

Figure 7.18 Normal probability paper showing a graph of $P[X \leqslant x]$ when X is distributed as $N(\mu, \sigma)$.

(*b*) Select a scale on the horizontal axis to accommodate all the observations.

(*c*) Plot the *modified* cumulative relative frequency $(i - \frac{1}{2})/n$ on the vertical scale against the value of the ith ordered observation on the horizontal scale.

(*d*) Examine the plot for departures from a straight-line pattern. Systematic departures indicate lack of normality.

Of course, we could use the usual cumulative relative frequency i/n here, but then the largest observation with a cumulative relative frequency of 1 would not fit in the graph. If the data set is large, the values can be plotted for only every fifth or tenth ordered observation together with the few extreme values. If the data are already grouped, the upper class boundaries and the corresponding cumulative relative frequencies can be graphed. The detailed discussion of frequency tables in Section 2.7 may be reviewed at this time.

Figure 7.19*a* illustrates a data set for which the assumption of a normal population is reasonable. The plot in Figure 7.19*b* indicates a serious departure from normality. The direction of departure is toward thicker tails than those of a normal distribution.

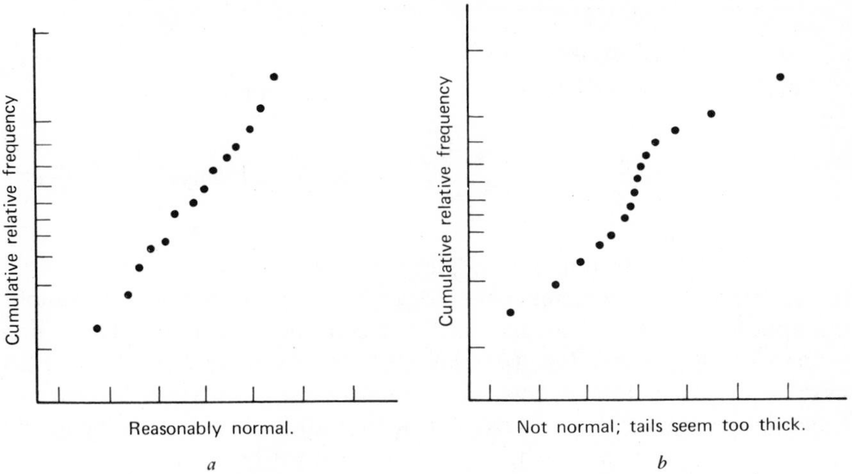

Figure 7.19 Plots on normal probability paper.

7.7 TRANSFORMING OBSERVATIONS TO ATTAIN NEAR NORMALITY

The versatility of the normal distribution as a model is greatly enhanced by the powerful technique of transforming the observations. Although histograms and dot diagrams often do not appear to be generated from an underlying bell-shaped curve, it is frequently possible to determine a transform of the original values that leads to a nearly symmetric graph. A valid application of many powerful techniques of statistical inference, especially those suited to small or moderate samples, requires that the population distribution be reasonably close to normal. When the sample measurements appear to have been taken from a population that departs drastically from normality, an appropriate conversion to a new variable may bring the distribution close to normal. Efficient techniques can then be safely applied to the converted data, whereas their application to the original data would have been questionable. Inferential methods requiring the assumption of normality are discussed in Chapter 8 and some subsequent chapters. The goal of our discussion here is to show how a transformation can improve the approximation to a normal distribution.

A transformation such as taking the square root of an observation can frequently be considered a change to a more natural measurement scale if the resulting graph has a simple bell-shaped form. There is no rule for determining the best transformation in a given situation. For any data set that does not have a symmetric histogram, we consider a variety of transformations.

Some useful transformations

make large values larger:	make large values smaller
x^3, x^2	$\sqrt{x}\,,\ \sqrt[4]{x}\,,\ \log_{10} x,\ \frac{1}{x}$

Recall that $\sqrt[4]{x}$ results from taking the square root of the square root. It is also important to remember that logarithms were widely used to simplify multiplication before the age of the electronic calculator. Their basic property is $\log_{10}(ab) = \log_{10}(a) + \log_{10}(b)$; here we use base 10, but any base will suffice. Fortunately, computers easily calculate and order the transformed values, so that several transformations in a list can be quickly tested. Note, however, that the observations must be positive if we intend to use $\sqrt{x}$, $\sqrt[4]{x}$, and $\log_{10}x$.

The selection of a good transformation is largely a matter of trial and error. If the data set contains a few large numbers that appear to be detached far to the right, $\sqrt{x}$, $\sqrt[4]{x}$, $\log_{10}x$ or negative powers that would pull these stragglers closer to the other data points should be considered.

Example 7.16 A forester randomly selects 49 plots in a large forest and determines the volume of timber measured in cords for each plot. The resulting data are given in Table 7.6 and the corresponding histogram appears in Figure 7.20*a*. The histogram exhibits a long tail to the right, so that it is reasonable to consider the transformations $\sqrt{x}$, $\sqrt[4]{x}$, $\log x$, and $1/x$. The most satisfactory result, obtained with

$$\text{Transformed observation} = \sqrt[4]{\text{Volume}}$$

is illustrated in Table 7.7 and in Figure 7.20*b*. The latter histogram is

TABLE 7.6
VOLUME OF TIMBER IN CORDS

39.3	14.8	6.3	.9	6.5
3.5	8.3	10.0	1.3	7.1
6.0	17.1	16.8	.7	7.9
2.7	26.2	24.3	17.7	3.2
7.4	6.6	5.2	8.3	5.9
3.5	8.3	44.8	8.3	13.4
19.4	19.0	14.1	1.9	12.0
19.7	10.3	3.4	16.7	4.3
1.0	7.6	28.3	26.2	31.7
8.7	18.9	3.4	10.0	

Courtesy of Professor Alan Ek.

TABLE 7.7

THE TRANSFORMED DATA $\sqrt[4]{\text{Volume}}$

2.50	1.96	1.58	.97	1.60
1.37	1.70	1.78	1.07	1.63
1.57	2.03	2.02	.91	1.68
1.29	2.26	2.22	2.05	1.34
1.64	1.60	1.51	1.70	1.56
1.37	1.70	2.59	1.70	1.91
2.07	2.10	1.93	1.17	1.86
2.11	1.79	1.36	2.02	1.44
1.00	1.66	2.31	2.26	2.37
1.72	2.09	1.36	1.78	

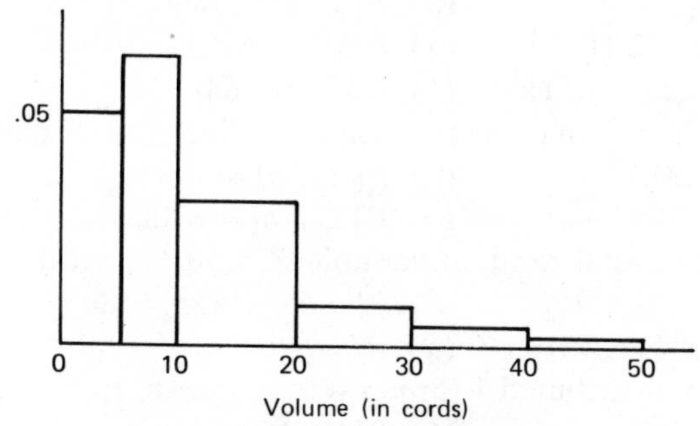

(a) Histogram of timber volume.

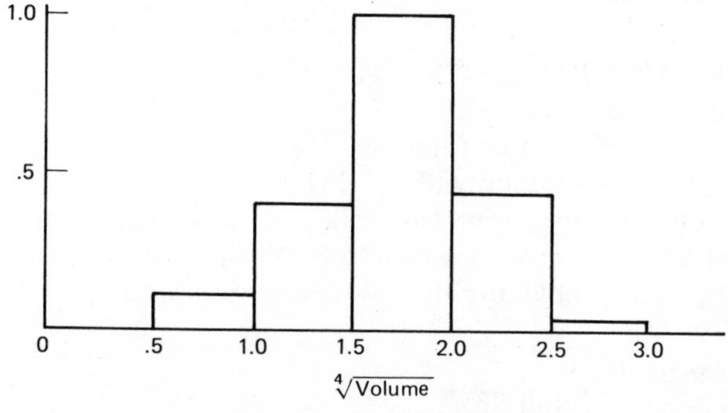

(b) Histogram of $\sqrt[4]{\text{volume}}$.

Figure 7.20 An illustration of the transformation technique.

relatively close to what would be expected from an underlying normal distribution. ■

The list of transformations could be extended to include all powers of x, say x^c. The question of selection then becomes somewhat more difficult. One method for selecting c is described in Box and Cox [1].

REFERENCES

1. Box, G. E. P., and Cox, D. R., "An analysis of transformations," *Journal of the Royal Statistical Society*, Series B, Vol. 26 (1964), 211–243.

EXERCISES

1. For a standard normal random variable Z, find:
   (*a*) $P[Z<.42]$ (*b*) $P[Z<-.42]$
   (*c*) $P[Z>1.69]$ (*d*) $P[Z>-1.69]$
   (*e*) $P[-1.2<Z<2.1]$ (*f*) $P[.05<Z<.8]$
   (*g*) $P[-1.62<Z<-.51]$ (*h*) $P[|Z|<1.64]$
2. If Z is a standard normal random variable, find b such that:
   (*a*) $P[Z<b]=.975$ (*b*) $P[Z<b]=.305$
   (*c*) $P[Z>b]=.025$ (*d*) $P[Z>b]=.8708$
3. For a standard normal random variable Z, find b such that:
   (*a*) $P[-b<Z<b]=.75$ (*b*) $P[-b<Z<b]=.80$
   (*c*) $P[-b<Z<b]=.99$ (*d*) $P[-b<Z<b]=.98$
4. If X is normally distributed with $\mu=100$ and $\sigma=8$, find:
   (*a*) $P[X<107]$ (*b*) $P[X<97]$
   (*c*) $P[X>110]$ (*d*) $P[X>90]$
   (*e*) $P[95<X<106]$ (*f*) $P[103<X<114]$
   (*g*) $P[88<X<100]$ (*h*) $P[60<X<108]$
5. If X is normally distributed with $\mu=100$ and $\sigma=8$, find b such that:
   (*a*) $P[X<b]=.975$ (*b*) $P[X<b]=.305$
   (*c*) $P[X>b]=.025$ (*d*) $P[X>b]=.8708$
6. Determine the quartiles of the:
   (*a*) Standard normal distribution.
   (*b*) Normal distribution with $\mu=220$ and $\sigma=25$.
7. Scores on a certain nationwide college entrance examination follow a normal distribution with a mean of 500 and a standard deviation of 100. Find the probability that a student will score:
   (*a*) Over 650.
   (*b*) Less than 250.
   (*c*) Between 325 and 675.

8. The error in measurements of blood sugar level by an instrument is normally distributed with a mean of .05 and a standard deviation of 1.5; that is, in repeated measurements, the distribution of the difference (Recorded level—true level) is $N(.05, 1.5)$.
   (*a*) What percentage of the measurements overestimate the true level?
   (*b*) Suppose that an error is regarded as serious when the recorded value differs from the true value by more than 2.8. What percentage of the measurements will be in serious error?
9. *Grading on a curve*: The scores on an examination are normally distributed with mean $\mu = 70$ and standard deviation $\sigma = 8$. Suppose that the instructor decides to assign letter grades according to the following scheme:

SCORES	GRADE
Less than $70 - 1.5\sigma$	F
$70 - 1.5\sigma$ to $70 - .5\sigma$	D
$70 - .5\sigma$ to $70 + .5\sigma$	C
$70 + .5\sigma$ to $70 + 1.5\sigma$	B
$70 + 1.5\sigma$ and above	A

   Find the percentage of students in each grade category.
   (NOTE: In practice, the sample quantities $\bar{x}$ and s, calculated from the scores of all students in a class, are used in place of μ and σ.)
10. An aptitude test administered to aircraft pilot trainees requires a series of operations to be performed in quick succession. Suppose that the time needed to complete the test is normally distributed with mean = 90 minutes and sd = 20 minutes.
   (*a*) To pass the test, a candidate must complete it within 80 minutes. What percentage of the candidates will pass the test?
   (*b*) If the top 5% of the candidates are to be given a certificate of commendation, how fast must a candidate complete the test to be eligible for a certificate?
11. A building materials store buys stone chips to use in the construction of decorative concrete blocks. The supplies arrive in mixed sizes and are then mechanically sorted according to size into three grades A, B, and C. It has already been determined that the size distribution of the stone chips can be approximated by a normal distribution with a mean of 135 and a standard deviation of 14. The grades correspond

to the following sizes:

GRADE	SIZE x
A	$150 \leqslant x \leqslant 160$
B	$115 \leqslant x < 150$
C	$x < 115$ or $x > 160$

Taking into account the cost of purchasing and handling and the selling price, suppose that the store's net profit per ton for each grade category is \$50 for grade A, \$25 for grade B, and $-\$5$ for grade C. Find the expected net profit from an incoming shipment of a ton of chips in mixed sizes.

*12. A dieter's daily lunch menu is a sandwich and a glass of milk. Suppose that the number of calories in the sandwich is normally distributed with mean = 200 and sd = 15 and that the number of calories in a glass of milk is normally distributed with mean = 80 and sd = 5.
   (*a*) What are the mean and the sd of the dieter's daily total caloric intake at lunch? What is the distribution?
   (*b*) If the dieter's New Year's resolution was not to consume more than 300 calories at lunch, in what percentage of days will the resolution be violated?

13. The bonding strength of a drop of plastic glue is normally distributed with mean = 100 pounds and sd = 8 pounds.
   (*a*) A broken plastic strip is repaired with a drop of this glue and then subjected to a test load of 98 pounds. What is the probability that the bonding will fail?
   (*b*) The strength test described in (*a*) is repeated 10 times. What is the probability of 6 or more failures?
   (*c*) A strength test with a load of 102 pounds is repeated 15 times. What is the probability of observing at most 5 failures?

14. Let X denote the number of successes in n Bernoulli trials with a success probability of p.
   (*a*) Find the exact probabilities of each of the following:
      (*i*) $X \leqslant 5$ when $n = 25, p = .4$.
      (*ii*) $11 \leqslant X \leqslant 17$ when $n = 20, p = .7$.
      (*iii*) $X \geqslant 11$ when $n = 16, p = .5$.
   (*b*) Use a normal approximation without the continuity correction for each situation in (*a*).
   (*c*) Use a normal approximation with the continuity correction for each situation in (*a*).

15. A basketball player makes 80% of her free throws. Given 120 chances to throw, what is the probability that she will make:
    (*a*) Less than 90 baskets?
    (*b*) 105 or more baskets?
    (*c*) exactly 100 baskets?
16. In a large midwestern university, 30% of the students live in apartments. If 200 students are randomly selected, find the probability that the number of them living in apartments will be between 50 and 75.
17. A particular program, say program A, previously drew 30% of the television audience. To determine if a recent rescheduling of the programs on a competing channel has adversely affected the audience of program A, a random sample of 400 viewers is to be asked whether or not they currently watch this program.
    (*a*) If the percentage of viewers watching program A has not changed, what is the probability that fewer than 105 out of a sample of 400 will be found to watch the program?
    (*b*) If the number of viewers of the program is actually found to be less than 105, will this strongly support the suspicion that the population percentage has dropped?
18. Suppose that 20% of the trees in a forest are infested with a certain type of parasite.
    (*a*) What is the probability that the number of trees having the parasite in a random sample of 300 will be between 49 and 71?
    (*b*) After sampling 300 trees, suppose that 72 trees are found to have the parasite. Does this contradict the hypothesis that the population proportion is 20%? Give reasons for your answer, using the terminologies of testing hypotheses.
19. *Normal approximation of the Poisson distribution*: When the parameter m of a Poisson distribution is large, a normal distribution with $\mu = m$ and $\sigma = \sqrt{m}$ provides a good approximation to the Poisson distribution. As in the binomial case, a correction for continuity improves the approximation.
    (*a*) Using the Poisson table, find the following probabilities:
        (*i*) $P[5 \leqslant X \leqslant 14]$ when $m = 9$.
        (*ii*) $P[X \leqslant 8]$ when $m = 9$.
    (*b*) Find the normal probabilities for (*i*) and (*ii*), using $\mu = 9$, and $\sigma = \sqrt{9}$, and a correction for continuity. Compare the results with the exact probabilities in (*a*).
20. Let the random variable X denote the number of days of a patient's stay in the intensive-care unit of a hospital after a particular operation. Considering the population of all patients, suppose that the

following probability distribution is assessed for X:

Number of days x	1	2	3
Probability $f(x)$	.3	.4	.3

(*a*) Calculate $\mu = E(X)$ and $\sigma = \text{sd}(X)$.

(*b*) Let the sample mean number of days of stay for a random sample of 2 patients be denoted by

$$\bar{X}_{(2)} = \frac{X_1 + X_2}{2}$$

Constructing the joint distribution of X_1 and X_2, obtain the probability distribution of $\bar{X}_{(2)}$. Verify that the mean and the standard deviation of this distribution are μ and $\sigma/\sqrt{2}$, respectively.

21. Referring to Exercise 20, let X_1, X_2, X_3, and X_4 denote the number of days of stay for a random sample of 4 patients, so that the sample mean is

$$\bar{X}_{(4)} = \frac{X_1 + X_2 + X_3 + X_4}{4}$$

Obtain the probability distribution of $\bar{X}_{(4)}$ and verify that its mean and its standard deviation are μ and $\sigma/\sqrt{4}$, respectively.
[*Hint*: Writing $Y_1 = \frac{X_1 + X_2}{2}$ and $Y_2 = \frac{X_3 + X_4}{2}$, gives us

$$\bar{X}_{(4)} = \frac{Y_1 + Y_2}{2}$$

Y_1 and Y_2 are independent, and each has the distribution obtained in Exercise 20*b*.]

22. Referring to Exercises 20 and 21, draw the probability histograms for the distributions of X, $\bar{X}_{(2)}$, and $\bar{X}_{(4)}$ on the same sheet of graph paper. Comment on their similarities to a normal distribution.

23. Suppose that the moisture content per pound of a dehydrated protein concentrate is normally distributed with a mean of 3.5 and a standard deviation of .5. A random sample of 16 specimens, each consisting of one pound of this concentrate, is to be tested. Letting $\bar{X}$ denote the sample mean of these measurements of moisture content:
(*a*) What is the distribution of $\bar{X}$? Is it the exact or an approximate distribution?
(*b*) What is the probability that:
(*i*) $\bar{X}$ will exceed 3.7?
(*ii*) $\bar{X}$ will be between 3.34 and 3.66?

24. *Quality Control*: A shoe factory owns a machine that cuts pieces from slabs of compressed rubber to be used as soles on a certain brand of men's shoe. The thickness measurements of these soles are normally distributed with the standard deviation $\sigma = .2$ millimeters. Occasionally, for some unforeseeable reason, the mean changes from its target setting of $\mu = 25$ millimeters. To be able to take timely corrective measures, such as readjusting the machine's setting, it is important to monitor product quality by measuring the thickness of a random sample of soles taken periodically from the machine's output. Suppose that the following plan is used to monitor the product quality. The thickness measurement for a random sample of 5 soles are observed, and the sample mean $\bar{X}$ is recorded. If $\bar{X} < 24.8$ or $\bar{X} > 25.2$, the machine is considered to be out of control. Production is then halted and the machine is readjusted.
   (*a*) When the true mean is $\mu = 25$ millimeters, what is the probability that a sample will indicate "out of control"?
   (*b*) Suppose that the true mean has changed to $\mu = 25.3$ millimeters. What is the probability that a sample will indicate "out of control"?

   (NOTE: In practical operations, *control charts* are plotted to show the $\bar{x}$ values at successive points of time to detect any drift in the process mean.)
25. A random sample of size 100 is taken from a population having a mean of 20 and a standard deviation of 5. The shape of the population distribution is unknown.
   (*a*) What can you say about the probability distribution of the sample mean $\bar{X}$?
   (*b*) Find the probability that $\bar{X}$ will exceed 20.75.
26. It has been found that with the standard dosage of a drug, the change in a person's uric-acid level has mean = 250 and standard deviation = 50. A new dosage of the drug is to be tried experimentally on 75 subjects, and the changes in their uric-acid levels are to be measured. The conjecture is that the new dosage will increase the population mean change of uric-acid level by 20 units but that it will not change the standard deviation.
   (*a*) Assuming that the conjecture is true, what is the probability that the sample mean for 75 subjects will be less than 260 or more than 280?
   (*b*) In addition to increasing the mean by 20 units, if the new dosage changes the standard deviation from 50 to 64, what is the probability that the sample mean will be less than 260 or more than 280?
27. The time required by workers to complete an assembly job has a

mean of 50 minutes and a standard deviation of 8 minutes. To spot check the workers' progress on a particular day, their supervisor intends to record the times 60 workers take to complete one assembly job apiece.

(*a*) What is the probability that the sample mean will be more than 52 minutes?

(*b*) If the sample mean is actually found to be 53 minutes, should the supervisor be concerned that a slow down is taking place in the assembly process? (Assume that the standard deviation did not change.)

28. Repeated measurements are made for the melting point of a new alloy. From the precision of the measuring device, it is known that the standard deviation of the measurements is 7 degrees. If the measurement is repeated 80 times, what is the probability that the sample mean will not deviate from the true melting point by more than 1.54 degrees?

CLASS PROJECTS

1. (*a*) Count the number of occupants X including the driver in each of 20 passing cars. Calculate the mean $\bar{x}$ of your sample.
   (*b*) Repeat (*a*) 10 times.
   (*c*) Collect the data sets of the individual car counts x from the entire class and construct a relative frequency histogram.
   (*d*) Collect the $\bar{x}$ values from the entire class (10 from each student) and construct a relative frequency histogram for $\bar{x}$ choosing appropriate class intervals.
   (*e*) Plot the two relative frequency histograms and comment on the closeness of their shapes to the normal distribution.
2. (*a*) Collect a sample of size 6 and compute $\bar{x}$, the sample median, s^2, and the sample range.
   (*b*) Repeat (*a*) 30 times.
   (*c*) Plot dot diagrams for the values of the four statistics in (*a*). These plots reflect the individual sampling distributions.
   (*d*) Compare the amount of variation in $\bar{X}$ and the median. Repeat for s and the sample range.

   In this exercise, you might record weekly soft-drink consumptions, sentence lengths or hours of sleep for different students. If a computer is available, you could use it to generate samples from several parent populations.

CHAPTER 8

Inferences About a Population

8.1 INTRODUCTION

Equipped with the idea of a probability distribution as a useful model for measurements and the associated concepts of sampling distributions, we are now ready to be introduced to the techniques of inference that have proved invaluable to data analyzers in modern times. When data are collected by sampling from a population, the most important objective of a statistical analysis is to draw inferences or generalities about that population from the partial information embodied in the sample data. Typically, our interest centers on learning about some numerical feature of the population, such as the proportion possessing a stated characteristic, the mean and the standard deviation of the population distribution, or some other numerical measures of center and variability. Any numerical feature of a probability distribution can be expressed in terms of the parameters that appear expressly in the formula for the distribution. In the context of statistical inference, it is usual to extend the use of the term parameter to include these functions of the basic parameters.

> Any numerical feature of a population distribution is called a *parameter*. *Statistical inference* deals with drawing generalizations about population parameters from an analysis of the sample data.

Before we collect a data set from which to draw inferences, we should consider the following points:

(*a*) Sample size and manner of sampling.
(*b*) Nature of inference desired.
(*c*) Strength and accuracy of the conclusions.

Although the question of sample size seems to be the first consideration, it is usually resolved only after a manner of sampling and a technique of inference have been selected. In this chapter, we introduce some inference techniques that are based on random samples and then discuss some methods for selecting the sample size.

The nature of the inference to be considered depends on the intent of the investigation. The two most important types of inferences are (*a*) *estimation of parameter*(*s*) and (*b*) *testing of statistical hypotheses*. Using either type of inference, the true value of a parameter is an unknown constant that can be correctly ascertained only by an exhaustive study of the population, if indeed that were possible. A realistic objective may be to obtain a guess or an estimate of the unknown true value or an interval of plausible values from the sample data and also to determine the accuracy of the procedure. This type of inference is called *estimation of parameters*. An alternative objective may be to decide whether or not the parameter lies in the specified range of values that corresponds to the investigator's conjecture. We have already introduced this type of inference, called *testing of statistical hypotheses*, in Chapter 6 in connection with testing proportions.

Example 8.1 To study the rate of growth of pine trees at an early stage, a nursery worker records 40 measurements of the heights of one-year-old red pine seedlings. This set of measurements appears in Table 8.1.

TABLE 8.1
HEIGHTS OF ONE YEAR OLD RED PINE SEEDLINGS MEASURED IN CENTIMETERS
(courtesy of Professor Alan Ek)

2.6	1.9	1.8	1.6	1.4	2.2	1.2	1.6
1.6	1.5	1.4	1.6	2.3	1.5	1.1	1.6
2.0	1.5	1.7	1.5	1.6	2.1	2.8	1.0
1.2	1.2	1.8	1.7	0.8	1.5	2.0	2.2
1.5	1.6	2.2	2.1	3.1	1.7	1.7	1.2

The methods outlined in Chapter 2 provide a means of describing these particular measurements. However, the purpose behind this investigation is to draw inferences not only about the heights of the pine seedlings included in the sample but also about the population of all possible one-year-old red pine seedlings. Idealizing the situation so that the observations represent a random sample from a probability distribution with a mean of μ, one goal of this study could be phrased "learning about μ." More specifically, the study could be conducted to:

(*a*) Estimate a single value for the unknown μ (*point estimation*).
(*b*) Determine an interval of plausible values for μ (*interval estimation*).
(*c*) Decide whether or not the mean height μ is 1.9 centimeters, which was found to be the mean from an extensive set of field trials conducted five years ago (*testing a hypothesis*). ■

Example 8.2 A government agency wishes to assess the prevailing rate of unemployment in a particular county. It is correctly felt that this assessment must be made quickly and effectively by sampling a small fraction of the labor force in the county and counting the number of persons currently unemployed. Suppose that a random sample of 500 persons is interviewed and that 41 are found to be unemployed.

The crucial stage is selecting a representative sample. Without pursuing this point here, we note as an example of difficulties that if a random sample is selected from a list of persons who filed income tax returns the previous year, newcomers to the county would not be included. Our current discussion concerns the nature of inference desired about the population parameter, or the proportion of the county's labor force that is currently unemployed. The inference required may be in the form of a point or an interval estimate of the value of this proportion. It may also be a test of the hypothesis that the unemployment rate in the county is not higher than the rate quoted in a federal report. ■

The central theme of this chapter constitutes a discussion of the three forms of statistical inference: (*a*) point estimation, (*b*) interval estimation, and (*c*) testing hypotheses. It is instructive to introduce the concept of point estimation before tackling the more difficult concepts associated with interval estimation. As we examine these three topics, it should become apparent that interval estimation, which provides an interval of plausible values, is ordinarily the most useful type of inference. We now explore these three areas of inference, to discover the essential concepts and reasoning that permit us to reach general conclusions on the basis of a sample.

8.2 POINT ESTIMATION OF A PARAMETER

The objective of point estimation is to produce a single number from the sample that is likely to be close to the unknown value of the parameter. For the purposes of general discussion, we use the symbol θ to denote a parameter that could typically be a proportion, mean, median, or some measure of variability. It is important to remember that θ has a constant numerical value, although its value is unknown to us. The available information is assumed to be in the form of a random sample $X_1, X_2, \ldots, X_n$ of size n taken from the population. We wish to formulate a function of the sample observations $X_1, X_2, \ldots, X_n$; that is, a statistic such that its value computed from the sample data would reflect the value of the population parameter as closely as possible.

A statistic intended for estimating a parameter θ is called a *point estimator* or, more simply, an *estimator* of θ, and it is commonly denoted by $\hat{\theta}$ or sometimes T. An estimator $\hat{\theta}$ is purely a function of the sample observations, and its value can be readily computed once the sample is observed. This numerical value is called a *point estimate* or an *estimate* of θ from the particular sample. Different realizations of the random sample usually result in different values of the statistic $\hat{\theta}$. Because $\hat{\theta}$ depends on the random variables that generate the sample, it too is a random variable with its own sampling distribution.

> An *estimator* $\hat{\theta}$ is a function of the sample observations that is used to estimate the unknown value of a parameter θ. The estimator $\hat{\theta}$ is a random variable with a probability distribution. When a random sample becomes available from the population and $\hat{\theta}$ is computed from the data set, the numerical value obtained is called an *estimate* of θ from the particular sample.

How do we choose an appropriate estimator? A few desirable properties that can help us make this choice may be mentioned here. First the distribution of an estimator should be centered in some sense at the value of the parameter to be estimated. Because expected value is a measure of the center of a distribution, a reasonable requirement for an estimator $\hat{\theta}$ may be $E(\hat{\theta})=\theta$. An estimator with this property is called an *unbiased estimator*.

> An estimator $\hat{\theta}$ is *unbiased* for a parameter θ if $E(\hat{\theta})=\theta$, whatever the true value of θ. If this property does not hold, $\hat{\theta}$ is a *biased* estimator.

The property of unbiasedness implies that use of the estimator $\hat{\theta}$ in repeated samplings will not result in a systematic overestimation or underestimation of θ. Although a desirable feature, unbiasedness does not indicate how close the estimator is likely to be to the parameter. Figure 8.1 illustrates a situation in which both the estimators T_1 and T_2 are unbiased but T_1 is more tightly distributed around θ than T_2. The estimator T_1 is likely to provide a more accurate estimate than T_2.

For a measure of the accuracy of estimation, we may consider the standard deviation of the sampling distribution of the estimator. When the distribution is nearly normal, this is indicative of its variability from θ and

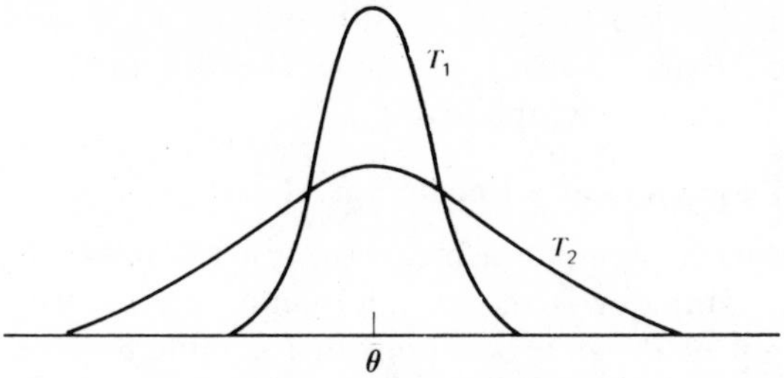

Figure 8.1 Distributions of two estimators T_1 and T_2, both of which are unbiased for θ.

the degree with which probability is concentrated. The smaller the standard deviation, the better the performance of the estimator. If, however, the sampling distribution is highly skewed or otherwise far from normal, another measure of the error in estimation may be preferable.

Based on the above discussion one method for selecting an estimator could be

> Select the unbiased estimator of θ that has the smallest variance, whatever the true value of θ. If one exists, it is called the *minimum variance unbiased estimator* of θ.

Unbiasedness should not be considered an overly important criterion in judging an estimator insofar as expectation is only one of several measures of center. There are examples of biased estimators that have even higher concentrations of probability near the parameter than the minimum variance unbiased estimator. However, for the problems within the scope of this text, most of the natural estimators suggested by common sense are unbiased.

Once an estimator has been chosen, its computation from a data set yields a single number—a point estimate. Without an assessment of accuracy, a single number quoted as an estimate may not serve a useful purpose. We must indicate the extent of variability in the distribution of the estimator. This is customarily done by stating the standard deviation of the estimator, which is also called the *standard error*.

> The standard deviation of the estimator $\hat{\theta}$ is called its *standard error* and is designated S.E. $(\hat{\theta})$.

Rather than develop the general situation in more detail here, we now amplify these ideas in terms of some specific problems of parameter estimation that are widely applicable.

8.2.1 Point Estimation of a Population Mean

When estimating a population mean from a random sample, perhaps the most intuitive estimator is the sample mean $\bar{X}$. For instance, to estimate the mean height of the pine seedlings in Example 8.1, we would naturally compute the mean of the sample measurements. Let us study the properties of this estimator that could lend support to this intuitive choice. The parameter θ in the present example is the population mean μ, and the estimator $\hat{\theta}$ is the sample mean $\bar{X}$. Repeating the results from Section 7.5

$$E(\bar{X}) = \mu, \quad \text{sd}(\bar{X}) = \frac{\sigma}{\sqrt{n}}$$

With large n, $\bar{X}$ is nearly normally distributed with mean μ and sd $\sigma/\sqrt{n}$.

The first two results show that $\bar{X}$ is an unbiased estimator of μ and that its standard error is $\sigma/\sqrt{n}$, where σ is the population standard deviation. To understand how closely $\bar{X}$ is expected to estimate μ, we now examine the third result, which is depicted in Figure 8.2. With a probability of .954, the estimator $\bar{X}$ will be within a distance $2\sigma/\sqrt{n}$ from the true parameter value μ. Prior to sampling we can make the statement: "With the high probability .954, the error of estimation $|\bar{X} - \mu|$ will not exceed the amount $2\sigma/\sqrt{n}$." Stemming from the approximate normal distribution of the estimator $\bar{X}$, the standard error $\sigma/\sqrt{n}$ can therefore be interpreted as indicative that we can be about 95.4% confident of controlling the error of estimation within ± 2(S.E.). Similarly, the probability is .997 that the error

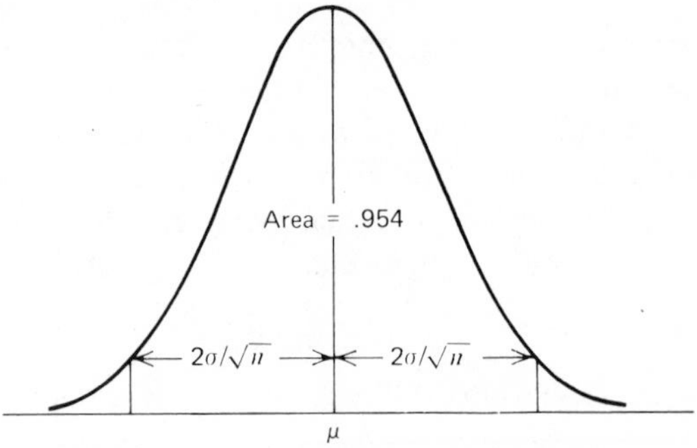

Figure 8.2 Approximate normal distribution of $\bar{X}$.

can be controlled within ± 3(S.E.). These probability statements can be rephrased by saying that when we are estimating μ by $\overline{X}$, the 95.4% error bound is $2\sigma/\sqrt{n}$ and the 99.7% error bound is $3\sigma/\sqrt{n}$.

A minor difficulty in computing the standard error of $\overline{X}$ remains, because this calculation involves the unknown population σ. Lacking a knowledge of σ, we can estimate the population standard deviation by the sample standard deviation

$$s=\sqrt{\frac{\sum_{i=1}^{n}(X_i-\overline{X})^2}{n-1}}$$

When n is large, the effect of approximating $\sigma/\sqrt{n}$ with $s/\sqrt{n}$ can be neglected, and the information provided by the estimate is still useful in small samples. The statistic $s/\sqrt{n}$ is an estimator of the standard error, and its multiples can be given as approximate probabilistic error bounds. We can now summarize in terms of the 95.4% error bound using 2(S.E.).

Point Estimation of the Mean

Parameter: population mean μ
Data: $X_1,\ldots,X_n$ (a random sample of size n)
Estimator: $\overline{X}$ (sample mean)
S.E. $(\overline{X}) = \sigma/\sqrt{n}$, estimated S.E. $(\overline{X}) = s/\sqrt{n}$
For large n, an approximate 95.4% error bound is $\pm 2s/\sqrt{n}$.

Example 8.3 From the data of Example 8.1, consisting of 40 measurements of the heights of one-year-old red pine seedlings, give a point estimate of the population mean height and state a 95.4% error bound.

The sample mean and the standard deviation computed from the 40 measurements in Table 8.1 are

$$\overline{x}=\frac{\sum x_i}{40}=1.715$$

$$s=\sqrt{\frac{\sum(x_i-\overline{x})^2}{39}}=\sqrt{.2254}=.475$$

A point estimate of the population mean height is $\bar{x}=1.715$ centimeters. Also

$$\text{Estimated S.E.}=\frac{s}{\sqrt{40}}=\frac{.475}{6.325}=.075$$

An approximate 95.4% error bound is then

$$\frac{2s}{\sqrt{40}}=.15 \text{ centimeters}$$

■

Caution

(*a*) Standard error should not be interpreted as the "typical" error in a problem of estimation as the word "standard" may suggest. For instance, when S.E. $(\bar{X})=.3$, we should not think that the error $(\bar{X}-\mu)$ is likely to be .3 but rather that, prior to observing the data, the probability is approximately .954 that the error will be within $\pm 2(\text{S.E.})=\pm .6$. The values that are covered when moving $\pm 2(\text{S.E.})$ from $\bar{X}$ may then be considered reasonable for the unknown quantity μ.

(*b*) An estimate and its variability are often reported in either of the forms: estimate $\pm$S.E. or estimate ± 2(S.E.). In reporting a numerical result such as 53.4 ± 4.6, we must specify whether 4.6 represents S.E., 2(S.E.), or some other multiple of the standard error.

At this point you may be asking the question: Is there a better estimator to estimate μ than old reliable $\bar{X}$? Example 8.4 helps motivate our answer.

Example 8.4 An engineer wishes to estimate the mean yield of a chemical process based on the yield measurements X_1, X_2, X_3 from three runs of an experiment. Consider the following two estimators of the mean yield μ:

$$T_1=\frac{X_1+X_2+X_3}{3} \quad \text{(the sample mean)}$$

$$T_2=\frac{X_1+2X_2+X_3}{4} \quad \text{(a weighted average of the observations)}$$

Which estimator should be preferred?

First we examine the property of unbiasedness. T_1 is simply the sample mean $\bar{X}$, which we know is unbiased.

$$E(T_2)=\tfrac{1}{4}(\mu+2\mu+\mu)=\mu$$

so that T_2 is also an unbiased estimator of μ. Denoting the population standard deviation by σ, we have

$$\text{Var}(T_1) = \frac{\sigma^2}{n} = \frac{\sigma^2}{3}$$

$$\text{Var}(T_2) = \frac{[\text{Var}(X_1) + \text{Var}(2X_2) + \text{Var}(X_3)]}{16}$$

$$= \frac{[\sigma^2 + 4\sigma^2 + \sigma^2]}{16}$$

$$= \frac{3\sigma^2}{8}$$

Because $\frac{1}{3} < \frac{3}{8}$, we have $\text{Var}(T_1) < \text{Var}(T_2)$ or $\text{S.E.}(T_1) < \text{S.E.}(T_2)$. Hence, $T_1 = \overline{X}$ is a better estimator of μ than T_2. ■

The observation that $\overline{X}$ has a smaller variance than a competing unbiased estimator is not accidental. In the context of Example 8.4, $\overline{X}$ has a smaller variance than any other unbiased weighted average of the observations. Moreover, $\overline{X}$ possesses the minimum variance property among all unbiased estimators under a great variety of population models.

8.2.2 Estimation of a Binomial Proportion

We now consider the type of problem that is illustrated in Example 8.2, where the parameter is the proportion p of a population having a specific characteristic. When n elements are randomly sampled from the population, the data will consist of the count X of the number of sampled elements possessing the characteristic. Common sense suggests the sample proportion

$$\hat{p} = \frac{X}{n}$$

as an estimator. When the sample size n is only a small fraction of the size of the population, as is usually the case, observations on n elements can be regarded as n independent Bernoulli trials with a success probability of p. As for the properties of this estimator, first we note that the sample count X has the binomial distribution $b(n,p)$, with mean np and variance npq where $q = (1 - p)$. Consequently

$$E(\hat{p}) = \frac{1}{n}E(X) = \frac{np}{n} = p$$

$$\text{Var}(\hat{p}) = \frac{1}{n^2}\text{Var}(X) = \frac{npq}{n^2} = \frac{pq}{n}$$

The first result shows that $\hat{p}$ is an unbiased estimator of p. In fact, $\hat{p}$ has a variance that is smaller than the variance of any other unbiased estimator. Second, the standard error of this estimator is given by

$$\text{S.E.}(\hat{p}) = \sqrt{\frac{pq}{n}}$$

and the estimated standard error can be obtained by substituting the sample estimate $\hat{p}$ for p and $\hat{q} = 1 - \hat{p}$ for q in the formula, or

$$\text{Estimated S.E.}(\hat{p}) = \sqrt{\frac{\hat{p}\hat{q}}{n}}$$

As we remarked in Section 7.5, when n is large, $\hat{p} = X/n$ is approximately normally distributed with mean p and standard deviation $\sqrt{pq/n}$. The normal approximation assures that prior to sampling the probability is approximately .954 that the error of estimation $|\hat{p} - p|$ will be less than $2 \times$(estimated S.E.).

Point Estimation of the Binomial Parameter

Parameter: Population proportion p

Data: X = Number having the characteristic out of a random sample of size n

Estimator: $\hat{p} = \dfrac{X}{n}$

$$\text{S.E.}(\hat{p}) = \sqrt{\frac{pq}{n}} \quad \text{and} \quad \text{estimated S.E.}(\hat{p}) = \sqrt{\frac{\hat{p}\hat{q}}{n}}$$

For large n, an approximate 95.4% error bound is $\pm 2\sqrt{\hat{p}\hat{q}/n}$.

Example 8.5 A large mail-order club that offers monthly specials wishes to test a possible special item. A trial mailing is sent to a random sample of 250 members selected from the list of over 9000 subscribers. Based on this sample mailing, 70 of the members decide to purchase the item. Give a point estimate of the proportion of club members that could be expected to purchase the item and attach a 95.4% error bound.

The number in the sample represents only a small fraction of the total membership, so that the count can be treated as if it were a binomial count.

Here $n = 250$ and $x = 70$, so that the estimate of the proportion of the population that would purchase the item is

$$\hat{p} = \frac{70}{250} = .28$$

$$\text{Estimated S.E.}(\hat{p}) = \sqrt{\frac{\hat{p}\hat{q}}{n}} = \sqrt{\frac{.28 \times .72}{250}} = .028$$

The estimated proportion is $\hat{p} = .28$, with a 95.4% error bound of $\pm .06$. ■

8.2.3 Point Estimation of Other Parameters

Mention of the point estimation of the population variance should be made here. We state without proof that the sample variance

$$s^2 = \frac{1}{n-1} \sum_{i=1}^{n} \left(X_i - \bar{X}\right)^2$$

satisfies $E(s^2) = \sigma^2$, whatever the value of σ^2, so that the sample variance s^2 is an unbiased estimator of σ^2. It is also true that $E(s) < \sigma$, so that the standard deviation is biased, but in large samples the bias is negligible and we continue to use s. The standard deviation of the point estimator s^2 is a little complicated. Moreover, because the distribution of s^2 is asymmetric, symmetric bounds using the standard error may be misleading as a measure of concentration of probability.

To close our discussion of point estimation we record that the population median or other percentiles can be estimated by using the corresponding sample percentiles.

8.3 ESTIMATION BY CONFIDENCE INTERVALS

Now that we are familiar to some degree with the formulation and structure for estimation procedures, we are ready to become acquainted with the technique called *confidence interval estimation*, which most statisticians consider to be the chief method of presenting evidence obtained from a sample. A point estimator calculated from the sample data provides a single number as an estimate of the parameter. This single number lies in the forefront even though a statement of accuracy in terms of the standard error is attached to it. An alternative approach to estimation is to extend the concept of error bound to produce an interval of values that is likely to contain the true value of the parameter. This is the concept underlying estimation by confidence intervals.

Formally, let $X_1, \ldots, X_n$ be a random sample and θ be an unknown population parameter. A confidence interval for θ is an interval (L, U) computed from the sample observations $X_1, \ldots, X_n$, such that prior to sampling, it includes the unknown true value of θ with a specified high probability. This probability, denoted by $1-\alpha$, is typically taken as .90, .95, or .99.

Let $(1-\alpha)$ be a specified high probability and L and U be functions of $X_1, \ldots, X_n$, such that

$$P[L < \theta < U] = 1 - \alpha$$

Then the interval (L, U) is called a $100(1-\alpha)\%$ *confidence interval* for the parameter, and $(1-\alpha)$ is called the *level of confidence* associated with the interval.

To clarify these concepts, we construct a confidence interval for a population mean μ when the sample size n is large and the standard deviation σ is known. The latter assumption is artificial, but it facilitates understanding the basic procedure. A more realistic formulation where σ is also unknown will be treated shortly.

A probability statement about the sample mean $\overline{X}$ based on the normal distribution provides the cornerstone for the development of confidence intervals. According to the central limit theorem, the distribution of $\overline{X}$ can be closely approximated by the $N(\mu, \sigma/\sqrt{n})$, where $\sigma/\sqrt{n}$ is a known number. This is a good approximation for large samples from nonnormal populations, but it holds exactly for all n, large or small, when the population distribution is normal. Consequently, our probability statements hold exactly for normal populations and as large sample approximations otherwise.

The normal table shows that the probability is .95 that a normal random variable will lie within 1.96 standard deviations of its mean. For $\overline{X}$,

$$P\left[\mu - 1.96\frac{\sigma}{\sqrt{n}} < \overline{X} < \mu + 1.96\frac{\sigma}{\sqrt{n}}\right] = .95$$

The event $[\mu - 1.96\sigma/\sqrt{n} < \overline{X}]$ is equivalent to $[\mu < \overline{X} + 1.96\sigma/\sqrt{n}]$, as we can see by adding the constant $1.96\sigma/\sqrt{n}$ to both sides of the first inequality. Also, because $[\overline{X} < \mu + 1.96\,\sigma/\sqrt{n}]$ is equivalent to $[\overline{X} - 1.96\sigma/\sqrt{n} < \mu]$, this probability statement can be alternatively expressed

$$P\left[\overline{X} - 1.96\frac{\sigma}{\sqrt{n}} < \mu < \overline{X} + 1.96\frac{\sigma}{\sqrt{n}}\right] = .95$$

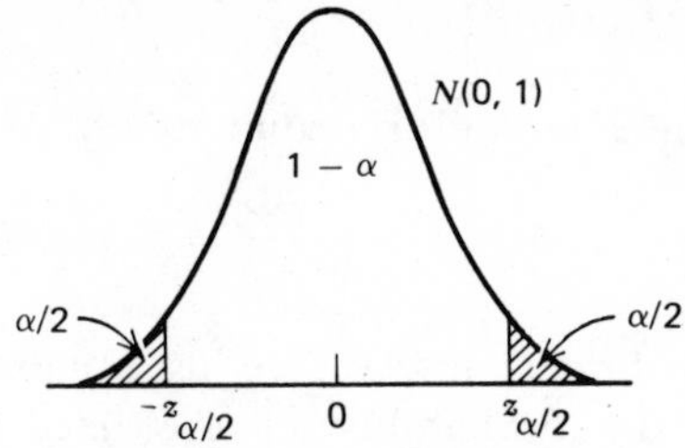

Figure 8.3 The notation $z_{\alpha/2}$.

If we now identify

$$L=\bar{X}-1.96\frac{\sigma}{\sqrt{n}} \quad \text{and} \quad U=\bar{X}+1.96\frac{\sigma}{\sqrt{n}}$$

the latter probability statement implies that, prior to sampling, the random interval (L, U) will include the parameter μ with a probability of .95. Because σ is assumed to be known, L and U can be computed as soon as the sample data are available.

More generally, when n is large and σ known, a $100(1-\alpha)\%$ confidence interval for μ is given by

$$\left(\bar{X}-z_{\alpha/2}\frac{\sigma}{\sqrt{n}},\quad \bar{X}+z_{\alpha/2}\frac{\sigma}{\sqrt{n}}\right)$$

where $z_{\alpha/2}$ denotes the upper $\alpha/2$ point of the standard normal distribution (i.e., the area to the right of $z_{\alpha/2}$ is $\alpha/2$, as shown in Figure 8.3). A few values of $z_{\alpha/2}$ obtained from the normal table appear in Table 8.2 for easy reference.

TABLE 8.2
VALUES OF $z_{\alpha/2}$

$1-\alpha$	.80	.85	.90	.95	.99
$z_{\alpha/2}$	1.28	1.44	1.64	1.96	2.58

Example 8.6 Given a random sample of 100 observations from a population for which μ is unknown and $\sigma=8$ suppose that the sample mean is found to be $\bar{x}=42.7$. Provide a 95% confidence interval for μ.

Here the sample size $n=100$ is large, so that the normal approximation to the distribution of $\bar{X}$ is justified. Using the observed value $\bar{x}=42.7$, a 95% confidence interval for μ becomes

$$\left(42.7-1.96\frac{8}{\sqrt{100}},\quad 42.7+1.96\frac{8}{\sqrt{100}}\right)=(41.1,\quad 44.3)$$

The result should be presented as

"A 95% confidence interval for μ computed from the observed sample is (41.1, 44.3)."

■

Referring to the confidence interval we just obtained in Example 8.6, we must *not* speak of the probability of the fixed interval (41.1, 44.3) covering the true mean μ: we must

$$\textit{never}\text{ write } P[41.1<\mu<44.3]=.95$$

because this expression contains no random variable within the brackets. The particular interval in Example 8.6 either does or does not cover μ, and we will never know which is the case.

To better understand the meaning of a confidence statement, we perform repeated samplings from a normal distribution with $\mu=100$ and $\sigma=10$. Ten samples of size 7 are selected, and a 95% confidence interval $\bar{x}\pm 1.96\times 10/\sqrt{7}$ is computed from each. These results are illustrated in Figure 8.4, where each vertical line segment represents one confidence interval. The midpoint of a line is the observed value of $\bar{X}$ for that particular sample. Also note that *all* of the intervals are of the same length $2\times 1.96\sigma/\sqrt{n}=14.8$. Of the ten intervals shown, nine cover the true value

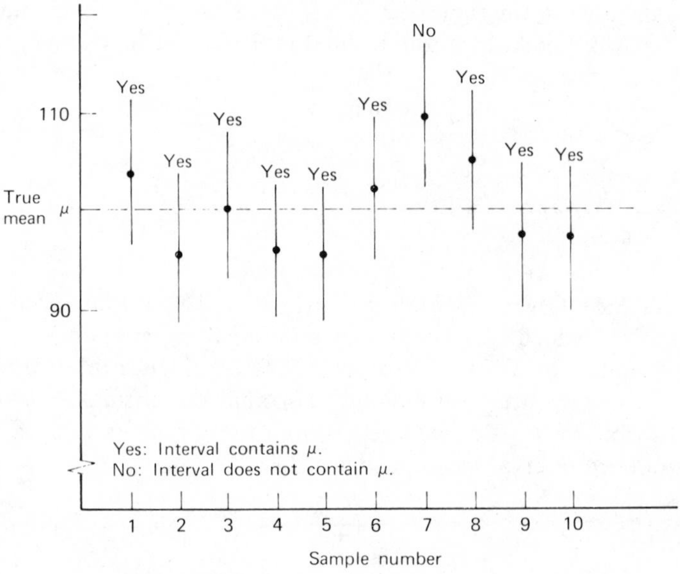

Figure 8.4 Interpretation of the confidence interval for μ.

of μ. This is not surprising, because the specified probability .95 represents the long-run relative frequency of these intervals crossing the dotted line through μ.

Because confidence interval statements are the most useful way to communicate information obtained from a sample, certain aspects of their formulation merit special emphasis to avoid any possible misunderstanding concerning their basic nature and properties. Stated in terms of a 95% confidence interval for μ, these are:

(*a*) A confidence interval $(\overline{X} - 1.96\sigma/\sqrt{n},\ \overline{X} + 1.96\sigma/\sqrt{n})$ is a random interval that attempts to cover the true values of the parameter μ.

(*b*) The probability

$$P\left[\overline{X} - 1.96\frac{\sigma}{\sqrt{n}} < \mu < \overline{X} + 1.96\frac{\sigma}{\sqrt{n}}\right] = .95$$

interpreted as the long-run relative frequency over many repetitions of sampling, asserts that about 95% of the intervals will cover μ.

(*c*) Once $\bar{x}$ is calculated from an observed sample, the interval $(\bar{x} - 1.96\sigma/\sqrt{n},\ \bar{x} + 1.96\sigma/\sqrt{n})$, which is a realization of the random interval, is presented as a 95% confidence interval for μ. Having determined a numerical interval, it is no longer sensible to speak about the probability of its covering a fixed quantity μ.

At this point one might protest "I am not really interested in repetitions of the same experiment." But if the confidence estimation techniques presented in this text are mastered and followed each time a problem of interval estimation arises, then over a lifetime approximately 95% of the intervals will correctly cover the true parameter. Of course, this is contingent on the validity of the assumptions underlying the techniques.

Finally, we do not view the center of a confidence interval as the best point estimate of the parameter. The philosophy of interval estimation does not attach preferential treatment to different points in a confidence interval but rather is directed toward determining a range of plausible values. Due to the symmetry of the approximate sampling distribution of $\overline{X}$, it so happens that the center of a confidence interval for μ is $\overline{X}$, which is also the best point estimator of μ. However, when a sampling distribution is not symmetric, the midpoint of a confidence interval need not be the best point estimate of the parameter, as the example of the estimation of σ^2 later in Section 8.7 will show.

8.3.1 Large Sample Confidence Interval for μ When σ Is Unknown

Having established the basic concepts underlying confidence interval statements, we now turn to the more realistic situation for which the population standard deviation σ is unknown. For large samples, the probability statement

$$P\left[\bar{X} - z_{\alpha/2}\frac{\sigma}{\sqrt{n}} < \mu < \bar{X} + z_{\alpha/2}\frac{\sigma}{\sqrt{n}}\right] = 1 - \alpha$$

is still correct, but because σ is unknown, the interval cannot be computed from the sample data and therefore cannot serve as a confidence interval. However, because n is large, replacing σ with its estimator s does not appreciably affect the probability statement. Summarizing:

Large sample confidence interval for μ

When n is large and the population σ is unknown, a $100(1-\alpha)\%$ confidence interval for μ is given by

$$\left(\bar{X} - z_{\alpha/2}\frac{s}{\sqrt{n}},\quad \bar{X} + z_{\alpha/2}\frac{s}{\sqrt{n}}\right)$$

where s is the sample standard deviation. No assumptions need be made regarding the form of the population other than finiteness of the variance σ^2.

Example 8.7 To estimate the average weekly income of restaurant waiters and waitresses in a large city, an investigator collects weekly income data from a random sample of 75 restaurant workers. The mean and the standard deviation are found to be \$127 and \$15, respectively. Compute (*a*) 90% and (*b*) 80% confidence intervals for the mean weekly income.

The sample size $n = 75$ is large, so that a normal approximation for the distribution of the sample mean $\bar{X}$ is appropriate. From the sample data, we know that

$$\bar{x} = \$127 \quad \text{and} \quad s = \$15$$

(*a*) With $1 - \alpha = .90$ we have $\alpha/2 = .05$, and $z_{\alpha/2} = 1.64$

$$1.64\frac{s}{\sqrt{n}} = \frac{1.64 \times 15}{\sqrt{75}} = 2.84$$

Hence, a 90% confidence interval for the population mean μ is

$$\left(\bar{x} - 1.64\frac{s}{\sqrt{n}},\quad \bar{x} + 1.64\frac{s}{\sqrt{n}}\right) = (127 - 2.84,\quad 127 + 2.84)$$

or approximately (124, 130)

This means that the investigator is 90% confident the mean income μ is in the interval of \$124 to \$130; that is, 90% of the time random samples of 75 waiters and waitresses produce intervals $\bar{x} \pm 1.64 s/\sqrt{75}$ that contain μ.

(*b*) With $1-\alpha=.80$, we have $\alpha/2=.10$, and $z_{.10}=1.28$

$$z_{.10}\frac{s}{\sqrt{n}}=\frac{1.28\times 15}{\sqrt{75}}=2.22$$

Hence, an 80% confidence interval for μ is

$$(127-2.22, \quad 127+2.22)=(125, 129)$$

Comparing the two results, the 80% confidence interval is shorter than the 90% interval. A shorter interval seems to give a more precise location for μ but suffers from a lower long-run frequency of being correct. ■

8.3.2 Confidence Intervals for Proportions

Arguments similar to those given above for the mean produce a large-sample confidence interval for the binomial parameter p. The interval is based on the estimator $\hat{p}$, the sample fraction of successes. According to our earlier discussion in Section 7.5, the estimator $\hat{p}$ is approximately normal with a mean of p and a standard deviation of $\sqrt{p(1-p)/n}$ when n is large and p is not too near 0 or 1. A slight complication is present here, because the unknown parameter p is present in both the mean and the standard deviation. This is easily overcome by estimating the standard deviation using $\sqrt{\hat{p}(1-\hat{p})/n}$. Again the effect of employing an estimated standard deviation in the confidence interval is negligible when n is large.

Large sample confidence interval for p

For large n, a $100(1-\alpha)\%$ confidence interval for p is given by

$$\left(\hat{p}-z_{\alpha/2}\sqrt{\frac{\hat{p}(1-\hat{p})}{n}}, \quad \hat{p}+z_{\alpha/2}\sqrt{\frac{\hat{p}(1-\hat{p})}{n}}\right)$$

Example 8.8 Using the data originally given in Example 8.2, where 41 people in a random sample of 500 persons from a county labor force were found to be unemployed, compute a 95% confidence interval for the rate of unemployment in the county.

The sample size $n=500$ is quite large, so that a normal approximation to the distribution of the sample proportion $\hat{p}$ is justified. Since $(1-\alpha)=.95$, we have $\alpha/2=.025$ and $z_{.025}=1.96$. The observed sample proportion is

$\hat{p}=41/500=.082$, so that

$$z_{.025}\sqrt{\frac{\hat{p}(1-\hat{p})}{n}}=1.96\sqrt{\frac{.082\times.918}{500}}=1.96\times.012=.024$$

Therefore, a 95% confidence interval for the rate of unemployment in the county is $.082\pm.024=(.058,.106)$, or (5.8%, 10.6%) in percentages.

This means that we can be 95% confident unemployment is between 5.8% and 10.6%, because our procedure will produce true statements 95% of the time. ■

8.4 CONFIDENCE INTERVAL FOR μ BASED ON SMALL SAMPLES (t DISTRIBUTION)

In Section 8.3, we learned to construct confidence intervals for a population mean μ based on the fact that the sample mean $\bar{X}$ is approximately distributed as $N(\mu, \sigma/\sqrt{n})$ when n is large. We did not need to make any additional assumptions about the form of the population distribution due to the versatility of the central limit theorem. But in many fields, especially those in which experimentation is costly or time consuming, inferences may have to be drawn from samples of limited size. When n is small (as a working rule, $n<30$ is usually regarded as small), we must be careful in invoking the normal distribution for $\bar{X}$. The distribution of $\bar{X}$ depends considerably on the shape of the population distribution. Therefore, no single inferential procedure can be expected to work for all kinds of population distributions. Here we present a method for constructing confidence intervals for μ *when it is reasonable to assume that the population distribution is normal.* It is important to remember that this procedure can be seriously misleading when n is small and the population distribution differs radically from a normal distribution.

If $X_1,\ldots,X_n$ is a random sample from a normal population $N(\mu,\sigma)$, the sample mean $\bar{X}$ is exactly distributed as $N(\mu, \sigma/\sqrt{n})$. If σ happens to be known, a $100(1-\alpha)\%$ confidence interval for μ is given by $\bar{X}\pm z_{\alpha/2}\sigma/\sqrt{n}$. This interval is obtained from the standard normal variable

$$Z=\frac{(\bar{X}-\mu)}{\sigma/\sqrt{n}}$$

for which $P[-z_{\alpha/2}<Z<z_{\alpha/2}]=1-\alpha$.

When σ is unknown, as is typically the case, an intuitive approach is to replace σ with the sample standard deviation s and hence to consider the ratio

$$t = \frac{(\overline{X} - \mu)}{s/\sqrt{n}}$$

Although replacing σ with s does not appreciably change the distribution in large samples, it does make a substantial difference if the sample is small. The new notation t is required because the random variable in the denominator increases the variance to a value greater than 1, so that the ratio is no longer standardized.

The distribution of the ratio t is known in statistical literature as "Student's t distribution." This distribution was first studied by a British chemist W. S. Gosset, who published his work in 1908 under the pseudoname "Student." By studying the behavior of a great many observed values of t under repeated random samplings, Gosset conjectured a mathematical form of the distribution that was later verified analytically.

Student's t distribution

If $X_1, \ldots, X_n$ is a random sample from a normal population $N(\mu, \sigma)$ and

$$\overline{X} = \frac{1}{n} \sum_{i=1}^{n} X_i \quad \text{and} \quad s = \sqrt{\frac{\sum_{i=1}^{n} (X_i - \overline{X})^2}{n-1}}$$

then the distribution of

$$t = \frac{(\overline{X} - \mu)}{s/\sqrt{n}}$$

is called *Student's t distribution with n − 1 degrees of freedom.*

The qualification "with $(n-1)$ degrees of freedom" is necessary, because with each different sample size or value of $(n-1)$ there is a different t distribution. Recall that for a sample size n the estimator s^2 of σ^2 is based on $(n-1)$ degrees of freedom (c.f. Section 2.5). The t distributions are symmetric around 0 but have tails that are thicker (i.e., more spread out) than the $N(0,1)$ distribution. Moreover, with increasing degrees of freedom the t distributions approach the $N(0,1)$ distribution. This conforms to our previous remark that for large n the ratio $\frac{(\bar{X}-\mu)}{s/\sqrt{n}}$ is approximately standard normal. The density curve for t with 5 degrees of freedom is plotted in Figure 8.5 along with the $N(0,1)$ curve.

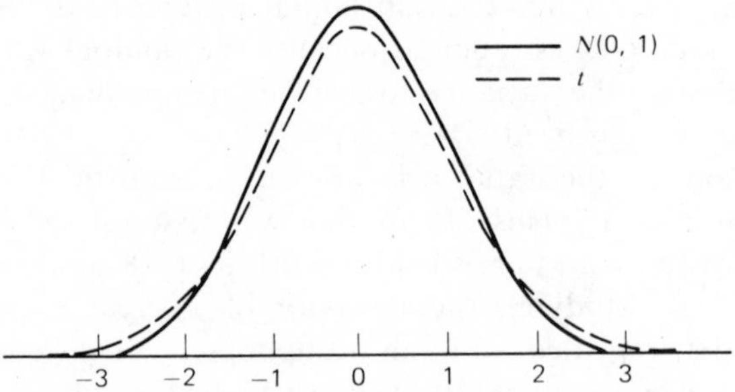

Figure 8.5 Comparison of $N(0,1)$ and t with d.f. = 5.

Appendix Table 5 gives the upper α points t_α for different values of α and the degrees of freedom (abbreviated d.f.). For example, with d.f. = 5, the upper .10 point of the t distribution is found from the table to be $t_{.10} = 1.476$. The curve is symmetric, so that the lower α point is simply $-t_\alpha$. The entries in the last row marked "d.f. = infinity" in Appendix Table 5 are exactly the percentage points of the $N(0,1)$ distribution.

Returning to the problem of determining a confidence interval for μ, we can conclude from the t distribution that

$$P\left[-t_{\alpha/2} < \frac{(\bar{X}-\mu)}{s/\sqrt{n}} < t_{\alpha/2}\right] = 1-\alpha$$

where $t_{\alpha/2}$ is the tabulated upper $\alpha/2$ point of the t distribution with $(n-1)$ degrees of freedom.

Rearranging terms within the brackets

$$P\left[\bar{X} - t_{\alpha/2}\frac{s}{\sqrt{n}} < \mu < \bar{X} + t_{\alpha/2}\frac{s}{\sqrt{n}}\right] = 1-\alpha$$

Therefore a $100(1-\alpha)\%$ confidence interval for μ is obtained from

$$\bar{X} \pm t_{\alpha/2}\frac{s}{\sqrt{n}}$$

If the population distribution is normal and if σ is unknown, a $100(1-\alpha)\%$ confidence interval for μ is

$$\bar{X} \pm t_{\alpha/2} \frac{s}{\sqrt{n}}$$

where $t_{\alpha/2}$ is the upper $\alpha/2$ point of the t distribution with d.f. $= n-1$.

The specified probability $(1-\alpha)$, whether it is .95, .99, or any other value, means that, prior to observation, the random interval will cover the true value of μ with probability $(1-\alpha)$. In the long run, this represents the approximate relative frequency of having μ included in the intervals obtained from repeated samplings. However, unlike the fixed length situation illustrated in Figure 8.4, the length of a confidence interval here is $2t_{\alpha/2}s/\sqrt{n}$, which is a random variable because it involves the sample standard deviation s. Figure 8.6 illustrates the realizations of these intervals from 10 samples of size 7, where $(1-\alpha)$ is taken to be .95 and the value of $t_{\alpha/2}$ with 6 d.f. is 2.447. In the first sample $\bar{x} = 103.88$ and $s = 7.96$, so that the interval is (96.52, 111.24).

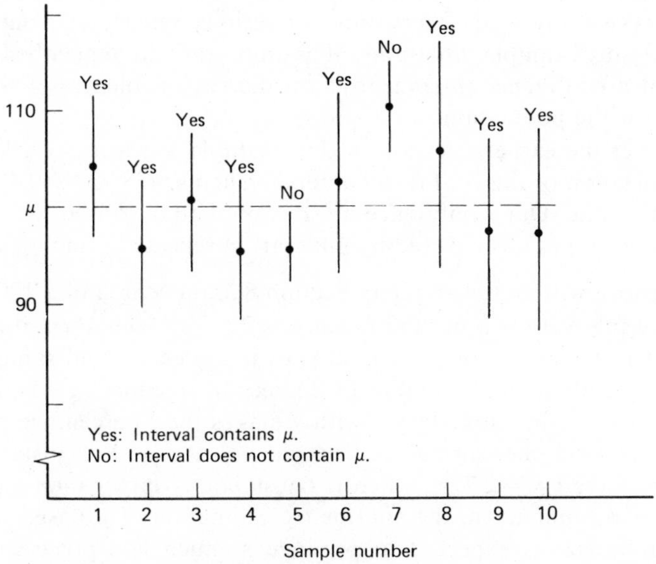

Figure 8.6 Behavior of confidence intervals based on the t distribution.

Example 8.9 A new alloy has been devised for use in a space vehicle. Tensile strength measurements are made on 15 pieces of the alloy, and the mean and the standard deviation of these measurements are found to be 39.3 and 2.6, respectively. Find a 90% confidence interval for the mean tensile strength of the alloy. Is μ included in this interval?

Assuming that strength measurements are normally distributed, a 90% confidence interval for the mean μ is given by

$$\bar{X} \pm t_{.05} \frac{s}{\sqrt{n}}$$

where $n = 15$ and consequently d.f. $= 14$. Consulting the t table, we find $t_{.05} = 1.761$. Hence a 90% confidence interval for μ computed from the observed sample is

$$39.3 \pm \frac{1.761 \times 2.6}{\sqrt{15}} = 39.3 \pm 1.18, \text{ or } (38.12, 40.48)$$

This means that we are 95% confident the tensile strength is between 38.12 and 40.48, because 95% of the intervals calculated in this manner will contain the true tensile strength.

When repeated independent measurements are made on the same material and any variation in the measurements is basically due to experimental error (possibly compounded by nonhomogeneous materials), the normal model is often found to be appropriate. It is still necessary to plot the individual data points (not given here) in a dot diagram on normal paper to reveal any wild observations or serious departures from normality. In all small-sample situations, it is important to remember that the validity of a confidence interval rests on the reasonableness of the model assumed for the population.

To answer the earlier question in this example, we will never know if a single realization of the confidence interval, such as (38.12, 40.48), covers the unknown μ. Our confidence in the method is based on the high percentage of times μ is covered by intervals in repeated samplings. ■

In comparison with our example in Section 8.3, the length of a $100(1-\alpha)\%$ confidence interval for a normal mean μ is $2z_{\alpha/2}\sigma/\sqrt{n}$ when σ is known, whereas it is $2t_{\alpha/2}s/\sqrt{n}$ when σ is unknown. Given a small sample size n and consequently a small number of degrees of freedom $(n-1)$, the extra variability caused by estimating σ with s makes the t percentage point $t_{\alpha/2}$ much larger than the normal percentage point $z_{\alpha/2}$. For instance, with d.f. $= 4$, we have $t_{.025} = 2.776$, which is considerably larger than $z_{.025} = 1.96$. Thus when σ is unknown, the confidence estimation of μ based on a very small sample size is expected to produce a much less precise inference (namely, a wide confidence interval) compared to the situation when σ is known. With increasing n, σ can be more closely estimated by s and the difference between $t_{\alpha/2}$ and $z_{\alpha/2}$ tends to diminish. The width of a confidence interval is then basically governed by the magnitude of $\sigma/\sqrt{n}$. The more variable a population, the larger the sample size required to produce a confidence interval of a desired length. This characteristic is discussed in greater detail in Section 8.9.

8.5 TESTING HYPOTHESES ABOUT A MEAN

Testing statistical hypotheses constitutes an alternative approach to inference when the primary goal is to determine whether a conjecture is supported or contradicted. Here estimating the numerical value of a parameter is not of direct interest, although implicitly an estimator may prove useful in evaluating the validity of a conjecture. In Chapter 6 we were already introduced to the basic concepts and terminologies that are applied in testing statistical hypotheses. A summary of the main steps in formulating and solving a testing problem is presented here, and a brief review of Chapter 6 may be advisable before reading further in this section.

For easy reference, the principal steps in hypothesis testing are:

(*a*) Identify the appropriate probability model and the parameter(s) about which hypotheses are made.
(*b*) Formulate the null hypothesis H_0 and the alternative hypothesis H_1.
(*c*) Choose the test statistic and determine the structure of the rejection region.
(*d*) Specify the sampling distribution of the test statistic under H_0 , and determine the rejection region for a specified α.
(*e*) Implement the test and draw a conclusion.

To delineate the basic steps in testing hypotheses about the mean of a population, we begin by considering a situation in which the sample size is small. A tentative assumption about the form of the population, namely, that it is normal, is again employed. The complete assumptions are that the observations form a random sample from a normal distribution $N(\mu,\sigma)$. Consequently, $\bar{X}$ is normal with distribution $N(\mu,\sigma/\sqrt{n})$, and based on our experience with estimation procedures, $\bar{X}$ should be a natural choice for a test statistic.

To avoid unnecessary complications as we introduce concepts, first we discuss a situation in which the population standard deviation σ is known to the investigator. This unrealistic assumption will be relaxed shortly. Suppose that the problem involves testing a one-sided hypothesis of the form

$$H_0: \mu \leqslant \mu_0 \quad \text{vs.} \quad H_1: \mu > \mu_0$$

where μ_0 is a specified number. The normal distribution $N(\mu, \sigma/\sqrt{n})$ of $\bar{X}$ is centered at μ, and the alternative hypothesis H_1 comprises larger values of μ than the null hypothesis H_0. As a visual aid, the shape of a normal distribution with a standard deviation of $\sigma/\sqrt{n}$ could be cut out of construction paper beforehand and slid to the left and right along the $\bar{x}$ scale so as to be centered at various values of μ. Comparing the situations when the distribution is centered at a μ value in H_0 to one so centered in H_1, it is apparent that large values of $\bar{X}$ have higher probabilities under H_1 than they do under H_0. The rejection of H_0 in favor of H_1 is therefore recommended if the observed value of $\bar{X}$ is too large. In other words, the structure of the rejection region should be

$$R: \quad \bar{X} \geqslant c$$

The boundary c of the rejection region must now be determined from a specified tolerance of the type I error probability α. The type I error probability is highest when $\mu = \mu_0$, the boundary point between H_0 and H_1, so that it is enough to determine c such that $P_{\mu_0}[\bar{X} \geqslant c] = \alpha$. When $\mu = \mu_0$, $\bar{X}$ has the distribution $N(\mu_0, \sigma/\sqrt{n})$ and the standardized variable

$$Z = \frac{(\bar{X} - \mu_0)}{\sigma/\sqrt{n}}$$

has the $N(0,1)$ distribution. The events

$$[\bar{X} \geqslant c] \quad \text{and} \quad \left[\frac{\bar{X} - \mu_0}{\sigma/\sqrt{n}} \geqslant \frac{c - \mu_0}{\sigma/\sqrt{n}}\right]$$

being equivalent, we obtain

$$P_{\mu_0}[\bar{X} \geqslant c] = P\left[Z \geqslant \frac{c - \mu_0}{\sigma/\sqrt{n}}\right]$$

Because $P[Z \geqslant z_\alpha] = \alpha$, the requirement for the type I error probability is satisfied by choosing c, so that

$$\frac{c - \mu_0}{\sigma/\sqrt{n}} = z_\alpha$$

This results in $c = \mu_0 + z_\alpha\ (\sigma/\sqrt{n}\ )$, and the rejection region of the level α test becomes

$$R: \quad \bar{X} \geqslant \mu_0 + z_\alpha \frac{\sigma}{\sqrt{n}}$$

From the equivalence of these two sets, the rejection region in terms of the test statistic $\bar{X}$ could alternatively be specified by employing the test statistic

$$Z = \frac{\bar{X} - \mu_0}{\sigma/\sqrt{n}}$$

as R: $Z \geqslant z_\alpha$. The latter version, Z, is called the standardized form of the test statistic, and the test is commonly called a *normal test* or a *Z test*. Its minor advantage lies in the fact that the rejection region can be directly read from the normal table.

When a problem concerns testing the null hypothesis H_0: $\mu \geqslant \mu_0$ vs. the alternative hypothesis H_1: $\mu < \mu_0$, small values of $\bar{X}$ or equivalently small values of the standardized test statistic Z should comprise the rejection region. Since $P[Z \leqslant -z_\alpha] = \alpha$, a level α test has the rejection region R: $Z \leqslant -z_\alpha$.

The above hypotheses are called *one-sided hypotheses*, because the values of the parameter μ under the alternative hypothesis lie on one side of those under the null hypothesis. The corresponding tests are called *one-sided tests* or *one-tailed tests*. By contrast, we can have a problem of testing two-sided hypotheses in the form

$$H_0: \mu = \mu_0 \quad \text{vs.} \quad H_1: \mu \neq \mu_0$$

Here H_0 is to be rejected if $\bar{X}$ is too far away from μ_0 in either direction; that is, if Z is too small or too large. For a level α test we divide the rejection probability α equally between the two tails and construct the rejection region

$$R: \quad Z \leqslant -z_{\alpha/2} \quad \text{or} \quad Z \geqslant z_{\alpha/2}$$

which can be expressed in more compact notation

$$R: \quad |Z| \geqslant z_{\alpha/2}$$

Example 8.10 From extensive records it is known that the duration of treating a disease by a standard therapy has a mean of 15 days and a standard deviation of 3 days. It is claimed that a new therapy can reduce the treatment time. To test this claim, the new therapy is to be tried on 70 patients and their times to recovery are to be recorded. Assume that the standard deviation for the new therapy is also 3 days.

Formulate the hypothesis and determine the rejection region of the test with level of significance $\alpha = .025$.

Let μ denote the population mean time to recovery for the new therapy. Because the aim of the investigation is to substantiate the assertion that $\mu < 15$, the appropriate formulation of the hypotheses is

$$H_0: \mu \geqslant 15 \quad \text{vs.} \quad H_1: \mu < 15$$

The sample size is $n = 70$, and it is assumed that $\sigma = 3$. Denoting the sample mean recovery time of 70 patients by $\overline{X}$, the rejection region of H_0 will consist of small values of $\overline{X}$, or of

$$Z = \frac{\overline{X} - \mu_0}{\sigma/\sqrt{n}} = \frac{\overline{X} - 15}{3/\sqrt{70}}$$

Since $z_{.025} = 1.96$, the test with a level of significance .025 can be expressed in either of the two forms

$$R: \quad Z \leqslant -1.96$$

$$R: \quad \overline{X} \leqslant 15 - \frac{1.96 \times 3}{\sqrt{70}} = 15 - .70 = 14.30$$ ■

Thus far we have focussed our attention on determining the rejection region for a test about μ, so that the maximum probability of rejecting H_0 when it is true is controlled at a specified level α. Translated into the long-run relative frequency, this means that under repeated samplings H_0 will not be falsely rejected in more than $100\alpha\%$ of the cases. Because a hypothesis testing problem is usually formulated to evaluate evidence for rejecting the null hypothesis, it is also interesting to know how often H_0 will be rejected when the alternative hypothesis H_1 is true. The corresponding probability, called *power*, can be calculated for a given value of μ in H_1. With regard to the problem of testing H_0: $\mu \leqslant \mu_0$ vs. H_1: $\mu > \mu_0$, where the rejection region of a level α test is

$$R: \quad \overline{X} \geqslant \mu_0 + z_\alpha \frac{\sigma}{\sqrt{n}}$$

the power of the test at an alternative $\mu_1 > \mu_0$ is

$$\gamma(\mu_1) = P_{\mu_1}\left[\overline{X} \geqslant \mu_0 + z_\alpha \frac{\sigma}{\sqrt{n}}\right]$$

Under $\mu = \mu_1$, $\overline{X}$ is normal $N(\mu_1, \sigma/\sqrt{n})$, so that

$$U = \frac{\overline{X} - \mu_1}{\sigma/\sqrt{n}} \qquad \text{is } N(0,1)$$

Here the notation U represents a standardized normal variable to avoid confusion with the standardized test statistic Z. The power is therefore given by

$$\gamma(\mu_1) = P\left[U \geqslant \frac{\mu_0 - \mu_1}{\sigma/\sqrt{n}} + z_\alpha \right]$$

and this probability can be computed using the normal table. Recall that $\beta(\mu_1) = 1 - \gamma(\mu_1)$ is the type II error probability at the alternative hypothesis μ_1. These concepts are illustrated in Figure 8.7. Because the true value of μ that prevails is never known, it is desirable to evaluate power at several different values of μ. A plot of the power as a function of μ, called

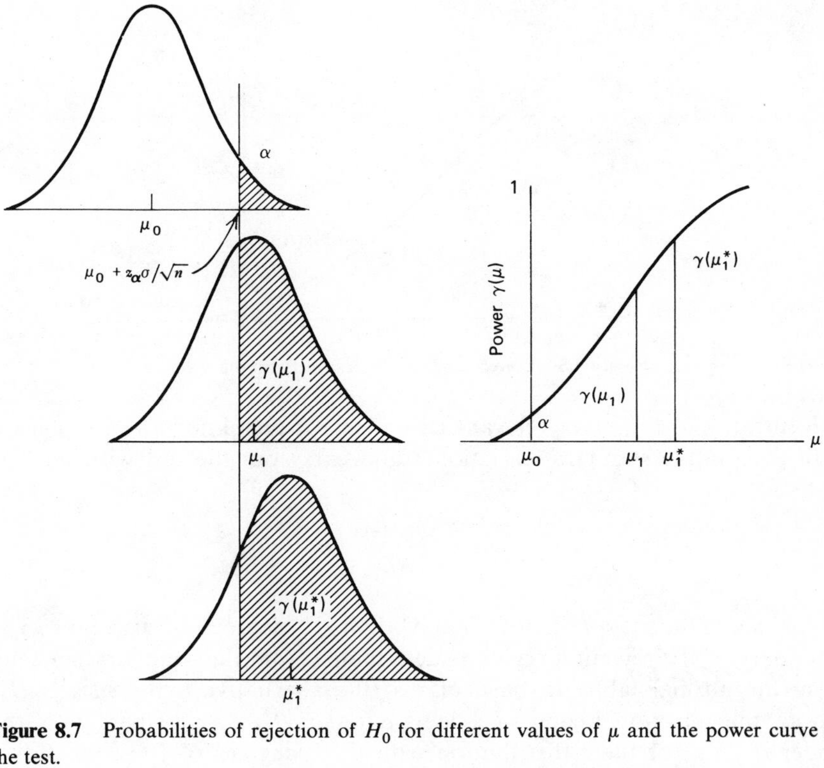

Figure 8.7 Probabilities of rejection of H_0 for different values of μ and the power curve of the test.

the *power curve*, gives a complete picture of the behavior of the test under the assumed normal model.

Example 8.11 When the new treatment in Example 8.10 is more effective than the standard therapy, the distribution of $\bar{X}$ is $N(\mu, 3/\sqrt{70})$ with $\mu < 15$. The test with $\alpha = .025$ obtained in that example rejects H_0: $\mu \geqslant 15$ if $\bar{X} \leqslant 14.30$. Obtain the power of this test at $\mu = 13.8$.

The power of the test at $\mu = 13.8$ is the probability of rejecting H_0 if the true value of μ is 13.8. In that event $\bar{X}$ is normally distributed with mean $= 13.8$ and sd $= \sigma/\sqrt{n} = 3/\sqrt{70}$. Using the normal table, the power is computed

$$\gamma(13.8) = P_{\mu=13.8}\left[\bar{X} \leqslant 14.30\right] = P\left[U \leqslant \frac{14.30 - 13.8}{3/\sqrt{70}}\right]$$

$$= P\left[U \leqslant 1.39\right] = .92$$

A similar calculation gives the additional values $\gamma(13.5) = .99$ and $\gamma(14.4) = .39$. The complete power curve appears in Figure 8.8. ■

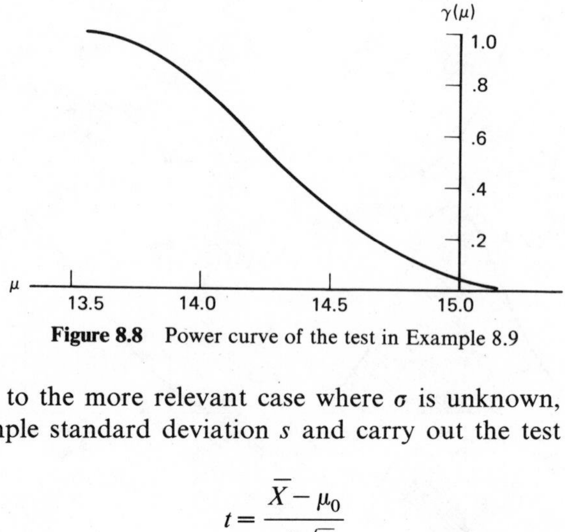

Figure 8.8 Power curve of the test in Example 8.9

Returning to the more relevant case where σ is unknown, we replace σ with the sample standard deviation s and carry out the test with the test statistic

$$t = \frac{\bar{X} - \mu_0}{s/\sqrt{n}}$$

which has Student's t distribution with $n-1$ degrees of freedom. The boundary of the rejection region is determined from the t table rather than from the normal table. In particular, if the alternative hypothesis is H_1: $\mu > \mu_0$, the rejection region of a level α test is R: $t \geqslant t_\alpha$, where t_α is the upper α point of the t distribution with $n-1$ degrees of freedom. For a two-sided alternative H_1: $\mu \neq \mu_0$ the rejection region is R: $|t| \geqslant t_{\alpha/2}$.

In testing the hypothesis H_0: $\mu = \mu_0$ for a *normal population when* σ *is unknown*, the test statistic is

$$t = \frac{(\bar{X} - \mu_0)}{s/\sqrt{n}}$$

which has Student's t distribution with $n-1$ degrees of freedom. The test is called *Student's t test* or, more simply, a *t test*.

Example 8.12 A city health department wishes to determine if the mean bacteria count per unit volume of water at a lake beach exceeds the safety level of 200. Researchers have collected 10 water samples of unit volume and have found the bacteria counts to be:

175,	190,	215,	198,	184
207,	210,	193,	196,	180

Do the data indicate a cause for concern?

Let μ denote the current mean bacteria count per unit volume of water and σ represent the standard deviation. Considering that failure to detect a violation of the safety level can result in serious consequences, formulation of the null and alternative hypotheses should be

$$H_0\colon \mu \geqslant 200 \quad \text{vs.} \quad H_1\colon \mu < 200$$

Since the counts are spread over a wide range, an approximation by a continuous distribution is not unrealistic for inference about the mean. Assuming that the measurements constitute a sample from a normal population and that the level of significance $\alpha = .01$, we employ the t test

$$t = \frac{\bar{X} - 200}{s/\sqrt{10}}$$

with the rejection region $t \leqslant -t_{.01}$ and d.f. = 9. From the t table we determine that $t_{.01}$ with d.f. = 9 is 2.82. Computations from the sample data yield

$$\bar{x} = 194.8$$

$$s = 13.14$$

$$t = \frac{194.8 - 200}{13.14/\sqrt{10}} = \frac{-5.2}{4.156} = -1.25$$

Because the observed value $t = -1.25$ is larger than -2.82, the null hypothesis is not rejected at $\alpha = .01$. On the basis of the data obtained from these 10 measurements, there does not seem to be strong evidence that the mean bacteria level is within the safety margin. ■

Unlike the normal test, which is applicable when σ is known, the power of a t test is more difficult to compute because it involves a generalization of the t distribution. The power depends on $(\mu - \mu_0)/\sigma$, that is, the difference $(\mu - \mu_0)$ measured in the scale of the population standard deviation. For a certain range of sample sizes, charts are provided in the *CRC Handbook of Tables for Probability and Statistics* [1, p. 284], from which the power of the t test can be read.

8.5.1 Large Sample Tests About μ

When we examine large samples (as a rule of thumb, say $n \geqslant 30$) the procedures discussed in this section are no longer heavily dependent on the assumption of a normal population. By virtue of the central limit theorem, $\bar{X}$ can be treated as an approximately normal $N(\mu, \sigma/\sqrt{n})$ variable whatever the form of the underlying population. Moreover, σ can be replaced by the sample standard deviation s without affecting the distribution to any appreciable extent. Thus an approximate test of H_0: $\mu = \mu_0$ can be performed by using the Z statistic

$$Z = \frac{\bar{X} - \mu_0}{s/\sqrt{n}}$$

and by consulting the normal table for the rejection region.

> When n is large and the population σ is unknown, a test of the null hypothesis H_0: $\mu = \mu_0$ is performed by using the normal test statistic
>
> $$Z = \frac{\bar{X} - \mu_0}{s/\sqrt{n}}$$
>
> No assumption is required as to the shape of the population distribution.

8.5.2 Large Sample Test About a Binomial p

In Chapter 6 we presented a detailed discussion of testing hypotheses for a binomial proportion in which we used the binomial table to determine the rejection region. Here we present a large sample test that is based essentially on the normal approximation to the binomial distribution. Denoting a population proportion by p, we consider testing H_0: $p=p_0$ vs. H_1: $p \neq p_0$ with a large number of trials n. Using X to denote the number of successes in n trials, the sample proportion

$$\hat{p}=\frac{X}{n}$$

is approximately normally distributed with a mean of p and a standard deviation of $\sqrt{pq}/\sqrt{n}$. Under the null hypothesis, p has the specified value p_0 and the distribution of $\hat{p}$ is approximately $N(p_0, \sqrt{p_0 q_0}/\sqrt{n})$. Consequently, the standardized statistic

$$Z=\frac{\hat{p}-p_0}{\sqrt{p_0 q_0}/\sqrt{n}}$$

has the $N(0,1)$ distribution. The alternative hypothesis is two-sided, so that the rejection region of a level α test is given by

$$R: |Z| \geqslant z_{\alpha/2}$$

For one-sided alternatives, we use a one-tailed test in exactly the same way we discussed earlier in connection with tests about μ. Because $\hat{p}=X/n$, we also have

$$Z=\frac{\dfrac{X}{n}-p_0}{\sqrt{p_0 q_0}/\sqrt{n}}=\frac{X-np_0}{\sqrt{np_0 q_0}}$$

which provides another equivalent formula for the Z test statistic.

Example 8.13 A five-year-old census recorded that 20% of the families in a large community lived below the poverty level. To determine if this percentage has changed, a random sample of 500 families is studied and 91 are found to be living below the poverty level. Does this finding indicate that the current percentage differs from the percentage of families earning incomes below the poverty level 5 years ago?

Let p denote the current population proportion of families living below the poverty level. Because we are seeking evidence to determine whether p is *different* from .20, we wish to test

$$H_0: p=.20 \quad \text{vs.} \quad H_1: p \neq .20$$

Because the sample size $n=500$ is large, we use the Z test statistic

$$Z=\frac{\hat{p}-.2}{\sqrt{.2\times.8/500}} \quad \text{or} \quad Z=\frac{X-500\times.2}{\sqrt{500\times.2\times.8}}$$

and the rejection region R: $|Z| \geqslant 1.96$ corresponds to $\alpha = .05$. The computed value of Z from the sample data is

$$z = \frac{(91/500) - .2}{\sqrt{.2 \times .8/500}} = \frac{.182 - .2}{.018} = -1.00$$

or, alternatively

$$z = \frac{91 - 500 \times .2}{\sqrt{500 \times .2 \times .8}} = \frac{91 - 100}{\sqrt{80}} = -1.00$$

Because $|z| = 1$ is smaller than 1.96, the null hypothesis is not rejected at $\alpha = .05$, and we conclude that the data do not provide strong evidence that a change in the percentage of families living below the poverty level has occurred. ■

8.6 RELATIONSHIP BETWEEN TESTS AND CONFIDENCE INTERVALS

At this point the careful reader should observe a similarity between the formulas we use in hypothesis testing and in estimation by a confidence interval. To clarify this connection, we consider the example of a $100(1-\alpha)\%$ confidence interval for μ given by

$$\bar{X} - t_{\alpha/2}\frac{s}{\sqrt{n}} < \mu < \bar{X} + t_{\alpha/2}\frac{s}{\sqrt{n}}$$

On the other hand, a rejection region of a level α test for H_0: $\mu = \mu_0$ vs. the two-sided alternative H_1: $\mu \neq \mu_0$ is

$$R: \left| \frac{\bar{X} - \mu_0}{s/\sqrt{n}} \right| \geqslant t_{\alpha/2}$$

The acceptance region of this test is obtained by reversing the inequality to obtain all the values of $\bar{X}$ that do not discredit μ_0, or

$$\text{Acceptance region: } \left| \frac{\bar{X} - \mu_0}{s/\sqrt{n}} \right| < t_{\alpha/2}$$

which can also be written

$$\text{Acceptance region: } \bar{X} - t_{\alpha/2}\frac{s}{\sqrt{n}} < \mu_0 < \bar{X} + t_{\alpha/2}\frac{s}{\sqrt{n}}$$

The latter expression shows that any given null hypothesis μ_0 will be accepted (more precisely, will not be rejected) at a level α if μ_0 lies within the $100(1-\alpha)\%$ confidence interval. Thus having established a $100(1-\alpha)\%$ confidence interval for μ, we know at once that all possible null hypotheses μ_0 lying outside this interval will be rejected at a level of significance α and that all those lying inside will not be rejected.

Example 8.14 A random sample of size $n=9$ from a normal population produced the mean $\bar{x}=8.3$ and the standard deviation $s=1.2$. Obtain a 95% confidence interval for μ and also test H_0: $\mu=8.5$ vs. H_1: $\mu\neq 8.5$ with $\alpha=.05$.

A 95% confidence interval has the form

$$\left(\bar{x}-t_{.025}\frac{s}{\sqrt{n}},\quad \bar{x}+t_{.025}\frac{s}{\sqrt{n}}\right)$$

where $t_{.025}=2.306$ corresponds to $n-1=8$ degrees of freedom. The interval then becomes

$$\left(8.3-2.306\frac{1.2}{\sqrt{9}},\quad 8.3+2.306\frac{1.2}{\sqrt{9}}\right)=(7.4,\quad 9.2)$$

In view of the equivalence just established, a test of $\mu=8.5$ vs. $\mu\neq 8.5$ can be implemented by noting that the value 8.5 lies within the 95% confidence interval, so that H_0 should not be rejected. Alternatively, a formal step-by-step solution can be based on the test statistic

$$\frac{\bar{X}-8.5}{s/\sqrt{n}}$$

The rejection region consists of both large and small values:

$$\text{Rejection region: } \left|\frac{\bar{X}-8.5}{s/\sqrt{n}}\right| \geqslant t_{.025}=2.306$$

Here the observed value is $\sqrt{9}\,|8.3-8.5|/1.2=.5$, so that the null hypothesis H_0: $\mu=8.5$ is not discredited. ∎

We now present a general statement concerning the connection between confidence estimation and tests of hypotheses with two-sided alternatives. This relationship indicates that these two methods of inference are really integrated in a common framework.

Relation between confidence intervals and tests

Let $L < \theta < U$ be a $100(1-\alpha)\%$ confidence interval for a parameter θ. Then the null hypothesis H_0: $\theta = \theta_0$ vs. H_1: $\theta \neq \theta_0$ will be rejected at a level of significance α if θ_0 falls outside the confidence interval:

$$(1 - \text{Level of significance of test}) = \text{Confidence level}$$

A confidence interval statement is regarded as a more comprehensive inference procedure than testing a single null hypothesis, because a confidence interval statement in effect tests many null hypotheses at the same time. Hypotheses testing still remains a useful method in inferential statistics, especially in complicated problems such as tests of independence and goodness of fit to be treated in Chapter 13. In these situations testing a hypothesis about some specific θ_0 is relatively easy, but it is difficult if not impossible to define a population parameter for expressing the alternatives.

8.7 INFERENCES ABOUT THE STANDARD DEVIATION σ OF A NORMAL POPULATION

Aside from inferences about the population mean, the population variability may also be of interest. In many industrial manufacturing processes the uniformity of some measurable characteristic is often a criterion of product quality, and the quality control engineer must ensure that the variability of the measurements does not exceed a specified limit. It may also be important to ensure sufficient uniformity of the inputed raw material for trouble-free operation of the machines. Apart from the record of a baseball player's batting average, information on the variability of the player's scores from one game to the next may be an indicator of reliability. We now consider inference procedures for the standard deviation σ of a population *under the assumption that the population distribution is normal*. In contrast to the inference procedures we developed in learning about μ, the usefulness of the methods to be examined here are extremely limited when this assumption is violated.

To make inferences about σ^2, the natural choice of a statistic is its sample analog, which is the sample variance

$$s^2 = \frac{\sum_{i=1}^{n} (X_i - \bar{X})^2}{n-1}$$

To estimate by confidence intervals and to test hypotheses, we must consider the sampling distribution of s^2. To do this, we introduce a new distribution, called the χ^2 distribution (read "chi-square distribution"), whose form depends on $n-1$.

> *χ^2 distribution*
>
> Let $X_1, \ldots, X_n$ be a random sample from a normal population, $N(\mu, \sigma)$. Then the distribution of
>
> $$\chi^2 = \frac{\sum_{i=1}^{n} (X_i - \bar{X})^2}{\sigma^2} = \frac{(n-1)s^2}{\sigma^2}$$
>
> is called a χ^2 *distribution with $n-1$ degrees of freedom.*

Unlike the normal or t distribution, the probability density curve of a χ^2 distribution is an asymmetric curve stretching over the positive side of the line and having a long right tail. The form of the curve depends on the value of the degrees of freedom. A typical χ^2 curve is illustrated in Figure 8.9.

Appendix Table 6 provides the upper α points of χ^2 distributions for various values of α and the degrees of freedom. As in both the cases of the t and the normal distributions, the upper α point χ^2_α denotes the χ^2 value such that the area to the right is α. The lower α point or 100αth percentile, read from the column $\chi^2_{1-\alpha}$ in the table, has an area $1-\alpha$ to the right. For

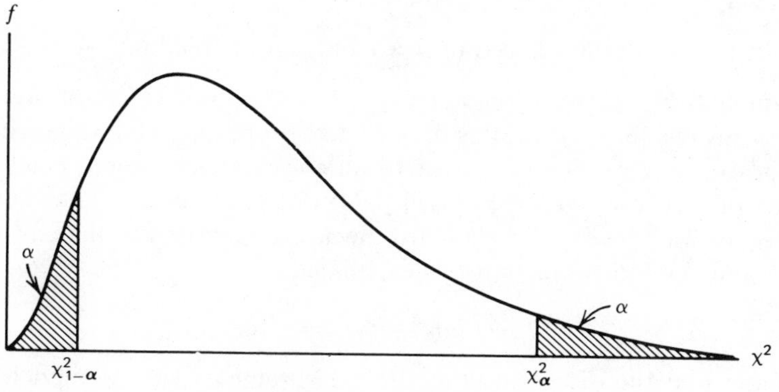

Figure 8.9 A χ^2 distribution.

example, the lower .05 point is obtained from the table by reading the $\chi^2_{.95}$ column, whereas the upper .05 point is obtained by reading the column $\chi^2_{.05}$.

Probability statements about the χ^2 distribution lead directly to confidence intervals, as we can see in the following construction of a 95% confidence interval for σ^2. Dividing the probability $\alpha=.05$ equally between the two tails of the χ^2 distribution and using the notations just explained we have

$$P\left[\chi^2_{.975}<\frac{(n-1)s^2}{\sigma^2}<\chi^2_{.025}\right]=.95$$

where the percentage points are read from the χ^2 table at d.f. $=n-1$. Because $(n-1)s^2/\sigma^2<\chi^2_{.025}$ is equivalent to $(n-1)s^2/\chi^2_{.025}<\sigma^2$ and $\chi^2_{.975}<(n-1)s^2/\sigma^2$ is equivalent to $\sigma^2<(n-1)s^2/\chi^2_{.975}$, the preceding probability statement can be rephrased

$$P\left[\frac{(n-1)s^2}{\chi^2_{.025}}<\sigma^2<\frac{(n-1)s^2}{\chi^2_{.975}}\right]=.95$$

This last statement, concerning a random interval covering σ^2, provides a 95% confidence interval for σ^2.

A confidence interval for σ can be obtained by taking the square root of the end points of the interval. For a confidence level .95, the interval for σ becomes

$$\left(s\sqrt{\frac{n-1}{\chi^2_{.025}}},\quad s\sqrt{\frac{n-1}{\chi^2_{.975}}}\right)$$

where the χ^2 values are based on $n-1$ degrees of freedom.

Example 8.15 A precision watchmaker wishes to learn about the variability of his product. To do this, he decides to obtain a confidence interval for σ based on a random sample of 10 watches selected from a much larger number of watches that have passed the final quality check. The deviations of these 10 watches from a standard clock are recorded at the end of one month and the following statistics calculated.

$$\bar{x}=.7 \text{ seconds} \qquad s=.4 \text{ seconds}$$

Assuming that the distribution of the measurements can be modeled as a normal distribution, find a 90% confidence interval for σ.

Here $n=10$, so that d.f. $=n-1=9$. The χ^2 table gives $\chi^2_{.95}=3.325$ and $\chi^2_{.05}=16.919$. Using the preceding formula, a 90% confidence interval for σ^2 is

$$\left(\frac{9\times(.4)^2}{16.919},\quad \frac{9\times(.4)^2}{3.325}\right)=(.085,\quad .433)$$

and the corresponding interval for σ is $(\sqrt{.085}\,,\ \sqrt{.433}\,)=(.29,\ .66)$. This means that the watchmaker can be 90% confident that σ is between .29 and .66 seconds, because 90% of the intervals calculated by this procedure in repeated samples will cover the true σ. ■

It is instructive to note that the midpoint of the confidence interval for σ^2 in Example 8.15 is not $s^2=.16$, which is the best point estimate. This is in sharp contrast to the confidence intervals for μ, and it serves to accent the difference in logic between interval and point estimation.

For a test of the null hypothesis H_0: $\sigma^2=\sigma_0^2$, it is natural to employ the test statistic s^2. If the alternative hypothesis is one-sided, say H_1: $\sigma^2>\sigma_0^2$, then the rejection region should consist of large values of s^2, or alternatively large values of the test statistic

$$\chi^2=\frac{(n-1)s^2}{\sigma_0^2}$$

This multiple of s^2 is perhaps the most convenient form of the test statistic, because its percentage points are directly attainable from the χ^2 table. The rejection region of a level α test is therefore

$$R:\ \frac{(n-1)s^2}{\sigma_0^2}\geqslant\chi^2_\alpha,\quad \text{d.f.}=n-1$$

For a two-sided alternative H_1: $\sigma^2\neq\sigma_0^2$, a level α rejection region is

$$R:\ \frac{(n-1)s^2}{\sigma_0^2}\leqslant\chi^2_{1-\alpha/2}\ \text{or}\ \frac{(n-1)s^2}{\sigma_0^2}\geqslant\chi^2_{\alpha/2}$$

Once again we must emphasize that the inference procedures for σ presented in this section are extremely sensitive to departures from a normal population.

8.8 THE EFFECT OF VIOLATION OF ASSUMPTIONS

The small-sample methods for both confidence-interval estimation and hypothesis testing presuppose that the sample is obtained from a normal population. Users of these methods would naturally ask:

(*a*) What method can be used to determine if the population distribution is nearly normal?
(*b*) What can go wrong if the population distribution is nonnormal?
(*c*) What procedures should be used if it is nonnormal?
(*d*) If the observations are not independent, is this serious?

(*a*) To answer the first question, we could examine a physical theory that might tentatively suggest that the population is normal or we could rely on the experience of other investigators who have extensively sampled similar populations and found them to be nearly normal. In either case we must still plot the data. The dot diagram or normal plot may indicate a wild observation or a striking departure from normality. Although a small-sample plot cannot provide convincing justification for normality, the investigator may feel more secure using the preceding inference procedures if no serious departures are indicated. Lacking sufficient observations to justify or to refute the normal assumption leads to a consideration of the second question.

(*b*) Confidence intervals and tests of hypotheses are based on the sampling distributions t and χ^2. The assumption of a normal population was invoked by statisticians who derived the forms of these distributions and then tabulated the areas under the curves. If the population is nonnormal, these forms will be incorrect and the actual percentage points may differ greatly from the tabulated values. When we say that $\bar{X} \pm t_{.025} s/\sqrt{n}$ is a 95% confidence interval for μ, the true probability that this random interval will contain μ may be, say, 75% or 99%. Likewise, the claimed control of type I error probability $\alpha = .05$ for a test may be violated, and the true probability may be double or even triple that amount. The direction and the extent of such disturbances will depend on how much the population distribution differs from a normal distribution. Regarding inferences about μ using the t statistic, the effect is not too serious if the sample size is at least moderately large (say 15). In larger samples such disturbances tend to disappear. Unfortunately, this is not true for inferences about σ^2 using the χ^2 distribution. Even those inferences drawn from large samples can be seriously misleading when the population distribution departs from normality. We express this by saying that inferences about σ^2 based on the χ^2 distribution are not "robust" against departures of the population distribution from normality, whereas inferences about μ using the t statistic are reasonably "robust." However, this qualitative discussion should not be considered a blanket endorsement for t. In small samples wild observations or distributions with long tails can produce misleading results.

(*c*) We cannot give a specific answer to the third question without knowing something about the nature of nonnormality. Dot diagrams or histograms of the original data may suggest some transformations that will bring the shape of the distribution closer to normality. If it is possible to obtain a transformation that leads to reasonably normal data plots, the problem can then be recast in terms of the transformed data. Alternatively, past experience with a similar population or a partial theory may suggest the use of a particular form of the probability model. The theory of optimal tests and confidence intervals has been developed for a wide variety of models, and users can benefit from consulting with experts in this field. Also, when no prior knowledge exists about the shape of the distribution, inference procedures that are valid under a wide family of population distributions can be examined. A few of these procedures will be discussed in Chapter 15. These techniques will not be as efficient as the procedures designed for specific distributions, but they can safeguard against the violation of assumptions that specify a form for the underlying distribution.

(*d*) A basic assumption throughout this chapter is that the sample is drawn at random, so that the observations are independently distributed. If the sampling is made in such a manner that the observations are dependent, however, all the inferential procedures we discussed here for small as well as large samples may be seriously in error. This applies to both the level of a test and a stated confidence level. Suppose that an investigator who is measuring the increase in blood pressure caused by the use of a drug includes 5 patients in an experiment and makes 4 successive measurements on each. This does *not* yield a random sample of size $5 \times 4 = 20$, because the 4 measurements made on each person are likely to be dependent. This type of sampling requires a more sophisticated mode of analysis. An investigator who is sampling opinions about a political issue may choose 100 families at random and record the opinions of both the husband and the wife in each family. This also does *not* provide a random sample of size $100 \times 2 = 200$, although it may be a convenient sampling method. When measurements are made closely together in time or distance there is a danger of losing independence, because adjacent observations are more likely to be similar than observations that are made farther apart. Because independence is the most crucial assumption, we must be constantly alert to detect such violations.

8.9 DETERMINATION OF THE SAMPLE SIZE

The formulas for standard error and confidence intervals discussed in Sections 8.2 and 8.3 have $\sqrt{n}$ in the denominator, so that with increasing sample size the standard error decreases and the confidence intervals become shorter. On the other hand, increasing the sample size is more costly and time consuming, both in the sampling operation and in processing the data. In the planning stage of an investigation an important practical decision must be made concerning the sample size required to achieve the desired protection against imprecision with the estimation procedure.

First we consider choosing the sample size to estimate a population mean. From the intended use of the estimate, the investigator should attempt to specify the amount of error that can be tolerated and, in addition, the assurance or probability with which this error bound can be expected to hold.

Suppose that the population standard deviation σ is known. Recall that the formula for a $100(1-\alpha)\%$ error bound for the estimation of μ by $\overline{X}$ is given by

$$z_{\alpha/2}\frac{\sigma}{\sqrt{n}}$$

To be $100(1-\alpha)\%$ sure that the error does not exceed an amount d, the investigator must have

$$z_{\alpha/2}\frac{\sigma}{\sqrt{n}}=d$$

This gives an equation in which n is unknown. Solving for n, we obtain

$$n=\left[\frac{z_{\alpha/2}\sigma}{d}\right]^2$$

which determines the required sample size. Of course, the solution is rounded to the next higher integer, because a sample size cannot be fractional.

Alternatively, the investigator may want to determine the required sample size so that a $100(1-\alpha)\%$ confidence interval for μ will have a specified length c. Because the length of the confidence interval $(\overline{X}-z_{\alpha/2}\sigma/\sqrt{n},\ \overline{X}+z_{\alpha/2}\sigma/\sqrt{n})$ is $2z_{\alpha/2}\sigma/\sqrt{n}$, the sample size n is determined from

$$2z_{\alpha/2}\frac{\sigma}{\sqrt{n}}=c \quad \text{or} \quad z_{\alpha/2}\frac{\sigma}{\sqrt{n}}=\frac{c}{2}$$

These two formulations are equivalent, as we can readily see by identifying $c/2$ with the specified error bound d.

We used the normal distribution for $\bar{X}$ to achieve this solution, so that it is valid if n is large or if the population distribution is normal. Otherwise, a *conservative* solution can be achieved with the aid of the Chebyshev inequality (see Mathematical Exercise 4.6, page 139). Thus

$$P\left[|\bar{X}-\mu|\leqslant d\right]\geqslant 1-\frac{\operatorname{var}(\bar{X})}{d^2}=1-\frac{\sigma^2}{nd^2}$$

Equating $(1-\sigma^2/nd^2)$ to the desired assurance $1-\alpha$, the sample size is given by

$$n=\frac{\sigma^2}{\alpha d^2}$$

To be $100(1-\alpha)\%$ sure that the error $|\bar{X}-\mu|$ does not exceed d, the *required sample size* is

$$n=\left[\frac{z_{\alpha/2}\sigma}{d}\right]^2$$

If n is small and the population is nonnormal, take

$$\frac{\sigma^2}{\alpha d^2}$$

as an upper bound for the sample size.

Note that knowledge of σ or at least a close approximation of σ is required to compute n. If σ is completely unknown, a small-scale preliminary sampling is necessary to obtain an estimate of σ to be used in the formula to compute n.

Example 8.16 A limnologist wishes to estimate the mean phosphate content per unit volume of a certain lake water. It is known from studies in previous years that the standard deviation has a fairly stable value $\sigma=4$. How many water samples must the limnologist analyze to be 90% certain that the error of estimation does not exceed 0.8?

Here $\sigma=4$ and $1-\alpha=.90$, so that $\alpha/2=.05$. The upper .05 point of the $N(0,1)$ distribution is $z_{.05}=1.64$. The tolerable error is $d=0.8$. Computing

$$n=\left[\frac{1.64\times 4}{.8}\right]^2=67.24$$

the required sample size is $n=68$. ■

The criterion for choosing n just discussed is to attain a specified precision while estimating the parameter μ. The problem of choosing the sample size may also arise in the context of testing hypotheses about μ. More specifically, we may wish to determine n, so that the level α test of H_0: $\mu=\mu_0$ will have a desired power γ at an alternative located at a specified distance d_0 from μ_0. The power depends on the distance measured in units of σ, or on the value of $\Delta=(\mu-\mu_0)/\sigma$. An analytical solution to this type of problem is somewhat complex, but charts of the required sample size n have been prepared and the interested reader is referred to [1], page 287.

8.9.1 Sample Size for the Estimation of p

A $100(1-\alpha)\%$ error bound for the estimation of p is given by $z_{\alpha/2}\sqrt{pq}\,/\sqrt{n}$, and the required sample size is obtained by replacing σ^2 with pq. We then obtain

$$n=pq\left[\frac{z_{\alpha/2}}{d}\right]^2$$

However, this solution is not readily usable, because it involves the parameter p we are seeking to estimate. We know that p ranges from 0 to 1, so that $p(1-p)$ increases from 0 to the maximum value $\frac{1}{4}$ at $p=\frac{1}{2}$ and then decreases to 0 (see Figure 8.10). The maximum possible value of pq is $\frac{1}{4}$, so that the solution n must satisfy

$$n\leqslant\frac{1}{4}\left[\frac{z_{\alpha/2}}{d}\right]^2$$

Without any prior knowledge of the approximate value of p, the choice of this maximum n will provide the desired protection. If the value of p is known to be roughly in the neighborhood of a value p^*, then n can be

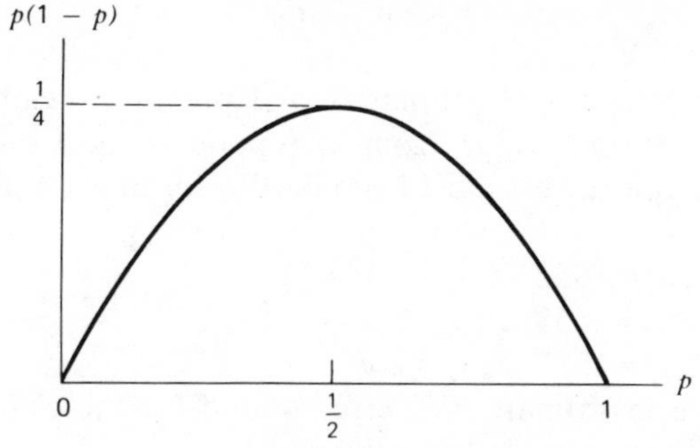

Figure 8.10 Values of $p(1-p)$.

determined from

$$n = p^*(1-p^*)\left[\frac{z_{\alpha/2}}{d}\right]^2$$

Example 8.17 A public health survey is to be designed to estimate the proportion p of a population having defective vision. How many persons should be examined if the public health commissioner wishes to be 98% certain that the error of estimation is within $\pm.05$ when (a) there is no knowledge about the value of p? (b) p is known to be about .3?

The tolerable error is $d = .05$. Also $(1-\alpha) = .98$, so that $\alpha/2 = .01$. From the normal table, we know that $z_{.01} = 2.33$. Therefore:

(a) Since p is unknown, the conservative value $\frac{1}{2}$ for p, gives an upper bound of

$$n = \frac{1}{4}\left[\frac{2.33}{.05}\right]^2 = 543$$

for the sample size.

(b) If $p^* = .3$, the required sample size is

$$n = (.3 \times .7)\left[\frac{2.33}{.05}\right]^2 = 456$$

■

REFERENCE

1. Beyer, W. H. (ed.), *Handbook of Tables for Probability and Statistics*, 2nd Ed., The Chemical Rubber Company, Cleveland, Ohio, 1968.

EXERCISES

1. Consider the problem of estimating a population mean μ based on a random sample of size n from the population. Compute a point estimate of μ and the estimated standard error in each of the following cases.
   (*a*) $n=70$, $\Sigma x_i=852$, $\Sigma(x_i-\bar{x})^2=215$.
   (*b*) $n=140$, $\Sigma x_i=1653$, $\Sigma(x_i-\bar{x})^2=464$.
   (*c*) $n=160$, $\Sigma x_i=1985$, $\Sigma(x_i-\bar{x})^2=475$.
2. Compute an approximate 95% error bound for the estimation of μ in each of the three cases in Exercise 1.
3. By what factor should the sample size be increased to reduce the standard error of $\bar{X}$ to:
   (*a*) $\frac{1}{2}$ its original value?
   (*b*) $\frac{1}{4}$ its original value?
4. Referring to Example 8.4, let

$$T_3=\frac{3X_1+X_2+X_3}{5} \qquad T_4=\frac{2X_1+X_2+2X_3}{5}$$

   be two other estimators of the mean yield.
   (*a*) Determine if these estimators are unbiased.
   (*b*) Compute the variances of these estimators and verify that both are larger than the variance of T_1.
5. An analyst wishes to estimate the market share captured by Brand X detergent; that is, the proportion of Brand X sales compared to the total sales of all detergents. From data supplied by several stores, the analyst finds that out of a total of 425 boxes of detergent sold, 120 were Brand X.
   (*a*) Estimate the market share captured by Brand X.
   (*b*) Compute the estimated 95% error bound. Also compute the 98% error bound.
6. A random sample of 2000 persons from the labor force of a large city are interviewed, and 165 of them are found to be unemployed.
   (*a*) Estimate the rate of unemployment based on the data.
   (*b*) Establish a 95% error bound for your estimate.
7. A zoologist wishes to estimate the mean blood sugar level of a species of animal when injected with a specified dosage of adrenaline. A sample of 55 animals of a common breed are injected with adrenaline, and their blood-sugar measurements are recorded in units of milligrams per 100 milliliters of blood. The mean and the standard deviation of these measurements are found to be 126.9 and 10.5, respectively.

(*a*) Give a point estimate of the population mean and find a 90% error bound.

(*b*) Determine a 90% confidence interval for the population mean.

8. An engineer wishes to estimate the mean setting time of a new gypsum cement mix used in highway spot repairs. From a record of the setting times for 100 spot repairs, the mean and the standard deviation are found to be 32 minutes and 4 minutes, respectively. Use these values to determine a 99% confidence interval for the mean setting time.
9. Referring to Exercise 5, determine a 95% confidence interval for the market share captured by Brand X detergent.
10. Referring to Exercise 6, compute a 98% confidence interval for the rate of unemployment.
11. To estimate the percentage of a species of rodent that carries a viral infection that produces mammary tumors, 128 rodents are histologically examined and 72 of them are found to be infected. Compute a 95% confidence interval for the population proportion.
12. Using the table of percentage points for the t distributions, find:

    (*a*) $t_{.05}$ when d.f. = 7.

    (*b*) $t_{.025}$ when d.f. = 12.

    (*c*) $t_{.01}$ when d.f. = 20.

    (*d*) $t_{.10}$ when d.f. = 8.
13. In an investigation on toxins produced by molds that infect corn crops, a biochemist prepares extracts of the mold culture with organic solvents and then measures the amount of the toxic substance per gram of solution. From 9 preparations of the mold culture, the following measurements of the toxic substance in milligrams are obtained: 1.2, .8, .6, 1.1, 1.2, .9, 1.5, .9, 1.0.

    (*a*) Calculate the mean $\bar{x}$ and the standard deviation s from the data.

    (*b*) Compute a 98% confidence interval for the mean weight of toxic substance per gram of mold culture. State the assumption you make about the population.
14. An experimenter studying the feasibility of extracting protein from seaweed to use in animal feed makes 18 determinations of the protein extract, each based on a different 50-kilogram sample of seaweed. The sample mean and the standard deviation are found to be 3.6 kilograms and .8 kilograms, respectively. Determine a 95% confidence interval for the mean yield of protein extract per 50 kilograms of seaweed.
15. In a lake pollution study, the concentration of lead in the upper sedimentary layer of a lake bottom is measured from 25 sediment samples of 1000 cubic centimeters each. The sample mean and the standard deviation of the measurements are found to be .38 and .06, respectively. Compute a 99% confidence interval for the mean con-

centration of lead per 1000 cubic centimeters of sediment in the lake bottom.

16. The supplier of a particular brand of vitamin pills claims that the average potency of these pills after a certain exposure to heat and humidity is at least 65. The distributor who wishes to buy these pills is interested in testing the supplier's claim. The distributor considers it a more serious mistake to buy a batch of pills with a mean retention of vitamin potency that is less than 65 than to reject a batch with a higher mean. From previous tests of similar vitamin pills, it is known that the potency measurements are approximately normally distributed with standard deviation $\sigma = 6$. To perform a quality inspection, the distributor intends to choose a random sample of 9 pills from a batch and measure their retention of potency after the specified exposure.
    (*a*) Formulate the hypotheses about the mean potency of a batch of pills μ and determine the rejection region of the test with $\alpha = .05$. (Assume that $\sigma = 6$.)
    (*b*) If $\mu = 67$, what is the probability of a type II error with this test?
    (*c*) Compute the power of the test at a few values of μ that range between 63 and 70 and plot the power curve.
17. Referring to the test in Exercise 16(a), suppose that the sample inspected from a particular batch yields the following data regarding potency: 63, 72, 64, 69, 59, 65, 66, 64, 65. Apply the test with these data and state your conclusions.
18. Again referring to Exercise 16, suppose that in addition to controlling α at .05 the distributor wants the rejection probability of a batch whose true mean potency is 68 to be less than .10.
    (*a*) Does the test in Exercise 16(a) meet this additional requirement?
    *(*b*) How should the sample size and the rejection region of the test be selected to meet this additional requirement?
19. A physical model suggests that the mean temperature increase in the water used as coolant in a compressor chamber should not be more than 5°C. Temperature increases in the coolant measured on 8 independent runs of the compressing unit revealed the following data: 6.4, 4.3, 5.7, 4.9, 6.5, 5.9, 6.4, 5.1.
    (*a*) Do the data contradict the assertion of the physical model? (Test at $\alpha = .05$.) State the assumption you make about the population.
    (*b*) Determine a 95% confidence interval for the mean increase of the temperature in the coolant.
20. The mean drying time of a brand of spray paint is known to be 90 seconds. The research division of the company that produces this paint contemplates that adding a new chemical ingredient to the paint will accelerate the drying process. To investigate this conjecture, the paint

with the chemical additions is sprayed on 15 surfaces and the drying times are recorded. The mean and the standard deviation computed from these measurements are 86 seconds and 4.5 seconds, respectively.
(*a*) Do these data provide strong evidence that the mean drying time is reduced by the addition of the new chemical?
(*b*) Construct a 98% confidence interval for the mean drying time of the paint with the chemical additive.

21. A car advertisement asserts that with the new collapsible bumper system, the average body repair cost for the damages sustained in a collision impact of 10 miles per hour does not exceed $300. To test the validity of this claim, 5 cars are crashed into a stone barrier at an impact force of 10 miles per hour and their subsequent body repair costs are recorded. The mean and the standard deviation are found to be $358 and $45, respectively. Do these data strongly contradict the advertiser's claim?
22. Rock specimens are excavated from a particular geological formation and are subjected to a chemical analysis to determine their percentage content of cadmium. After analyzing 25 specimens, the mean and the standard deviation are found to be 10.2 and 3.1, respectively. Suppose that a commercial extraction of this mineral will be economically feasible if the mean percentage content is at least 8.
(*a*) Do the data strongly support the feasibility of commercial extraction? (Test at $\alpha = .01$.)
(*b*) Construct a 99% confidence interval for the mean percentage content of cadmium in the geological formation.
23. Packs of a certain brand of cigarettes carry the specification that the average nicotine content is .60 milligrams per cigarette. A government testing agency chemically analyzes a random sample of 100 of these cigarettes and finds that the sample mean and the standard deviation are .63 milligrams and .11 milligrams, respectively.
(*a*) Assuming $\alpha = .05$, construct a test to determine if these results provide strong evidence that the mean nicotine content is higher than the specified value.
(*b*) What is the smallest level of significance α that will lead to a rejection of the null hypothesis on the basis of the data?
24. In a study to determine whether stimulants produce hyperactivity, a random sample of 55 purebred rats are subjected to an intravenous injection of 10 micrograms of morphine. During a period of 15–25 minutes after the injection, each rat is tested for hyperactivity, which is measured on a scale that assigns scores to particular types of reactions. The mean score of hyperactivity is 14.9 and the standard deviation is 2.8.

(*a*) Do these data strongly support the experimenter's conjecture that the population mean score is higher than 14?

(*b*) Compute a 98% confidence interval for the population mean level of hyperactivity produced by the stated dosage of the stimulant.

25. It is claimed that a new treatment is more effective than the standard treatment for prolonging the lives of terminal cancer patients. The standard treatment has been in use for a long time, and from records in medical journals the mean survival period is known to be 4.2 years. The new treatment is administered to 80 patients, and their duration of survival is recorded. The sample mean and the standard deviation are found to be 4.5 years and 1.1 years, respectively. Is the claim supported by these results? Test at $\alpha = .05$.

26. A marketing manager wishes to determine if lemon flavor and almond flavor in a dishwashing liquid are equally liked by consumers. Out of 250 consumers interviewed, 145 expressed their preference for the lemon flavor and the remaining 105 preferred the almond flavor.

    (*a*) Do these data provide strong evidence that there is a difference in popularity between the two flavors? (Test with $\alpha = .05$).

    (*b*) Construct a 95% confidence interval for the population proportion of consumers who prefer the almond flavor.

27. A recent survey conducted by the U.S. Census Bureau states that the national unemployment rate is 7.8%. A social research group takes a random sample of 1600 persons from the labor force population in a particular county and finds that 96 of them are unemployed. Is this strong evidence that the unemployment rate in the county is lower than the national figure?

28. Refer to Exercise 25. Researchers in cancer therapy often report only the number of patients who survive for a specified period of time after treatment rather than the patients' actual survival times. Suppose that 30% of the patients who undergo the standard treatment are known to survive 5 years. The new treatment is administered to 100 patients, and 38 of them are still alive after a period of 5 years.

    (*a*) With this type of data, formulate the hypotheses for testing the validity of the claim that the new treatment is more effective than the standard therapy.

    (*b*) Test with $\alpha = .05$ and state your conclusions.

29. A genetic model suggests that 80% of the plants grown from a cross between two given strains of seeds will be of the dwarf variety. After breeding 200 of these plants, 64 were not of the dwarf variety.

    (*a*) Does this observation strongly contradict the genetic model?

    (*b*) Construct a 95% confidence interval for the true proportion of dwarf plants obtained from the given cross.

30. Referring to Exercise 15, consider the problem of testing H_0: $\mu = \mu_0$ vs. H_1: $\mu \neq \mu_0$ at $\alpha = .01$, where μ denotes the population mean concentration of lead per 1000 cubic centimeters of sediment. What values of μ_0 will fail to be rejected on the basis of the data given in Exercise 15?
31. Referring to Exercise 26, denote the population proportion of consumers preferring the almond flavor by p and consider the problem of testing H_0: $p = p_0$ vs. H_1: $p \neq p_0$ at the 5% level of significance. From your answer to Exercise 26(*b*), give the range of values of p_0 for which the null hypothesis will fail to be rejected.
32. Using the table of percentage points for the χ^2 distributions, find:
    (*a*) The upper 5% point when d.f. = 5.
    (*b*) The upper 1% point when d.f. = 9.
    (*c*) The lower 2.5% point when d.f. = 16.
    (*d*) The lower 1% point when d.f. = 10.
33. Plastic sheets produced by a machine are periodically monitored for possible fluctuation in thickness. Uncontrollable heterogeneity in the viscosity of the liquid mold makes some variation in thickness measurements unavoidable. However, if the true standard deviation of thickness exceeds 1.5 millimeters, there is cause to be concerned about product quality. Thickness measurements in millimeters of 10 specimens produced on a particular shift resulted in the following data: 226, 228, 226, 225, 232, 228, 227, 229, 225, 230.
    (*a*) Do the data substantiate the suspicion that the process variability exceeded the stated level on this particular shift? (Test at $\alpha = .05$.) State the assumption you make about the population distribution.
    (*b*) Construct a 95% confidence interval for the true standard deviation of the thickness of sheets produced on this shift.
34. Referring to Exercise 22, one other indicator of the quality of an ore is the uniformity of its mineral content. Suppose that the ore quality is considered satisfactory if the true standard deviation σ of the percentage content of cadmium does not exceed 4. Using the data given in Exercise 22:
    (*a*) Construct a test to determine if there is strong evidence that σ is smaller than 4.
    (*b*) Construct a 98% confidence interval for σ.
35. Refer to Exercise 8. Prior to sampling, the engineer wishes to determine the sample size, (i.e., the number of observations of the setting time with the new mix) required to attain a desired precision in estimating the mean. From experience with other cement mixes, the engineer expects the standard deviation of the measurements to be approximately 5. How many observations of the setting times with the new mix should the engineer collect to be 95% certain that the error of estimation of the true mean does not exceed 1 minute?

36. In a large-scale, cost-of-living survey undertaken last January, weekly grocery expenses for families with 1 or 2 children were found to have a mean of \$70 and a standard deviation of \$15. To investigate the current situation, a random sample of families with 1 or 2 children is to be chosen and their last week's grocery expenses are to be recorded.
   (*a*) How large a sample should be taken if one wants to be 95% sure that the error of estimation of the population mean grocery expense per week for families with 1 or 2 children does not exceed \$2?
   (*b*) A random sample of 100 families is actually chosen, and from the data of their last week's grocery bills, the mean and the standard deviation are found to be \$75 and \$12, respectively. Construct a 98% confidence interval for the current mean grocery expense per week for the population of families with 1 or 2 children.
37. In a survey of the prevalence of drug use, randomly sampled high-school students are to be asked if they have ever used any hard drugs, such as LSD, Dexedrine, or heroin. Their responses are to be completely confidential.
   (*a*) How many students should be included in the sample if the investigator wishes to be 90% certain that the error of estimation of the proportion of hard-drug users does not exceed .02? (The investigator feels that about 10% of the students have tried hard drugs at least one time.)
   (*b*) Suppose that a random sample of 400 students is actually taken and that 52 of them report they have used hard drugs. Based on this information, construct a 95% confidence interval for the proportion of hard-drug users among high-school students.
38. In a psychological experiment, individuals are permitted to react to a stimulus in one of two ways, say A or B. The experimenter wishes to estimate the proportion of persons exhibiting reaction A, which we denote by p. How many persons should be included in the experiment to be 90% confident that the error of estimation is within .04 if the experimenter:
   (*a*) Knows that p is about .2.
   (*b*) Has no idea about the value of p.
39. A researcher in a heart and lung center wishes to estimate the rate of incidence of respiratory disorders among middle-aged males who have been smoking more than two packs of cigarettes per day during the last 5 years. How large a sample should the researcher select to be 95% confident that the error of estimation of the proportion of the population afflicted with respiratory disorders does not exceed .03? (The true value of p is expected to be near .15.)

40. To estimate the proportion of car owners who purchase more than \$150,000 of liability coverage in their automobile insurance policies, a random sample of 400 car owners is taken. If 56 of them are found to have chosen this extent of coverage:
    (*a*) Construct a 95% confidence interval for the proportion being estimated.
    (*b*) How large a sample should be chosen to estimate the proportion with a 95% error bound of .008?
41. A national safety council wishes to estimate the proportion of automobile accidents that involve pedestrians. How large a sample of accident records must be examined to be 98% certain that the estimate does not differ from the true proportion by more than .04? (The council believes that the true proportion is below .25.)
42. Henry Cavendish (1731–1810) provided direct experimental evidence of Newton's law of universal gravitation, which specifies the force of attraction between two masses. In an experiment with known masses determined by weighing, the measured force can also be used to calculate a value for the density of the earth. The values of the earth's density from Cavendish's renowned experiment in time order by column are:

5.36	5.62	5.27	5.46
5.29	5.29	5.39	5.30
5.58	5.44	5.42	5.75
5.65	5.34	5.47	5.68
5.57	5.79	5.63	5.85
5.53	5.10	5.34	

(These data were published in *Philosophical Transactions*, Vol. 17, 1798, p. 469.) Find a 99% confidence interval for the density of the earth.

MATHEMATICAL EXERCISES

1. *Cauchy-Schwarz inequality*. A useful result says that for any $a_1,\ldots,a_n$ and $b_1,\ldots,b_n$

$$\left(\sum_{i=1}^{n} a_i b_i\right)^2 \leqslant \left(\sum_{i=1}^{n} a_i^2\right)\left(\sum_{i=1}^{n} b_i^2\right)$$

Establish this inequality.
[*Hint*: Note that, for any z,

$$0 \leqslant \sum_{i=1}^{n} (a_i z + b_i)^2 = z^2\left(\sum_{i=1}^{n} a_i^2\right) + 2z\left(\sum_{i=1}^{n} a_i b_i\right) + \left(\sum_{i=1}^{n} b_i^2\right)$$

For the quadratic function of z to be $\geqslant 0$, we must have

$$\left(\sum_{i=1}^{n} a_i b_i\right)^2 - \left(\sum_{i=1}^{n} a_i^2\right)\left(\sum_{i=1}^{n} b_i^2\right) \leqslant 0 \qquad]$$

2. Let $Y_1, Y_2, \ldots, Y_n$ be a random sample from a population with mean μ and variance σ^2. Show that among all linear combinations,

$$a_1 Y_1 + a_2 Y_2 + \cdots + a_n Y_n$$

which are unbiased estimators of μ, the sample mean $\bar{Y} = \frac{1}{n} Y_1 + \frac{1}{n} Y_2 + \cdots + \frac{1}{n} Y_n$ has the smallest variance.

[*Hint*: To be unbiased, $\mu = E(a_1 Y_1 + \cdots + a_n Y_n) = a_1 \mu + \cdots + a_n \mu = (a_1 + \cdots + a_n)\mu$ for all possible values of μ. Therefore $a_1 + \cdots + a_n = 1$. Also,

$$\text{Var}(a_1 Y_1 + \cdots + a_n Y_n) = (a_1^2 + \cdots + a_n^2)\sigma^2$$

By the Cauchy-Schwarz inequality in Mathematical Exercise 1.

$$\left(\sum_{i=1}^{n} a_i^2\right)\left[\sum_{i=1}^{n} \frac{1}{n^2}\right] \geqslant \left[\sum_{i=1}^{n} a_i \frac{1}{n}\right]^2 = \frac{1}{n^2}\left(\sum_{i=1}^{n} a_i\right)^2 = \frac{1}{n^2}.$$

or $\sum_{i=1}^{n} a_i^2 \geqslant 1/n$. The choice $a_1 = \cdots = a_n = 1/n$ gives this minimum.]

3. *Combining two measurements of unequal precision*: Suppose that you have two measurements X and Y, both having means $\mu = E(X) = E(Y)$ but different variances σ_X^2 and σ_Y^2. When the variances are equal, we can combine the observations by taking the mean of the two observations as our estimator of μ. However, if σ_Y^2 is much larger than σ_X^2, clearly Y should not receive as much weight as X. Assume that X and Y are independent and that their variances are known.
   (*a*) In order that the combined estimator $aX + bY$ will be unbiased for μ, show that the constants a and b must satisfy the relation

   $$b = 1 - a$$

   Then for any arbitrary choice of the constant a, the estimator

   $$aX + (1-a)Y$$

   is unbiased for μ.

(*b*) Among all these unbiased estimators, find the one with the smallest variance. In other words, find the value of a in terms of the known σ_X^2 and σ_Y^2 such that $\text{Var}[aX+(1-a)Y]$ is minimized.

This situation would arise, for example, if we used two different meters to measure water pressure in a pipe and one were more accurate than the other. Check your answer in (*b*) to be sure that it reduces to $(X+Y)/2$ if the variances are equal.

[*Hint*: Show that

$$\text{Var}\left[aX+(1-a)Y\right]=a^2\sigma_X^2+(1-a)^2\sigma_Y^2$$

$$=\left(\sigma_X^2+\sigma_Y^2\right)\left[a^2-2a\frac{\sigma_Y^2}{\sigma_X^2+\sigma_Y^2}\right]+\sigma_Y^2$$

$$=\left(\sigma_X^2+\sigma_Y^2\right)\left[a-\frac{\sigma_Y^2}{\sigma_X^2+\sigma_Y^2}\right]^2+\frac{\sigma_X^2\sigma_Y^2}{\sigma_X^2+\sigma_Y^2}$$

Then the value of a that minimizes the variance is $\sigma_Y^2/(\sigma_X^2+\sigma_Y^2)$.]

4. *An unbiased estimator of* σ^2: Let $X_1, X_2, \ldots, X_n$ be a random sample from a population with a mean of μ and a variance of σ^2. Prove that the sample variance

$$s^2=\frac{1}{n-1}\sum_{i=1}^{n}\left(X_i-\bar{X}\right)^2$$

is an unbiased estimator of σ^2.

[*Hint*: Verify the following facts:

$$\sum_{i=1}^{n}\left(X_i-\bar{X}\right)^2=\sum_{i=1}^{n}X_i^2-n\bar{X}^2$$

$$E\left(X_i^2\right)=\text{Var}(X_i)+\left[E\left(X_i\right)\right]^2=\sigma^2+\mu^2$$

$$E\left(\bar{X}^2\right)=\frac{\sigma^2}{n}+\mu^2$$

and use them to calculate $E(s^2)$.]

5. If T is an unbiased estimator of θ, prove that T^2 must be a biased estimator of θ^2, except for the trivial case when $\text{Var}(T)=0$. [*Hint*: $E(T^2)=\text{Var}(T)+[E(T)]^2$.]
6. Use your results from mathematical Exercises 4 and 5 to show that the sample standard deviation s is not an unbiased estimator of σ.

CHAPTER 9

Comparing Two Treatments

9.1 INTRODUCTION

In virtually every area of human activity, search is continually underway to develop new modes of action or to modify and revise existing techniques. The exploration of new methods, whether in the field of industrial production, conservation of resources, health care, or public education, is not just the pastime of curious scientific minds but plays an indispensable role in meeting current societal needs. To comparatively study two or more competing techniques, we must perform experiments, collect informative data, and reach inferences from experimental evidence. The impact of statistical methods is immediately felt, because they provide a logical basis for the evaluation of evidence as well as for planning an effective process of data collection. Confining ourselves to the context of a comparison between two techniques or treatments, we introduce the basic statistical concepts in this chapter while emphasizing the important issues, including:

(*a*) The manner of conducting the experiment to obtain a valid basis for comparison.
(*b*) Methods of analyzing data to test if a real difference exists and to estimate the magnitude of the difference.

Example 9.1 Agricultural field trials: To ascertain if a new strain of seeds developed by a genetic cross actually produces a higher yield per acre compared to a current major variety, field trials must be performed by planting each variety under appropriate farming conditions. A record of crop yields from the field trials will form the data base for making a comparison between the two varieties. It may also be desirable to compare the varieties from standpoints of other factors, such as disease resistance and fertilizer requirements. An outstanding breakthrough providing food for millions, now known as the "green revolution," was accomplished by the extensive coordinated efforts of many scientists in undertaking comparative field trials with newer hybrids. ■

Example 9.2 Drug evaluation: Pharmaceutical researchers strive to synthesize chemicals to improve their efficiency in curing diseases. New chemicals may result from educated guesses concerning potential biological reactions, but evaluations must be based on a test of their effects on diseased animals or human beings. To test the effectiveness of a drug in controlling tumors in mice, several mice of an identical breed may be taken as experimental subjects. After infecting them with cancer cells, some will be subsequently treated with the drug and others will remain untreated as a control group. After a period of time, the size of the tumor will be recorded for each mouse in the control group as well as in the treated group. The data of tumor sizes for the two groups will then provide a basis for comparison in determining if the drug is effective in checking tumor growth. ■

Example 9.3 Monitoring advertising claims: The public is constantly bombarded with commercials that claim the superiority of one product brand in comparison to others. When such comparisons are founded on sound experimental evidence, they serve to educate the consumers. Not infrequently, however, misleading advertising claims are made due to insufficient experimentation, faulty analysis of data, or even direct manipulation of experimental results. Government agencies and consumer groups must be prepared to verify the comparative quality of products by using adequate data collection procedures and proper methods of statistical analysis. ■

These are only three examples of the comparison between two kinds of substances, techniques, products, or modes of action. In a general discussion the common statistical term *treatments* is often used to refer to the things that are being compared. In performing an experiment the basic units that are exposed to one treatment or another are called *experimental units* or *experimental subjects*, and the characteristic that is recorded after the application of a treatment to a subject is called the *response*. For instance, the two treatments involved in a comparative study may be two varieties of seeds, the use or nonuse of a drug, two training programs, or two geographical locations of farms. The experimental subjects may be garden plots, mice, children, or farmers, and the responses may be yield, size of tumor, score on a skill test, or income, respectively. The goal of our discussion may then be stated as a comparison of the response measurements between two groups of subjects, each of which receives one of the two treatments. Alternatively, each set of response measurements can be considered a sample from a population, and we can then speak in terms of a comparison between two population distributions.

The manner in which subjects are chosen and assigned to treatments is called an *experimental design* or a *sampling design*. A carefully planned design is crucial to a successful experiment that produces a valid comparison between two treatments. Two fundamental features of a well-planned comparative experiment are (*a*) randomization of subjects between treatments, and (*b*) pairing or matching like experimental units to reduce interference from extraneous factors. These two ideas are discussed in Sections 9.3 and 9.4, respectively.

A design for the sampling procedure determines the structure of the data, and the method of inference must be geared to the particular data structure. Two basic types of designs and their consequent data structures are examined in this chapter. In the first, a random sample is available from the population corresponding to treatment 1 that is independent of a random sample from the population corresponding to treatment 2. In a later case, experimental subjects are chosen in matched pairs, so that one member of each pair receives treatment 1 and the other receives treatment 2. An understanding of these two basic data structures and their underlying assumptions is essential to the selection of a correct method of inferential analysis.

9.2 INDEPENDENT RANDOM SAMPLES FROM TWO POPULATIONS

With the objective of drawing a comparison between two populations or, synonymously, between two treatments, we examine the situation in which the data are in the form of realizations of random samples of size n_1 from population 1 and size n_2 from population 2. The data are the measurements of response associated with the following experimental design. A collection of n_1+n_2 subjects are randomly divided into two groups of sizes n_1 and n_2. Each member of the first group receives treatment 1, and each member of the second group receives treatment 2.

Sample	*Summary statistics*
$X_1, X_2, \ldots, X_{n_1}$ from population 1	$\bar{X}=\frac{1}{n_1}\sum_{i=1}^{n_1} X_i, \quad s_1^2=\frac{\sum_{i=1}^{n_1}(X_i-\bar{X})^2}{n_1-1}$
$Y_1, Y_2, \ldots, Y_{n_2}$ from population 2	$\bar{Y}=\frac{1}{n_2}\sum_{i=1}^{n_2} Y_i, \quad s_2^2=\frac{\sum_{i=1}^{n_2}(Y_i-\bar{Y})^2}{n_2-1}$

Specifically, within the framework of statistical language, we are interested in making inferences about the parameter, which is

$$(\text{mean of population 1}) - (\text{mean of population 2}) = \mu_1 - \mu_2$$

Any confidence statements or tests of hypotheses must be based on assumptions regarding the structure of the underlying distributions.

Structure of data: independent random samples

(*a*) $X_1, X_2, \ldots, X_{n_1}$ is a random sample of size n_1 from population 1 whose mean is denoted by μ_1 and whose variance is denoted by σ_1^2.

(*b*) $Y_1, Y_2, \ldots, Y_{n_2}$ is a random sample of size n_2 from population 2 whose mean is denoted by μ_2 and whose variance is denoted by σ_2^2.

(*c*) $X_1, X_2, \ldots, X_{n_1}$ are independent of $Y_1, Y_2, \ldots, Y_{n_2}$. In other words, the response measurements under one treatment are unrelated to the response measurements under the other treatment.

These are the only assumptions required when the sample sizes n_1 and n_2 are both large. Not surprisingly, more distributional structure is required to formulate appropriate inference procedures for small samples. We now introduce the small-sample inference procedures that are valid under the following assumptions about the population distributions. Naturally the practical usefulness of such procedures in given situations depends on how closely these assumptions are realized.

9.2.1 Inferences From Small Samples

Further assumptions when sample sizes are small

(*a*) Both distributions are normal.

(*b*) The population variances σ_1^2 and σ_2^2 are equal.

The second assumption is somewhat artificial, but we reserve comment until after the approach to inference has been completely described.

In summary:

Small-sample assumptions:

(*a*) $X_1, X_2, \ldots, X_{n_1}$ is a random sample from $N(\mu_1, \sigma)$.

(*b*) $Y_1, Y_2, \ldots, Y_{n_2}$ is a random sample from $N(\mu_2, \sigma)$.

(NOTE: σ is the same for both distributions.)

(*c*) $X_1, X_2, \ldots, X_{n_1}$ and $Y_1, Y_2, \ldots, Y_{n_2}$ are independent.

What statistic should be considered in drawing inferences about $\mu_1 - \mu_2$? From our discussion in Chapter 8, we know that $\overline{X}$ has mean μ_1 and variance σ_1^2/n_1 and that $\overline{Y}$ has mean μ_2 and variance σ_2^2/n_2. Therefore

$$E(\overline{X} - \overline{Y}) = E(\overline{X}) - E(\overline{Y}) = \mu_1 - \mu_2$$

In other words, $\overline{X} - \overline{Y}$ is an unbiased estimator of $\mu_1 - \mu_2$. Moreover, because $\overline{X}$ and $\overline{Y}$ are independent, the covariance term is zero and

$$\mathrm{Var}(\overline{X} - \overline{Y}) = \mathrm{Var}(\overline{X}) + \mathrm{Var}(\overline{Y}) = \frac{\sigma_1^2}{n_1} + \frac{\sigma_2^2}{n_2}$$

For small samples we assume a common population variance, say σ^2. In this case

$$\mathrm{Var}(\overline{X} - \overline{Y}) = \sigma^2\left(\frac{1}{n_1} + \frac{1}{n_2}\right)$$

The common variance σ^2 can be estimated by combining information provided by both samples. Specifically, the sum $\Sigma_{i=1}^{n_1}(X_i - \overline{X})^2$ incorporates $n_1 - 1$ pieces of information about σ^2, in view of the constraint that the deviations $X_i - \overline{X}$ sum to zero. Independently of this, $\Sigma_{i=1}^{n_2}(Y_i - \overline{Y})^2$ comprises $n_2 - 1$ pieces of information about σ^2. These two quantities can then be combined to obtain a *pooled estimate* of the common σ^2. The proper divisor is the sum of the component degrees of freedom, or $(n_1 - 1) + (n_2 - 1) = n_1 + n_2 - 2$.

Pooled estimator of the common σ^2

$$s^2_{\text{pooled}} = \frac{\sum_{i=1}^{n_1}\left(X_i-\bar{X}\right)^2 + \sum_{i=1}^{n_2}\left(Y_i-\bar{Y}\right)^2}{n_1+n_2-2} = \frac{(n_1-1)s_1^2+(n_2-1)s_2^2}{n_1+n_2-2}$$

Example 9.4 To illustrate the calculation of pooled sample variance, consider these two samples:

Sample from population 1: 8, 5, 7, 6, 9, 7
Sample from population 2: 2, 6, 4, 7, 6

The sample means are

$$\bar{x} = \frac{\Sigma x_i}{6} = \frac{42}{6} = 7$$

$$\bar{y} = \frac{\Sigma y_i}{5} = \frac{25}{5} = 5$$

Further

$$(6-1)s_1^2 = \Sigma(x_i-\bar{x})^2$$
$$= (8-7)^2+(5-7)^2+(7-7)^2+(6-7)^2+(9-7)^2+(7-7)^2 = 10$$

$$(5-1)s_2^2 = \Sigma(y_i-\bar{y})^2$$
$$= (2-5)^2+(6-5)^2+(4-5)^2+(7-5)^2+(6-5)^2 = 16$$

The individual sample variances are $s_1^2 = 10/5 = 2$ and $s_2^2 = 16/4 = 4$.
To continue with the calculation of pooled variance

$$s^2_{\text{pooled}} = \frac{(n_1-1)s_1^2+(n_2-1)s_2^2}{n_1+n_2-2} = \frac{10+16}{6+5-2} = 2.89$$

which is closer to 2 than 4 because the first sample-size is larger. ■

Assuming normal populations, we know that $\bar{X}$ is distributed as $N(\mu_1, \sigma/\sqrt{n_1})$ and that $\bar{Y}$ is distributed as $N(\mu_2, \sigma/\sqrt{n_2})$. The statistic $\bar{X}-\bar{Y}$, the difference between two independent normal variables, is also normally distributed with mean $= \mu_1 - \mu_2$ and standard deviation $= \sigma\sqrt{(1/n_1)+(1/n_2)}$. Consequently

$$\frac{(\bar{X}-\bar{Y})-(\mu_1-\mu_2)}{\sigma\sqrt{\dfrac{1}{n_1}+\dfrac{1}{n_2}}} \quad \text{is} \quad N(0,1)$$

Employing the pooled estimator $\sqrt{s^2_{\text{pooled}}}$ for σ in this $N(0,1)$ variable, we obtain a Student's t variable that is basic to inferences about $\mu_1-\mu_2$.

$$\frac{(\bar{X}-\bar{Y})-(\mu_1-\mu_2)}{s_{\text{pooled}}\sqrt{\dfrac{1}{n_1}+\dfrac{1}{n_2}}}$$

has Student's t distribution, with n_1+n_2-2 degrees of freedom.

We can now obtain confidence intervals for $(\mu_1-\mu_2)$ as we did for the single-sample problem discussed in Chapter 8, by taking

Estimate of parameter $\pm$ (Estimated standard deviation) $\times$ (t value)

Confidence interval for $\mu_1-\mu_2$

A $100(1-\alpha)\%$ confidence interval for $\mu_1-\mu_2$ is given by

$$\bar{X}-\bar{Y} \pm t_{\alpha/2}s_{\text{pooled}}\sqrt{\frac{1}{n_1}+\frac{1}{n_2}}$$

where

$$s^2_{\text{pooled}} = \frac{(n_1-1)s_1^2+(n_2-1)s_2^2}{n_1+n_2-2}$$

and $t_{\alpha/2}$ is the upper $\alpha/2$ point of the t distribution with d.f. $= n_1+n_2-2$.

Our intent is still to stress confidence intervals, because they provide intervals of plausible values for the unknown parameter and constitute the most comprehensive form of inference. Because tests about the parameter $\mu_1 - \mu_2$ are sometimes of interest, we note the important details of such tests here.

Testing $H_0: \mu_1 - \mu_2 = \delta_0$

Test statistic:

$$t = \frac{\bar{X} - \bar{Y} - \delta_0}{s_{\text{pooled}}\sqrt{\frac{1}{n_1} + \frac{1}{n_2}}}, \quad \text{d.f.} = n_1 + n_2 - 2$$

Alternative hypothesis:	Level α rejection region:
$H_1: \mu_1 - \mu_2 \neq \delta_0$	$\lvert t \rvert > t_{\alpha/2}$
$H_1: \mu_1 - \mu_2 > \delta_0$	$t > t_\alpha$
$H_1: \mu_1 - \mu_2 < \delta_0$	$t < -t_\alpha$

Example 9.5 A feeding test is conducted on a herd of 25 milking cows to compare two diets, one of dewatered alfalfa and the other of field-wilted alfalfa. Dewatered alfalfa has an economic advantage in that its mechanical processing produces a liquid protein-rich by-product that can be used to supplement the feed of other animals. A sample of 12 cows randomly selected from the herd are fed dewatered alfalfa; the remaining 13 cows are fed field-wilted alfalfa. From observations made over a three-week period, the average daily milk production in pounds recorded for each cow is:

Field-wilted alfalfa	44, 44, 56, 46, 47, 38, 58, 53, 49, 35, 46, 30, 41
Dewatered alfalfa	35, 47, 55, 29, 40, 39, 32, 41, 42, 57, 51, 39

(*a*) Obtain a 95% confidence interval for the difference in mean daily milk yield per cow between the two diets. (*b*) Do the data strongly indicate that the milk yield is less with dewatered alfalfa than with field-wilted alfalfa? (Test at $\alpha = .05$.)

The dot diagrams of these data, plotted in Figure 9.1, give the appearance of approximately equal amounts of variation.

We assume that the milk-yield data for both field-wilted and dewatered

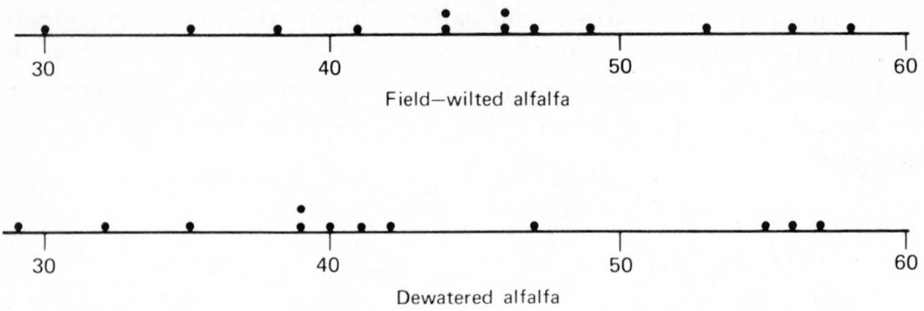

Figure 9.1 Dot diagrams of the milk-yield data in Example 9.5.

alfalfa are random samples from normal populations with means of μ_1 and μ_2, respectively, and with a common variance of σ^2. Computations from these data provide the summary statistics

$$\text{Field-wilted alfalfa:} \quad \bar{x}=45.15, \quad \Sigma(x_i-\bar{x})^2=767.69, \quad s_1^2=64.0$$

$$\text{Dewatered alfalfa:} \quad \bar{y}=42.25, \quad \Sigma(y_i-\bar{y})^2=840.25, \quad s_2^2=76.4$$

The sample sizes are $n_1=13$ and $n_2=12$. The pooled sample variance is

$$s^2_{\text{pooled}}=\frac{\Sigma(x_i-\bar{x})^2+\Sigma(y_i-\bar{y})^2}{n_1+n_2-2}=\frac{767.69+840.25}{23}=69.9$$

(*a*) With a 95% confidence level $\alpha/2=.025$, and consulting the t table we find that $t_{.025}=2.069$ with d.f. $=n_1+n_2-2=23$. Thus a 95% confidence interval for $\mu_1-\mu_2$ is

$$\bar{x}-\bar{y}\pm t_{.025}s_{\text{pooled}}\sqrt{\frac{1}{n_1}+\frac{1}{n_2}}=45.15-42.25\pm 2.069\times 8.36\sqrt{\frac{1}{13}+\frac{1}{12}}$$

$$=2.90\pm 6.92 \quad \text{or} \quad (-4.02, 9.82)$$

This means that we can be 95% confident the mean yield from field-wilted alfalfa can be as little as 4.02 pounds lower or as much as 9.82 pounds higher than for dewatered alfalfa.

(*b*) To test $H_0: \mu_1 = \mu_2$ vs. $H_1: \mu_1 > \mu_2$, we will employ the test statistic

$$t = \frac{\bar{x} - \bar{y}}{s_{\text{pooled}} \sqrt{\frac{1}{n_1} + \frac{1}{n_2}}}$$

Assuming $\alpha = .05$, the one-sided rejection region is $R: t > t_{.05}$. The observed value of t is then

$$t = \frac{45.15 - 42.25}{8.63\sqrt{\frac{1}{13} + \frac{1}{12}}} = \frac{2.90}{3.45} = .84$$

For d.f. $= 23$, the tabled value $t_{.05} = 1.714$ which is larger than the observed value. Hence the null hypothesis is not rejected with $\alpha = .05$. ■

9.2.2 Inferences From Large Samples

When both sample sizes n_1 and n_2 are greater than 25 or 30, the assumptions concerning small samples can be greatly relaxed. It is no longer necessary to assume that the parent distributions are normal, because the central limit theorem assures that $\bar{X}$ is distributed approximately as $N(\mu_1, \sigma_1/\sqrt{n_1})$ and that $\bar{Y}$ is approximately distributed as $N(\mu_2, \sigma_2/\sqrt{n_2})$. Subtracting the mean and dividing by the standard deviation, produces the result

$$\frac{\bar{X} - \bar{Y} - (\mu_1 - \mu_2)}{\sqrt{\frac{\sigma_1^2}{n_1} + \frac{\sigma_2^2}{n_2}}} \qquad \text{approximately} \quad N(0, 1)$$

Because n_1 and n_2 are both large, the approximation remains valid if σ_1^2 and σ_2^2 are replaced by their sample estimators $s_1^2 = \sum_{i=1}^{n_1}(X_i - \bar{X})^2/(n_1 - 1)$ and $s_2^2 = \sum_{i=1}^{n_2}(Y_i - \bar{Y})^2/(n_2 - 1)$. The assumption of equal population variances is not required in inferences derived from large samples. The primary formulas for confidence intervals and tests of hypotheses are noted here. The choice of a one-sided or a two-sided rejection region depends on the type of alternative hypothesis.

Large sample inferences for $\mu_1 - \mu_2$

An approximate $100(1-\alpha)\%$ confidence interval for $\mu_1 - \mu_2$ is given by

$$\bar{X} - \bar{Y} \pm z_{\alpha/2} \sqrt{\frac{s_1^2}{n_1} + \frac{s_2^2}{n_2}}$$

where $z_{\alpha/2}$ is the upper $\alpha/2$ point of $N(0, 1)$.

In Testing $H_0: \mu_1 - \mu_2 = \delta_0$, the test statistic is

$$Z = \frac{\bar{X} - \bar{Y} - \delta_0}{\sqrt{\frac{s_1^2}{n_1} + \frac{s_2^2}{n_2}}}$$

which has approximately a $N(0, 1)$ distribution under H_0.

The most striking difference from the small-sample solutions is that inferences drawn from large samples are valid over a much broader spectrum of underlying population distributions. Large-sample solutions only require the distribution of the sample means, not the original populations, to be nearly normal, and this is ensured by the central limit theorem. Moreover, the artificial assumption $\sigma_1 = \sigma_2$ that is used in small-sample solutions is avoided in large-sample situations.

Example 9.6 Rural and urban students are to be compared on the basis of their scores on a nationwide musical aptitude test. Two random samples of sizes 90 and 100 are selected from rural and urban seventh grade students. The summary statistics from the test scores are:

	Rural	Urban
Sample size	90	100
Mean	76.4	81.2
Standard deviation	8.2	7.6

(*a*) Establish a 98% confidence interval for the difference in population mean scores between urban and rural students. (*b*) Given a 2% level of significance, construct a test to determine if there is a significant difference between the population mean scores.

We denote the population mean scores of seventh grade students in rural and urban areas by μ_1 and μ_2, respectively. From the sample data we know that $n_1=90$, $n_2=100$, $\bar{x}=76.4$, $\bar{y}=81.2$, $s_1=8.2$, $s_2=7.6$.

(*a*) Since $1-\alpha=.98$, we have $\alpha/2=.01$ and $z_{.01}=2.33$. Thus a 98% confidence interval for $\mu_1-\mu_2$ is

$$\bar{x}-\bar{y}\pm z_{.01}\sqrt{\frac{s_1^2}{n_1}+\frac{s_2^2}{n_2}}$$

$$=76.4-81.2\pm 2.33\sqrt{\frac{(8.2)^2}{90}+\frac{(7.6)^2}{100}}$$

$$=-4.8\pm 2.7 \quad \text{or} \quad (-7.5,-2.2)$$

We conclude, with 98% confidence, that the mean of the urban scores is at least 2.2 units higher and can be as much as 7.5 units higher than the mean of the rural scores. Our 98% confidence derives from the fact that approximately 98% of the intervals, calculated in this manner from repeated samples of sizes 90 and 100, will cover the true mean difference $\mu_1-\mu_2$.

(*b*) Because the confidence interval does not include zero, the null hypothesis $H_0:\mu_1=\mu_2$ is rejected at level $\alpha=.02$ in favor of the alternative $H_1:\mu_1\neq\mu_2$. ■

9.3 RANDOMIZATION AND ITS ROLE IN INFERENCE

Now that we have presented the details as to how to implement statistical procedures in drawing inferences about the difference between two population means, we concentrate on some important questions regarding the design of the experiment or data-collection procedure. The manner in which experimental subjects are chosen for the two treatment groups can be crucial to the success of an inference or a comparison drawn about the treatments. For example, suppose that a remedial-reading instructor has developed a new teaching technique and is permitted to use the new method to instruct half the pupils in a class. The instructor might choose the most alert or the students who are more promising in some other way, leaving the weaker students to be taught in the conventional manner. Clearly, a comparison between the reading achievements of these two groups would not just be a comparison of two teaching methods. A similar fallacy can result in comparing the nutritional quality of a new school lunch package if the new diet is given to a group of children suffering from malnutrition and the conventional diet is given to a group of children who are already in good health. A comparison of health improvement between these groups is likely to reveal a much more dramatic performance for the new diet than is due to its inherent quality. Such a faulty experimental design actually occurred in a nutritional study of milk among Scottish school children a few years ago.

When the choice of which experimental unit is to receive each treatment is under our control, steps must be taken to ensure logical support for any inference concerning a comparison between the two treatments. At the core of a valid comparison lies the principle of impartial selection, which is given the statistical name *randomization*. Because human minds are potentially partial, the principle of randomization is based on the idea that a choice of the experimental units for one treatment or the other must be left to a chance mechanism that does not favor one particular selection over any other. The choice must not be left to the discretion of the experimenter, no matter how balanced or impartial a person he or she may appear to be.

Suppose that a comparative experiment is to be run with n experimental units, of which n_1 units are to be assigned to treatment 1 and the remaining $n_2 = n - n_1$ units are to be assigned to treatment 2. When little is known about the relative magnitude of the variability of responses under the two treatments or when the variances are expected to be nearly the same, it is advisable to make the sample sizes n_1 and n_2 nearly equal. Recalling the rule of combinations discussed in Section 3.5, we know that there are $\binom{n}{n_1}$ possible choices of n_1 units to assign to treatment 1. The principle of randomization tells us that the n_1 units for treatment 1 must be chosen at random from the available collection of n units; that is, in a manner such that all $\binom{n}{n_1}$ possible choices are equally likely to be selected.

> *Randomization procedure for comparing two treatments*
>
> From the available $n = n_1 + n_2$ experimental units, choose n_1 units at random to receive treatment 1 and assign the remaining n_2 units to treatment 2. The random choice entails that all $\binom{n}{n_1}$ possible selections are equally likely to be chosen.

As a mode of implementation of the random selection, which is conceptually simple although it is not the most convenient practical method, we can label n available units with numbers from 1 to n. Then n identical cards, each marked with a different number from 1 to n, can be placed in a container, shuffled thoroughly, and n_1 cards can be drawn blindfolded. The experimental units corresponding to the numbers on the selected cards can be assigned to treatment 1, and the remaining units can be assigned to treatment 2. A quicker and more efficient means of random sampling can be accomplished by using random number tables, to be discussed in Chapter 16.

Although randomization is not a difficult concept, it is one of the most fundamental principles of a good experimental design. Randomization safeguards against the introduction of bias in choosing subjects for the two treatment groups and thereby greatly enhances the credibility of a comparison. Uncontrolled sources of variation that enter through the preexisting conditions of the experimental units tend to cause variability in the responses that is not an effect of the treatments. Randomization guarantees that these external sources have the same chance of helping the response of treatment 1 as they do of helping the response of treatment 2. Any systematic effects of uncontrolled variables, such as age, strength, resistance, or intelligence, are chopped up or confused in their attempt to influence the treatment responses, although the presence of such effects causes more variability of response within each group.

> Randomization prevents uncontrolled sources of variation from influencing the responses in a systematic manner.

The reader may wonder why the principle of randomization, which is essential during the data-collection stage, did not appear in the prescription for confidence intervals and tests detailed in Section 9.2. This is because we have made the idealistic assumptions that the response observations constitute independent random samples and that the only factor that changes systematically between the two sets is the treatment. Actually, the process of randomization makes this assumption plausible, and thus it enhances the validity of our inference procedures.

As another example of the manner in which randomization provides protection against misleading comparisons, suppose that a chemical engineer is interested in determining how a new temperature setting affects the quality of paper produced, during a particular stage of the production cycle. Also suppose that unknown to the experimenter the new temperature setting has no effect on quality, but that the quality drifts over time, as shown in Figure 9.2.

If the new temperature setting is tried only at the later times, say 4 hours or more, and the old setting is tried only at the earlier times, the mean quality with the new setting would appear to be substantially higher than the mean quality with the old setting. The engineer would then reach the *incorrect* conclusion that the new temperature setting improved product

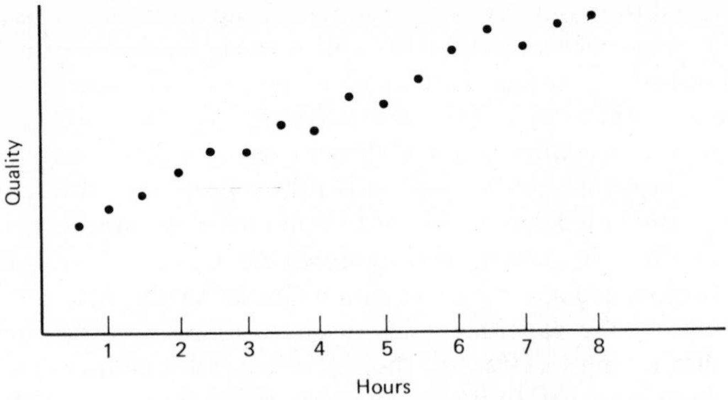

Figure 9.2 Quality drifting over time.

quality. This mistake can be avoided if each setting is tried at randomly selected times. Then some relatively high as well as some low product quality will be observed for each setting.

In many situations, a random selection of experimental subjects in the true sense of the term may not be practically feasible. This is usually the case in many clinical trials, where a doctor cannot be asked to practice a different therapy on two randomly chosen groups of patients. One therapy may have to be uniformly practiced in a clinic in one locality and another clinic may be chosen for the second treatment. Before comparing the treatments from two sets of data from two different places, the data must be carefully examined for other factors that could influence the therapeutic results, such as general health, age, or socioeconomic conditions. In the presence of a systematic variation of external conditions between the two treatment groups, an interpretation of a comparison of treatments can be dubious. A similar difficulty in interpretation arises when comparing the crime rates of cities before and after a new law. Aside from a package of criminal laws, other factors such as poverty, inflation, and unemployment play a significant role in the prevalence of crime. As long as these contingent factors cannot be regulated during the observation period, caution should be exercised in crediting the new law if a decline in the crime rate is observed or in discrediting the new law if an increase in the crime rate is observed.

9.4 PAIRED COMPARISONS

In comparing two treatments, it is desirable that the experimental units be as homogeneous as possible, so that a difference in responses between the two groups can be attributed to differences in treatment. If some identifiable conditions that can influence the response are allowed to vary over the units in an uncontrolled manner, they would introduce a large variability in the measurements and could obscure a real difference in treatment effects. On the other hand, the requirement of homogeneity may impose a severe limitation on the number of subjects available for a comparative experiment. To compare two analgesics, for example, it would be impractical to look for a sizable number of patients who are of the same sex, age, and general health condition and who have the same severity of pain. Aside from the question of practicality, we would rarely want to confine a comparison to such a narrow group. A broader scope of inference can be attained by applying the treatments on a variety of patients of both sexes and different age groups and health conditions.

The concept of *matching* or *blocking* is fundamental to providing a compromise between the two conflicting requirements of homogeneity and diversity of experimental units. The procedure consists of choosing units in groups or blocks so that the units in each block are homogeneous and the units in different blocks are dissimilar. Some of the units within each block are assigned to treatment 1; the others are assigned to treatment 2. This process preserves the effectiveness of a comparison within each block and permits a diversity of conditions to exist in different blocks. For example, in studying how two different environments influence the learning capacities of preschoolers, it is desirable to remove the effect of heredity: ideally this is accomplished by working with twins.

Here we discuss a simple form of blocking where each block consists of a pair of like experimental units: one of which receives treatment 1 and the other receives treatment 2. Of course, the treatments must be allotted to each pair randomly to avoid selection bias. This design is called *sampling by matched pairs* or a *paired comparison*.

Paired comparison

Pair or block:	Experimental units:	
1	2	1
2	1	2
3	1	2
⋮	⋮	⋮
n	2	1

Units in each pair are alike, whereas units in different pairs may be dissimilar. In each pair, a unit is chosen at random to receive treatment 1; the other unit receives treatment 2.

In making a paired comparison the response of an experimental unit can be viewed as composed of (*a*) an effect contributed by the conditions prevailing in the block, (*b*) a treatment effect, and (*c*) an unexplained chance component. By taking the difference between the two observations in a block, we can filter out the common block effect. These differences then permit us to focus on the effects of treatments that are freed from undesirable sources of variation.

> *Pairing (or blocking)*
>
> Pairing like experimental units according to some identifiable characteristic(s) serves to remove this source of variation from the experiment.

The structure of the observations in a paired comparison is given below, where X and Y denote the responses to treatment 1 and treatment 2, respectively. The difference between the responses in each pair is recorded in the last column, and the summary statistics are also presented.

Structure of data for a paired comparison

Pair	Treatment 1	Treatment 2	Difference
1	X_1	Y_1	$D_1 = X_1 - Y_1$
2	X_2	Y_2	$D_2 = X_2 - Y_2$
⋮	⋮	⋮	⋮
n	X_n	Y_n	$D_n = X_n - Y_n$

The pairs $(X_1, Y_1), (X_2, Y_2), \ldots, (X_n, Y_n)$ are independent.

Summary statistics:

$$\bar{D} = \frac{1}{n}\sum_{i=1}^{n} D_i, \quad s_D^2 = \frac{\sum_{i=1}^{n}(D_i - \bar{D})^2}{n-1}$$

Although the pairs (X_i, Y_i) are independent, X_i and Y_i within the ith pair, will usually be dependent. In fact, if the pairing of experimental units is effective, we would expect X_i and Y_i to be relatively large or small together. Expressed in another way, we would expect (X_i, Y_i) to have a high positive correlation. Because the differences $D_i = X_i - Y_i$, $i = 1, 2, \ldots, n$ are freed from the block effects, it is reasonable to assume that they constitute a random sample from a population with mean $= \delta$ and variance $= \sigma_D^2$, where δ represents the mean difference of the treatment effects. In other words

$$E(D_i) = E(X_i - Y_i) = \delta$$

$$\text{Var}(D_i) = \text{Var}(X_i - Y_i) = \sigma_D^2, \quad i = 1, \ldots, n$$

If the mean difference δ is zero, then the two treatments can be considered equivalent. A positive δ signifies that treatment 1 has a higher mean response than treatment 2. Considering $D_1, \ldots, D_n$ to be a single random sample from a population, we can immediately apply the techniques discussed in Chapter 8 to learn about the population mean δ.

Small sample inference about the mean difference δ

Assume that the differences $D_i = X_i - Y_i$ are independent with a $N(\delta, \sigma_D)$ distribution. Let

$$\bar{D} = \sum_{i=1}^{n} D_i / n \quad \text{and} \quad s_D = \sqrt{\sum_{i=1}^{n} \left(D_i - \bar{D}\right)^2 / (n-1)} \ .$$

Then:

(*a*) *A* $100(1-\alpha)\%$ *confidence interval for δ is given by*

$$\bar{D} \pm t_{\alpha/2} s_D / \sqrt{n}$$

where t corresponds to $n-1$ degrees of freedom.

(*b*) A test of $H_0 : \delta = \delta_0$ is based on the test statistic

$$t = \frac{\bar{D} - \delta_0}{s_D / \sqrt{n}}, \quad \text{d.f.} = n-1$$

As we learned in Chapter 8, the assumption of an underlying normal distribution can be relaxed when the sample size is large. Applying the central limit theorem to the differences $D_1, \ldots, D_n$ suggests a nearly normal distribution of $(\bar{D} - \delta)/(s_D/\sqrt{n}\,)$ when n is large, say greater than 30. The inferences can then be based on the percentage points of the $N(0, 1)$ distribution or, equivalently, on the percentage points of the t distribution, with the degrees of freedom marked "infinity."

Example 9.7 A medical researcher wishes to determine if a pill has the undesirable side effect of reducing the blood pressure of the user. The study involves recording the initial blood pressures of 15 college-age women. After they use the pill regularly for six months, their blood pressures are again recorded. The researcher wishes to draw inferences about the effect of the pill on blood pressure from the observations given in Table 9.1.

TABLE 9.1
BLOOD-PRESSURE MEASUREMENTS BEFORE AND AFTER USE OF PILL

SUBJECT	1	2	3	4	5	6	7	8	9	10	11	12	13	14	15
Before (x)	70	80	72	76	76	76	72	78	82	64	74	92	74	68	84
After (y)	68	72	62	70	58	66	68	52	64	72	74	60	74	72	74
$d=(x-y)$	2	8	10	6	18	10	4	26	18	−8	0	32	0	−4	10

(Courtesy of a family planning clinic)

Here each subject represents a block generating a pair of measurements: one before using the pill and the other after using the pill. The paired differences $d_i = x_i - y_i$ are computed in the last row of Table 9.1, and the following summary statistics are calculated:

$$\bar{d} = \frac{\sum_{i=1}^{15} d_i}{15} = 8.80, \quad s_D = \sqrt{\frac{\sum_{i=1}^{15} (d_i - \bar{d})^2}{14}} = 10.98$$

Assuming that the paired differences constitute a random sample from a normal population $N(\delta, \sigma_D)$, a 95% confidence interval for the mean difference δ is given by

$$\bar{d} \pm t_{.025} \frac{s_D}{\sqrt{15}}$$

where $t_{.025}$ is based on d.f. = 14. From the t table we find $t_{.025} = 2.145$. The 95% confidence interval can then be computed as

$$8.80 \pm 2.145 \times \frac{10.98}{\sqrt{15}} = 8.80 \pm 6.08 \quad \text{or} \quad (2.72, 14.88)$$

This means that we are 95% confident the mean reduction for subjects is between 2.72 and 14.88.

Do these data substantiate the claim that use of the pill reduces blood pressure? Because the confidence interval includes only positive values of δ, a reduction of blood pressure is strongly indicated. We may wish to formally test the null hypothesis $H_0:\delta=0$ vs. $H_1:\delta>0$. Assuming $\alpha=.05$, the one-sided rejection region is $t>t_{.05}=1.761$. The observed value of the test statistic

$$t=\frac{\bar{d}}{s_D/\sqrt{n}}=\frac{8.80}{10.98/\sqrt{15}}=\frac{8.80}{2.84}=3.10$$

falls in the rejection region. Consequently H_0 is rejected in favor of H_1 at a 5% level of significance.

To be certain that the pill causes the reduction in blood pressure, it is advisable to measure the blood pressures of the same subjects once again after they have stopped using the pill for a period of time. This amounts to performing the experiment in reverse order to check the findings of the first stage. ■

Example 9.7 is a typical before–after situation. Data gathered to determine the effectiveness of a safety program or an exercise program would have the same structure. In such cases there is really no way to choose how to order the experiments within a pair. The before situation must precede the after situation. If something other than the institution of the program causes performance to improve, the improvement will be incorrectly credited to the program. However, when the order of the application of treatments can be determined by the investigator, something can be done about such systematic influences. Suppose that a coin is flipped to select the treatment for the first unit in each pair and that the other treatment is then applied to the second unit. Any change in uncontrolled variables then has an equal chance of helping the performance of treatment 1 or treatment 2. After eliminating an identified source of variation by pairing, we return to randomization in an attempt to reduce the systematic effects of any uncontrolled sources of variation.

Randomization with pairing

After pairing, a coin should be flipped to determine which treatment should be applied to the first unit in a pair.

As before, this restricted randomization within each pair chops up or confuses any systematic influences, that we are unable to control.

9.5 CHOOSING BETWEEN INDEPENDENT SAMPLES AND PAIRED SAMPLES

At the stage of planning an experiment to compare two treatments, we must often choose between designing two independent samples or designing a sample with paired observations. To provide guidelines to help us make this decision, some comments about the pros and cons of these two sampling methods are in order here. Because a paired sample with n pairs of observations contains $2n$ measurements, a comparable situation would be two independent samples with n observations in each. Noting that $\bar{D}=\bar{X}-\bar{Y}$, the confidence intervals for the difference between treatment effects using either sampling method have the common structure

$$(\bar{X}-\bar{Y})\pm t_{\alpha/2}\left[\text{estimated sd of } (\bar{X}-\bar{Y})\right]$$

However, d.f. $=n-1$ for t using a paired sample, whereas d.f. $=2n-2$ using independent samples. The formulas for the estimated standard deviation are also different for the two sampling methods. Because the length of a confidence interval is determined by these two components, we now examine their behavior under the two competing sampling processes.

Paired sampling results in a loss of degrees of freedom and consequently, in a larger value of $t_{\alpha/2}$. For instance, a paired sample with $n=10$ has $t_{.05}=1.833$ with d.f. $=9$. But the t value associated with independent samples each of size 10, is $t_{.05}=1.734$ with d.f. $=18$. Thus if all other factors remain equal, then a loss of degrees of freedom tends to make confidence intervals larger for paired samples. Likewise, in testing hypotheses a loss of degrees of freedom for the t test results in a loss of power.

The merit of paired sampling emerges when we turn our attention to the second component. Recall that in independent sampling, treatments 1 and 2 are applied to randomly chosen groups of experimental units. If some identifiable condition of the experimental units that can appreciably influence the response is allowed to vary over the two groups in an uncontrolled manner, a marked variability will be produced in the measurements for each group. Consequently, a large standard deviation of $\bar{X}-\bar{Y}$ will result. If units are paired so that the interfering factor is held constant, or nearly so, between members of each pair, the treatment responses X and Y within each pair will be equally affected by this factor. If the prevailing condition in a pair causes the X measurement to be large, it will also cause the corresponding Y measurement to be large and vice versa. A statistical way of stating this is to say that "a good pairing of units results in X and Y having a positive covariance." Since

$$\text{Var}(X-Y)=\text{Var}(X)+\text{Var}(Y)-2\,\text{Cov}(X,Y)$$

the variance of the difference $X-Y$ will be smaller in the case of an effective pairing than it will be in the case of independent random variables. This reduction in variance causes $\bar{D}=\bar{X}-\bar{Y}$ to have a smaller standard deviation in the case of paired sampling and the estimated standard deviation to be typically smaller as well. With an effective pairing, the reduction in the standard deviation usually more than compensates for the loss of degrees of freedom.

In Example 9.7 concerning the effect of a pill in reducing blood pressure, we noted that a number of important factors (age, weight, height, general health, etc.) affect a person's blood pressure. By measuring the blood pressure of the same person before and after use of the pill these influencing factors can be held nearly constant for each pair of measurements. Independent samples of one group of persons using the pill and a separate control group of persons not using the pill may produce a greater variability in blood-pressure measurements if all the persons selected are not similar in age, weight, height, and general health.

As a further illustration we consider an experiment comparing the wear of two brands of bicycle tires under conditions of normal usage. Using the scheme of independent sampling, a random sample of n_1 bicycles are mounted with one brand of tire and another random sample of n_2 bicycles are mounted with the other brand. Suppose that the amount of wear is to be measured for each tire after 1000 miles of riding. To simplify this discussion we consider recording a single measurement per bicycle, which we assume is the average wear on the front and back tires. The quantity $\bar{X}-\bar{Y}$ can then be computed and with the formula for the standard deviation under independent sampling, used to draw a comparison between the quality of the two brands. Unfortunately, extraneous factors such as road conditions, riders' weight, and riding habits affect the wear of a tire. When allowed to vary in an uncontrolled manner, these factors will confound the measurements of quality difference. Although randomization helps distribute these effects equally between $\bar{X}$ and $\bar{Y}$,the extraneous disturbances make the standard deviation of the wear measurements large and thereby reduce the precision of inference. A more effective comparison can be obtained by mounting one tire of each brand on each bicycle, so that the interfering factors remain constant for every pair of measurements. This experimental design is nothing but paired sampling. Note that the two brands must be randomly assigned to the front and back wheels of each bicycle for a valid comparison using the t statistic.

Paired sampling is preferable to independent sampling only when an appreciable reduction in variability can be anticipated by means of pairing. When the experimental units are already homogeneous or when their heterogeneity cannot be linked to identifiable factors, an arbitrary pairing may fail to achieve a reduction in variance. The loss of degrees of freedom will then make a paired comparison less precise.

9.6 COMPARISON OF TWO BINOMIAL PROPORTIONS

We now turn to statistical inferences concerning a comparison between the rates of incidence of a characteristic in two populations. The unknown proportion of elements possessing the particular characteristic in population 1 and in population 2 are denoted by p_1 and p_2, respectively. Our aims are to construct confidence intervals for the parameter $(p_1 - p_2)$ and to test the null hypothesis of no difference $H_0: p_1 = p_2$. Comparing infant mortality in two ethnic groups, the unemployment rates in rural and urban populations, and the proportion of substandard items produced by two competing manufacturing processes are only three examples of this type. Inherent in this concept are also comparisons between two treatments where the response of an experimental subject falls into one of two possible categories that we may technically call "success" and "failure." The success rates for the two treatments can then be identified as the two population proportions p_1 and p_2.

The major distinction between the present inference problem and problems treated in preceding sections is that the sample data here involve binomial counts rather than measurements on a scale. Specifically, a random sample of size n_1 is taken from population 1 and the number of successes is denoted by X. An independent random sample of size n_2 is taken from population 2 and the number of successes is denoted by Y. Because a sample proportion is the basic quantity used to draw inferences about a population proportion, we naturally consider the statistic $\hat{p}_1 - \hat{p}_2$ to be a starting point for making inferences about $p_1 - p_2$.

Structure of inference

Parameter: $p_1 - p_2 =$ (Proportion in population 1) $-$ (Proportion in population 2)

Sample information: Based on independent random samples from the populations

X = Number of successes in n_1 trials from population 1

Y = Number of successes in n_2 trials from population 2

Sample proportions: $\hat{p}_1 = \dfrac{X}{n_1}, \quad \hat{p}_2 = \dfrac{Y}{n_2}$

Recalling the results from Section 8.2, the mean and the variance of the sample proportions are

$$E(\hat{p}_1)=p_1 \qquad\qquad E(\hat{p}_2)=p_2$$

$$\text{Var}(\hat{p}_1)=\frac{p_1(1-p_1)}{n_1} \qquad\qquad \text{Var}(\hat{p}_2)=\frac{p_2(1-p_2)}{n_2}$$

Given the fact that $\hat{p}_1$ and $\hat{p}_2$ are independent, the mean, variance, and standard deviation of the difference $\hat{p}_1-\hat{p}_2$ are then given by

$$E(\hat{p}_1-\hat{p}_2)=p_1-p_2$$

$$\text{Var}(\hat{p}_1-\hat{p}_2)=\frac{p_1(1-p_1)}{n_1}+\frac{p_2(1-p_2)}{n_2}$$

$$\text{sd}(\hat{p}_1-\hat{p}_2)=\sqrt{\frac{p_1(1-p_1)}{n_1}+\frac{p_2(1-p_2)}{n_2}}$$

The first result shows that $\hat{p}_1-\hat{p}_2$ is an unbiased estimator of p_1-p_2. An estimate of the standard deviation can be obtained by replacing p_1 and p_2 with their respective sample estimates $\hat{p}_1$ and $\hat{p}_2$. Moreover, for large sample sizes n_1 and n_2, the statistic $(\hat{p}_1-\hat{p}_2)$ is approximately normally distributed, so that

$$\frac{(\hat{p}_1-\hat{p}_2)-(p_1-p_2)}{\sqrt{\dfrac{\hat{p}_1(1-\hat{p}_1)}{n_1}+\dfrac{\hat{p}_2(1-\hat{p}_2)}{n_2}}} \quad \text{is approximately} \quad N(0,1)$$

Therefore, in the case of large samples a $100(1-\alpha)\%$ confidence interval for (p_1-p_2) can be readily constructed from this normal approximation. Summarizing:

> An approximate $100(1-\alpha)\%$ confidence interval for p_1-p_2 in large samples is
>
> $$(\hat{p}_1-\hat{p}_2)\pm z_{\alpha/2}\sqrt{\frac{\hat{p}_1(1-\hat{p}_1)}{n_1}+\frac{\hat{p}_2(1-\hat{p}_2)}{n_2}}$$

To test the null hypothesis $H_0: p_1=p_2$, we denote the unspecified common population proportion by p. Under the null hypothesis, the statistic

$\hat{p}_1 - \hat{p}_2$ is approximately normally distributed with

$$E(\hat{p}_1 - \hat{p}_2) = 0$$

$$\text{sd}(\hat{p}_1 - \hat{p}_2) = \sqrt{p(1-p)} \sqrt{\frac{1}{n_1} + \frac{1}{n_2}}$$

The unknown parameter p involved in the standard deviation must now be estimated by pooling information from the two samples. The proportion of successes in the combined sample provides

$$\text{Pooled estimate} \quad \hat{p} = \frac{X+Y}{n_1+n_2}$$

$$\text{Estimated sd of}(\hat{p}_1 - \hat{p}_2) = \sqrt{\hat{p}(1-\hat{p})} \sqrt{\frac{1}{n_1} + \frac{1}{n_2}}$$

Because n_1 and n_2 are large, the statistic

$$Z = \frac{\hat{p}_1 - \hat{p}_2}{\sqrt{\hat{p}(1-\hat{p})} \sqrt{\dfrac{1}{n_1} + \dfrac{1}{n_2}}}$$

is approximately $N(0,1)$ under the null hypothesis. Depending on the alternative, a one- or a two-sided rejection region can be formulated in terms of the standardized test statistic Z.

Example 9.8 A proponent of innovative teaching methods wishes to compare the effectiveness of teaching English by the traditional classroom lecture system and by the extensive use of audio-visual aids. To do so, 100 students are selected at random from a class of 250 and assigned to audio-visual instruction. The remaining 150 students are taught English in classroom lectures. At the end of the term, all 250 students are given a test; the number of students from each group who passed the test is recorded in Table 9.2. (*a*) Find a 95% confidence interval for the difference between success rates for the two methods of instruction. (*b*) Do the data strongly support that a better passing rate is achieved using the classroom lecture technique than is achieved using the audio-visual method?

TABLE 9.2

	AUDIO-VISUAL INSTRUCTION	CLASSROOM LECTURE
Pass	63	107
Fail	37	43
Total	100	150

(*a*) Denoting the population proportion of passes using the lecture method and the audio-visual method by p_1 and p_2, respectively, we compute

$$\hat{p}_1 = \frac{107}{150} = .71$$

$$\hat{p}_2 = \frac{63}{100} = .63$$

$$\hat{p}_1 - \hat{p}_2 = .08$$

$$\text{Estimated standard deviation} = \sqrt{\frac{.71 \times .29}{150} + \frac{.63 \times .37}{100}}$$

$$= \sqrt{.00137 + .00233}$$

$$= \sqrt{.00370} = .06$$

From the normal table we know that $z_{.025} = 1.96$, so that an approximate 95% confidence interval for $p_1 - p_2$ is

$$(.08 - 1.96 \times .06,\ .08 + 1.96 \times .06) = (-.04,\ .20)$$

This means that we are 95% confident the difference in probabilities $p_1 - p_2$ lies in the interval $-.04$ to .20. Approximately 95% of the intervals calculated in this manner will contain the true difference.

(*b*) Here we want to test the null hypothesis $H_0: p_1 = p_2$ vs. the alternative $H_1: p_1 > p_2$. The pooled estimate of p, the common value of $p_1 = p_2$ under H_0 is

$$\hat{p} = \frac{107 + 63}{150 + 100} = \frac{170}{250} = .68$$

The computed value of the test statistic Z is

$$Z = \frac{\hat{p}_1 - \hat{p}_2}{\sqrt{\hat{p}(1 - \hat{p})}\sqrt{\dfrac{1}{n_1} + \dfrac{1}{n_2}}} = \frac{.08}{\sqrt{.68 \times .32}\sqrt{\dfrac{1}{150} + \dfrac{1}{100}}} = 1.33$$

Assuming a level of significance $\alpha = .05$, the one-sided rejection region is $Z \geqslant 1.64$. Because the observed value of Z is smaller, the null hypothesis is not rejected at $\alpha = .05$. The claim that classroom method is superior to audio-visual instruction is not well substantiated by the observed data.

It should be emphasized here that the validity of our inferential procedure is crucially dependent on the random assignment of students between the two methods of instruction. Without randomization the comparison could obviously be manipulated in a desired direction. For instance, students with high GPAs could be assigned to one method of instruction and students with low GPAs could be assigned to the other. Also, if the capacity of the audio-visual facilities permits, a more effective comparison is achieved with samples of equal size. ■

9.7 COMPARING THE VARIANCES OF TWO NORMAL POPULATIONS

Occasionally, it is of interest to an experimenter to compare the variability of two populations, especially when variability is considered to be an important indicator of the performance of a treatment. Examples of variability in single populations have already been cited in Section 8.7. A statistical procedure for comparing the variances of two populations relies on this set of assumptions:

Assumptions

(*a*) $X_1, \ldots, X_{n_1}$ is a random sample from a normal population $N(\mu_1, \sigma_1)$.

(*b*) $Y_1, \ldots, Y_{n_2}$ is a random sample from a normal population $N(\mu_2, \sigma_2)$

(*c*) The two samples are independent.

Inferences about the population variances is naturally based on the sample variances

$$s_1^2 = \frac{\sum_{i=1}^{n_1} \left(X_i - \overline{X}\right)^2}{n_1 - 1}$$

$$s_2^2 = \frac{\sum_{i=1}^{n_2} \left(Y_i - \overline{Y}\right)^2}{n_2 - 1}$$

In particular, we may wish to test the null hypothesis that the population variances are equal. This can be stated

$$H_0 : \frac{\sigma_1^2}{\sigma_2^2} = 1$$

The question is whether s_1^2/s_2^2 is too far from 1 to be explained by chance. Note that unlike a comparison between two means, which is phrased in terms of the difference $\mu_1-\mu_2$, a comparison between variances is formulated using the ratio σ_1^2/σ_2^2. This is because a sampling distribution that is instrumental to the present inference situation involves the parameters σ_1^2 and σ_2^2 only through the ratio.

F distribution

If $X_1,\ldots,X_{n_1}$ and $Y_1,\ldots,Y_{n_2}$ are independent random samples from two normal populations $N(\mu_1,\sigma_1)$ and $N(\mu_2,\sigma_2)$, respectively, then the distribution of

$$F=\frac{s_1^2/\sigma_1^2}{s_2^2/\sigma_2^2}=\frac{\sum_{i=1}^{n_1}\left(X_i-\bar{X}\right)^2/(n_1-1)\sigma_1^2}{\sum_{i=1}^{n_2}\left(Y_i-\bar{Y}\right)^2/(n_2-1)\sigma_2^2}$$

is called an *F distribution with* d.f. $=(n_1-1,n_2-1)$.

An F distribution has a long tail to the right. Its shape is governed by the quantities n_1-1 and n_2-1, which are the degrees of freedom associated with s_1^2 and s_2^2, respectively. We use the symbol $F_\alpha(\nu_1,\nu_2)$ to represent the upper $100\alpha\%$ point of an F distribution having (ν_1,ν_2) degrees of freedom. Appendix Table 7 lists these percentage points for $\alpha=.05$ and $\alpha=.10$ for various pairs (ν_1,ν_2). For instance, the upper 5% point of an F distribution with d.f. $=(8,15)$ is $F_{.05}(8,15)=2.64$.

To determine a lower percentage point, note the interesting feature that the reciprocal F^* of an F ratio $F^*=1/F$ has the same structure as an F ratio, but the degrees of freedom are interchanged.

If F has an F distribution with d.f. $=(\nu_1,\nu_2)$, then the reciprocal $F^*=1/F$ has an F distribution with d.f. $=(\nu_2,\nu_1)$.

Suppose that we wish to find the lower 5% point of an F distribution with d.f. $=(8,15)$. In other words we are looking for the number a such that

$$P\left[F\leqslant a\right]=.05$$

Writing $F^* = 1/F$, the inequality $F \leqslant a$ is equivalent to $1/F \geqslant 1/a$, or $F^* \geqslant 1/a$. Hence, the number a is such that $P(F^* \geqslant 1/a) = .05$. The new variable F^* has the F distribution with d.f. $= (15, 8)$, and $1/a$ is its upper 5% point. Consulting Appendix Table 7 we can see that $1/a = 3.22$, which gives $a = 1/3.22 = .311$. This example illustrates the general rule:

$$\text{Lower } 100\alpha\% \text{ point of } F \text{ with d.f.} = (\nu_1, \nu_2)$$

$$= \frac{1}{\text{Upper } 100\alpha\% \text{ point of } F \text{ with d.f.} = (\nu_2, \nu_1)}$$

Returning to the problem of testing the null hypothesis $H_0: \sigma_1^2/\sigma_2^2 = 1$, note that under H_0 the F ratio just defined reduces to the statistic

$$F = \frac{s_1^2}{s_2^2}$$

which has an F distribution with d.f. $= (n_1 - 1, n_2 - 1)$. If the alternative hypothesis is $\sigma_1^2 > \sigma_2^2$ (that is, $H_1: \sigma_1^2/\sigma_2^2 > 1$), then large values of s_1^2/s_2^2 should constitute the rejection region. Given a level of significance α, the rule is then:

$$\text{Reject } H_0: \frac{\sigma_1^2}{\sigma_2^2} = 1 \text{ in favor of } H_1: \frac{\sigma_1^2}{\sigma_2^2} > 1$$

$$\text{if} \quad \frac{s_1^2}{s_2^2} \geqslant F_\alpha(n_1 - 1, n_2 - 1).$$

If the alternative hypothesis is $H_1: \sigma_1^2/\sigma_2^2 < 1$, we can simply interchange the roles of the two samples and use the test statistic s_2^2/s_1^2, with large values of this ratio comprising the rejection region.

$$\text{Reject } H_0: \frac{\sigma_1^2}{\sigma_2^2} = 1 \text{ in favor of } H_1: \frac{\sigma_1^2}{\sigma_2^2} < 1$$

$$\text{if} \quad \frac{s_2^2}{s_1^2} \geqslant F_\alpha(n_2 - 1, n_1 - 1).$$

In the case of a two-sided alternative $H_1: \sigma_1^2/\sigma_2^2 \neq 1$, the rejection region consists of both tails of the F distribution. Dividing the significance probability α equally between the two tails, the rule is then:

$$\text{Reject } H_0: \frac{\sigma_1^2}{\sigma_2^2} = 1 \text{ in favor of } H_1: \frac{\sigma_1^2}{\sigma_2^2} \neq 1$$

$$\text{if} \quad \frac{s_1^2}{s_2^2} \geqslant F_{\alpha/2}(n_1 - 1, n_2 - 1)$$

$$\text{or} \quad \frac{s_1^2}{s_2^2} \leqslant \frac{1}{F_{\alpha/2}(n_2 - 1, n_1 - 1)}.$$

A $100(1-\alpha)\%$ confidence interval for the variance ratio σ_1^2/σ_2^2 can be constructed by again referring to the F distribution of

$$\frac{s_1^2/\sigma_1^2}{s_2^2/\sigma_2^2} = \frac{\sigma_2^2}{\sigma_1^2}\frac{s_1^2}{s_2^2}$$

Noting that

$$P\left[\frac{1}{F_{\alpha/2}(n_2 - 1, n_1 - 1)} \leqslant \frac{\sigma_2^2}{\sigma_1^2}\frac{s_1^2}{s_2^2} \leqslant F_{\alpha/2}(n_1 - 1, n_2 - 1)\right] = 1 - \alpha$$

and rearranging the inequalities within the brackets:

A $100(1-\alpha)\%$ confidence interval for $\dfrac{\sigma_1^2}{\sigma_2^2}$ is given by

$$\frac{s_1^2}{s_2^2}\frac{1}{F_{\alpha/2}(n_1 - 1, n_2 - 1)} \leqslant \frac{\sigma_1^2}{\sigma_2^2} \leqslant \frac{s_1^2}{s_2^2} F_{\alpha/2}(n_2 - 1, n_1 - 1)$$

Example 9.9 The uniformity in fiber density of knitting yarns manufactured by two spinning machines are to be compared. Each observation consists of taking a 100-yard length of yarn at random and measuring its weight. Suppose that 12 determinations of weight are made from the output of machine A and that 10 are made from the output of machine B. The standard deviations of weight measurements are found to be 2.3 for machine A and 1.5 for machine B. (*a*) Is there strong evidence that product variability is higher for machine A than machine B? (*b*) Construct a 90% confidence interval for the ratio of the population standard deviations.

We assume that the weight measurements are random samples from normal distributions, and we let σ_A^2 and σ_B^2 denote the population variances corresponding to machines A and B, respectively.

(*a*) We want to test $H_0: \sigma_A^2/\sigma_B^2 = 1$ vs. $H_1: \sigma_A^2/\sigma_B^2 > 1$. The sample sizes and the sample variances are

$$n_A = 12 \qquad\qquad n_B = 10$$
$$s_A^2 = (2.3)^2 = 5.29 \qquad\qquad s_B^2 = (1.5)^2 = 2.25$$

The calculated value of the F statistic is

$$F = \frac{s_A^2}{s_B^2} = \frac{5.29}{2.25} = 2.35, \quad \text{d.f.} = (11, 9)$$

The rejection region is determined from the upper tail of the F distribution. Taking $\alpha = .05$ and consulting the F table, we find that $F_{.05}(11,9) = 3.10$. Because the observed value of F is smaller, H_0 is not rejected at $\alpha = .05$.

(*b*) A 90% confidence interval for the ratio σ_A^2/σ_B^2 is given by

$$\left(\frac{s_A^2}{s_B^2}\frac{1}{F_{.05}(11,9)}, \quad \frac{s_A^2}{s_B^2}F_{.05}(9,11)\right)$$
$$= \left(\frac{2.35}{3.10}, \quad 2.35 \times 2.90\right)$$
$$= (.76, 6.81)$$

Consequently, a 90% confidence interval for σ_A/σ_B is

$$(\sqrt{.76}\,, \sqrt{6.81}\,) = (.87, 2.61)$$

We are 90% confident that σ_A may be as small as $.87\sigma_B$ or as large as $2.61\sigma_B$. Our 90% confidence derives from the property that, approximately, 90% of the intervals, calculated in this manner from repeated samples of sizes 12 and 10, will contain the true ratio σ_A/σ_B. ■

Caution The testing and confidence procedures described in this section are heavily dependent on the assumption of normality. Grossly misleading inferences can result if the population distributions differ substantially from the normal distributions. With moderate or large samples, the normality assumption should be checked by plotting the data on normal probability paper. If drastic departures from the normal distribution occur, the appropriate normalizing transformations should be applied to the data before the techniques discussed here are employed.

*9.8 RELATIONSHIPS OF χ^2, t, AND F TO THE NORMAL DISTRIBUTION

The sampling distributions χ^2, t, and F are basic to many statistical inference procedures. This section systematically presents their definitions and interrelationships. The material is slightly demanding mathematically, but persistent study here will prove beneficial later when these distributions reappear as basic sampling distributions in new settings.

> *Definition 1: Chi-square (χ^2) distribution*
>
> If $Z_1, Z_2, \ldots, Z_k$ are independent random variables, each having the standard normal distribution, then the distribution of
>
> $$\chi_k^2 = Z_1^2 + Z_2^2 + \ldots + Z_k^2$$
>
> is called a χ^2 distribution with d.f. $= k$.

In our notation the suffix of a χ^2 variable indicates the number of degrees of freedom. The χ^2 distributions possess the property of additivity, which is formalized below.

Additivity Property of χ^2

If $\chi_{k_1}^2$ and $\chi_{k_2}^2$ are two independent χ^2 variables, then their sum $\chi_{k_1}^2 + \chi_{k_2}^2$ has a χ^2 distribution with d.f. $= k_1 + k_2$.

This property is derived from the structure

$$\chi_{k_1}^2 = Z_1^2 + \ldots + Z_{k_1}^2, \chi_{k_2}^2 = Z_1'^2 + \ldots + Z_{k_2}'^2$$

where all Z_i and Z_i' are independent $N(0, 1)$ variables. The sum $\chi_{k_1}^2 + \chi_{k_2}^2$ is then the sum of $k_1 + k_2$ squares of independent $N(0, 1)$ variables, and Definition 1 implies that this sum has a $\chi_{k_1+k_2}^2$ distribution.

For a random sample $X_1, X_2, \ldots, X_n$ from a normal population $N(\mu, \sigma)$, the standardized random variables

$$Z_i = \frac{X_i - \mu}{\sigma}, \quad i = 1, 2, \ldots, n$$

are independent $N(0, 1)$. Applying Definition 1, we can then conclude that

$$\sum_{i=1}^{n} Z_i^2 = \frac{1}{\sigma^2} \sum_{i=1}^{n} (X_i - \mu)^2$$

has a χ_n^2 distribution. We now relate this quantity to the sample variance s^2, which is given by $(n-1)s^2 = \sum_{i=1}^{n}(X_i - \bar{X})^2$. Squaring each side of

$$X_i - \mu = (X_i - \bar{X}) + (\bar{X} - \mu)$$

we obtain

$$(X_i - \mu)^2 = (X_i - \bar{X})^2 + (\bar{X} - \mu)^2 + 2(\bar{X} - \mu)(X_i - \bar{X})$$

Summing both sides over $i = 1, \ldots, n$ and noting that $\sum_{i=1}^{n}(X_i - \bar{X}) = 0$, we obtain the relation

$$\sum_{i=1}^{n} (X_i - \mu)^2 = \sum_{i=1}^{n} (X_i - \bar{X})^2 + n(\bar{X} - \mu)^2$$

which yields

$$\frac{1}{\sigma^2} \sum_{i=1}^{n} (X_i - \mu)^2 = \frac{(n-1)s^2}{\sigma^2} + \frac{n(\bar{X} - \mu)^2}{\sigma^2}$$

Now $\bar{X}$ is distributed as $N(\mu, \sigma/\sqrt{n})$, so that $\sqrt{n}(\bar{X} - \mu)/\sigma$ is $N(0, 1)$. Hence, by Definition 1, $n(\bar{X} - \mu)^2/\sigma^2$ is χ_1^2. Moreover, the random variable $\bar{X}$ is independent of s^2 (a fact we merely state without proof). The left-hand side of the last equality given above has been shown to have a χ_n^2 distribution. The right-hand side of this equality is a sum of two independent terms, and the term $n(\bar{X} - \mu)^2/\sigma^2$ has a χ_1^2 distribution. This provides an intuitive justification for the mathematical fact that

$$\frac{(n-1)s^2}{\sigma^2} \quad \text{is distributed as } \chi_{n-1}^2$$

Definition 2: t distribution

Let the random variables Z and χ_k^2 be independent, with Z distributed as $N(0,1)$ and χ_k^2 having a χ^2 distribution with d.f. $= k$. Then the distribution of

$$t_k = \frac{Z}{\sqrt{\chi_k^2/k}} \text{ or descriptively } \frac{N(0,1)}{\sqrt{\text{Chi-square/d.f.}}}$$

is defined as *the Student's t distribution with* d.f. $= k$.

Referring to a random sample $X_1, \ldots, X_n$ from a normal population $N(\mu, \sigma)$, we have already noted that $Z = \sqrt{n}\,(\bar{X} - \mu)/\sigma$ is $N(0,1)$ and that it is independent of $(n-1)s^2/\sigma^2$, which is χ_{n-1}^2. Applying Definition 2

$$\frac{\sqrt{n}\,(\bar{X} - \mu)/\sigma}{\sqrt{\dfrac{(n-1)s^2/\sigma^2}{n-1}}} = \frac{(\bar{X} - \mu)}{s/\sqrt{n}}$$

is distributed as t_{n-1}. This fact has been used to make inferences about μ.

Given two independent random samples $X_1, \ldots, X_{n_1}$ from $N(\mu_1, \sigma)$ and $Y_1, \ldots, Y_{n_2}$ from $N(\mu_2, \sigma)$, we know that

$$Z = \frac{(\bar{X} - \bar{Y}) - (\mu_1 - \mu_2)}{\sigma\sqrt{\dfrac{1}{n_1} + \dfrac{1}{n_2}}} \quad \text{is} \quad N(0,1)$$

Applying the same line of reasoning, $(n_1 - 1)s_1^2/\sigma^2$ is distributed as $\chi_{n_1-1}^2$ and $(n_2 - 1)s_2^2/\sigma^2$ is distributed as $\chi_{n_2-1}^2$. These variables are independent, and by the additivity property of the χ^2 distribution their sum

$$\frac{(n_1 - 1)s_1^2 + (n_2 - 1)s_2^2}{\sigma^2} = \frac{(n_1 + n_2 - 2)s_{\text{pooled}}^2}{\sigma^2}$$

is distributed as $\chi_{n_1+n_2-2}^2$.

Moreover, this χ^2 variable is independent of Z because $\bar{X}$, $\bar{Y}$, s_1^2, and s_2^2 are all independent. Applying Definition 2 gives us

$$\frac{\dfrac{(\bar{X}-\bar{Y})-(\mu_1-\mu_2)}{\sigma\sqrt{\dfrac{1}{n_1}+\dfrac{1}{n_2}}}}{\sqrt{\dfrac{(n_1+n_2-2)s_{\text{pooled}}^2}{\sigma^2(n_1+n_2-2)}}}=\frac{(\bar{X}-\bar{Y})-(\mu_1-\mu_2)}{s_{\text{pooled}}\sqrt{\dfrac{1}{n_1}+\dfrac{1}{n_2}}}$$

distributed as Student's t with d.f. $=n_1+n_2-2$.

> *Definition 3: F distribution*
>
> If $\chi^2_{k_1}$ and $\chi^2_{k_2}$ are two independent χ^2 variables with k_1 and k_2 degrees of freedom, respectively, then the distribution of
>
> $$F_{k_1,k_2}=\frac{\chi^2_{k_1}/k_1}{\chi^2_{k_2}/k_2} \quad \text{or descriptively} \quad \frac{\text{Chi-square/d.f.}}{\text{Chi-square/d.f.}}$$
>
> is called an *F distribution with* d.f. $=(k_1,k_2)$.

For two independent random samples $X_1,\ldots,X_{n_1}$ from $N(\mu_1,\sigma_1)$ and $Y_1,\ldots,Y_{n_2}$ from $N(\mu_2,\sigma_2)$, we know that

$$\frac{(n_1-1)s_1^2}{\sigma_1^2} \quad \text{is distributed as} \quad \chi^2_{n_1-1}.$$

$$\frac{(n_2-1)s_2^2}{\sigma_2^2} \quad \text{is distributed as} \quad \chi^2_{n_2-1}.$$

These two χ^2 variables are independent. Applying Definition 3, we can then conclude that

$$\frac{(n_1-1)s_1^2/\sigma_1^2(n_1-1)}{(n_2-1)s_2^2/\sigma_2^2(n_2-1)}=\frac{s_1^2/\sigma_1^2}{s_2^2/\sigma_2^2}$$

has an F distribution with d.f. $=(n_1-1,n_2-1)$. We used this result to make inferences about σ_1^2/σ_2^2 in Section 9.7.

Relation Between t and F

Because the square of an $N(0, 1)$ variable is distributed as χ_1^2, the square of a t variable has the structure

$$t_k^2 = \frac{Z^2}{\chi_k^2/k} = \frac{\chi_1^2/1}{\chi_k^2/k}$$

Hence, t_k^2 has the same distribution as $F_{1,k}$.

EXERCISES

1. To compare two programs for training industrial workers to perform a skilled job, 20 workers are included in an experiment. Of these 10 are selected at random to be trained by method 1; the remaining 10 workers are to be trained by method 2. After completion of training, all the workers are subjected to a time-and-motion test that records the speed of performance of a skilled job. The following data are obtained:

Time (in minutes)

Method 1	15	20	11	23	16	21	18	16	27	24
Method 2	23	31	13	19	23	17	28	26	25	28

   (*a*) Can you conclude from the data that the mean job time is significantly less after training with method 1 than after training with method 2? (Test with $\alpha = .05$.)
   (*b*) State the assumptions you make for the population distributions.
   (*c*) Construct a 95% confidence interval for the population mean difference in job times between the two methods.

2. The peak oxygen intake per unit of body weight, called the "aerobic capacity," of an individual performing a strenuous activity is a measure of work capacity. For a comparative study, measurements of aerobic capacities are recorded for a group of 20 Peruvian Highland natives and for a group of 10 U.S. lowlanders acclimatized as adults in high altitudes. The following summary statistics were obtained from the data [Source: A. R. Frisancho, *Science* **187** (1975), 317]:

	Peruvian natives	U.S. subjects acclimatized
Mean	46.3	38.5
Standard deviation	5.0	5.8

(*a*) Do the data provide a strong indication of a difference in mean aerobic capacity between the highland natives and the acclimatized low landers?

(*b*) Construct a 98% confidence interval for the mean difference in aerobic capacity between the two groups.

3. To determine how an experimental dose of a dental anesthesia affects male and female patients, random samples of 15 male and 16 female patients are selected and their reaction times are recorded in minutes. The mean and the standard deviations obtained from the data sets are:

	Male	Female
Mean	4.8	4.4
sd	.8	.9

(*a*) Devise a test with $\alpha = .1$ to determine whether there is a significant difference in the mean reaction times between the males and females.

(*b*) Construct a 95% confidence interval for the difference between the mean reaction times of males and females.

(*c*) Construct a 99% confidence interval for the mean reaction time of each group individually.

4. It is anticipated that a new instructional method will more effectively improve the reading ability of elementary-school children than the standard method currently in use. To test this conjecture, 16 children are divided at random into two groups of 8 each. One group is instructed using the standard method and the other group is instructed using the new method. The children's scores on a reading test are found to be:

Reading test scores

Standard	65	70	76	63	72	71	68	68
New	75	80	72	77	69	81	71	78

(*a*) Analyze the data and make an inference about the difference of the population mean scores achieved using the standard and new methods of instruction.

(*b*) State the assumptions you have made for your analysis.

5. Specimens of brain tissue are collected by performing autopsies on 9 schizophrenic patients and 9 control patients of comparable ages. A certain enzyme activity is measured for each subject in terms of the amount of a substance formed per gram of tissue per hour. The following means and standard deviations are calculated from the data

[SOURCE: R. J. Wyatt et al., *Science* **187** (1975), 369]:

	Control subjects	Schizophrenic subjects
Mean	39.8	35.5
sd	8.16	6.93

(*a*) Test to determine if the mean activity is significantly lower for the schizophrenic subjects than for the control subjects.

(*b*) Construct a 95% confidence interval for the mean difference in enzyme activity between the two groups.

6. Two brands of furniture polish are to be compared with respect to the area of coverage achieved with a fixed quantity of polish. For the experiment, 7 one-quart cans of brand A and 5 one-quart cans of brand B are used. The surface area covered by each quart of polish is measured in square feet, and computations from the data sets yield these results:

	Brand A	Brand B
Mean area	106	90
sd	5.8	4.7

(*a*) Do these results support the claim that brand A covers a larger area than brand B?

(*b*) Construct a 95% confidence interval for the mean coverage by one quart of polish of each brand.

7. An investigation is undertaken to determine how the administration of a growth hormone affects the weight gain of pregnant rats. Weight gains during gestation are recorded for 6 control rats and for 6 rats receiving the growth hormone. The following summary statistics are obtained [SOURCE: V. R. Sara et al., *Science* **186** (1974), 446]:

	Control rats	Hormone rats
Mean	41.8	60.8
sd	7.6	16.4

(*a*) State the assumptions about the populations and test to determine if the mean weight gain is significantly higher for the rats receiving the hormone than for the rats in the control group.

(*b*) Do the data indicate that you should be concerned about the possible violation of any assumption? If so, which one?

8. To compare the effectiveness of Isometric and Isotonic exercise methods in abdominal reduction, 20 potbellied business executives are

included in an experiment: 10 are selected at random and assigned to one exercise method; the remaining 10 are assigned to the other exercise method. After five weeks, the reductions in abdomen measurements are recorded in centimeters, and the following results obtained:

	Isometric method A	Isotonic method B
Mean	2.5	3.1
sd	0.8	1.0

(*a*) Do these data support a claim that the Isotonic method is more effective?

(*b*) Construct a 95% confidence interval for $\mu_B - \mu_A$.

9. To compare the age at first marriage of females in two ethnic groups A and B, a random sample of 100 ever-married females is taken from each group and the ages at first marriage are recorded. The mean and the standard deviation are found to be:

	A	B
Mean	18.5	20.7
sd	5.8	6.3

Use hypothesis testing and confidence interval methods to make inferences about any difference between the two ethnic groups regarding age at first marriage.

10. In June 1975, chemical analyses were made of 85 water samples (each of unit volume) taken from various parts of a city lake, and the measurements of chlorine content were recorded. During the next two winters, the use of road salt was substantially reduced in the catchment areas of the lake. In June 1977, 110 water samples were analyzed and their chlorine contents recorded. Calculations of the mean and the standard deviation for the two sets of data were:

	Chlorine content	
	1975	1977
Mean	18.3	17.8
sd	1.2	1.8

(*a*) Do the data provide strong evidence that there is a reduction of average chlorine level in the lake water in 1977 compared to the level in 1975?

(*b*) Establish a 98% confidence interval for the difference between the mean chlorine levels in 1975 and 1977.

(*c*) Construct a 95% confidence interval for the mean chlorine level in June 1977.

11. A national equal employment opportunities committee is conducting an investigation to determine if women employees are as well paid as their male counterparts in comparable jobs. Random samples of 75 males and 64 females in junior academic positions are selected, and the following calculations are obtained from their salary data:

	Male	Female
Mean	\$11,530	\$10,620
sd	780	750

Construct a 95% confidence interval for the difference between the mean salaries of males and females in junior academic positions.

12. An investigation is conducted to determine if the mean age of welfare recipients differs between two cities A and B. Random samples of 75 and 100 welfare recipients are selected from city A and city B, respectively, and the following computations are made:

	City A	City B
Mean	38	43
sd	6.8	7.5

(*a*) Do the data provide strong evidence that the mean ages are different in city A and city B? (Test at $\alpha = .02$.)

(*b*) Construct a 98% confidence interval for the difference in mean ages between A and B.

(*c*) Construct a 98% confidence interval for the mean age for city A and city B individually.

13. A sociologist wishes to compare the fertility rates of women in two tribal sects A and B of eastern Africa. From each sect a random sample of 100 women in the age group 50–60 years is selected, and the number of children born to each is recorded. The following frequency distributions are obtained:

		Number of children 0	1	2	3	4	5	6	7	8	Total
frequency	A	6	14	18	25	19	11	5	2	3	100
	B	0	3	8	18	30	19	15	5	2	100

(*a*) Calculate the mean and the standard deviation for each frequency distribution.

(*b*) Do the data indicate a significant difference in the mean number of children born to women in the two sects?

(*c*) Construct a 98% confidence interval for the difference between the population means.

14. Two methods of psychotherapy for the treatment of manic depressive patients are being compared on the basis of case records supplied by a number of psychiatrists. From the records of 65 patients treated by method 1, the mean and the standard deviation for the duration of treatment are found to be 123 days and 21 days, respectively. From the records of 53 patients treated by method 2, the mean and the standard deviation are found to be 132 days and 30 days, respectively. Do these data indicate a significant difference between the mean duration of treatments using the two methods?

15. Referring to Exercise 7, suppose that you are asked to design an experiment to study the effect of a hormone injection on the weight gain of pregnant rats during gestation. If you decide to inject 6 of the 12 rats available for the experiment, and to retain the other 6 as controls,

(*a*) Briefly explain why it is important to randomly divide the rats into the two groups. What might be wrong with the experimental results if you choose to treat your favorite 6 rats with the hormone?

(*b*) Suppose that the 12 rats are tagged with serial numbers from 1 through 12 and that 12 marbles identical in appearance are also numbered from 1 through 12. How can you use this numbering system to randomly select the rats in the treatment and control groups?

16. Referring to Exercise 8:

(*a*) Aside from the type of exercise method, identify a few other factors that are likely to have an important effect on the amount of reduction accomplished in a five-week period.

(*b*) What role does randomization play in achieving a valid comparison between the two exercise methods?

(*c*) If you were to design this experiment, describe how you would divide the 20 business executives into two groups [refer to Exercise 15(*b*)].

17. Two methods of memorizing difficult material are being tested to determine if one produces better retention. Nine pairs of students are included in the study. The students in each pair are matched according to I.Q. and academic background and then are assigned to the two methods at random. A memorization test is given to all the students,

and the following scores are obtained:

	Pair								
	1	2	3	4	5	6	7	8	9
Method A	90	86	72	65	44	52	46	38	43
Method B	85	87	70	62	44	53	42	35	46

At $\alpha = .05$, test to determine if there is a significant difference in the effectiveness of the two methods.

18. A blanching process currently used in the canning industry consists of treating vegetables with a large volume of boiling water before canning. A newly developed method, called Steam Blanching Process (SBP), is expected to remove less vitamins and minerals from vegetables, because it is more of a steam wash than a flowing water wash. Ten batches of string beans from different farms are to be used to compare the SBP and the standard process. One-half of each batch of beans is treated with the standard process; the other half of each batch is treated with the SBP. Measurements of the vitamin content per pound of canned beans are:

	Batches									
	1	2	3	4	5	6	7	8	9	10
SBP	35	48	65	33	61	54	49	37	58	65
Standard	33	40	55	41	62	54	40	35	59	56

(*a*) Do the data provide strong evidence that SBP removes less vitamins in canned beans than the standard method of blanching?
(*b*) Construct a 98% confidence interval for the difference between the mean vitamin contents per pound using the two methods of blanching.

19. A study is to be made of the relative effectiveness of two kinds of cough medicines in increasing sleep. Six people with colds are given medicine A the first night and medicine B the second night. Their hours of sleep each night are recorded. The data are:

	Subject					
	1	2	3	4	5	6
Medicine A	4.8	4.1	5.8	4.9	5.3	7.4
Medicine B	3.9	4.2	5.0	4.9	5.4	7.1

(*a*) Establish a 95% confidence interval for the mean increase in hours of sleep from medicine B to medicine A.

(*b*) How and what would you randomize in this study? Briefly explain your reasoning for randomization.

20. Five pairs of tests are conducted to compare two methods of making rope. Each sample batch contains enough hemp to make two ropes. The tensile strength measurements are:

	Test				
	1	2	3	4	5
Method 1	14	12	18	16	15
Method 2	16	15	17	16	14

(*a*) Treat the data as five paired observations, and calculate a 95% confidence interval for the mean difference in tensile strengths between ropes made by the two methods.

(*b*) Repeat the calculation of a 95% confidence interval treating the data as independent random samples.

(*c*) Briefly discuss the conditions under which each type of analysis would be appropriate.

21. It is claimed that an industrial safety program is effective in reducing the loss of working hours due to factory accidents. The following data are collected concerning the weekly loss of working hours due to accidents in 6 plants both before and after the safety program is initiated.

	Plant					
	1	2	3	4	5	6
Before	12	29	16	37	28	15
After	10	28	17	35	25	16

State any assumption that you make. Then devise a suitable test of the hypothesis and determine if the data support the claim.

22. An experiment is conducted to determine if the use of a special chemical additive with a standard fertilizer accelerates plant growth. Ten locations are included in the study. At each location two plants growing in close proximity are treated: one is given the standard fertilizer; the other is given the standard fertilizer with the chemical additive. Plant growth after four weeks is measured in centimeters, and

the following data are obtained:

	Location									
	1	2	3	4	5	6	7	8	9	10
Without additive	20	31	16	22	19	32	25	18	20	19
With additive	23	34	15	21	22	31	29	20	24	23

Do the data substantiate the claim that use of the chemical additive accelerates plant growth? State the assumptions that you make and devise an appropriate test of the hypothesis.

23. Measurements of the left-hand and right-hand gripping strengths of 10 left-handed writers are recorded:

	Person									
	1	2	3	4	5	6	7	8	9	10
Left hand	140	90	125	130	95	121	85	97	131	110
Right hand	138	87	110	132	96	120	86	90	129	100

(*a*) Do the data provide strong evidence that people who write with the left hand have a greater gripping strength in the left hand than they do in the right hand?

(*b*) Construct a 90% confidence interval for the mean difference.

24. Sarah claims that her "miracle bait" is a more effective lure for panfish than the old-fashioned lure that Bill uses. Sarah and Bill went on 12 fishing expeditions in the same boat last summer and kept the following day-by-day record of the number of panfish they caught:

	Days											
	1	2	3	4	5	6	7	8	9	10	11	12
Sarah	8	27	7	9	18	15	13	19	3	12	18	12
Bill	13	20	2	9	19	12	10	23	0	11	15	10

Do these statistics support Sarah's argument sufficiently to convice Bill to switch to her "miracle bait," which is somewhat more expensive than the bait Bill is currently using?

25. A business firm wishes to be certain that a recent radio-TV advertisement of its brand of pantyhose was effective in increasing sales. The marketing division selects 48 department stores from the area covered by the advertisement. For each store, the numbers of the firm's brand of pantyhose sold a month before the advertisement and a month after

the advertisement are recorded. The increase of sales is calculated for each store, and the mean and the standard deviation for these data of increases are found to be 120.5 and 198.7, respectively.

(*a*) From this information, what can you conclude about the effectiveness of the advertisement in increasing sales?

(*b*) Construct a 95% confidence interval for the mean increase in sales per store following the advertisement.

26. Referring to Exercise 22, suppose that the two plants at each location are chosen from a row stretching in the East–West direction. In designing this experiment, you must decide which of the two plants at each location—the one in the East or the one in the West—is to be given the chemical additive.

(*a*) Explain how by repeatedly tossing a coin you can randomly allocate the treatment to the plants at the 10 locations.

(*b*) Perform the randomization by actually tossing a coin 10 times.

27. In each of the following cases, how would you select the experimental units and conduct the experiment?

(*a*) Compare the mileage obtained from two gasolines; 16 cars are available for the experiment.

(*b*) Test two varnishes; 12 birch boards are available for the experiment.

(*c*) Compare two methods of teaching basic ice skating; 40 seven-year-old boys are available for the experiment.

28. In a poll taken among college students, 295 out of 500 fraternity men and 64 out of 100 nonfraternity men favored a certain proposition. Estimate the difference in the fractions favoring the proposition and place a 99.7% bound on the error of estimation.

29. An antibiotic for pneumonia was injected into 100 patients with kidney malfunctions (called uremic patients) and into 100 patients with no kidney malfunctions (called normal patients). Some allergic reaction developed in 38 of the uremic patients and in 21 of the normal patients.

(*a*) Do the data provide strong evidence that the rate of incidence of allergic reaction to the antibiotic is higher in uremic patients than it is in normal patients?

(*b*) Construct a 95% confidence interval for the difference between the population proportions.

30. Referring to Exercise 23, suppose that 50 left-handed writers are included in an experiment and that the only information recorded is that 31 have greater strength in the left hand than in the right hand. Formulate and test the appropriate hypothesis to answer the question raised in Exercise 23(*a*).

31. A study (courtesy of R. Golubjatnikov) is undertaken to compare the rates of prevalence of CF antibody to Parainfluenza I virus among boys and girls in the age group 5–9 years. Among 113 boys tested, 34 are found to have the antibody; among 139 girls tested, 54 have the antibody. Do the data provide strong evidence that the rate of prevalence of the antibody is significantly higher in girls than in boys?

32. Referring to Exercise 9, suppose that 100 females sampled from each group are asked the question: "Did you get married before you were 19?" and that the following responses are obtained:

	A	*B*
Yes	62	29
No	38	71

(*a*) Formulate the hypotheses and perform the test with $\alpha = .05$.

(*b*) Construct a 98% confidence interval for p_A, the population proportion of ever-married females in group *A* who married before age 19.

(*c*) Establish a 95% confidence interval for the difference $p_A - p_B$.

33. An experiment is conducted to compare the viability of seeds with and without a cathodic protection, which consists of subjecting the seeds to a negatively charged conductor. Seeds of a common type are randomly divided in two batches of 250 each. One batch is given cathodic protection, and the other is retained as the control group. Both batches are then subjected to a common high temperature to induce artificial aging. Subsequently, all the seeds are soaked in water and left to germinate. It is found that 25% of the control seeds and 10% of the cathodically protected seeds fail to germinate.

(*a*) Do the data provide strong evidence that the cathodic protection permits a higher germination rate in seeds subjected to artificial aging?

(*b*) Construct a 98% confidence interval for the difference $p_T - p_C$, where p_T and p_C represent the germination rates of cathodically treated seeds and control seeds, respectively.

34. Using the table of percentage points for the *F* distributions, find:

(*a*) The upper 5% point when d.f. = (7, 9).

(*b*) The upper .10 point when d.f. = (3, 8).

(*c*) The lower 5% point when d.f. = (12, 7).

(*d*) The lower .10 point when d.f. = (4, 8).

35. Referring to Exercise 1, devise a test to determine if the variabilities of job times are significantly different for the two training methods, at a .10 level of significance. State the assumption that is required for your test procedure.

36. Referring to Exercise 5:
    (*a*) Do the data provide strong evidence that the enzyme activity in the control subjects has a higher standard deviation than the enzyme activity in schizophrenic patients?
    (*b*) Construct a 90% confidence interval for the ratio of the two population standard deviations.
37. Referring to Exercise 7, would you conclude from the data that the variability of weight gains in the hormone-injected rats is significantly higher than it is in the control rats?
38. An experiment is conducted to compare the precision of two brands of mercury detectors in measuring mercury concentration in the air. During the noon hour of one day in a downtown city area, 7 measurements of the mercury concentration are made with brand *A* instrument and 6 measurements are made with brand *B* instrument. The measurements recorded in units of micrograms per cubic meter of air are:

Brand *A*	.95	.82	.78	.96	.71	.86	.99
Brand *B*	.89	.91	.94	.91	.90	.89	

    (*a*) Do the data provide strong evidence that brand *B* measures mercury concentration in the air more precisely than brand *A*?
    (*b*) Construct a 90% confidence interval for the ratio of the standard deviations of measurements made with brand *A* and brand *B*.
39. One aspect of a study of sex differences involves the play behavior of monkeys during the first year of life (courtesy of H. Harlow, U. W. Primate Laboratory). Six male and six female monkeys are observed in groups of four families during several ten-minute test sessions. The mean total number of times each monkey initiates play with another age mate is recorded:

Males	3.64	3.11	3.80	3.58	4.55	3.92
Females	1.91	2.06	1.78	2.00	1.30	2.32

    (*a*) Plot the observations.
    (*b*) Establish a 95% confidence interval for the difference in population means.
40. In the early 1970s, students started a phenomenon called "streaking." Within a two-week period following the first streaking sighted on campus, a standard psychological test was given to a group of 19 males who were admitted streakers and to a control group of 19 males who were nonstreakers. S. Stoner and M. Watman reported the following data [*Psychology* **11**:4 (1975) 14–16], regarding scores on a test de-

signed to determine extroversion:

Streakers	Nonstreakers
$\bar{x} = 15.26$	$\bar{y} = 13.90$
$s_1 = 2.62$	$s_2 = 4.11$

(*a*) Construct a 95% confidence interval for the difference in population means. Does there appear to be a difference between the two groups?

(*b*) Repeat (*a*) without pooling.

(*c*) It may be true that those who admit to streaking differ from those who do not admit to streaking. In light of this possibility, what criticism can be made of your analytical conclusion?

41. J. J. Thomson (1856–1940), discovered the electron while investigating the basic nature of cathode rays. In laboratory experiments Thomson isolated negatively charged particles for which he could determine the mass–charge ratio. This ratio appeared to be constant over a wide variety of experimental conditions and to be a characteristic of these new particles. Thomson obtained the following results with two different cathode ray tubes, using air as the gas [SOURCE: *Philosophical Magazine* **44**:5 (1897), 293]:

	Mass–charge ratio (10^7)						
Tube 1	.57	.34	.43	.32	.48	.40	.40
Tube 2	.53	.47	.47	.51	.63	.61	.48

(*a*) Construct a 95% confidence interval for the difference in the means of the two tubes. Do the two tubes appear to produce consistent results?

(*b*) Establish a 99% confidence interval for the mean mass–charge ratio, treating Thomson's two sets of measurements as one sample of size 14.

CHAPTER 10

Regression Analysis: Simple Linear Relation

10.1 OBJECTIVES OF REGRESSION ANALYSIS

In many areas of scientific investigation the variation in the experimental measurements of a variable is caused to a great extent by other related variables whose magnitudes change over the course of the experiment. By explicitly incorporating the data of these influential variables into the statistical analysis, it is often possible to assess the nature of the relationship and then to utilize this information to improve the description of and inferences about the variable of primary interest. Probing the relationship among variables is also important in that the value of one variable can be predicted from observations on the other variables or even controlled and optimized by manipulating influencing factors.

Regression analysis is a body of statistical methods dealing with the formulation of mathematical models that depict relationships among variables, and the use of these modeled relationships for the purpose of prediction and other statistical inferences. How such a powerful tool of modern statistics was christened with the unlikely name "regression" demands an explanation. Historically, the word "regression" was first used in its present technical context by Sir Francis Galton, who analyzed the heights of sons and the average heights of their parents. From his observations, Galton concluded that sons of very tall (or short) parents were generally taller (shorter) than the average but not as tall (short) as their parents. This result was published in 1885 under the title "Regression Toward Mediocrity in Hereditary Stature." In this context the term "regression" connoted that the sons' heights tended toward the average rather than to more extreme values.

From this origin, the term *regression analysis* has developed to the point that it currently encompasses the analysis of data involving two or more variables when the objective is to discover the nature of their relationship and then to explore it for the purposes of prediction. The study of relationships among variables is prevalent in many fields of scientific endeavor. The advertising manager of a firm is interested in the relationship between money spent for advertising and the corresponding increase in sales. A major concern in radiation therapy is the extent of cellular damage induced by the duration and intensity of exposure. To forecast floods, a hydrologist must study the rate of river discharge measured at a particular site in relation to recent precipitation and suitably time-lagged discharge rates at sites located upstream. In a study of political awareness, a sociologist may wish to relate the percentage of people who are eligible to vote with such socioeconomic factors as age structure, education level, and average income. The personnel departments of business firms often explore the connection between employee job performance and the preliminary evaluation based on interview scores at the time of hiring. In view of the diversity of the nature of possible relationships among variables, a broad classification and some examples should help to outline the domain and scope of regression analysis.

Deterministic Relation

Variables are often related by some law that can be expressed by a precise mathematical function. Some universally recognized theoretical basis justifies the functional form, and any deviation of the observations from this relation is considered to be an experimental error.

For example, if x dollars are deposited in a savings account that earns $100r\%$ interest compounded annually and if y denotes the amount in this account in n years, then y is related to x, r, and n by an exact formula

$$y = x(1+r)^n$$

which is the law of compound interest. As a second example, the time t that it takes a metal ball to reach the surface of the earth when it is dropped from a height h is related to h by the physical law of gravitation

$$t = \sqrt{\frac{2h}{g}}$$

where g is the gravitational constant. Galileo originally postulated that h was proportional to t^2 and left it to later experimenters to obtain accurate estimates of g. Current values are such that, for most purposes, further experimentation and analysis of data are unnecessary. These cases are excluded from the domain of regression analysis.

Semideterministic Relation

In other cases a well-established theory prescribes a form for the law relating the variables but not the particular values of the parameters that appear in the relation. To learn about these parameters, we must perform experiments. The limited precision of the measuring device, uncontrollable perturbations of experimental conditions, and other factors introduce experimental errors in the data that normally cause some variation about the true relation.

Example 10.1 Each gas has its own specific heat ratio γ. If we conduct an experiment in which there is no variation in heat as we change the volume V and measure the pressure P, the ideal gas law specifies that

$$PV^{\gamma} = \text{constant}$$

where the specific heat ratio γ must be estimated from the experimental data on P and V.

Two other examples of this type are estimating the rate constants in a known form of rate equation for a chemical reaction, and estimating the thermal expansion of a new alloy. ■

In some situations a partial theoretical background suggests some plausible form for the relationship, but the theoretical basis is not precise or universally acceptable. Moreover, additional fluctuation is often produced by uncontrolled variables that are not included in the relation.

Example 10.2 Suppose that a factory manufactures items in batches and that the production manager wishes to relate the production cost of a batch y to the size of the batch x. A certain component of the cost is practically constant, regardless of the batch size x, at least within a realistic range of variation of x. Building costs and administrative and supervisory salaries are included in this category, and we denote this fixed cost collectively by F. A second component is directly proportional to the number of units produced. For example, both the raw materials and the labor required to produce the product are included in this category. Let c denote the variable cost of producing one item. In the absence of any other factors we can then expect to have a deterministic cost–size relation of the form

$$y = F + cx$$

However, we must consider a third component of cost whose magnitude is somewhat unpredictable in nature. Machines occasionally break down in the course of production, resulting in varying amounts of idle time and repair costs. Fluctuations in the quality of the raw materials may also occur, resulting in a slowdown of the production process. Thus a deterministic relation can be masked by chance components that result from these and other undiscovered factors. Consequently, the relationship between y and x must be investigated by a statistical analysis of the cost-batch size data. ■

Empirical Relation

In contrast to the above situations many natural phenomena are comprised of mutually related variables, or one variable that is dependent on a number of influencing or causal variables, where the relation is not governed by a precise physical law. A plot of some observed values of these variables on graph paper depicts, in a rather crude form, a relation intertwined with chance fluctuation. The following are a few examples of situations where the form of the underlying relationship is completely unknown. After gaining sufficient knowledge about an empirical relation it may be possible for the investigator to formulate a theory that leads to a mathematical formula and hence to the semi-deterministic case.

Example 10.3 To combat automobile pollution, research is underway to determine the chemical composition of a gasoline additive that will improve emission quality. One aspect of the investigation is to study the relationship between the amount of a particular additive and the reduction in the emission of nitrogen oxides. Other ingredients may also produce some effects, but their amounts are to remain fixed during the study. Several new Mustang automobiles are selected as the experimental units. The amount of nitrogen oxides in the exhaust of each car is measured first without the additive and then with a specified amount x of the additive. The reduction of nitrogen oxides in the exhaust is taken to be the response y. Due to the complexity of chemical reactions and of the internal conditions of the automobile engine, a deterministic formula of the relationship between y and x is beyond present knowledge. ■

Example 10.4 Suppose that the yield of tomato plants y in an agricultural experiment is to be studied in relation to the dosage x of a certain fertilizer, while other contributing factors such as irrigation and soil dressing are to remain as constant as possible. The experiment consists of applying different dosages of the fertilizer, over the range of interest, in different plots and then recording the tomato yield from these plots. Different dosages of fertilizer will typically produce different yields, but the relationship is not expected to follow a precise mathematical formula. Aside from unpredictable chance variations, the underlying form of the relation in this case cannot be determined from any theoretical basis. ■

Example 10.5 The aptitude of a newly trained operator for performing a skilled job depends on both the duration of the training period and the nature of the training program. To evaluate the effectiveness of the training program, we must conduct an experimental study of the relation between growth in skill or learning y and duration of training x. It is unlikely that this relation will be deterministic, due to the simple fact that no two human beings are exactly alike. Yet an analysis of the data of the two variables could help us to assess the nature of the relation and to utilize it in evaluating and designing such a training program. ■

These examples illustrate the domain of the application of regression analysis in the rather simple context of determining how one variable is related to one other variable. In more complex situations several variables may be interrelated or one variable of major interest may depend on several influencing variables, and studies of such relationships require the observation and analysis of all these variables. In Example 10.5 the growth of learning may be studied in relation to I.Q., the score on an initial aptitude test, the amount of classroom and laboratory training received, etc. Similarly, the yield of a chemical process may be studied in relation to several variables, such as the temperature maintained in the system, the initial concentration of the ingredients, or the rate of cooling. The usefulness of regression analysis extends to these multivariable problems. It provides the methods for building models for such relationships, estimating the unknown parameters, determining which variables are important and which are redundant, and finally employing these models in prediction and control.

10.2 A SIMPLE REGRESSION PROBLEM

To exhibit the basic concepts, we begin with an experiment to determine the relation between two variables x and y; x acts as an independent variable whose values are controlled by the experimenter, whereas y is dependent on x and is subject to uncontrollable sources of error.

> The *independent* or *controlled* variable is also called the *predictor variable* and is denoted by x. The *effect* or *response variable* is denoted by y.

The dependence of y on x is unidirectional so that we are primarily concerned with situations in which the values of x are set with no appreciable error. (Cases in which both x and y are beyond the control of the experimenter and can only be observed by random sampling are discussed later in this chapter.) For a more concrete discussion we suppose that n Mustang automobiles are used in the experiment described in Example 10.3. The quantity of nitrogen oxides that each car emits is first measured without the additive. Then a specified amount x of the additive is used in a full tank and the emission of nitrogen oxides is measured again. The reduction in the amount of nitrogen oxides can then be recorded as the response variable y. The data can be arranged as in Table 10.1.

TABLE 10.1
FORM OF DATA

Amount of additive x	$x_1 \quad x_2 \quad x_3 \cdots x_n$
Reduction in nitrogen oxides y	$y_1 \quad y_2 \quad y_3 \cdots y_n$

For numerical illustration, we consider the data given in Table 10.2 as the observations obtained in an experiment with $n = 10$ cars. The amount of additive x and the reduction in nitrogen oxides y are measured in some suitable units. Seven different levels of x are included in the experiment, and some of these levels are repeated for more than one car. A glance at the table shows us that y generally increases with x but it is difficult to say much more about the form of the relation simply by looking at this tabular data.

TABLE 10.2
THE AMOUNT OF ADDITIVE AND REDUCTION INNITROGEN OXIDES DATA FROM 10 CARS

Amount of additive x	1	1	2	3	4	4	5	6	6	7
Reduction in nitrogen oxides y	2.1	2.5	3.1	3.0	3.8	3.2	4.3	3.9	4.4	4.8

10.3 PLOTTING THE DATA

In studying the relationship between two variables, the logical first step is to plot the data as points on graph paper. The resulting figure, called a *scatter diagram*, indicates whether the points are clustered around a straight line or some curve and also provides a visual impression of the extent of variation about the line or curve. In most situations there is no a priori theoretical relation known to apply so that the information portrayed by the scatter diagram is useful in the search for an appropriate mathematical model.

The scatter diagram of the observations in Table 10.2 appears in Figure 10.1. This scatter diagram reveals that the relationship is approximately linear in nature; that is, the points seem to cluster around a straight line. Because a linear relation is the simplest relationship to handle mathematically, we present the details of the statistical regression analysis for this case. Other situations can often be reduced to this case by applying suitable transformations to one or both variables.

We conclude this section with the reminder:

> *First step in the analysis*
>
> In investigating the relationship between two variables, plotting a *scatter diagram* is an important preliminary step that must be made before undertaking a formal statistical analysis. A scatter diagram provides insight into the nature of the relationship exhibited by the data.

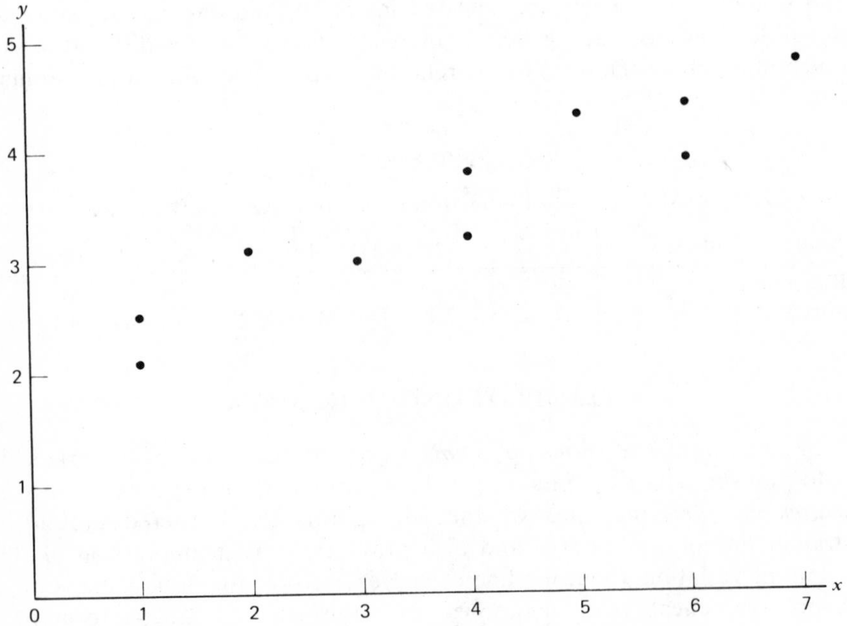

Figure 10.1 Scatter diagram of the data given in Table 10.2.

10.4 A STRAIGHT-LINE REGRESSION MODEL

Recall that if the relation between y and x is exactly a straight line, then the variables are connected by the formula

$$y = \alpha + \beta x$$

where α indicates the intercept of the line with the y axis and β represents the slope of the line, or the change in y per unit change in x (see Figure 10.2).

In a nondeterministic situation it is therefore reasonable to postulate that an underlying linear relation is being masked by random disturbances or experimental errors. Given this viewpoint, we can formulate the following linear regression model as a tentative representation of the mode of relationship between y and x. We then proceed with our statistical analysis.

Statistical Model

We assume that the response Y_i is related to the value x_i of the controlled variable by

$$Y_i = \alpha + \beta x_i + e_i, \quad i = 1, \ldots, n$$

where

(*a*) $x_1, x_2, \ldots, x_n$ are the set values of the controlled variable x that the experimenter has selected for the study.

(*b*) $e_1, \ldots, e_n$ are the unknown error components that are superimposed on the true linear relation. These are *unobservable random variables*, which we assume are independently and normally distributed with a mean of zero and an unknown variance of σ^2.

(*c*) The parameters α and β, which together locate the straight line, are unknown.

According to this model each observation Y_i that corresponds to level x_i of the controlled variable is a random sample of one observation from the normal distribution with mean $= \alpha + \beta x_i$ and standard deviation $= \sigma$. One interpretation of this is that as we attempt to observe the true value on the line, nature adds the random error e to this quantity. This error structure is illustrated in Figure 10.3, which shows a few normal distributions of Y.

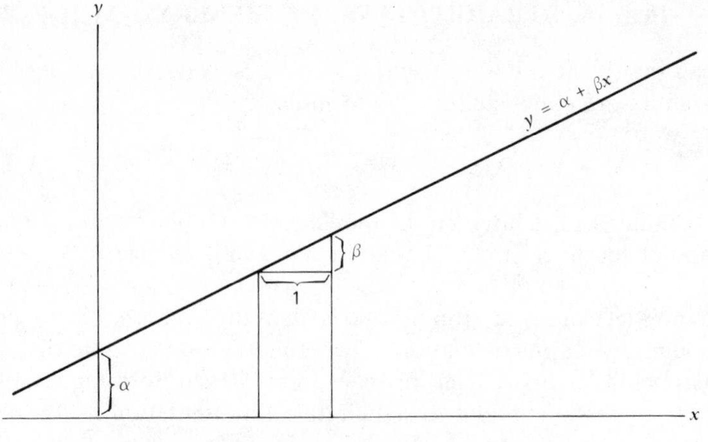

Figure 10.2 Graph of straight line $y = \alpha + \beta x$.

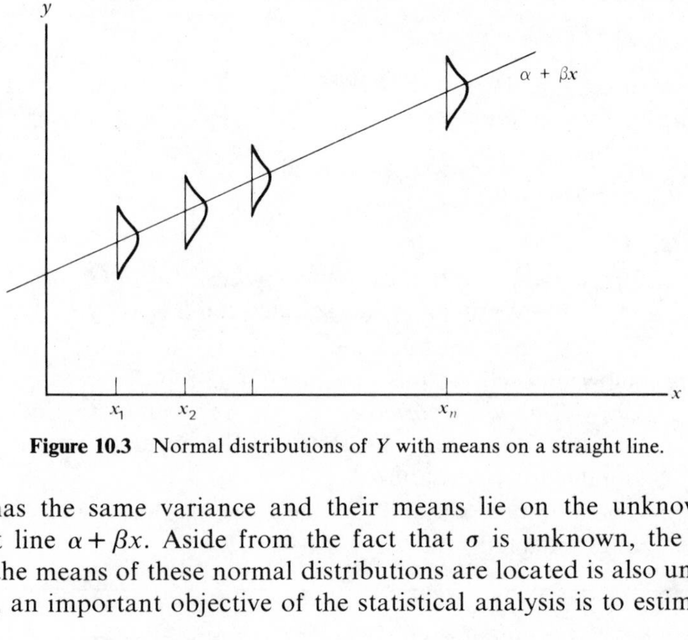

Figure 10.3 Normal distributions of Y with means on a straight line.

Each has the same variance and their means lie on the unknown true straight line $\alpha + \beta x$. Aside from the fact that σ is unknown, the line on which the means of these normal distributions are located is also unknown. In fact, an important objective of the statistical analysis is to estimate this line.

10.5 THE METHOD OF LEAST SQUARES

If we tentatively assume that the preceding formulation of the model is correct, we can proceed to estimate the regression line and to solve a few related inference problems. The problem of estimating the regression parameters α and β can be viewed as fitting the best straight line on the scatter diagram. One simple method is to slide a transparent ruler over the diagram to visually determine a straight line that closely fits the data. Although it is simple to use, this method of "eye estimation" has some serious drawbacks. Being a subjective procedure, it leaves no scope for statistical inferences such as the construction of confidence intervals and tests of hypotheses. Second, it cannot be used in a study of more than two variables, because a scatter diagram cannot be plotted in such a case. Even with two variables the method is not easy to use if the relation appears to be a curve rather than a line. The *method of least squares*, which we explain here, is an objective, efficient method of estimating the regression parameters and its application is not limited to the straight-line model.

Suppose that an arbitrary line $y = a + bx$ is drawn on the scatter diagram, as it is in Figure 10.4. At the value x_i of the controlled variable, the y value predicted by this line is $a + bx_i$, whereas the observed y value is y_i. The discrepancy between these values is $(y_i - a - bx_i) = d_i$, which is the vertical distance of the point from the line.

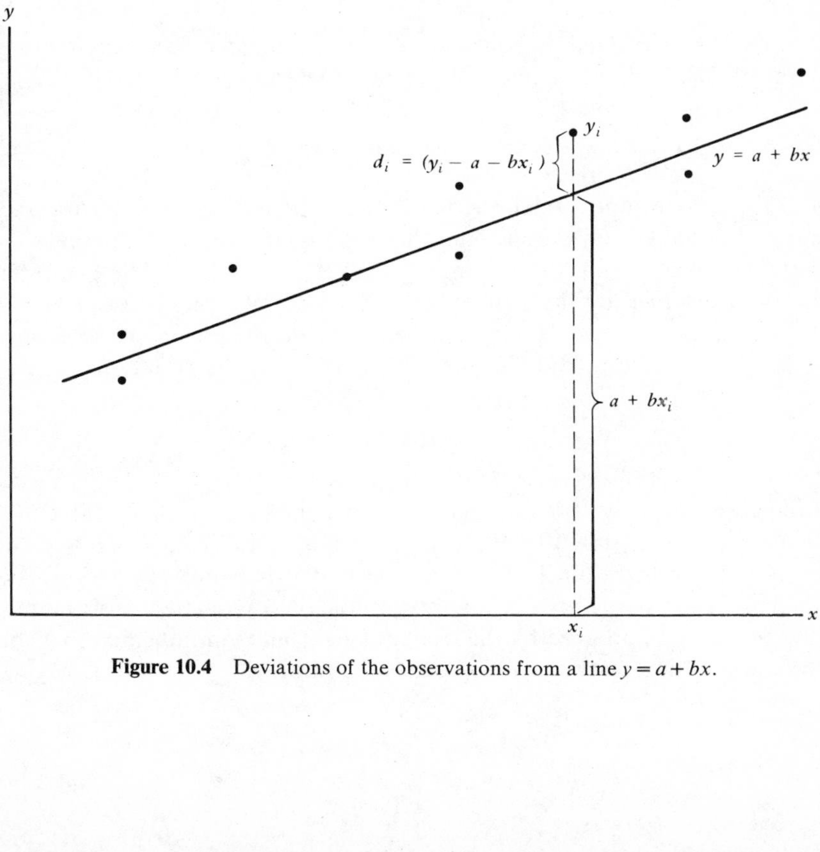

Figure 10.4 Deviations of the observations from a line $y = a + bx$.

Considering such discrepancies at all the n points, we take

$$D=\sum_{1}^{n} d_i^2=\sum_{1}^{n}(y_i-a-bx_i)^2$$

as an overall measure of the discrepancy of the observed points from the fitted line. The magnitude of D obviously depends on the line that is drawn; in other words, it depends on a and b, the two quantities that determine the line. A good fit will make D as small as possible. We now state the basic principle we are following here, first in general terms to illustrate its versatility in applications to other models.

The Principle of Least Squares

The principle of least squares consists of determining the values for the unknown parameters to minimize the overall discrepancy. The overall discrepancy D is defined as

$$D=\sum(\text{Observed response}-\text{Predicted response})^2$$

where the predicted response involves the unknown parameters of the model. The parameter values thus determined are called the *least squares estimates*.

For the straight-line model, the predicted response is $a+bx_i$ corresponding to the observed response y_i, and the least squares principle involves the determination of a and b to minimize $D=\sum_1^n(y_i-a-bx_i)^2$. The quantities a and b determined by this principle are denoted by $\hat{\alpha}$ and $\hat{\beta}$, respectively, and are called the *least squares estimates* of the regression parameters α and β. The best fitting straight line is then given by the equation

$$\hat{y}=\hat{\alpha}+\hat{\beta}x$$

To illustrate the idea of least squares, we use the data given in Table 10.2 to compute D in Table 10.3 for two choices of a and b: $a=0$, $b=1$, and $a=2$, $b=.5$. From Table 10.3 we can see that it is better to choose $a=2$, $b=.5$ than $a=0$, $b=1$, because the magnitude of D is smaller. Fortunately, we do not have to proceed by the trial and error method indicated in Table 10.3 to obtain the least squares estimates $\hat{\alpha}$ and $\hat{\beta}$.

TABLE 10.3
COMPUTATION OF THE DISCREPANCY D FOR TWO CHOICES a AND b

		$a=0,\quad b=1$			$a=2,\quad b=.5$		
x	y	$a+bx$	$d=(y-a-bx)$	d^2	$a+bx$	d	d^2
1	2.1	1	1.1	1.21	2.5	−.4	.16
1	2.5	1	1.5	2.25	2.5	0	0
2	3.1	2	1.1	1.21	3.0	.1	.01
3	3.0	3	0	0	3.5	−.5	.25
4	3.8	4	−.2	.04	4.0	−.2	.04
4	3.2	4	−.8	.64	4.0	−.8	.64
5	4.3	5	−.7	.49	4.5	−.2	.04
6	3.9	6	−2.1	4.41	5.0	−1.1	1.21
6	4.4	6	−1.6	2.56	5.0	−.6	.36
7	4.8	7	−2.2	4.84	5.5	−.7	.49
				$D=17.65$			$D=3.20$

An analytical solution is available for least squares estimation in the straight-line regression model. For simplicity in the presentation of this solution we introduce a few basic notations:

Basic notations

$$\bar{x}=\frac{1}{n}\Sigma x_i,\qquad \bar{y}=\frac{1}{n}\Sigma y_i$$

$$S_x^2=\Sigma(x_i-\bar{x})^2=\Sigma x_i^2-n\bar{x}^2$$

$$S_y^2=\Sigma(y_i-\bar{y})^2=\Sigma y_i^2-n\bar{y}^2$$

$$S_{xy}=\Sigma(x_i-\bar{x})(y_i-\bar{y})=\Sigma x_iy_i-n\bar{x}\bar{y}$$

The quantities $\bar{x}$ and $\bar{y}$ are the sample means of the x and y values; S_x^2 and S_y^2 are the sums of squared deviations from the means. We have already become familiar with these expressions in connection with the definition of sample variance. Here, however, the x values are fixed by the experimenter. S_{xy} is the sum of the cross products of deviations, and its second form can be derived from the first by writing $(x_i-\bar{x})(y_i-\bar{y})=x_iy_i-\bar{x}y_i-\bar{y}x_i+\bar{x}\bar{y}$ and summing term by term.

*10.6 DERIVATION OF THE LEAST SQUARES ESTIMATES

According to the principle of least squares, we must determine the quantities a and b such that $D=\Sigma(y_i-a-bx_i)^2$ is minimized. First we write

$$y_i-a-bx_i=(y_i-\bar{y})-b(x_i-\bar{x})+(\bar{y}-a-b\bar{x})$$

Squaring both sides we then obtain

$$(y_i-a-bx_i)^2=(y_i-\bar{y})^2+b^2(x_i-\bar{x})^2+(\bar{y}-a-b\bar{x})^2$$
$$-2b(x_i-\bar{x})(y_i-\bar{y})-2b(x_i-\bar{x})(\bar{y}-a-b\bar{x})+2(y_i-\bar{y})(\bar{y}-a-b\bar{x})$$

We now sum both sides over $i=1,\ldots,n$ and note that the last two terms in the right-hand side of the formula disappear after summation, because $\Sigma(x_i-\bar{x})=0$ and $\Sigma(y_i-\bar{y})=0$. Hence we have

$$D=S_y^2+b^2S_x^2+n(\bar{y}-a-b\bar{x})^2-2bS_{xy}$$

We can now rearrange terms and complete a square with

$$D=n(\bar{y}-a-b\bar{x})^2+b^2S_x^2-2bS_{xy}+S_y^2$$

$$=n(\bar{y}-a-b\bar{x})^2+\left(b^2S_x^2-2bS_{xy}+\frac{S_{xy}^2}{S_x^2}\right)+S_y^2-\frac{S_{xy}^2}{S_x^2}$$

$$=n(\bar{y}-a-b\bar{x})^2+\left(bS_x-\frac{S_{xy}}{S_x}\right)^2+\left(S_y^2-\frac{S_{xy}^2}{S_x^2}\right)$$

The last term does not involve a and b. The first two terms can be reduced to the smallest value of zero if we set

$$b=\frac{S_{xy}}{S_x^2} \quad \text{and} \quad a=\bar{y}-b\bar{x}$$

10.7 LEAST SQUARES REGRESSION LINE

The derivation presented in Section 10.6 provides formulas for the least squares estimates:

Least squares estimate of α: $\hat{\alpha} = \bar{y} - \hat{\beta}\bar{x}$

Least squares estimate of β: $\hat{\beta} = \dfrac{S_{xy}}{S_x^2}$

The estimates $\hat{\alpha}$ and $\hat{\beta}$ can then be used to locate the best fitting line:

Least squares regression line: $\hat{y} = \hat{\alpha} + \hat{\beta}x$

Employing the least squares estimates, the minimum value of D is

$$S_y^2 - \frac{S_{xy}^2}{S_x^2} = S_y^2 - \hat{\beta}^2 S_x^2$$

This quantity is called the *residual sum of squares* or the *sum of squares due to error* and is denoted by SSE. Thus

The *residual sum of squares* or the *sum of squares due to error* is

$$\text{SSE} = S_y^2 - \hat{\beta}^2 S_x^2 = \sum_{i=1}^{n} (y_i - \hat{\alpha} - \hat{\beta}x_i)^2$$

where the second form follows from the fact that this is the overall discrepancy of the observations around the fitted regression line $\hat{y} = \hat{\alpha} + \hat{\beta}x$. The individual deviations of the observations y_i from the fitted line are called the *residuals*. In addition to providing an alternative calculation for SSE, the residuals play a central role in checking the assumptions of the model.

Residuals $= y_i - \hat{\alpha} - \hat{\beta}x_i\,, \quad i = 1, 2, \ldots, n$

In applying the least squares method to a given data set, it is convenient first to compute the basic quantities $\bar{x}$, $\bar{y}$, S_x^2, S_y^2, and S_{xy} introduced in Section 10.5. Then the preceding formulas can be used to obtain the least squares regression line, the residuals, and the value of SSE. Computations for the data given in Table 10.2 are illustrated in Table 10.4.

TABLE 10.4

COMPUTATIONS FOR THE LEAST SQUARES LINE AND SSE

	x	y	x^2	y^2	xy	$\hat{\alpha}+\hat{\beta}x$	Residual
	1	2.1	1	4.41	2.1	2.387	−.287
	1	2.5	1	6.25	2.5	2.387	.113
	2	3.1	4	9.61	6.2	2.774	.326
	3	3.0	9	9.00	9.0	3.161	−.161
	4	3.8	16	14.44	15.2	3.548	.252
	4	3.2	16	10.24	12.8	3.548	−.348
	5	4.3	25	18.49	21.5	3.935	.365
	6	3.9	36	15.21	23.4	4.322	−.422
	6	4.4	36	19.36	26.4	4.322	.078
	7	4.8	49	23.04	33.6	4.709	.091
Total	39	35.1	193	130.05	152.7		.007

$$\bar{x}=3.9, \quad \bar{y}=3.51 \qquad \hat{\beta}=\frac{15.81}{40.9}=.387$$

$$S_x^2=193-10(3.9)^2=40.9 \qquad \hat{\alpha}=3.51-.387\times 3.9=2.00$$

$$S_y^2=130.05-10(3.51)^2=6.85 \qquad \text{SSE}=6.85-\frac{(15.81)^2}{40.9}=.74$$

$$S_{xy}=152.7-10(3.9)(3.51)=15.81$$

The equation of the line fitted by the least squares method is then

$$\hat{y}=2.00+.387x$$

Figure 10.5 shows a plot of the data along with the least squares regression line.

The deviations $y_i-2.00-.387x_i$ of the observed y values from the fitted line are computed in the last column of Table 10.4. It is necessary to retain additional decimal places in this residual column to perform the alternate calculation of SSE as the sum of squares of residuals. From this second form we obtain

$$\text{SSE}=(-.287)^2+(.113)^2+(.326)^2+\ldots+(.091)^2=.7376$$

The difference between .7376 and .74 is due to rounding. Theoretically, the sum of the residuals should be zero, and the difference between the sum .007 and zero is also due to rounding. This form of calculation may be better suited for electronic computers. An alternative form for the calculation of $\hat{\alpha}$ and $\hat{\beta}$ is given in Exercise 6 at the end of this chapter.

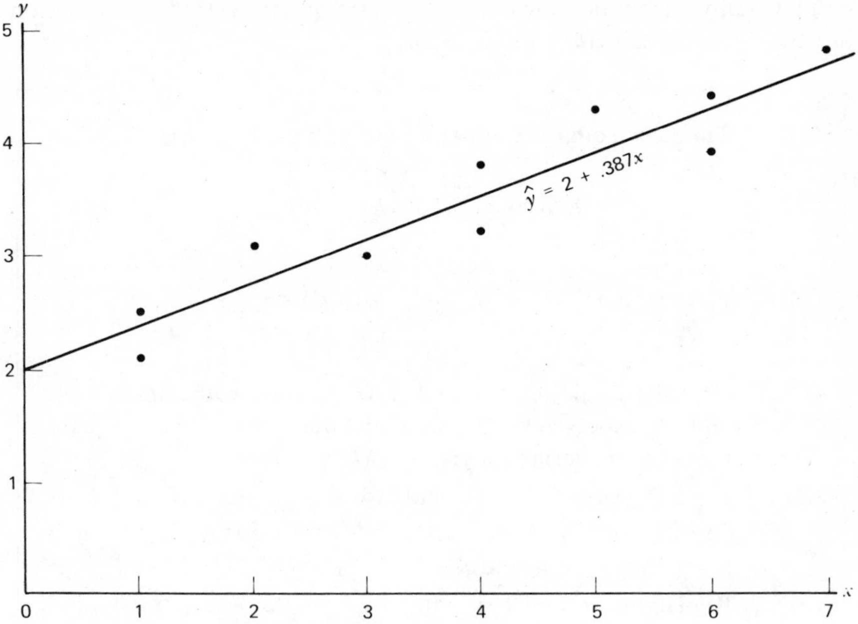

Figure 10.5 The least squares regression line for the data given in Table 10.2.

10.8 PROPERTIES OF THE LEAST SQUARES ESTIMATORS

It is important to remember that the line $\hat{y} = \hat{\alpha} + \hat{\beta}x$ obtained by the principle of least squares is an *estimate*, based on the sample data, of the unknown true regression line $y = \alpha + \beta x$. In our automobile emissions problem (originally Example 10.3), the estimated line is $2 + .387x$. This suggests that each unit of the additive improves the mean reduction in nitrogen oxides by .387. If, for example, $x = 3.2$ units of additive are tried, we can also use the fitted regression line to calculate the estimated amount of reduction as $2 + .387(3.2) = 3.24$. Two questions concerning these estimates naturally arise at this point:

(*a*) In light of the value .387 for $\hat{\beta}$, could $\beta = 0$ so that y does not depend on x? What are plausible values for β?

(*b*) How much uncertainty should be attached to the estimate $2 + .387(3.2) = 3.24$ of the point $\alpha + \beta(3.2)$ on the true regression line?

To answer these and related questions, we must know something about the sampling distributions of the least squares estimators $\hat{\alpha}$ and $\hat{\beta}$. To avoid lengthy algebraic calculations, we simply state these distributions and their properties here without proof.

(*a*) The least squares estimators are unbiased; that is

$$E(\hat{\alpha})=\alpha \quad \text{and} \quad E(\hat{\beta})=\beta$$

(*b*) $\operatorname{Var}(\hat{\alpha})=\sigma^2\left[\dfrac{1}{n}+\dfrac{\bar{x}^2}{S_x^2}\right]$ and $\operatorname{Var}(\hat{\beta})=\dfrac{\sigma^2}{S_x^2}$

(*c*) The distributions of $\hat{\alpha}$ and $\hat{\beta}$ are normal with means of α and β, respectively; the standard deviations are the square roots of the variances given in (*b*).

(*d*) $s^2=\text{SSE}/(n-2)$ is an unbiased estimator of σ^2. Also, $(n-2)s^2/\sigma^2$ is distributed as χ^2 with d.f. $=n-2$, and it is independent of $\hat{\alpha}$ and $\hat{\beta}$.

(*e*) Replacing σ^2 in (*b*) with its sample estimate s^2 and considering the square roots of the variances, we obtain the estimated standard errors of $\hat{\alpha}$ and $\hat{\beta}$:

$$\text{Estimated standard error of } \hat{\alpha}=s\sqrt{\frac{1}{n}+\frac{\bar{x}^2}{S_x^2}}$$

$$\text{Estimated standard error of } \hat{\beta}=\frac{s}{S_x}$$

(*f*) $\dfrac{S_x(\hat{\beta}-\beta)}{s}$ has a *t* distribution with d.f. $=n-2$

$\dfrac{(\hat{\alpha}-\alpha)}{s\sqrt{\dfrac{1}{n}+\dfrac{\bar{x}^2}{S_x^2}}}$ has a *t* distribution with d.f. $=n-2$

Incidentally, we mention here without proof a strong point in favor of the principle of least squares estimation in addition to its intuitive plausibility. The estimators $\hat{\alpha}$ and $\hat{\beta}$ are not only unbiased, but they also have the smallest possible variances among all unbiased estimators. In other words,

$\hat{\alpha}$ and $\hat{\beta}$ are the best unbiased estimators. This mathematical property is often given as a justification for the least squares method.

10.9 IMPORTANT INFERENCE PROBLEMS

We now consider how to test hypotheses, construct confidence intervals, and make predictions in the context of a linear regression model.

10.9.1 Inference Concerning the Slope β

In a regression analysis problem it may be of special interest to determine whether the expected response does or does not vary with the magnitude of the controlled variable. According to the model, the expected response is related to the level of the controlled variable by

$$E(Y|x) = \alpha + \beta x$$

where $Y|x$ means that the response Y corresponds to the given level x of the controlled variable. The expected response $\alpha + \beta x$ does not change with a change in x if and only if $\beta = 0$. We can therefore test the null hypothesis $H_0: \beta = 0$ against a one- or a two-sided alternative, depending on the nature of the relation that is anticipated. Referring to property (f), given in Section 10.8, the null hypothesis $H_0: \beta = 0$ is to be tested by:

$$H_0: \beta = 0: \quad \text{test statistic} \quad t = \frac{S_x \hat{\beta}}{s}, \quad \text{d.f.} = n - 2$$

Example 10.6 Do the data given in Table 10.2 constitute strong evidence that the additive reduces the amount of nitrogen oxides over the range covered in the study?

To answer this question, we consider testing $H_0: \beta = 0$ vs. $H_1: \beta > 0$. Because $s^2 = \text{SSE}/(n-2) = .74/8 = .0925$, the value of the test statistic is

$$t = \frac{S_x \hat{\beta}}{s} = \sqrt{\frac{40.9}{.0925}} \times .387 = 8.14, \quad \text{d.f.} = 10 - 2 = 8$$

With d.f. = 8, the upper 5% tabulated value of t is 1.860. The observed t value is therefore highly significant, and H_0 is rejected, reflecting a significant reduction in nitrogen oxides when the additive is used. ■

A warning is in order here concerning the interpretation of the test of $H_0: \beta = 0$. If H_0 is not rejected, we may be tempted to conclude that y does not depend on x. Such an unqualified statement may be erroneous. First, the absence of a linear relation has only been established over the range of the x values in the experiment. There is no basis on which to draw conclusions about the relation for x values that lie outside the observed range. Second, the interpretation of lack of dependence is valid only if our model formulation is correct. If the scatter diagram depicts a relation on a curve but we inadvertently formulate a linear model and test $H_0:\beta=0$, the conclusion of accepting H_0 should be interpreted to mean "no linear relation" rather than "no relation." We elaborate on this point further in Chapter 11. Our present viewpoint is to assume that the model is correctly formulated and to discuss the various inference problems associated with it.

More generally, we may test whether or not β is equal to some postulated value β_0, not necessarily zero.

The test for

$$H_0:\beta=\beta_0 \quad \text{is based on}$$

$$t=\frac{S_x(\hat{\beta}-\beta_0)}{s}, \quad \text{d.f.}=n-2$$

In addition to testing hypotheses, we can provide a confidence interval for the parameter β using the t distribution. For instance,

95% confidence interval for β:

$$\hat{\beta}\pm t_{.025}\frac{s}{S_x}$$

where $t_{.025}$ is the upper 2.5% point of the t distribution with d.f. $=n-2$.

Example 10.7 For the reduction in nitrogen-oxides data given in Table 10.2, a 95% confidence interval for β is given by

$$.387\pm 2.306\frac{.304}{6.4}=.387\pm .110 \quad \text{or} \quad (.277, .497)$$

This means that we are 95% confident that by adding one extra unit of the additive we will attain a mean reduction in nitrogen oxides of between .277 and .497. ■

10.9.2 Inference About α

Although somewhat less important in practice, inference procedures similar to the ones outlined in Section 10.9.1 can be provided for the parameter α using the t distribution with d.f. $=n-2$, stated for $\hat{\alpha}$ in property (f) in Section 10.8. For example,

90% confidence interval for α:

$$\hat{\alpha} \pm t_{.05}\, s \sqrt{\frac{1}{n} + \frac{\bar{x}^2}{S_x^2}}$$

Further, for testing

H_0: $\alpha = \alpha_0$, test is based on

$$t = \frac{(\hat{\alpha} - \alpha_0)}{s \sqrt{\dfrac{1}{n} + \dfrac{\bar{x}^2}{S_x^2}}}, \quad \text{d.f.} = n - 2$$

10.9.3 Prediction of the Mean Response for a Specified x Value

The most important objective in a regression study may be to employ the fitted regression model in estimating the expected response corresponding to a specified level of the controlled variable. For example, we may want to estimate the expected reduction in nitrogen oxides for a specified amount x^* of this additive, in all Mustangs, by employing the model of a linear relationship. According to the linear model examined in Section 10.4, the expected response at a value x^* of the controlled variable x is given by $E(Y|x^*) = \alpha + \beta x^*$. An unbiased estimator of this is $\hat{\alpha} + \hat{\beta}x^*$, because $\hat{\alpha}$ and $\hat{\beta}$ are unbiased estimators of α and β, respectively. In other words, the point on the fitted regression line corresponding to the value x^* of x provides an unbiased estimate of the expected response. Further properties of the estimator are:

To estimate $E(Y|x^*) = \alpha + \beta x^*$, use the estimator $\hat{\alpha} + \hat{\beta}x^*$:

Estimated standard error: $s \sqrt{\dfrac{1}{n} + \dfrac{(x^* - \bar{x})^2}{S_x^2}}$

Distribution: $$\frac{\hat{\alpha} + \hat{\beta}x^* - (\alpha + \beta x^*)}{s \sqrt{\dfrac{1}{n} + \dfrac{(x^* - \bar{x})^2}{S_x^2}}}$$

has a t distribution with d.f. $= n - 2$.

This t distribution can be used to construct confidence intervals or hypothesis tests in the usual manner.

95% confidence interval for the expected response $E(Y|x^*)=\alpha+\beta x^*$ is

$$\hat{\alpha}+\hat{\beta}x^* \pm t_{.025}\, s\sqrt{\frac{1}{n}+\frac{(x^*-\bar{x})^2}{S_x^2}}$$

To test the hypothesis that $E(Y|x^*)=\mu_0$, some specified value, we use

$$t=\frac{\hat{\alpha}+\hat{\beta}x^*-\mu_0}{s\sqrt{\frac{1}{n}+\frac{(x^*-\bar{x})^2}{S_x^2}}}, \qquad \text{d.f.}=n-2$$

Example 10.8 Again consider the reduction in nitrogen oxides data given in Table 10.2 and the calculations for the regression analysis given in Table 10.4. The fitted regression line is

$$\hat{y}=2+.387x$$

The expected reduction corresponding to the amount $x^*=4$ of the additive is estimated by

$$\hat{\alpha}+\hat{\beta}x^*=2+.387\times 4=3.548$$

$$\text{Estimated standard error}=s\sqrt{\frac{1}{10}+\frac{(4-3.9)^2}{40.9}}$$

$$=.304\times .317=.096$$

A 95% confidence interval for the mean reduction in nitrogen oxides at $x^*=4$ is therefore

$$3.548 \pm t_{.025}\times .096=3.548\pm 2.306\times .096$$

$$=3.548\pm .22 \quad \text{or} \quad (3.33,\ 3.77)$$

We are 95% confident of achieving a reduction in nitrogen oxides between 3.33 and 3.77.

Suppose that we also wish to estimate the mean reduction at $x^* = 7.5$. Following the same steps, the point estimate is

$$\hat{\alpha} + \hat{\beta}x^* = 2 + .387 \times 7.5 = 4.90$$

$$\text{Estimated standard error} = .304\sqrt{\frac{1}{10} + \frac{(7.5-3.9)^2}{40.9}}$$

$$= .304 \times .646 = .196$$

A 95% confidence interval is

$$4.90 \pm 2.306 \times .196 = 4.90 \pm .45 \quad \text{or} \quad (4.45,\ 5.35)$$ ■

The formula for the standard error of prediction shows that when x^* is close to $\bar{x}$, the standard error is smaller than it is when x^* is far removed from $\bar{x}$. This is confirmed by Example 10.8, where the standard error of prediction at $x^* = 7.5$ can be seen to be more than twice as large as the value at $x^* = 4$. Consequently, the confidence interval for the former is also larger. Thus we can conclude that, in general, prediction is more precise near the mean $\bar{x}$ than it is for values of the x variable that lie far from the mean.

Moreover, the formula for the standard error of prediction reflects the strength of regression techniques, because for x^* at the center $\bar{x}$ of the x values it equals σ^2/n. In other words, regression methods not only enable us to determine how x explains y, but an estimate of the mean $\alpha + \beta\bar{x}$ has the same variance it would have if all the observations were made using that single amount of additive $\bar{x}$ for all trials.

Caution. Extreme caution should be exercised in extending a fitted regression line to make long-range predictions far away from the range of x values covered in the experiment. Not only does the confidence interval

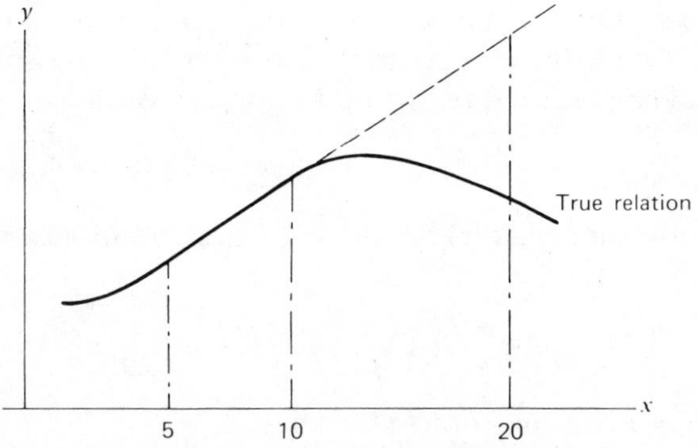

Figure 10.6 Danger in long-range prediction.

become so wide that predictions based on it can be extremely unreliable, but an even greater danger exists. If the pattern of the relationship between the variables changes drastically at a distant value of x, the data provide no information with which to detect such a change. Figure 10.6 illustrates this situation. We would observe a good linear relationship if we experimented with x values in the 5–10 range, but if the fitted line were extended to estimate the response at $x^* = 20$, then our estimate would drastically miss the mark.

10.9.4 Prediction of a Single Response for a Specified x Value

Suppose that we give a specified amount x^* of the additive to a *single* Mustang car and we want to predict the reduction in nitrogen oxides. This problem is different from the one considered in Section 10.9.3, where we were interested in estimating the mean reduction for the population of *all* Mustangs for an amount x^* of the additive. The prediction is still determined from the fitted line; that is, the predicted value of the response is $\hat{\alpha} + \hat{\beta}x^*$ as it was in the preceding case. However, the standard error of the prediction here is inflated, because a single observation is more uncertain than the mean of a large number of observations. We now give the formula of the estimated standard error for this case.

> The estimated standard error when predicting a single observation y at a given x^* is
>
> $$s\sqrt{1 + \frac{1}{n} + \frac{(x^* - \bar{x})^2}{S_x^2}}$$

The formula for the confidence interval must be modified accordingly. ■

Example 10.9 Once again, consider the reduction in nitrogen oxides data given in Table 10.2. A new trial is to be made on a single Mustang with $x^* = 4.5$. The predicted reduction in nitrogen oxides is

$$\hat{\alpha} + \hat{\beta}x^* = 2 + .387 \times 4.5 = 3.74$$

and a 95% confidence interval for the reduction in nitrogen oxides in this Mustang is

$$3.74 \pm 2.306 \times .304\sqrt{1 + \frac{1}{10} + \frac{(4.5 - 3.9)^2}{40.9}}$$

$$= 3.74 \pm .74 \quad \text{or} \quad (3.00, 4.48)$$

This means that we are 95% confident this new car will have a reduction in nitrogen oxides between 3.00 and 4.48 if 4.5 units of the additive are used. This is true because 95% of the intervals calculated in this manner from repeated samples will include the new measurement. ■

10.10 WHAT TO DO IF THE PREDICTOR VARIABLE CANNOT BE CONTROLLED BY THE EXPERIMENTER

The analysis of the linear regression model presented in the preceding sections is based on the assumption that the independent variable x is nonrandom. The experimenter chooses the values of x to be included in the experiment and then observes the random variable y corresponding to these chosen levels of x. This type of procedure is appropriate in most controlled experiments, such as studies of the relation between weight gain y and carbohydrate intake x, crop yield y and fertilizer dosage x, or reaction time y and magnitude of stimulation impulse x. The levels of the causal variable x are preset by the experimenter over a realistic range of values, and the response y at a fixed level of x is considered a random variable with a normal distribution, as illustrated earlier in Figure 10.3.

In other experimental studies involving two variables x and y, even though x may be considered the causal variable that influences the response y, the experimenter may not be able to make controlled choices of the x values. In the population of experimental units, both x and y are regarded as random variables governed by some joint probability distribution. The experimenter selects a random sample of n experimental units and observes the pairs of values $(x_1,y_1), (x_2,y_2), \ldots, (x_n,y_n)$. The objective of the experiment is still to use this type of data to formulate a model for predicting the expected value of y when the value of x becomes known. For example, the administrator of an educational program may wish to study the relationship between the scores x that participants achieve on an admissions test and their subsequent performance y in the program. A prediction model of y from x would be useful in planning admissions standards that would be consistent with the program objective as well as in forecasting performance in the program. The data set would consist of the x scores and y scores for a random sample of n participants; obviously, the experimenter would have no control over the values of x in this situation.

When x and y are both random variables that have been observed by random sampling, the mathematical formulation of a prediction model and the associated conditions are based on a *bivariate* normal probability distribution which is briefly discussed in Chapter 12. However, *as long as x is viewed as the causal variable that influences y and the objective of sampling* *is to make predictions about y from the value of x, the operational steps of analysis are the same as those discussed in preceding sections*. In other words, the observed values of x are regarded as the fixed levels at which the experiment has been performed, and the results are interpreted as conditional on the observed set of x values. In contrast, special techniques must be used in situations where appreciable measurement errors are involved in recording x values as well as the y values.

10.11 REMARKS ABOUT THE MODEL

The basic statistical inference procedures associated with a linear regression model with one independent variable are described in Section 10.9. However, it is important to remember that the validity of these procedures and the usefulness of the conclusions are constrained by the assumptions made in the model formulation. A regression study is not conducted by performing a few routine hypothesis tests and constructing confidence intervals for parameters on the basis of the formulas given in Section 10.9. Such conclusions can be *seriously* misleading if the assumptions made in the model formulation are grossly incompatible with the data. It is therefore essential to check the data carefully for indications of any violation of the assumptions. To review, the assumptions involved in the formulation of our straight-line model are briefly stated again here:

(*a*) The underlying relation is linear.
(*b*) Independence of errors.
(*c*) Constant variance.
(*d*) Normal distribution.

Of course, when the general nature of the relationship between y and x forms a curve rather than a straight line, the predictions obtained from fitting a straight-line model to the data may produce nonsensical results. Often a suitable transformation of the data reduces a nonlinear relation to one that is approximately linear in form. A few simple transformations are discussed in Chapter 11. Violating the assumption of independence is perhaps the most serious error, because this can drastically distort the conclusions drawn from the t tests and the confidence statements associated with interval estimation. The implications of assumptions (*c*) and (*d*) were illustrated earlier in Figure 10.3. If the scatter diagram shows different amounts of fluctuations in the y values at different levels of x, then the assumption of constant variance may have been violated. Here again, an appropriate transformation of the data often helps to stabilize the variance. Lastly, using the t distribution in hypothesis testing and confidence interval estimation is valid as long as the errors are approximately normally distributed. A moderate departure from normality does not impair the conclusions, however, especially when the data set is large. In other words, a violation of assumption (*d*) alone is not as serious as a violation of any of the other assumptions. Methods of checking the adequacy of an assumed model are also discussed in Chapter 11.

We have yet to address ourselves to one important aspect of the regression problem: what values $x_1, \ldots, x_n$ should be included in the experiment? The interval $[a,b]$ where the x values should be located and the number of observations to be collected are typically determined from the problem under investigation. A question of *experimental design* is how to choose the n values of x in the interval $[a,b]$ to obtain the most precise inferences. Because all the standard error formulas given in Section 10.8 contain $S_x^2 = \Sigma(x_i - \bar{x})^2$ in the denominator, making S_x^2 as large as possible would provide the most precise estimates. When n is even, the maximum value of S_x^2 is realized by placing half of the x values at each of the two end points of the interval $[a,b]$. Accordingly, to achieve maximum precision in estimation, the experiment should be repeated at only the two distinct values of x that are the two extremes of the range of interest. This recommendation is impractical, however, because we rarely know a priori that the straight-line model is correct, and it is impossible to determine if the true relation is linear when all the y observations correspond to only two distinct x values. To determine whether a straight-line model or some other model is appropriate, it is essential to spread the x values over the range of interest rather than to locate all of them at the two end points of the interval.

EXERCISES

1. Plot the line $y = 3x + 2$ on graph paper by locating the points for $x = 0$, 1, 2.
2. Given the linear relationship $3x + 2y - 12 = 0$:
   (*a*) Find the intercept and slope.
   (*b*) Graph the line.
   (*c*) What is the value of y corresponding to $x = 1$?
   (*d*) What is the value of x corresponding to $y = 3$?
3. Identify the independent variable x and the dependent variable y in each of the following situations:
   (*a*) A training director wishes to study the relationship between the duration of training for new recruits and their performance in a skilled job.
   (*b*) A chemist wishes to study the relationship between the drying time of a paint and the concentration of a chemical additive that is supposed to accelerate the drying process.
   (*c*) The aim of a study is to relate the carbon monoxide level in blood

samples from smokers with the average number of cigarettes they smoke per day.

(*d*) An agronomist wishes to investigate the growth rate of a fungus in relation to the level of humidity in the environment.

(*e*) A market analyst wishes to relate the expenditures incurred in promoting a product in test markets and the subsequent amount of product sales.

4. Given these five pairs of (x,y) values:

x	1	2	3	4	5
y	.9	2.1	2.5	3.3	3.8

(*a*) Plot the points on graph paper.

(*b*) From a visual inspection, draw a straight line that appears to fit the data well.

(*c*) Compute the least squares estimates $\hat{\alpha}$ and $\hat{\beta}$ and draw the fitted line.

5. Referring to Exercise 4:

(*a*) Compute the residuals and verify that they sum to zero.

(*b*) Compute the sum of squares due to error by (*i*) adding the squares of the residuals, and (*ii*) using the formula $\text{SSE} = S_y^2 - \hat{\beta}^2 S_x^2$.

(*c*) Estimate σ^2.

6. *An alternative form for the calculation of* $\hat{\alpha}$ *and* $\hat{\beta}$: When using an electronic computer, it is usually easier to calculate the means $\bar{x}$ and $\bar{y}$, then the individual deviations from the mean $x_i - \bar{x}$ and $y_i - \bar{y}$, and finally their squares and product, which are summed to obtain S_x^2, S_y^2, and S_{xy}. These calculations are illustrated in the following table for the reduction in nitrogen-oxides data given in Table 10.2:

x_i	y_i	$x_i - \bar{x}$	$y_i - \bar{y}$	$(x_i - \bar{x})^2$	$(y_i - \bar{y})^2$	$(x_i - \bar{x})(y_i - \bar{y})$
1	2.1	−2.9	−1.41	8.41	1.9881	4.089
1	2.5	−2.9		8.41	1.0201	2.929
2	3.1	−1.9	−.41		.1681	
3	3.0	−.9	−.51	.81		.459
4	3.8	.1	.29	.01	.0841	.029
4	3.2	.1	−.31	.01	.0961	−.031
5	4.3	1.1	.79	1.21	.6241	.869
6	3.9		.39		.1521	.819
6	4.4	2.1		4.41	.7921	1.869
7	4.8	3.1	1.29	9.61	1.6641	
$\bar{x} = 3.9$	$\bar{y} = 3.51$	0 (check)	0 (check)	$S_x^2 =$	$S_y^2 =$	$S_{xy} =$

Fill in the missing values and compute the estimates $\hat{\alpha}$ and $\hat{\beta}$.

7. Demonstrate your familiarity with least squares calculations by fitting a straight line to the pairs of values:

x	1	1	2	4
y	7	8	6	3

(*a*) By the method given in Table 10.4.
(*b*) By the method illustrated in Exercise 6.

8. A morning newspaper lists the following used-car prices for a foreign compact, with age x measured in years and selling price y measured in thousands of dollars:

x	1	2	2	3	3	4	6	7	8	10
y	2.45	1.80	2.00	2.00	1.70	1.20	1.15	.69	.60	.47

(*a*) Plot the scatter diagram.
(*b*) Determine the equation of the least squares regression line and draw this line on the scatter diagram.
(*c*) Construct a 95% confidence interval for the slope of the regression line.

9. Referring to Exercise 8:
(*a*) From the fitted regression line, determine the predicted value for the average selling price of a five-year-old model compact and construct a 95% confidence interval.
(*b*) Determine the predicted value for a five-year-old compact to be listed in next week's paper. Construct a 90% confidence interval.
(*c*) Is it justifiable to predict the selling price of a 20-year-old compact from the fitted regression line? Give reasons for your answer.

10. Referring to the reduction in nitrogen-oxides data given in Table 10.2, use the fitted regression line to predict the expected value of y that corresponds to $x=0$. Is this prediction reasonable? Why or why not?

11. In an experiment designed to determine the relationship between the dose of a compost fertilizer x and the yield of a crop y, the following summary statistics are recorded:

$$n=15, \qquad \bar{x}=10.8, \qquad \bar{y}=122.7$$
$$S_x^2=70.6, \qquad S_y^2=98.5, \qquad S_{xy}=68.3$$

Assume a linear relationship.
(*a*) Find the equation of the least squares regression line.
(*b*) Compute the error sum of squares and estimate σ^2.
(*c*) Do the data contradict the experimenter's conjecture that over the range of x values covered in the study, the average increase in yield per unit increase in the compost dose is at least 1.5?

(*d*) Construct a 95% confidence interval for the expected yield corresponding to $x = 12$.

12. Some chemists interested in a property of plutonium called solubility, which depends on temperature, reported the following measurements [J. C. Mailen, et al., *Journal of Chemical and Engineering Data* (1971), Vol 16, No. 1, 69] for a plutonium powder (P_uF_3) in a molten mixture ($2LiF - BeF_2$), where y is $-\log_{10}$(solubility) and x is 1000/(temperature °C).

x	1.68	1.74	1.85	1.92	1.99	1.82	1.69	1.60	1.52
y	.33	.41	.57	.65	.77	.57	.35	.18	.14

(*a*) Find the least squares fit of a straight line.
(*b*) Establish a 95% confidence interval for β. Does solubility depend on temperature over the range of the experiment?
(*c*) Predict the mean value of $-\log_{10}$(solubility) at 714°C or $x = 1.40$. Construct a 95% confidence interval.
(*d*) Construct a 90% confidence interval for a new measurement to be made at $x = 1.8$.

13. Many college students obtain college degree credits by demonstrating their proficiency on exams developed as part of the College Level Examination Program (CLEP). Based on their scores on the College Qualification Tests (CQT), it would be helpful if students could predict their scores on a corresponding portion of the CLEP exam. For the following data (courtesy of R. Johnson) where x = Total CQT score and y = mathematical CLEP:

x	170	147	166	125	182	133	146	125	136	179
y	698	518	725	485	745	538	485	625	471	798

x	174	128	152	157	174	185	171	102	150	192
y	645	578	625	558	698	745	611	458	538	778

(*a*) Plot the scatter diagram and find the least squares fit of a straight line.
(*b*) Calculate the residuals.
(*c*) Construct a 95% confidence interval for the CLEP score of a student who obtains a CQT score of 150.
(*d*) Repeat (*c*) with $x = 175$ and $x = 195$.

14. An engineer interested in studying the thermal properties of ductile cast iron measures its cooling rate y in °F per hour during a heat-treatment stage. The engineer also records the data of

x_1 = The number of graphite spheroids present in a square millimeter area of the cast iron.

x_2 = The percentage of pearlite present in the cast iron.

(Courtesy of A. Tosh.)

x_1	346	272	276	350	345	304	311	338	344	346	272	276
x_2	1	1	1	1	1	1	1	1	1	2	2	2
y	45	20	20	65	65	20	30	30	55	55	30	30

x_1	239	254	238	304	311	338	344	346	272	276	239	254
x_2	2	2	2	2	2	2	2	3	3	3	3	3
y	20	20	20	30	45	45	65	65	45	45	30	30

x_1	304	311	338	239	254	238	304	338
x_2	3	3	3	4	4	4	4	4
y	45	55	55	45	45	30	55	65

As a first step in the analysis, disregard the variable x_2 and study the suitability of the simple linear regression model

$$Y = \alpha + \beta x_1 + e$$

In particular:

(*a*) Plot the scatter diagram of y vs. x_1.

(*b*) Determine the equation of the fitted straight line.

(*c*) Test to determine if the data provide strong evidence that the cooling rate increases as the count of graphite spheroids increases.

(*d*) Construct a 95% confidence interval for the average cooling rate corresponding to $x_1 = 300$.

15. *Testing the equality of the slopes of two regression lines*: Suppose that an experimenter wishes to determine how the response y is influenced by the dosage x of each of two comparable treatments. Treatment 1 is administered to n_1 subjects in different dosages, and their response measurements are recorded. Similarly, treatment 2 is administered to another independent group of n_2 subjects, and their responses are recorded. The data structure is then:

Dosage x_1 of treatment 1	x_{11}	x_{12}	$\cdots$	x_{1i}	$\cdots$	x_{1n_1}
Response y_1	y_{11}	y_{12}	$\cdots$	y_{1i}	$\cdots$	y_{1n_1}

Dosage x_2 of treatment 2	x_{21}	x_{22}	$\cdots$	x_{2i}	$\cdots$	x_{2n_2}
Response y_2	y_{21}	y_{22}	$\cdots$	y_{2i}	$\cdots$	y_{2n_2}

Assume that a linear relationship is appropriate for each treatment. Thus:

$$\text{Treatment 1:} \quad Y_{1i} = \alpha_1 + \beta_1 x_{1i} + e_{1i}, \quad i = 1, \dots, n_1$$

$$\text{Treatment 2:} \quad Y_{2i} = \alpha_2 + \beta_2 x_{2i} + e_{2i}, \quad i = 1, \dots, n_2$$

In addition to the standard assumptions of a linear regression model, we further assume here that the error variances are equal, or

$$\text{Var}(e_{1i}) = \text{Var}(e_{2i}) = \sigma^2$$

It is often of practical interest to test the null hypothesis that the two regression lines have equal slopes; that is, $H_0: \beta_1 = \beta_2$. Graphically, this is equivalent to the hypothesis that these two lines are parallel. The first step of analysis involves fitting separate regression lines to the two data sets and then calculating the individual error sum of squares. Let $\hat{\beta}_1$ and $\hat{\beta}_2$ denote the least squares estimators of β_1 and β_2, respectively, and let SSE(1) and SSE(2) denote the corresponding error sum of squares. Further, denoting

$$S_{x_1}^2 = \sum_{i=1}^{n_1} (x_{1i} - \bar{x}_1)^2, \qquad S_{x_2}^2 = \sum_{i=1}^{n_2} (x_{2i} - \bar{x}_2)^2$$

the statistic $(\hat{\beta}_1 - \hat{\beta}_2)$ is normally distributed with a mean of $\beta_1 - \beta_2$ and a variance of

$$\sigma^2 \left(\frac{1}{S_{x_1}^2} + \frac{1}{S_{x_2}^2} \right)$$

A pooled estimator of σ^2 is given by

$$\hat{\sigma}^2 = s_{\text{pooled}}^2 = \frac{\text{SSE}(1) + \text{SSE}(2)}{n_1 + n_2 - 4}, \quad \text{d.f.} = n_1 + n_2 - 4$$

and the ratio

$$\frac{(\hat{\beta}_1 - \hat{\beta}_2) - (\beta_1 - \beta_2)}{s_{\text{pooled}} \sqrt{(1/S_{x_1}^2) + (1/S_{x_2}^2)}}$$

has a Student's t distribution with d.f. $= n_1 + n_2 - 4$. A test of $H_0: \beta_1 = \beta_2$ can then be based on the statistic

$$t = \frac{\hat{\beta}_1 - \hat{\beta}_2}{s_{\text{pooled}} \sqrt{(1/S_{x_1}^2) + (1/S_{x_2}^2)}}, \quad \text{d.f.} = n_1 + n_2 - 4$$

Construct a $100(1-\alpha)\%$ confidence interval for $\beta_1 - \beta_2$.

16. Two preparations *A* and *B* of insulin are being studied to determine their effects on reducing the blood-sugar level in rats. Thirteen rats of an identical breed are randomly divided into two groups of 7 and 6 and injected with different dosages of insulins *A* and *B*, respectively. Reductions in their blood-sugar levels are:

Dosage of *A*	.20	.25	.25	.30	.40	.50	.50
Reduction in blood sugar	30	26	40	35	54	56	65

Dosage of *B*	.20	.25	.30	.40	.40	.50
Reduction in blood sugar	23	24	42	49	55	70

(*a*) Plot scatter diagrams for both data sets on the same graph paper, using two different marking symbols.
(*b*) Determine the least squares regression lines for both data sets and plot them on the same graph.
(*c*) Do the data provide strong evidence that the reduction rates in blood-sugar levels with increasing insulin dosages are different for *A* and *B*?
(*d*) Construct a 95% confidence interval for the difference in the slopes of the two regression lines.

17. Robert Boyle (1627–1691) performed a famous experiment to prove that pressure × volume is constant for a gas at constant temperature. Today, this is known as Boyle's Law:

Pressure is increased on the air in the short leg of a J-shaped tube by pouring mercury into the open top of the long side. The volume of air in the short leg equals h (cross section). Set y = height of mercury, adjusted for the pressure of the atmosphere on the open end, and $x = 1/h$. If air obeys the law, y and x should follow a straight-line relationship.

(SOURCE: *The Laws of Gases*, Edited by Carl Barus (1899), Harper and Brothers Publishers.)

h	48	46	44	42	40	38	36	34	32	30	28	26	24
y	$29\frac{2}{16}$	$30\frac{9}{16}$	$31\frac{15}{16}$	$33\frac{8}{16}$	$35\frac{5}{16}$	37	$39\frac{5}{16}$	$41\frac{10}{16}$	$44\frac{3}{16}$	$47\frac{1}{16}$	$50\frac{5}{16}$	$54\frac{5}{16}$	$58\frac{13}{16}$

h	23	22	21	20	19	18	17	16	15	14	13	12
y	$61\frac{5}{16}$	$64\frac{1}{16}$	$67\frac{1}{16}$	$70\frac{11}{16}$	$74\frac{2}{16}$	$77\frac{14}{16}$	$82\frac{12}{16}$	$87\frac{14}{16}$	$93\frac{1}{16}$	$100\frac{7}{16}$	$107\frac{13}{16}$	$117\frac{9}{16}$

(*a*) Plot the data to see if a straight-line relationship appears reasonable.
(*b*) Fit a straight line $Y_i = \alpha + \beta x_i + e_i$, using the least squares method.

(*c*) Construct a 95% confidence interval for α. Is zero a plausible value?

MATHEMATICAL EXERCISES

1. Show that $E(\hat{\alpha}) = \alpha$ and $E(\hat{\beta}) = \beta$.
   [*Hint*: Note that $\hat{\beta}$ is of the form

$$b_1 Y_1 + b_2 Y_2 + \ldots + b_n Y_n \quad \text{with} \quad b_i = \frac{x_i - \bar{x}}{S_x^2}$$

   and $\hat{\alpha}$ is of the form

$$a_1 Y_1 + a_2 Y_2 + \ldots + a_n Y_n \quad \text{with} \quad a_i = \frac{1}{n} - \bar{x} b_i$$

   (To emphasize that the y's are random variables, we use Y_i in place of y_i.) Apply the rule for expectation of a sum, using $E(Y_i) = \alpha + \beta x_i$.]
2. Show that

   (*a*) $\text{Var}(\hat{\alpha}) = \sigma^2 \left[\frac{1}{n} + \frac{\bar{x}^2}{S_x^2} \right]$

   (*b*) $\text{Var}(\hat{\beta}) = \frac{\sigma^2}{S_x^2}$

   [*Hint*: The variance of Y_i is

$$\text{Var}(\alpha + \beta x_i + e_i) = \text{Var}(e_i) = \sigma^2 \quad \text{all} \quad i$$

   Now use the expression for the variance of a sum.]
3. Establish that

$$\text{Cov}(\hat{\alpha}, \hat{\beta}) = \frac{-\bar{x}}{S_x^2} \sigma^2$$

   [*Hint*: $\text{Var}(\hat{\alpha} + \hat{\beta}) = \text{Var}(\hat{\alpha}) + \text{Var}(\hat{\beta}) + 2\,\text{Cov}(\hat{\alpha}, \hat{\beta})$. Alternatively, $\text{Var}(\hat{\alpha} + \hat{\beta}) = \text{Var}[(a_1 + b_1) Y_1 + \ldots + (a_n + b_n) Y_n]$.]
4. If the model is written $Y = \alpha_1 + \gamma(x - \bar{x}) + e$, show that $\text{Cov}(\hat{\alpha}_1, \hat{\gamma}) = 0$.
5. Establish that

$$\text{Var}(\hat{Y} | x^*) = \text{Var}(\hat{\alpha} + \hat{\beta} x^*) = \sigma^2 \left[\frac{1}{n} + \frac{(x^* - \bar{x})^2}{S_x^2} \right]$$

[*Hint*: Write $\text{Var}(\hat{\alpha}) + x^{*2}\text{Var}(\hat{\beta}) + 2x^*\text{Cov}(\hat{\alpha}, \hat{\beta})$ and refer to the previous exercises.]

6. For a new observation at x^*, show that

$$\text{Var}(\hat{\alpha} + \hat{\beta}x^* + e) = \sigma^2\left[1 + \frac{1}{n} + \frac{(x^* - \bar{x})^2}{S_x^2}\right]$$

CHAPTER 11

Regression Analysis: Model Checking and Multiple Linear Regression

11.1 GENERAL ATTITUDE TOWARD A STATISTICAL MODEL

It is of primary importance to be aware of the fact that a regression analysis is *not* completed by fitting a model by least squares, by providing confidence intervals, and by testing various hypotheses. These steps tell only half the story: the statistical inferences that *can* be made when the postulated model is *adequate*. In most studies in the social and natural sciences, the relations among variables are empirical in form, so that we can never be sure that a particular model is correct. Therefore, we should adopt the following strategy:

(*a*) *Tentatively entertain* a model.
(*b*) Obtain least squares estimates and compute the residuals.
(*c*) Criticize the model by examining the residuals.

In most investigations step (*c*) suggests methods of appropriately modifying the model. Returning to stage (*a*), the modified model is then entertained, and this *iteration* is continued until a model is obtained for which the data do not seem to contradict the assumptions made about the model.

11.2 EXAMINING THE RESIDUALS

We now discuss techniques for criticizing the model, that apply not only to the straight-line model but also to all the extensions we consider later in this chapter. Once a model is fitted by least squares, all the information on variation that cannot be explained by the model is contained in the residuals

$$\hat{e}_i = y_i - \hat{y}_i, \quad i = 1, 2, \ldots, n$$

where y_i is the observed value and $\hat{y}_i$ denotes the corresponding value predicted by the fitted model. For example, in the case of a simple linear regression model, $\hat{y}_i = \hat{\alpha} + \hat{\beta}x_i$.

Recall from our examination of the straight-line model in Chapter 10 that we have made the assumptions of independence, constant variance, and a normal distribution for the error components e_i. The inference procedures are based on these assumptions. When the model is correct, the residuals can be considered as estimates of the errors e_i which are distributed as $N(0, \sigma)$.

To determine the merits of the tentatively entertained model, we can examine the residuals by plotting them on graph paper. Then if we recognize any systematic pattern formed by the plotted residuals, we can suspect that some assumptions regarding the model are invalid. There are many ways to plot the residuals, depending on what aspect is to be examined. We mention a few of these here to illustrate the techniques. A more comprehensive discussion can be found in Chapter 3 of Draper and Smith [1].

Histogram or Dot Diagram

To picture the overall behavior of the residuals, we can plot a *histogram* with suitable class intervals for a large number of observations or a dot diagram on a line for fewer observations. For example, in a dot diagram like the one in Figure 11.1*a*, the pattern of the data seems similar to a

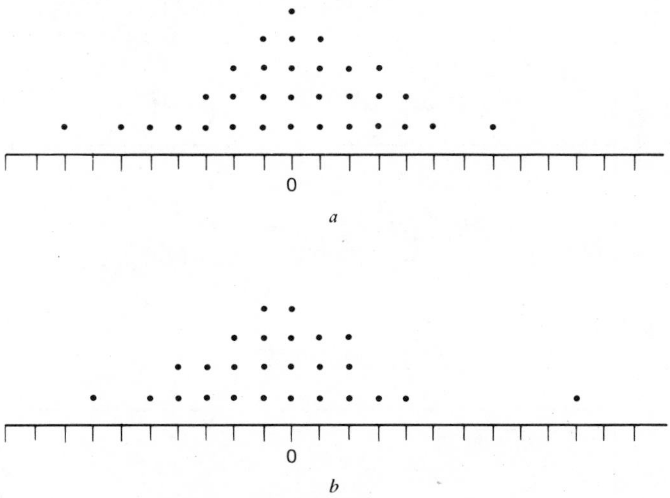

Figure 11.1 Dot diagram of residuals.

sample from a normal population and there do not appear to be any "wild" observations. In contrast, Figure 11.1*b* illustrates a situation in which the distribution appears to be quite normal except for a single residual that lies far to the right of the other residuals. The circumstances that produced the associated observation demand serious consideration to ensure that this observation does not exert undue influence on the fitting of the model.

Residual vs. Predicted Value

A plot of the residuals $\hat{e}_i$ vs. the predicted value $\hat{y}_i$ often helps to detect the inadequacies of an assumed relation or a violation of the assumption of constant error variance. Figure 11.2 illustrates some typical phenomena. If

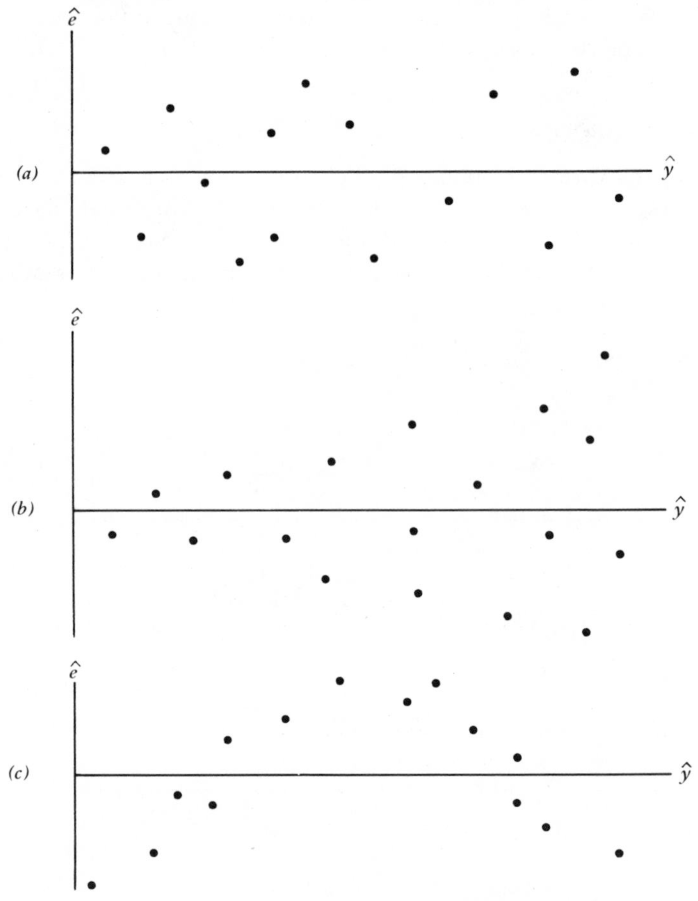

Figure 11.2 Plot of residuals vs. predicted values.

the points form a horizontal band around zero, as in Figure 11.2*a*, then no abnormality is indicated. If the width of the band increases noticeably with increasing values of $\hat{y}$, as in Figure 11.2*b*, this indicates that the error variance σ^2 tends to increase with an increasing level of response. We would then suspect the validity of the assumption of constant variance in the model. There are ways of transforming the data to stabilize the variance, but we do not discuss them here. Figure 11.2*c* shows that the residuals form a systematic pattern. Instead of being randomly distributed around the $\hat{y}$-axis, they tend first to increase steadily and then to decrease. This would lead us to suspect that the model is inadequate and that a squared-term or some other nonlinear model should be considered.

Residual vs. Time Order

We mentioned in Chapter 10 that the most serious violation of assumptions occurs when the errors e_i are *not* independent. Lack of independence frequently occurs in business and economic applications, where the observations are collected in a time sequence with the intention of using regression techniques to predict future trends. In many other experiments, trials are conducted successively in time. In any event, whenever time varies substantially in the course of an experiment, a plot of the residuals vs. *time order* often detects a violation of the assumption of independence. For example, the plot in Figure 11.3 exhibits a systematic pattern in that a string of high values is followed by a string of low values. This indicates that consecutive residuals are (positively) correlated, and we would suspect a violation of the independence assumption. Independence can also be checked by plotting the successive pairs $(\hat{e}_i, \hat{e}_{i-1})$, where $\hat{e}_1$ indicates the residual from the first y value observed, $\hat{e}_2$ indicates the second, and so on. Independence is suggested if the scatter diagram is a patternless cluster, whereas points clustered along a line suggest a lack of independence between adjacent observations.

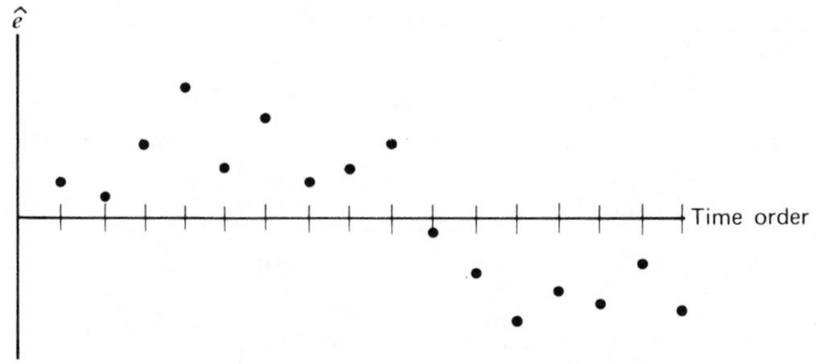

Figure 11.3 Plot of residuals vs. time order.

It is important to remember that our confidence in statistical inference procedures is related to the validity of the assumptions about them. A mechanically made inference may be misleading if some model assumption is grossly violated. An examination of the residuals is an important part of regression analysis, because it helps to detect any inconsistency between the data and the postulated model. If no abnormalities are exposed in this process, then we can consider the model adequate and proceed with the relevant inferences. Otherwise, we must search for a more appropriate model.

A description of the standard techniques employed in handling the data when certain assumptions appear to be violated would lead us far beyond the scope of this book. Instead, we choose the more modest goal of stressing the importance of checking the assumptions and present some descriptive methods to employ in studying the residuals.

11.3 FURTHER CHECKS ON THE STRAIGHT-LINE MODEL

We now determine the adequacy of the straight-line model by examining how much of the variation in the values of the response variable can be explained by the fitted model. After obtaining the least squares estimates $\hat{\alpha}$ and $\hat{\beta}$, we can view any observed y_i as consisting of these two components:

y_i	$=$	$(\hat{\alpha}+\hat{\beta}x_i)$	$+$	$(y_i-\hat{\alpha}-\hat{\beta}x_i)$
Observed y value		Explained by linear relation		Residual or deviation from linear relation

In an ideal situation where all the points lie exactly on the line, the residuals are all zero, and the y values are completely accounted for or *explained* by the linear dependence on x.

As an overall measure of the discrepancy or the variation from linearity we can consider the sum of squares of the residuals

$$\text{SSE} = \sum (y_i - \hat{\alpha} - \hat{\beta}x_i)^2 = S_y^2 - \hat{\beta}^2 S_x^2$$

the last form of which was derived in Section 10.7. Recall that the abbreviation SSE stands for *sum of squares due to error*, which is also called the *residual sum of squares*. The total variability of the y values is reflected in the sum of squares

$$S_y^2 = \sum (y_i - \bar{y})^2$$

of which $\text{SSE} = S_y^2 - \hat{\beta}^2 S_x^2$ forms a part, the other part being $\hat{\beta}^2 S_x^2$. Motivated by the decomposition of the observation y_i just given, we now consider a decomposition of the variability of the y values:

S_y^2	$=$	$\hat{\beta}^2 S_x^2$	$+$	SSE
Total SS of y		SS explained by linear relation		Residual SS (unexplained)

The first term on the right-hand side of this equation is called the *sum of squares due to linear regression*. If the straight-line model is considered to provide a good fit for the data, then SS due to linear regression should comprise a major portion of S_y^2 and leave only a small part for SSE. In an ideal situation in which all points lie on the line, SSE is zero. S_y^2 can then be completely explained by the fact that the x values vary during the experiment, and the linear relationship between y and x is solely responsible for the variability in the y values.

As an index of how well the straight-line model fits, it is then reasonable to consider the proportion

$$r^2 = \frac{\hat{\beta}^2 S_x^2}{S_y^2} = \frac{\text{SS due to linear regression}}{\text{Total SS of } y}$$

where r^2 represents the proportion of the y variability explained by the linear relation with x. Recall that $\hat{\beta}$ is given by

$$\hat{\beta} = \frac{S_{xy}}{S_x^2}$$

so that r^2 can also be written

$$r^2 = \frac{S_{xy}^2}{S_x^2 S_y^2}$$

Incidentally, we mention that $r = S_{xy}/S_x S_y$ is called the *sample correlation coefficient* between the x and y values observed in the data. This coefficient is so important in a related context that it is the topic of Chapter 12. The square of the correlation coefficient serves as a measure of the closeness of the relation to linearity.

Example 11.1 We now consider the adequacy of the straight-line model for the reduction in nitrogen-oxides data originally given in Table 10.2.

The results obtained for the data

x	1	1	2	3	4	4	5	6	6	7
y	2.1	2.5	3.1	3.0	3.8	3.2	4.3	3.9	4.4	4.8

are

$$S_x^2 = 40.9 \qquad S_y^2 = 6.85 \qquad S_{xy} = 15.81$$

$$\text{Fitted line:} \quad \hat{y} = 2.00 + .387x$$

How much of the variability in y is explained by the linear regression model?

To answer this question, we compute

$$r^2 = \frac{S_{xy}^2}{S_x^2 S_y^2} = \frac{249.956}{(40.9)(6.85)} = .89$$

This means that 89% of the variability in y is explained by linear regression, and the linear model seems satisfactory in this respect. ■

When the value of r^2 is small, we can only conclude that a straight-line relation does not give a good fit for the data. Such a case may arise due to the following reasons:

(*a*) There is little relation between the variables in the sense that the scatter diagram fails to exhibit any pattern, as illustrated in Figure 11.4*a*. In this case, any alternative regression model other than the one we have used is not likely to reduce the SSE or to explain a substantial part of S_y^2.

(*b*) There is a prominent relation but it is nonlinear in nature; that is, the scatter is banded around a curve rather than a line. The part of S_y^2 that is explained by straight-line regression is small because the model is inappropriate. Some other relationship may improve the fit substantially. Figure 11.4*b* illustrates such a case, where the SSE itself contains a part which is due to lack of fit of the linear model but which can be accounted for by fitting a suitable curve to the data. The other part of the SSE, called the *pure error* SS, reflects the inherent variability of the y values in repeated observations when x is not varied.

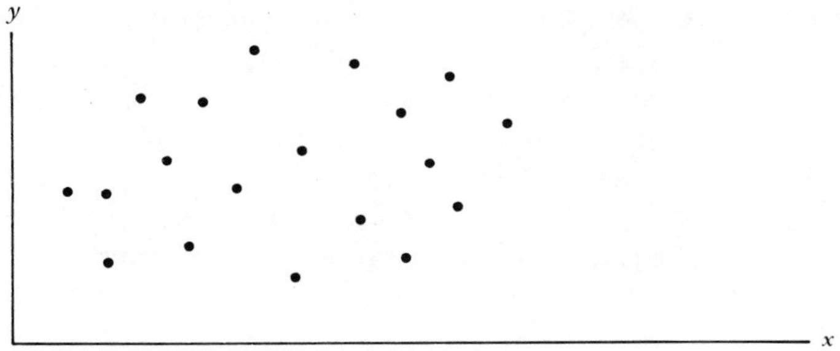

(*a*) No relation.

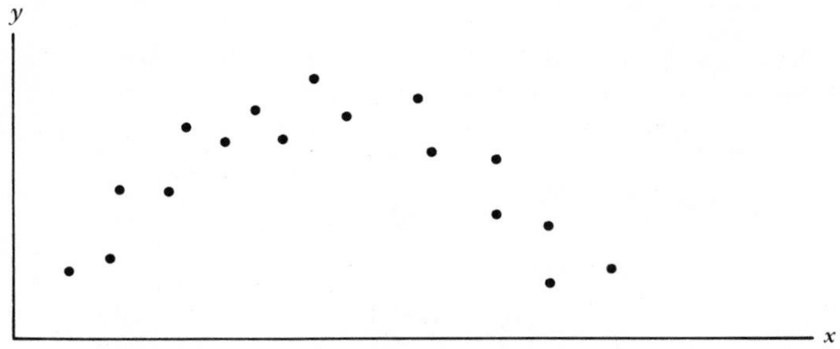

(*b*) A nonlinear relation.

Figure 11.4 Scatter diagram patterns.

The lack of fit and the pure error components of SSE cannot be separated if the experiment is conducted only once for each different level of x. To understand the intrinsic variability of the y values we must design the experiment so that repeated y observations are available, at least for some of the x values. The data given in Example 11.1 illustrate a design in which the values $x=1,4,6$ are repeated. With such a design, it is possible to separate the two sources contributing to SSE and to test *lack of fit of the linear model*. The principle underlying the formulation of this type of test is explained in Section 11.4, where a numerical example is provided. The important point to note is that some repeated experimental trials are necessary to test for lack of fit.

*11.4 TEST OF LACK OF FIT BASED ON REPEATED RUNS

In this section we describe the method of performing a lack of fit test in terms of fitting the straight-line model. This concept can be extended to any fitted model, but for the first exposure it is probably best to work with the most familiar model. Suppose that k different levels $x_1, \ldots, x_k$ of the independent variable are included in the experiment and that the y observations are repeated $n_1, \ldots, n_k$ times, respectively, at these levels of x. Table 11.1 indicates the nature of the resulting data. In each row of the table, the x value is fixed, so that the variability of the y values is solely due to the error component whose variance is σ^2. From each row we can compute the sum of squared deviations from the mean, which provides an estimate of σ^2 when it is divided by the degrees of freedom. For example

$$S_1 = \sum_{j=1}^{n_1} (y_{1j} - \bar{y}_1)^2, \quad \text{d.f.} = n_1 - 1$$

where $\bar{y}_1$ is the mean of the y values in the first row of Table 11.1.

The sum of squared deviations and the degrees of freedom are presented in the last two columns of the table. The total of these sums of squares

$$\text{SS}_{pe} = \sum_1^k S_i$$

is called the *pure error sum of squares*, and the associated number of degrees of freedom is $\sum_1^k (n_i - 1) = n - k$, where n is the total number of y observations. The mean square MS_{pe} defined by

$$\text{MS}_{pe} = \frac{\text{SS}_{pe}}{n-k} = \text{mean square for pure error}$$

provides an unbiased estimate of the error variance σ^2, irrespective of any model of the relationship of y with x.

TABLE 11.1
PATTERN OF DATA HAVING REPEATED RUNS

DISTINCT x VALUES	REPEATED y VALUES	MEANS	SS	d.f.
x_1	$y_{11}\ y_{12}\ \cdots\ y_{1n_1}$	$\bar{y}_1$	S_1	$n_1 - 1$
x_2	$y_{21}\ y_{22}\ \cdots\ y_{2n_2}$	$\bar{y}_2$	S_2	$n_2 - 1$
$\vdots$	$\cdots$	$\vdots$	$\vdots$	$\vdots$
x_k	$y_{k1}\ y_{k2}\ \cdots\ y_{kn_k}$	$\bar{y}_k$	S_k	$n_k - 1$
		Total	SS_{pe}	$n - k \quad n = \sum_1^k n_i$

Taking the n data points, we can fit a regression line and compute the residual sum of squares SSE as usual by

$$\mathrm{SSE}=S_y^2-\hat{\beta}^2 S_x^2, \quad \text{d.f.}=n-2$$

Figure 11.5 shows the breakdown of this SSE into two components: the SS_{pe} just described and the remaining $\mathrm{SS}_L=\mathrm{SSE}-\mathrm{SS}_{pe}$, which is called the *lack of fit sum of squares*. The associated degrees of freedom are decomposed in a similar manner to give d.f. $=k-2$ for lack of fit.

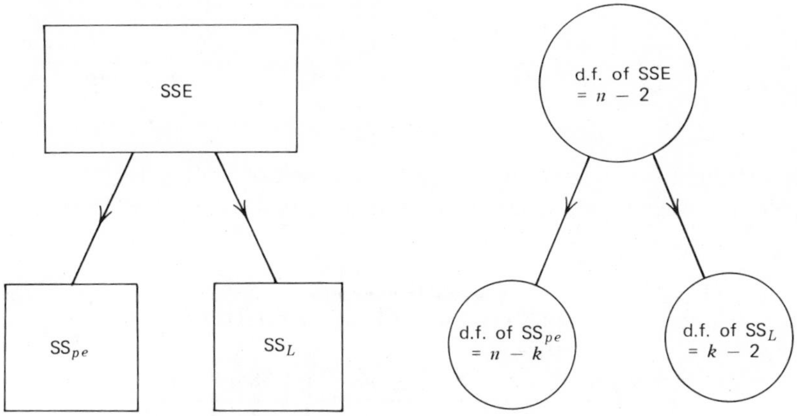

Figure 11.5 Breakdown of SSE and degrees of freedom.

Now let us call

$$\mathrm{MS}_L=\frac{\mathrm{SS}_L}{k-2}=\textit{mean square due to lack of fit}.$$

If the linear model is correct, the MS_L also estimates the error variance σ^2, just as the MS_{pe} does. On the other hand, if a nonlinear model is appropriate, then the MS_L overestimates σ^2 and is likely to be significantly larger than the MS_{pe}. Thus a test of lack of fit for the linear model can be performed by computing the F ratio

$$F=\frac{\mathrm{MS}_L}{\mathrm{MS}_{pe}}=\frac{(\mathrm{SSE}-\mathrm{SS}_{pe})/(k-2)}{\mathrm{SS}_{pe}/(n-k)}, \quad \text{d.f.}=(k-2,\ n-k)$$

If this F test shows significance, we conclude that the lack of fit for the straight-line model is significant and we should then seek an alternate model for the relationship. If F is not significant, little useful purpose can be served by applying a more complicated model. This fact by itself does not ensure that a straight-line model is adequate. The adequacy of a linear model is determined by the magnitude of r^2, as we explained earlier.

Example 11.2 Five levels of the independent variable x are included in an experiment, some of which are repeated to provide the 11 data points given in Table 11.2.

TABLE 11.2
DATA HAVING REPEATED RUNS

x	2	2	2	3	3	4	5	5	6	6	6
y	4	3	8	18	22	24	24	18	13	10	16

Fit a straight-line regression and perform a test of lack of fit.

We rearrange the data as shown in the first two columns of Table 11.3.

TABLE 11.3
COMPUTATIONS FOR LACK OF FIT TEST

x	y	$\bar{y}$	SS	d.f.
2	4, 3, 8	5	14	2
3	18, 22	20	8	1
4	24	24	0	0
5	24, 18	21	18	1
6	13, 10, 16	13	18	2
		Total	58	6

For $x=2$, the mean of y is 5 and the sum of squares is computed

$$(4-5)^2+(3-5)^2+(8-5)^2=14$$

Similar calculations are performed for each row in Table 11.3 to obtain

$$SS_{pe}=14+8+18+18=58$$

$$\text{d.f.}=2+1+1+2=6$$

Using the formulas given in Chapter 10 to fit a straight line to the 11 pairs of (x,y) values, we obtain $\hat{\beta}=1.786$, $\hat{\alpha}=7.41$, $S_y^2=571$, and SSE$=481$, with d.f.$=9$. Therefore, $SS_L=481-58=423$ with d.f.$=9-6=3$. Thus our measure of lack of fit

$$F = \frac{423/3}{58/6} = 14.6, \quad \text{d.f.} = (3,6)$$

is highly significant, because the tabulated 5% point of F is 4.76 with d.f. $= (3,6)$. This demonstrates a significant lack of fit for the straight-line model. The scatter diagram of the data given in Table 11.2 is plotted in Figure 11.6. The conclusion we can draw from the lack of fit test is evidently substantiated by a visual inspection of this scatter diagram, which indicates the presence of a curvilinear relationship. It would be instructive at this point to plot the residuals to see how they indicate the presence of lack of fit. ■

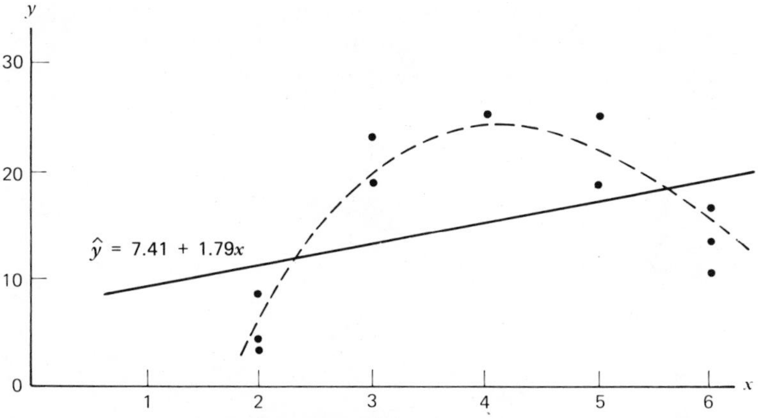

Figure 11.6 Scatter diagram of the data given in Table 11.2 and the fitted line. The relationship is better represented by the dotted curve.

In the next section, we consider the possibility of transforming certain relations so that they become linear.

11.5 NONLINEAR RELATIONS AND LINEARIZING TRANSFORMATIONS

Previously, we studied situations where the underlying relationship between a dependent variable y and an independent variable x could reasonably be formulated in terms of a linear regression model. In addition to its simplicity, a straight-line model is particularly appealing because simple, well-behaved statistical inference procedures are readily available.

However, the discussion relating to these procedures should not give the impression that it is appropriate to apply a straight-line model to most sets of real-life data. In a great many situations a plot of the data in a scatter diagram indicates that a relationship, although present, is far from linear. This can be established on a statistical basis by checking that the observed value of r^2 is small when fitting a straight line or, by conducting the lack of fit test just described in Section 11.4. Statistical procedures for handling nonlinear relationships are more complicated than those for handling linear relationships, with the exception of a specific type of model called the *polynomial regression model*, which is discussed in the next section. In some situations, however, it may be possible to transform the variables x and/or y in such a way that the new relationship is close to being linear. A linear regression model can then be formulated in terms of the transformed variables, and the appropriate analysis can be based on the transformed data. This analysis should include an examination of the residuals of the transformed model, because the assumptions of independent normal errors with constant variance now apply to the transformed model.

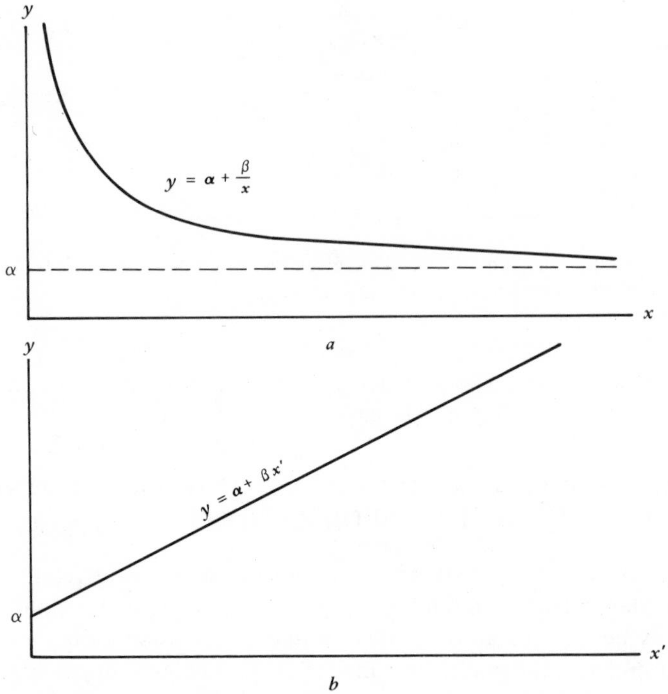

Figure 11.7 Transformation $x' = 1/x$ linearizes $y = \alpha + \beta/x$.

To illustrate this line of reasoning, consider the problem of determining the relationship between growth in skill y and duration of training x originally described in Section 10.1. More specifically, suppose that x is the number of hours of training that a machinist receives and that y is the time the machinist requires to set up a piece of complicated machinery. The time to complete the job y is expected to decrease with increasing training x, but at some point a level is expected to be reached below which y cannot be reduced despite additional training. Perhaps we can consider a relation of the form

$$y = \alpha + \frac{\beta}{x}$$

whose graph, shown in Figure 11.7a, exhibits the prescribed behavior. This relationship is quite nonlinear.

If instead of x we consider the transformed variable $x' = 1/x$, then our model becomes $y = \alpha + \beta x'$, which is a straight line as shown in Figure 11.7b. We can then begin to analyze the data on the growth of skill by performing a simple linear regression analysis of y and the new x variable, which is the reciprocal of the number of hours of job training.

A few common nonlinear models and their corresponding linearizing transformations are given in Table 11.4.

In some situations a specific nonlinear relation is obviously suggested either by the data or by a theoretical consideration. Even when initial information about the form is lacking, a study of the scatter diagram often indicates the appropriate linearizing transformation. It can be helpful to plot the points (x_i, y_i) on various types of graph paper, such as semilog or

TABLE 11.4
SOME NONLINEAR MODELS AND THEIR LINEARIZING TRANSFORMATIONS

	NONLINEAR MODEL	TRANSFORMATION		TRANSFORMED MODEL		
(a)	$y = ae^{bx}$	$y' = \log_e y$,	$x' = x$	$y' = \alpha + \beta x'$,	$\alpha = \log_e a$,	$\beta = b$
(b)	$y = ax^b$	$y' = \log y$,	$x' = \log x$	$y' = \alpha + \beta x'$,	$\alpha = \log a$,	$\beta = b$
(c)	$y = \dfrac{1}{a+bx}$	$y' = \dfrac{1}{y}$,	$x' = x$	$y' = \alpha + \beta x'$,	$\alpha = a$,	$\beta = b$
(d)	$y = \dfrac{1}{(a+bx)^2}$	$y' = \dfrac{1}{\sqrt{y}}$,	$x' = x$	$y' = \alpha + \beta x'$,	$\alpha = a$,	$\beta = b$
(e)	$\dfrac{1}{y} = a + \dfrac{b}{1+x}$	$y' = \dfrac{1}{y}$,	$x' = \dfrac{1}{1+x}$	$y' = \alpha + \beta x'$,	$\alpha = a$,	$\beta = b$
(f)	$y = a + b\sqrt{x}$	$y' = y$,	$x' = \sqrt{x}$	$y' = \alpha + \beta x'$,	$\alpha = a$,	$\beta = b$

double-log paper, to see if any transformed relations are close to being linear. For instance, relation (*a*) in Table 11.4 is graphed as a straight line on semilog paper. Sometimes when the scatter diagram exhibits a relationship on a curve in which the y values increase too fast in comparison with the x values, a plot of $\sqrt{y}$ or some other fractional power of y can help to linearize the relation. This situation is illustrated in Example 11.3. Some analytical guidelines for determining the appropriate power in this type of transformation are available, but a discussion of them is beyond the scope of this book.

Example 11.3 To determine the maximum stopping ability of cars when their breaks are fully applied, 10 cars are to be driven each at a specified speed and the distance each requires to come to a complete stop is to be measured. The various initial speeds selected for each of the 10 cars and the stopping distances recorded are given in Table 11.5. The scatter diagram for the data appears in Figure 11.8.

TABLE 11.5
DATA ON SPEED AND STOPPING DISTANCE

Initial speed x	20	20	30	30	30	40	40	50	50	60
Stopping distance y	16.3	26.7	39.2	63.5	51.3	98.4	65.7	104.1	155.6	217.2

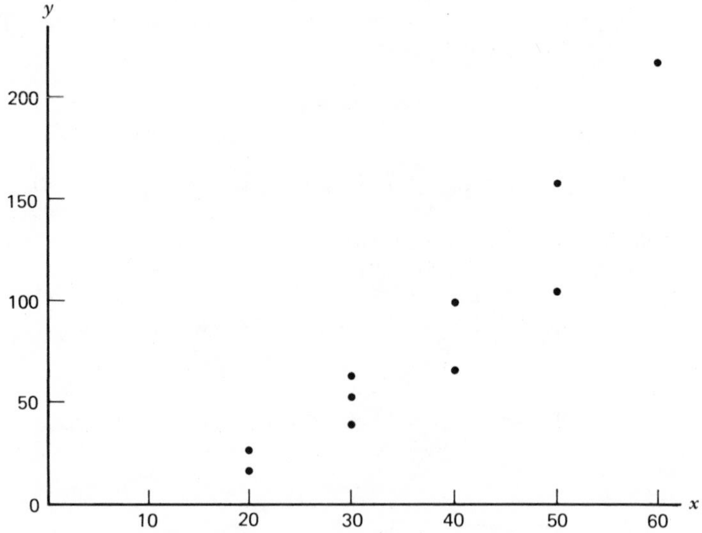

Figure 11.8 Scatter diagram of the data given in Table 11.5.

The relation deviates from a straight line most markedly in that y increases at a much faster rate at large x than at small x. This suggests that we can try to linearize the relation by plotting $\sqrt{y}$ or some other fractional power of y with x. A plot of $\sqrt{y}$ yields the transformed data given in Table 11.6, and the scatter diagram for these data, which exhibits an approximate linear relation, appears in Figure 11.9.

TABLE 11.6
DATA ON SPEED AND SQUARE ROOT OF STOPPING DISTANCE

x	20	20	30	30	30	40	40	50	50	60
$y' = \sqrt{y}$	4.037	5.167	6.261	7.969	7.162	9.920	8.106	10.203	12.474	14.738

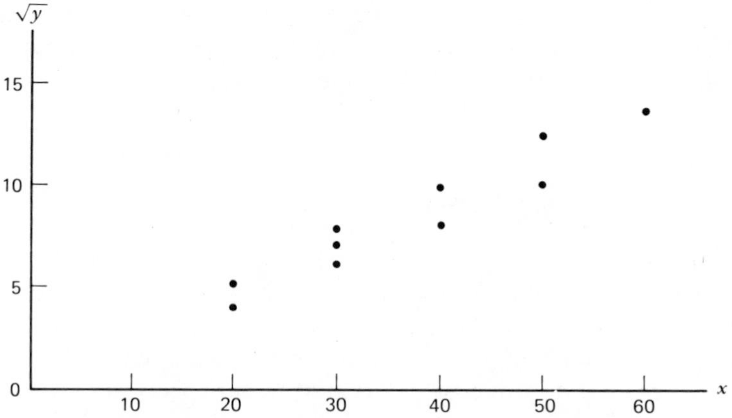

Figure 11.9 Scatter diagram of the data given in Table 11.6.

With the aid of a standard computer program for regression analysis, the following results are obtained:

$$\bar{x} = 37 \qquad \bar{y}' = 8.604$$
$$S_x^2 = 1610 \qquad S_{y'}^2 = 97.773 \qquad S_{xy'} = 381.621$$
$$\hat{\alpha} = -.166 \qquad \hat{\beta} = .237$$

Thus the equation of the fitted line is

$$\hat{y}' = -.166 + .237x$$

The proportion of the y' variation that is explained by the straight-line model is

$$r^2 = \frac{(381.621)^2}{(1610)(97.773)} = .92$$ ■

To review we must remember that all inferences about the transformed model are based on the assumptions of a linear relation and independent normal errors with constant variance. Before we can trust these inferences, this transformed model must be scrutinized to determine whether any serious violation of these assumptions may have occurred.

> The pattern revealed by the scatter diagram often indicates the presence of a nonlinear relationship. It is sometimes possible to obtain transformations of the original data, so that the relationship of the new variables is approximately linear. When this is possible, we can then perform the usual straight-line regression analysis and draw inferences from the transformed model. The assumptions concerning the error structure of the transformed variables should be checked by the usual methods.

11.6 MULTIPLE LINEAR REGRESSION

Easily the most important single situation leading to the modification of the straight-line model is the following. After carrying out the usual linear regression analysis between y and x, we may obtain a small value of r^2 but an inspection of the scatter diagram or lack of fit test may fail to discredit the linear relationship due to a large error variance. Moreover, further examination of both the experimental method and the data-collection process may reveal the existence of causal variables other than x that influenced the response variable y but that were ignored in the simple linear regression analysis. If the variation in these influencing variables has been uncontrolled during the experiment, it may have obscured the true relation between y and x by inflating the error variance σ^2. In the reduction in nitrogen-oxides example originally discussed in Chapter 10, if influencing factors such as average speed, length of idling time and surrounding temperature are allowed to vary during the experiment, a scatter diagram of y vs. x may show wide fluctuations and the SSE may be quite large. Any available information about these additional factors can be used to improve the prediction.

Therefore, to obtain a useful prediction model as well as unbiased and efficient parameter estimators, we must record data relevant to all variables that are known to influence the response variable y and incorporate them explicitly in the regression analysis. To be more specific, suppose that the response variable y in an experiment is expected to be influenced by three causal variables x_1, x_2, and x_3 and that the data relevant to all these causal variables are recorded with the measurements of y. By analogy with the simple linear regression model, we can then tentatively formulate the simple model of the relationship of y with x_1, x_2, and x_3

$$Y_i = \alpha + \beta_1 x_{i1} + \beta_2 x_{i2} + \beta_3 x_{i3} + e_i, \quad i = 1, \dots, n$$

where x_{i1}, x_{i2}, and x_{i3} are the fixed values of the three independent variables in the ith experimental trial and y_i is the corresponding response. The error components e_i are assumed to be independent normal variables with mean $= 0$ and variance $= \sigma^2$. The parameters α, β_1, β_2, and β_3 are unknown fixed quantities.

This model suggests that aside from the error components, the response varies linearly with each of the independent variables when the other two remain fixed. Consequently, this response function is a *plane* with two predictors and a *hyperplane* with more than two predictors. Due to the presence of more than one predictor variable, this model is called a *multiple regression model*.

Although the scatter diagram cannot be plotted, the principle of least squares is useful in estimating the regression parameters. For this model, we are required to vary α, β_1, β_2, and β_3 simultaneously to minimize the sum of squared deviations

$$\sum_{i=1}^{n} (y_i - \alpha - \beta_1 x_{i1} - \beta_2 x_{i2} - \beta_3 x_{i3})^2$$

The least squares estimates $\hat{\alpha}$, $\hat{\beta}_1$, $\hat{\beta}_2$, and $\hat{\beta}_3$ can be verified to be the solutions to the following *normal equations*, which are extensions of the equations, for the least squares solutions for the straight-line model, given in Section 10.7:

$$\hat{\beta}_1 S_{x_1}^2 + \hat{\beta}_2 S_{x_1x_2} + \hat{\beta}_3 S_{x_1x_3} = S_{x_1y}$$

$$\hat{\beta}_1 S_{x_1x_2} + \hat{\beta}_2 S_{x_2}^2 + \hat{\beta}_3 S_{x_2x_3} = S_{x_2y}$$

$$\hat{\beta}_1 S_{x_1x_3} + \hat{\beta}_2 S_{x_2x_3} + \hat{\beta}_3 S_{x_3}^2 = S_{x_3y}$$

$$\hat{\alpha} = \bar{y} - \hat{\beta}_1 \bar{x}_1 - \hat{\beta}_2 \bar{x}_2 - \hat{\beta}_3 \bar{x}_3$$

where $S^2_{x_1}$, $S_{x_1x_2}$, etc. are the sums of squares and cross products of the variables in the suffix and can be computed just as they are in a straight-line regression model, because a maximum of two variables can enter any S. Methods are available for interval estimation and hypotheses testing, testing for lack of fit of the model and for determining which of the x variables are important enough to include in the model and which ones can be omitted. In principle, these methods are similar to those used in the simple regression model, but the algebraic formulas required become more and more involved as the number of x variables increases. A full discussion of the statistical analyses of these models is beyond the scope of this book. However, the examination of residuals is exactly the same for the multiple regression model as it is for the straight-line regression model and provides an important verification of models that are fit by least squares on a computer.

We conclude this section with an example illustrating the method of calculation for fitting a linear model that has two predictor variables.

Example 11.4 We are interested in studying the systolic blood pressure y in relation to weight x_1 and age x_2 in a class of males of approximately the same height. From 13 subjects preselected according to weight and age, we obtain the data listed in the first three columns of Table 11.7. Fit the multiple regression model $Y_i = \alpha + \beta_1 x_{i1} + \beta_2 x_{i2} + e_i$ to the data.

TABLE 11.7
DATA AND CALCULATIONS FOR MULTIPLE REGRESSION ANALYSIS IN EXAMPLE 11.4

x_1	x_2	y	x_1^2	x_2^2	x_1x_2	x_1y	x_2y	$\hat{y}$
152	50	120	23,104	2,500	7,600	18,240	6,000	119.89
183	20	141	33,489	400	3,660	25,803	2,820	140.52
171	20	124	29,241	400	3,420	21,204	2,480	127.59
165	30	126	27,225	900	4,950	20,790	3,780	125.38
158	30	117	24,964	900	4,740	18,486	3,510	117.84
161	50	129	25,921	2,500	8,050	20,769	6,450	129.58
149	60	123	22,201	3,600	8.940	18,327	7,380	120.91
158	50	125	24,964	2,500	7,900	19,750	6,250	126.35
170	40	132	28,900	1,600	6,800	22,440	5,280	135.02
153	55	123	23,409	3,025	8,415	18,819	6,765	123.09
164	40	132	26,896	1,600	6,560	21,648	5,280	128.56
190	40	155	36,100	1,600	7,600	29,450	6,200	156.57
185	20	147	34,225	400	3,700	27,195	2,940	142.67

$\sum x_1 = 2159$ $\quad \sum x_1^2 = 360{,}639$ $\quad \sum x_1x_2 = 82{,}335$ $\quad \sum x_2 y = 65{,}135$

$\sum x_2 = 505$ $\quad \sum x_2^2 = 21{,}925$ $\quad \sum x_1 y = 282{,}921$

$\sum y = 1694$

A multiple regression analysis is usually conducted with the aid of standard computer programs that provide both the least squares estimates and the residuals. However, for the purposes of illustration, the details of the types of calculations involved are presented in Table 11.7. All the entries beyond the third column were calculated on a computer.

From the column totals in Table 11.7, it follows that $S_{x_1x_2} = 82{,}335 - 13(166.077)(38.846) = -1533.8$ etc. so that the normal equations become

$$2078.9\hat{\beta}_1 - 1533.8\hat{\beta}_2 = 1586.7$$

$$-1533.8\hat{\beta}_1 + 2307.7\hat{\beta}_2 = -670.38$$

$$\hat{\alpha} = 130.308 - 166.077\hat{\beta}_1 - 38.846\hat{\beta}_2$$

Multiplying the first equation *by* 1533.8 and the second equation by 2078.9, we eliminate $\hat{\beta}_1$ by summing the resulting equations. Solving for $\hat{\beta}_2$ yields $\hat{\beta}_2 = .425$, and either of the first two equations then gives $\hat{\beta}_1 = 1.077$. Finally, the third equation gives $\hat{\alpha} = -65.10$. The last column in Table 11.7 lists the predicted values derived from the preceding least squares estimates, or

$$\hat{y} = -65.10 + 1.077x_1 + .425x_2$$

This means that the mean blood pressure will increase by 1.077 if weight x_1 increases by one unit and age x_2 remains fixed. Similarly, a one-year increase in age with weight held fixed will only increase the mean blood pressure by .425. It would be instructive at this point to calculate the residuals to see if any systematic patterns can be determined. ■

Example 11.4 illustrates the analysis of a multiple regression model with two predictor variables x_1 and x_2. The analysis of a straight-line regression model with a single predictor variable x has already been detailed in Chapter 10 and a few linearizing transformations were subsequently examined in Section 11.5. Even for the single predictor case, the scatter diagram may exhibit a relationship on a curve for which a suitable linearizing transformation cannot be constructed. An alternative method of handling such a nonlinear relation is to include terms with higher powers of x in the model $Y = \alpha + \beta x + e$. In this instance, by including the second power of x, we obtain the model

$$Y_i = \alpha + \beta_1 x_i + \beta_2 x_i^2 + e_i, \quad i = 1, \ldots, n$$

which states that aside from the error components e_i, the response y is a *quadratic function* (or a *second-degree polynomial*) of the independent variable x. Such a model is called a *polynomial regression model* of y with x, and the highest power of x that occurs in the model is called the *degree* or the *order* of the polynomial regression. It is interesting to note that the analysis of a polynomial regression model does not require any special techniques other than those used in multiple regression analysis. By identifying x and x^2 as the two variables x_1 and x_2, respectively, this second-degree polynomial model reduces to the form of a multiple regression model

$$Y_i = \alpha + \beta_1 x_{i1} + \beta_2 x_{i2} + e_i, \quad i = 1, \ldots, n$$

where $x_{i1} = x_i$ and $x_{i2} = x_i^2$. In fact, both of these types of models and many more types are special cases of a general class called *linear models*, in which the response y is expressed

$$Y_i = l_{i0}\alpha + l_{i1}\beta_1 + \cdots + l_{ik}\beta_k + e_i$$

This relation is linear in the parameters $(\alpha, \beta_1, \ldots, \beta_k)$, and the coefficients l of these parameters are known from the values of the predictor variables selected for the experiment. These coefficients may be the values of different causal variables, as is the case in multiple regression, or the values of different powers of the same causal variable, as is the case in polynomial regression. There may even be both kinds of variables in a model. Because all these models possess the important property of being linear in the parameters, the format of the analysis remains the same as that illustrated in Example 11.4.

*11.7 FURTHER ASPECTS OF MULTIPLE REGRESSION

By virtue of its wide applicability, the multiple linear regression model plays a prominent role in the portfolio of a statistician. Although a complete analysis cannot be given here, certain aspects of multiple regression merit further attention. The examples in this section elucidate the following points:

(*a*) If an important variable is not included in the model, the estimators will not be unbiased.
(*b*) An individual regression coefficient β_i must be carefully interpreted.
(*c*) If the experimental design is so poor that one input variable can be obtained by adding constant multiples of the other variables, then an indeterminacy arises.

Example 11.5 The estimators are not unbiased if a straight line is fitted when the true response also depends on another variable. Suppose that the true mean is $4+3x_1+2x_2$ where

x_1	1	-3	2
x_2	2	3	4

but for some reason the investigator only records x_1 and fits $Y=\alpha+\beta x_1+e$. The least squares estimators then become

$$\hat{\alpha}=\bar{y}-\hat{\beta}\bar{x}_1=\bar{y}$$

$$\hat{\beta}=\frac{S_{x_1y}}{S_{x_1}^2}=\frac{1}{14}(y_1-3y_2+2y_3)$$

with $E(Y_1)=4+3(1)+2(2)=11$, $E(Y_2)=4+3(-3)+2(3)=1$, and $E(Y_3)=18$. Consequently,

$$E(\hat{\alpha})=\frac{1}{3}\left[E(Y_1)+E(Y_2)+E(Y_3)\right]=\frac{11+1+18}{3}=10$$

$$E(\hat{\beta})=\frac{1}{14}\left[E(Y_1)-3E(Y_2)+2E(Y_3)\right]=\frac{1}{14}[11-3+36]=\frac{22}{7}$$

The expected response at x_1^*, actually at (x_1^*,x_2^*), is

$$E\left(\hat{\alpha}+\hat{\beta}x_1^*\right)=10+\frac{22}{7}x_1^*$$

not $4+3x_1^*+2x_2^*$, which is the true value. The simple straight-line model is insufficient; the more complicated model is required. ■

Example 11.6 Interpretation of regression coefficients: In the multiple regression model $\alpha+\beta_1x_1+\beta_2x_2=4+3x_1+2x_2$, the value $\beta_1=3$ describes the influence of x_1 when x_2 is *fixed*. This response is illustrated in Figure 11.10. If x_2 is fixed, then a straight-line relationship exists between $E(Y)$ and x_1.

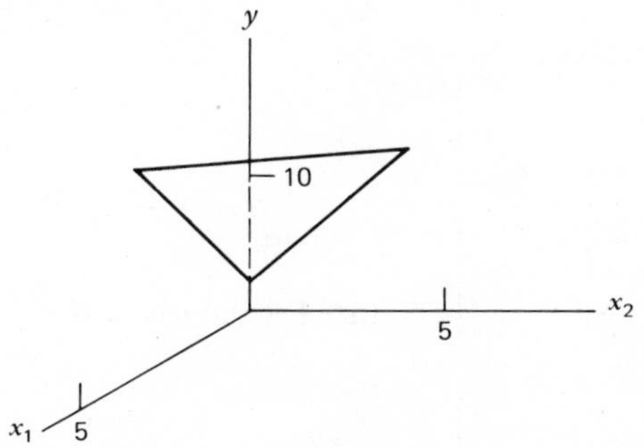

Figure 11.10 The expected response plane $4+3x_1+2x_2$.

To consider the estimated regression, suppose that we obtain $3.5+3.6x_1+2.8x_2$ based on 20 observations. Although 3.6 describes the estimated influence of x_1 with x_2 held fixed, we cannot conclude that x_1 is more important than x_2. In fact, if (x_1,x_2) usually occurs with the second component more than twice the first, the reverse may well be true. We can also use R^2, subsequently defined in Exercise 18, to determine the relative importance of x_1 and x_2. ■

Example 11.7 A case of poor design in which the coefficients are not unique: Suppose that the expected response $E(Y)=4+3x_1+2x_2$ and that $x_2=2x_1$ for all trials. The x_1,x_2 values are:

x_1	1	3	−2	2
x_2	2	6	−4	4
$4+3x_1+2x_2$	11	25	−10	18
$4+5x_1+x_2$	11	25	−10	18

Here $4+3x_1+2x_2=4+3x_1+2x_1+x_2=4+5x_1+x_2$, so that we cannot distinguish $\beta_1=3$, $\beta_2=2$ from $\beta_1=5$, $\beta_2=1$, or from any other number of choices. Not surprisingly, this indeterminancy also applies to the estimates, and both the terms must not be retained in the model. ■

We have already mentioned that most least squares analyses of multiple linear regression models are carried out with the aid of a computer. Programs for implementing the analysis all require the investigator to provide the values of the response y_i and the p input variables $x_{i1},\ldots,x_{ip}$ for each trial $i=1,2,\ldots,n$. Writing $1\cdot\alpha$, so that 1 is the known value of an extra "dummy" input variable corresponding to α, the model is

Observation — Input variables — Error

$$\underset{\text{Observation}}{Y_i} = \underbrace{1\cdot\alpha + x_{i1}\beta_1 + x_{i2}\beta_2 + \cdots + x_{ip}\beta_p}_{\text{Input variables}} + \underset{\text{Error}}{e_i}$$

The basic quantities can be arranged in the form of these arrays, which are denoted by boldface letters:

Observation — Input variables

$$\mathbf{y}=\begin{bmatrix} y_1 \\ y_2 \\ \vdots \\ y_i \\ \vdots \\ y_n \end{bmatrix}, \quad \mathbf{X}=\begin{bmatrix} 1 & x_{11} & \cdots & x_{1p} \\ 1 & x_{21} & \cdots & x_{2p} \\ \vdots & \vdots & & \vdots \\ 1 & x_{i1} & \cdots & x_{ip} \\ \vdots & \vdots & & \vdots \\ 1 & x_{n1} & \cdots & x_{np} \end{bmatrix} \quad \text{values for the } i\text{th trial}$$

Only the arrays **y** and **X** are required to obtain the least squares estimates of $\alpha, \beta_1, \ldots, \beta_p$ that minimize

$$\sum_{i=1}^{n} (y_i - \alpha - x_{i1}\beta_1 - \cdots - x_{ip}\beta_p)^2$$

The input array **X** is called the *design matrix*.

In the same vein, setting

$$\mathbf{e} = \begin{bmatrix} e_1 \\ e_2 \\ \cdot \\ \cdot \\ \cdot \\ e_n \end{bmatrix} \quad \text{and} \quad \boldsymbol{\beta} = \begin{bmatrix} \alpha \\ \beta_1 \\ \cdot \\ \cdot \\ \cdot \\ \beta_p \end{bmatrix}$$

we can write the model in the suggestive form

$$\underset{\text{Observation}}{\mathbf{y}} = \underset{\substack{\text{Design}\\ \text{matrix}}}{\mathbf{X}} \quad \underset{\text{Parameters}}{\boldsymbol{\beta}} \quad \underset{\text{Error}}{+\mathbf{e}}$$

which forms the basis for a thorough but more advanced treatment of regression.

REFERENCES

1. Draper, N. R., and Smith, H., *Applied Regression Analysis*, John Wiley & Sons, New York, 1965.

EXERCISES

1. A least squares fit of a straight line produces the following residuals:

$\hat{y}$	11.3	14.8	18.4	22.0	25.5	27.0	31.2	32.7	34.1
Residuals	−.3	.2	−5.4	2.0	.5	5.0	−9.2	6.3	12.9

$\hat{y}$	36.2	39.8	43.4	45.5	46.2	46.9	46.9
Residuals	−4.2	−14.8	7.6	−1.5	16.8	1.1	−16.9

Plot these residuals against the predicted values and determine if any assumption appears to be violated.

2. Referring to the reduction in nitrogen-oxides data originally given in Table 10.2 and to the calculations given in Table 10.4 for fitting a

straight line, plot the residuals against the predicted values and determine whether any assumption appears to be violated.

3. A second-degree polynomial $\hat{y} = \hat{\alpha} + \hat{\beta}_1 x + \hat{\beta}_2 x^2$ is fitted to a response y, and the following predicted values and residuals are obtained:

$\hat{y}$	4.01	5.53	6.21	6.85	8.17	8.34	8.81	9.62
Residuals	.28	−.33	−.21	.24	−.97	.46	.79	−1.02

$\hat{y}$	10.05	10.55	10.77	10.77	10.94	10.98	10.98
Residuals	1.35	−1.55	.63	1.73	−2.14	1.92	−1.18

Do the assumptions appear to be violated?

4. The following predicted values and residuals are obtained in an experiment conducted to determine the degree to which the yield of an important chemical in the manufacture of penicillin is dependent on sugar concentration (the time order of the experiments is given in parentheses):

Predicted	2.2(9)	3.1(6)	2.5(13)	3.3(1)	2.3(7)	3.6(14)	2.6(8)
Residual	−1	−2	3	−3	−1	5	0

	2.5(3)	3.0(12)	3.2(4)	2.9(11)	3.3(2)	2.7(10)	3.2(5)
	0	3	−2	2	−5	0	1

(*a*) Plot the residuals against the predicted values and also against the time order.

(*b*) Do the basic assumptions appear to be violated?

5. An experimenter obtains the following residuals after fitting a quadratic expression in x:

$x=1$	$x=2$	$x=3$	$x=4$	$x=5$
−.1	1.3	−.1	0	−.2
0	−.2	−.3	.2	0
−.2	−.1	.1	−.1	−.2
.6	−.3	.4	0	−.2
−.1	.1	−.1	−.2	−.3
		.1		−.1

Do the basic assumptions appear to be violated?

6. Referring to Exercise 10.12 (page 362):

(*a*) Calculate the residuals.

(*b*) Plot the residuals graphically and determine if the assumptions of a linear regression model appear to be violated.

7. Referring to Exercise 10.13 (page 362), plot the residuals and check any possible violations of the model assumptions.
8. An interested student used the method of least squares to fit the straight line $\hat{y}=264.3+18.77x$ to gross national product y in real dollars. The results for 26 recent years, $x=1,2,\ldots,26$ appear below. Which assumption(s) for a linear regression model appear to be seriously violated by the data? (NOTE: Regression methods are usually not appropriate for this type of data.)

Year	1	2	3	4	5	6	7
y	309.9	323.7	324.1	355.3	383.4	395.1	412.8
$\hat{y}$	283.1	301.9	320.6	339.4	358.2	376.9	395.7
Residual	26.8	21.8	3.5	15.9	25.2	18.2	17.1

Year	8	9	10	11	12	13	14
y	407	438	446.1	452.5	447.3	475.9	487.7
$\hat{y}$	414.5	433.2	452.0	470.8	489.5	508.3	527.1
Residual	−7.5	4.8	−5.9	−18.3	−42.2	−32.4	−39.4

Year	15	16	17	18	19	20	21
y	497.2	529.8	551	581.1	617.8	658.1	675.2
$\hat{y}$	545.8	564.6	583.4	602.1	620.9	639.7	658.4
Residual	−48.6	−34.8	−32.4	−21	−3.1	18.4	16.8

Year	22	23	24	25	26
y	706.6	725.6	722.5	745.4	790.7
$\hat{y}$	677.2	696.0	714.7	733.5	752.3
Residual	29.4	29.6	7.8	11.9	38.4

9. Given $S_x^2=10.1$, $S_y^2=16.5$, and $S_{xy}=9.3$, determine the proportion of variation in y that is explained by linear regression.
10. A calculation shows that $S_x^2=92$, $S_y^2=457$, and $S_{xy}=160$. Determine the proportion of variation in y that is explained by linear regression.
11. Referring to Exercise 10.12, determine the proportion of variation in the y values that is explained by the linear regression of y on x. What is the correlation coefficient between the x and y values?

*12. Perform a lack of fit test after fitting a straight line to the following data:

x	1	1	1	2	3	3	4	5	5
y	9	7	8	10	15	12	19	24	21

*13. Referring to the data given in Exercise 10.8 (page 361), although there is a paucity of replicates, show how to conduct a test of lack of fit for a straight-line model.

14. A forester seeking information on basic tree dimensions obtains the following measurements of the diameters 4.5 feet above the ground and the heights of 12 sugar maple trees (courtesy of A. Ek). The forester wishes to determine if the diameter measurements can be used to predict tree height.

Diameter x (inches)	.9	1.2	2.9	3.1	3.3	3.9	4.3	6.2	9.6	12.6	16.1	25.8
Height y (feet)	18	26	32	36	44.5	35.6	40.5	57.5	67.3	84	67	87.5

(*a*) Plot the scatter diagram and determine if a straight-line relation is appropriate.

(*b*) Determine an appropriate linearizing transformation. In particular, try $x' = \log x$, $y' = \log y$.

(*c*) Fit a straight-line regression to the transformed data.

(*d*) What proportion of variability is explained by the fitted model?

15. A genetic experiment is undertaken to study the competition between two types of female Drosophila melanogaster in cages with one male genotype acting as a substrate. The independent variable x is the time spent in cages, and the dependent variable y is the ratio of the numbers of type 1 to type 2 females. The following data (courtesy of C. Denniston) are recorded:

Time x (days)	No. type 1	No. type 2	$y = \dfrac{\text{No. type 1}}{\text{No. type 2}}$
17	137	586	.23
31	278	479	.58
45	331	167	1.98
59	769	227	3.39
73	976	75	13.01

(*a*) Plot the scatter diagram of y vs. x and determine if a linear model of relation is appropriate.

(*b*) Determine if a linear relation is plausible for the transformed data $y' = \log_{10} y$.

(*c*) Fit a straight-line regression to the transformed data.

16. Carry out the necessary computations to obtain the least squares estimates of the parameters in the multiple regression model $Y = \alpha + \beta_1 x_1 + \beta_2 x_2 + e$, given:

x_1	2	2	3	4	4
x_2	1	1	0	0	3
y	12	10	9	13	20

17. Referring to Exercise 10.14 (page 362):

(*a*) Find the least squares fit of the multiple regression model $Y = \alpha + \beta_1 x_1 + \beta_2 x_2 + e$ to predict the cooling rate of cast iron, using the number of graphite spheroids x_1 and the pearlite content x_2 as predictors.

(*b*) What is the predicted cooling rate corresponding to $x_1 = 350$ and $x_2 = 3$?

*18. *Multiple correlation coefficient R*: Recall that in fitting a simple linear regression $Y = \alpha + \beta x + e$ by least squares, the quantity $r^2 = 1 - \text{SSE}/S_y^2$ measures the proportion of variability in the y values that can be explained by the fitted line. Extending this concept to the case of fitting a multiple regression model, we define the square of the multiple correlation R as the proportion of variability in y that is explained by the fitted model. In particular, suppose that the model is $Y = \alpha + \beta_1 x_1 + \beta_2 x_2 + \beta_3 x_3 + e$ with 3 predictors x_1, x_2, and x_3. After obtaining the least squares estimates $\hat{\alpha}$, $\hat{\beta}_1$, $\hat{\beta}_2$, and $\hat{\beta}_3$, we can compute the residuals $y_i - \hat{y}_i$, where

$$\hat{y}_i = \hat{\alpha} + \hat{\beta}_1 x_{i1} + \hat{\beta}_2 x_{i2} + \hat{\beta}_3 x_{i3}, \quad i = 1, \ldots, n$$

The residual sum of squares $\text{SSE} = \sum (y_i - \hat{y}_i)^2$ represents that part of the total variation in y values $S_y^2 = \sum (y_i - \bar{y})^2$ that is not explained by the fitted model. The square of the sample multiple correlation of y on (x_1, x_2, x_3) is defined as

$$R_{y\cdot 123}^2 = 1 - \frac{\text{SSE}}{S_y^2} = \frac{\hat{\beta}_1 S_{x_1 y} + \hat{\beta}_2 S_{x_2 y} + \hat{\beta}_3 S_{x_3 y}}{S_y^2}$$

which represents the proportion of variability in y that is explained by the predictor variables x_1, x_2, and x_3 using the multiple regression model. This definition applies to any number p of predictor variables. Incidentally, R is the ordinary correlation coefficient between $(y_i, \hat{y}_i)$, $i=1,\ldots,n$. The term *coefficient of determination* is often used to describe the square of the multiple correlation. In particular, when $p=1$, R^2 is the same as r^2.

(*a*) Referring to Exercise 17, compute the multiple correlation coefficient for the prediction of the cooling rate, using x_1 and x_2 as predictors.

(*b*) Comment on the improvement of fit due to the addition of the extra variable x_2 in the model by comparing $R^2_{y\cdot 12}$ and $R^2_{y\cdot 1}$.

19. Using the method of least squares, fit a quadratic model $Y=\alpha+\beta_1 x + \beta_2 x^2 + e$ to the following data:

x	-2	-1	0	1	2
y	.4	1.3	2.2	2.5	3.0

[*Hint*: Follow the scheme of computations given in Example 11.4, using x in place of x_1 and x^2 in place of x_2.]

*20. Again referring to Exercise 10.13:

(*a*) Fit a quadratic model $y=\alpha+\beta_1 x+\beta_2 x^2+e$ to the data for CLEP scores y and CQT scores x.

(*b*) Use the fitted regression to predict the expected CLEP score when $x=160$.

(*c*) Compute r^2 for fitting a line, and the square of the multiple correlation for fitting a quadratic expression. Interpret these values and comment on the improvement of fit.

*21. *A test for the presence of one type of serial correlation*: In a multiple regression model, for instance one with three predictor variables

$$Y_i=\alpha+\beta_1 x_{i1}+\beta_2 x_{i2}+\beta_3 x_{i3}+e_i\,, \quad i=1,2,\ldots,n$$

it is assumed that the errors e_i are independently distributed as $N(0,\sigma)$. However, the errors may be associated in time, so that those adjacent in time have the correlation ρ. Further, if the special model holds in which errors two time units apart have correlation ρ^2, those three units apart $\rho^3,\ldots$, those k units apart ρ^k, it is possible to test for independence. This test, called the *Durbin-Watson test*, is based on the statistic

$$d=\frac{\sum_{t=2}^{n}(u_t-u_{t-1})^2}{\sum_{t=1}^{n}u_t^2}$$

where $u_t = y_t - \hat{y}_t$, $t = 1, \ldots, n$ are the residuals arranged in time order. The distribution of d is complicated. However, two critical values d_L and d_U are tabulated in Durbin, J. and Watson, G. S., Testing for Serial Correlation in Least Squares Regression, II [*Biometrika* (1951), Vol. 38, 159–78] for different sample sizes n and for several numbers of predictor variables in the model. In testing H_0: $\rho = 0$ vs. H_1: $\rho \neq 0$, the procedure has an unusual structure that enables us to conclude that:

(*i*) d is significant if $d < d_L$ or $4 - d < d_L$

(*ii*) d is not significant if $d > d_U$ and $4 - d > d_U$

(*iii*) The test is inconclusive otherwise.

To test the two-sided alternative at level α, both d_L and d_U are obtained from the above mentioned table with $\alpha/2$. One-sided tests can be constructed by noting that $d < d_L$ indicates a positive correlation and that $4 - d < d_L$ indicates a negative correlation. Although correlation among the errors e_i is a serious violation of assumptions, it usually requires 20 or more observations to detect even moderately large values of ρ in this special model.

(*a*) Calculate d for the data on gross national product given in Exercise 8.

(*b*) Referring to Exercise 10.14 and to Exercise 17 here, calculate d for the multiple regression of the cooling rate y on x_1 and x_2, assuming the data are presented in time order.

*22. (*Continuation of Exercise* 21) Durbin and Watson give the pairs of values ($\alpha = .05$)

$(d_L, d_U) = (1.30, 1.46)$ for 26 observations and 1 predictor.

$(d_L, d_U) = (1.31, 1.57)$ for 32 observations and 2 predictors.

Using these values, test for lack of independence:

(*a*) In Exercise 21(*a*).

(*b*) In Exercise 21(*b*).

23. Robert A. Millikan (1865–1953) devised a method to observe the motion of a single drop of water or oil under the influence of both electric and gravitational fields. Millikan's method helped to establish the discreteness of charge and produced the first accurate measurements of the charge e of an electron. The droplet under observation carried multiple charges, and direct calculations based on voltage, time of fall, and so on, provided an estimate of total charge. Below we record the results for water drops falling two spaces by converting each data set of voltage, time measurements, and other factors to a multiple-charge value [SOURCE: *Philosophical Magazine*, **19** (1910),

(209–28)]. Originally, Millikan averaged some values before converting to charge.

x (No. of e's)	Observations ($10^9 \times$ Charge)
3	1.392, 1.392, 1.398, 1.368, 1.368, 1.368, 1.345
4	1.768, 1.768, 1.910, 1.768, 1.746, 1.746, 1.886, 1.768, 1.768, 1.768
5	2.471, 2.471, 2.256, 2.256, 2.471
2	.944, .992
6	2.981, 2.688

(*a*) Fit a straight-line relationship to the data using the method of least squares.
(*b*) Construct a 95% confidence interval for β, the charge on the electron.
(*c*) Test H_0: $\alpha = 0$.
(*d*) Examine the residuals.

24. Find the pure error sum of squares in Exercise 23 and test for lack of fit.
25. In an experiment (courtesy of W. Burkholder) involving stored-product beetles (*Trogoderma glabrum*) and their sex-attractant pheromone, the pheromone is placed in a pit-trap in the centers of identical square arenas. Marked beetles are then released along the diagonals of each square at various distances from the pheromone source. After 48 hours, the pit-traps are inspected. Control pit-traps containing no pheromone captured no beetles.

Release distance (centimeters)	No. of beetles captured out of 8
6.25	5, 3, 4, 6
12.5	5, 2, 5, 4
25	4, 5, 3, 0
50	3, 4, 2, 2
100	1, 2, 2, 3

(*a*) Plot the original data with y = number of beetles captured. Repeat with $x = \log_e$(distance).
(*b*) Fit a straight line by least squares to the appropriate graph in (*a*).
(*c*) Construct a 95% confidence interval for β.
(*d*) Establish a 95% confidence interval for the mean at a release distance of 18 cm.

*26. Suppose the least squares regression line becomes

$$\hat{y}=4+3x_1+2x_2$$

(a) If x_1 is increased one unit with x_2 held fixed, find the estimated mean increase in Y.

(b) If x_2 is increased by one unit with x_1 held fixed, determine the estimated mean increase in Y. What is the answer if x_2 is increased by 10 units?

*27. Show that the model $Y=\alpha+\beta_1x_1+\beta_2x_2+e$ has an indeterminancy if x_1 and x_2 take the pairs of values

x_1	1	2	3	5
x_2	1	4	7	13

[*Hint*: $x_2=3x_1-2$ so $\alpha+\beta_1x_1+\beta_2x_2=(\alpha-2\beta_2)+(\beta_1+3\beta_2)x_1$].

*28. When fitting a quadratic model to the data in Exercise 15

(a) write, completely, the design matrix **X**

(b) write the arrays for the parameters $\boldsymbol{\beta}$, errors **e** and the observations **y**.

*29. Write the design matrix **X** for fitting a straight-line model to the data in Exercise 10.12 page 362.

MATHEMATICAL EXERCISES

1. In fitting a straight-line model $Y_i=\alpha+\beta x_i+e_i$ the sample correlation coefficient between the observed values y_i and the corresponding predicted values $\hat{y}_i=\hat{\alpha}+\hat{\beta}x_i$, $i=1,\ldots,n$, is defined as

$$r_{y,\hat{y}}=\frac{\sum(y_i-\bar{y})(\hat{y}_i-\bar{y})}{\sqrt{\sum(y_i-\bar{y})^2}\sqrt{(\hat{y}_i-\bar{y})^2}}$$

Show that $r_{y,\hat{y}}=|r|$, where r is the sample correlation coefficient between (x_i,y_i), $i=1,\ldots,n$, as defined in Section 11.3.

$$\left[\textit{Hint}:\ \sum(y_i-\bar{y})(\hat{y}_i-\bar{y})=\hat{\beta}\sum(y_i-\bar{y})(x_i-\bar{x})=\hat{\beta}S_{xy}\right.$$

$$\left.\sum(\hat{y}_i-\bar{y})^2=\beta^2\sum(x_i-\bar{x})^2=\hat{\beta}^2S_x^2\right]$$

CHAPTER 12

Correlation: A Measure of Linear Relationship

12.1 INTRODUCTION

Regression methods are appropriate when a random variable Y depends on a causal variable X that is often controlled by the experimenter and the analysis is conducted to determine the effect of X or its ability to predict Y. In contrast, the experimenter's primary goal may be to study the strength of the relationship between two random variables, neither of which can be singled out as the cause of the other. For instance, the two variables may be the girth X and the height Y of oak trees. A helmet manufacturer may want to know how head length X and breadth Y vary together in the population of potential customers, whereas a garment manufacturer may be interested in waist size X and inseam Y. In a study of social mobility, we may wish to relate the number of years of schooling for father X and son Y. In each of these events neither of the variables under consideration are controlled in advance of the collection of data; both are treated as random variables. The data set consisting of measurements of X and Y made on a sample of n experimental subjects can then be viewed as a bivariate random sample $(X_1, Y_1), \ldots, (X_n, Y_n)$, where the different pairs are independent. From this perspective, a study of the relationship between the variables is effected by *correlation analysis*.

> *Structure of observations*
>
> In a random sample of n experimental subjects, observations on the variables X and Y are denoted by
>
> $$(X_1, Y_1), (X_2, Y_2), \ldots, (X_n, Y_n)$$
>
> where each pair has the same bivariate distribution and the different pairs are independent.

The first step in the study of a relationship should now be familiar—plot the observations on a graph. The scatter diagram provides a useful aid in discerning the nature of the relationship.

Example 12.1 In a study of social mobility, information about occupational status is collected from a random sample of 43 persons and their fathers. The occupational status is recorded on the Duncan scale, which is widely used by sociologists. The data set is presented in Table 12.1, and the scatter diagram is plotted in Figure 12.1. The diagram seems to indicate an increasing relationship between father's status X and son's status Y, but the data points fluctuate too widely to indicate a strong relationship.

TABLE 12.1
DATA OF THE OCCUPATIONAL STATUS OF FATHER (x) AND SON (y) RECORDED ON THE DUNCAN SCALE

x	y	x	y
22	13	53	65
14	49	14	14
14	72	25	25
14	44	37	31
68	44	53	64
12	19	19	17
32	17	14	18
22	13	15	18
19	22	49	47
14	14	36	18
44	21	21	41
18	15	14	15
61	66	18	44
82	67	53	72
14	44	44	37
18	13	24	44
44	16	87	45
32	15	61	19
72	40	19	15
86	17	44	50
26	31	19	41
65	65		

Courtesy of R. Hauser and W. Sewell.

■

In addition to examining the scatter diagram, it is therefore advisable to compute a descriptive measure that will reflect the strength of the existing relationship in much the same way as the measures of location and spread we examined in Chapter 2. Our discussion focuses on an index called the

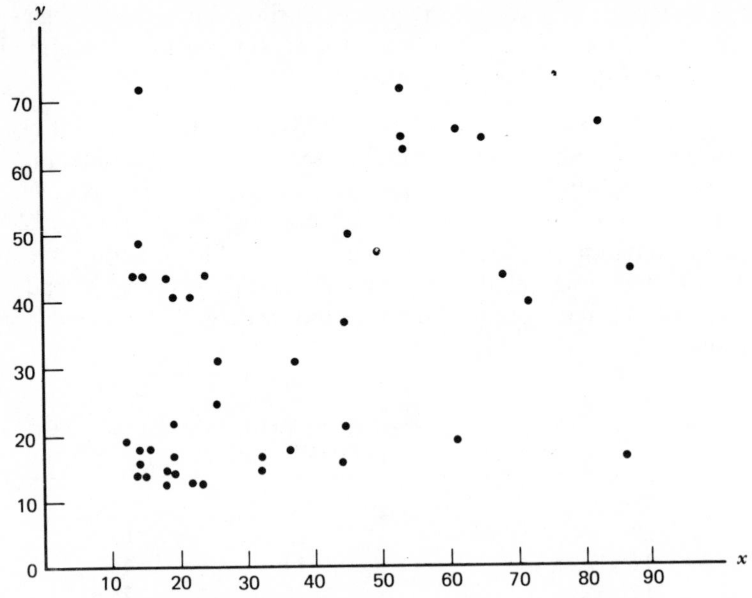

Figure 12.1 Scatter diagram of the data in Table 12.1.

correlation coefficient, which is designed to measure the closeness of the relationship to a straight-line form.

12.2 THE SAMPLE CORRELATION COEFFICIENT

One simple kind of association between the variables X and Y produces pairs of values or, graphically, points that scatter about a straight line. A small amount of scatter about a line indicates strong association; a large amount of scatter is a manifestation of weak association. A numerical measure of this relationship is called the *sample correlation coefficient* or, sometimes, *Pearson's product moment correlation coefficient*. The sample correlation coefficient r is given by:

> *Sample correlation coefficient*
>
> $$r = \frac{\sum_{i=1}^{n} (X_i - \bar{X})(Y_i - \bar{Y})}{\sqrt{\left[\sum_{i=1}^{n} (X_i - \bar{X})^2\right]\left[\sum_{i=1}^{n} (Y_i - \bar{Y})^2\right]}}$$
>
> where $(X_1, Y_1), \ldots, (X_n, Y_n)$ are the n pairs of observations.

Recall from Section 4.6 that the *population correlation coefficient* for a bivariate distribution is defined

$$\operatorname{Corr}(X, Y) = \frac{\operatorname{Cov}(X, Y)}{\sigma_X \sigma_Y}$$

which is usually denoted by ρ. The statistic r is a sample analog of ρ, as we can see by replacing the population parameters by their sample analogs. That is, replace $\operatorname{Cov}(X, Y)$ with $\sum (X_i - \bar{X})(Y_i - \bar{Y})/(n-1)$, σ_X^2 with $\sum (X_i - \bar{X})^2/(n-1)$, and σ_Y^2 with $\sum (Y_i - \bar{Y})^2/(n-1)$. Therefore, the sample correlation coefficient r can be considered an estimator of the population correlation ρ.

Another form of r, sometimes useful in hand calculation, is

$$r = \frac{S_{xy}}{\sqrt{S_x^2 S_y^2}} = \frac{\sum_{i=1}^{n} X_i Y_i - \left(\sum_{i=1}^{n} X_i\right)\left(\sum_{i=1}^{n} Y_i\right)/n}{\sqrt{\left[\sum_{i=1}^{n} X_i^2 - \left(\sum_{i=1}^{n} X_i\right)^2/n\right]\left[\sum_{i=1}^{n} Y_i^2 - \left(\sum_{i=1}^{n} Y_i\right)^2/n\right]}}$$

where the symbols

$$S_{xy} = \sum_{i=1}^{n} (X_i - \bar{X})(Y_i - \bar{Y}) \qquad S_x^2 = \sum_{i=1}^{n} (X_i - \bar{X})^2 \qquad S_y^2 = \sum_{i=1}^{n} (Y_i - \bar{Y})^2$$

are identical to those given in Chapter 10.

Example 12.2 Compute the correlation coefficient r for the five pairs of observations

x	5	1	4	3	2
y	0	4	2	0	−1

using each of the two formulas.

The calculations using the first formula appear in Table 12.2, and those using the second formula appear in Table 12.3.

TABLE 12.2
CALCULATION OF r USING THE FIRST FORMULA

x	y	$x-\bar{x}$	$y-\bar{y}$	$(x-\bar{x})^2$	$(y-\bar{y})^2$	$(x-\bar{x})(y-\bar{y})$
5	0	2	−1	4	1	−2
1	4	−2	3	4	9	−6
4	2	1	1	1	1	1
3	0	0	−1	0	1	0
2	−1	−1	−2	1	4	2
$\bar{x}=3$	$\bar{y}=1$			$\sum(x-\bar{x})^2 = 10$	$\sum(y-\bar{y})^2 = 16$	$\sum(x-\bar{x})(y-\bar{y}) = -5$

$$r=\frac{\sum(x_i-\bar{x})(y_i-\bar{y})}{\sqrt{\left[\sum(x_i-\bar{x})^2\right]\left[\sum(y_i-\bar{y})^2\right]}}=\frac{-5}{\sqrt{10\times 16}}=-.395$$

TABLE 12.3
CALCULATION OF r USING THE SECOND FORMULA

x	y	xy	x^2	y^2
5	0	0	25	0
1	4	4	1	16
4	2	8	16	4
3	0	0	9	0
2	−1	−2	4	1
$\sum x=15$	$\sum y=5$	$\sum xy=10$	$\sum x^2=55$	$\sum y^2=21$

$$r=\frac{\sum x_iy_i-\left(\sum x_i\right)\left(\sum y_i\right)/n}{\sqrt{\left[\sum x_i^2-\left(\sum x_i\right)^2/n\right]\left[\sum y_i^2-\left(\sum y_i\right)^2/n\right]}}$$

$$=\frac{10-(15)(5)/5}{\sqrt{\left[55-(15)^2/5\right]\left[21-(5)^2/5\right]}}=\frac{-5}{\sqrt{10\times 16}}=-.395 \quad \blacksquare$$

Because r is a sample analog of the population correlation ρ, the properties of ρ discussed in Section 4.6 are also shared by r. In particular, r can assume values between -1 and 1. The value $r=1$ occurs only when all the data points lie perfectly on a straight line with a positive slope; $r=-1$ is also a perfect linear relation in which the line has a negative slope. A value of r close to either of these extremes corresponds to a tight clustering of data points around a straight line, constituting a strong linear relation. With lesser amounts of cluster about a straight line, r assumes smaller numerical values, and $r=0$ is interpreted as an absence of linear relation. The five scatter diagrams in Figure 12.2 illustrate the correspondence between the extent of the linearity of a relation and the values of r.

Another interpretation of the sample correlation coefficient can be made in the context of fitting a straight line to the data by least squares. As we saw in Section 11.3, the proportion of variability in the y values that can be explained by a linear relation is precisely r^2. Thus with $r=.9$, 81% of the variability in the y values is explained by a linear relation; with $r=.5$, only 25% is explained.

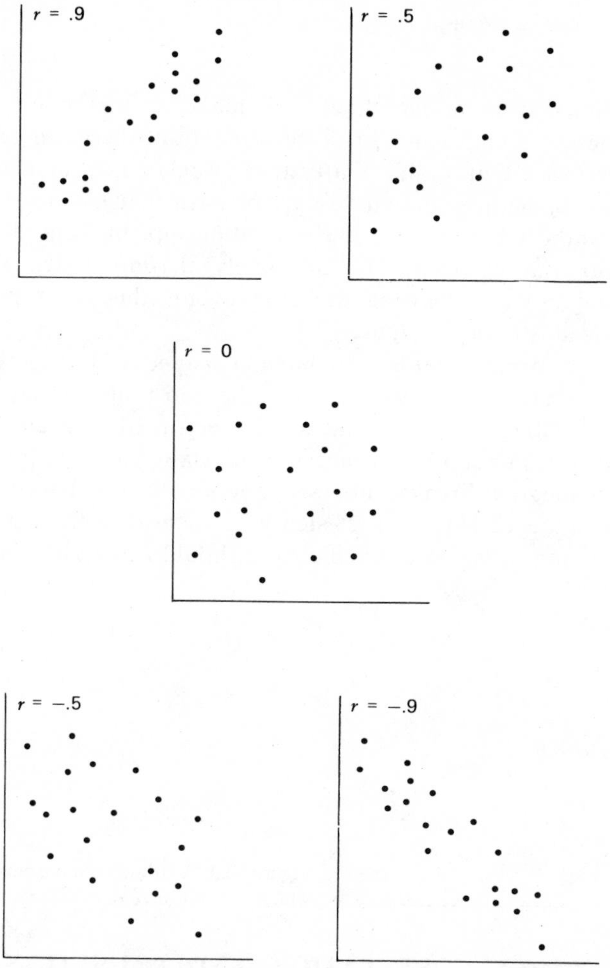

Figure 12.2 Correspondence between the values of r and the amount of scatter.

Properties of r

(*a*) r must lie between -1 and $+1$.

(*b*) The numerical value of r measures the strength of the linear relation, and the sign of r indicates the direction of the relationship.

(*c*) r^2 is the proportion of variability in the y values that is explained by a straight line fitted by least squares.

(*d*) r remains unchanged if the x values are changed to $ax+b$ and the y values are changed to $cy+d$, where a and c have the same sign.

An important reminder should be made at this point. The sample correlation coefficient r measures the strength of the *linear* relationship. It may be the case that X and Y are strongly related but that the relation is curvilinear. Sometimes the curve may be such that r is nearly zero, which correctly indicates a lack of linear relationship, but this does not mean there is no relationship at all. Fortunately, the initial step of plotting the scatter diagram and viewing the data avoids this latter pitfall. Such a situation is illustrated in Figure 12.3.

When the overall pattern is banana-shaped (see Figure 12.4*a*), the sample correlation r is usually not the most appropriate measure of association. Other measures that are better suited to such situations, are presented in Chapter 15. No measure of relationship is appropriate when the scatter diagram breaks into two or more clusters. Faced with separate clusters (Figure 12.4*b*), the first step is to determine the underlying cause for the division of the data. Each cluster should be handled individually.

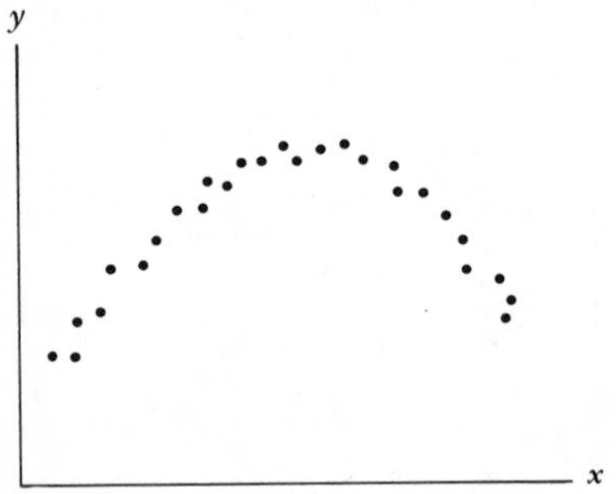

Figure 12.3 A strong relationship along a curve for which r is almost zero.

12.3 CORRELATION AND CAUSATION

Historically, analysts have often jumped to unjustified conclusions by mistaking an observed correlation for a cause–effect relationship. A high

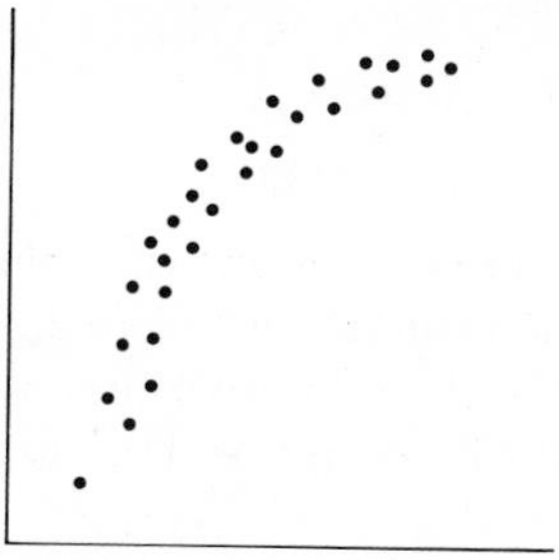

Relationship not linear.

a

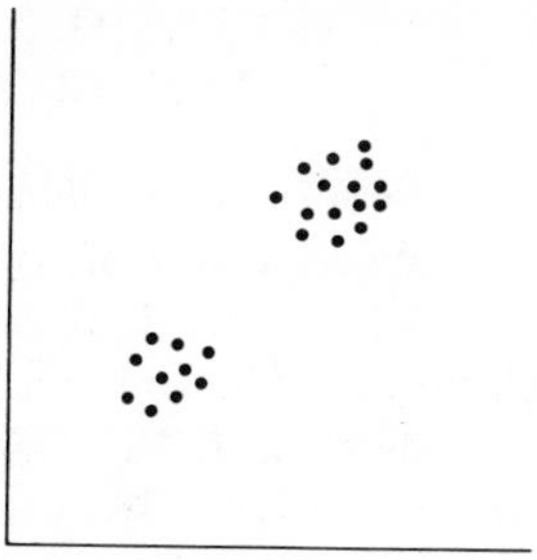

Really samples from two populations.

b

Figure 12.4 Cases where r is inappropriate.

sample correlation coefficient does not necessarily signify any causal relation between two variables. A frequently quoted example concerns an observed high positive correlation between the number of storks sighted and the number of births in a European city. Hopefully, no one today would conclude that this evidence indicates that storks bring babies or, worse yet, that killing storks would control population growth.

The observation that two variables tend to simultaneously vary in a certain direction does not imply the presence of a direct relationship between them. If we record the monthly number of homicides x and the monthly number of religious meetings y for several cities of widely varying sizes, the data will probably indicate a high positive correlation. It is the fluctuation of a third variable (namely, the city population) that causes x and y to vary in the same direction, despite the fact that x and y may be unrelated or even negatively related. Picturesquely, the third variable, which in this example, is actually causing the observed correlation between crime and religious meetings, is referred to as a *lurking variable*. The false correlation that it produces is called *spurious correlation*. Sometimes it becomes more a matter of common sense than of statistical application to determine if an observed correlation can be practically interpreted or if it is spurious. At other times, the presence of spurious correlation in studies in which the variables were initially believed to be correlated on logical grounds has sparked serious and continuing controversies.

WARNING: An observed correlation may be *spurious*; that is, it may be due to a third unknown causal variable.

When using the correlation coefficient as a measure of relationship, we must be careful to avoid the possibility that a lurking variable is affecting any of the variables under consideration.

12.4 A POPULATION MODEL FOR CORRELATION: THE BIVARIATE NORMAL

The *bivariate normal distribution* is a widely used population model for dealing with observations on two continuous random variables X and Y. In this model each individual variable is normally distributed. The population correlation ρ is the only parameter in the joint distribution in addition to the two means and the two standard deviations that appear in the marginal distributions of X and Y.

With two random variables, a probability density is a hill in three dimensions that describes the manner in which probability is spread over the plane of (x,y) values. The mathematical formula for the bivariate normal density is rather inconsequential to our discussion, but some illustrations should prove illuminating here. Three bivariate normal distributions are shown in Figure 12.5. All of these distributions have the same means, and both components X and Y have the same standard deviations. The differences in appearance are exclusively due to differences in the population correlation ρ. With a high positive or negative ρ, the probability surface tends to form a sharp ridge. The sign of ρ determines the orientation of the ridge; a positive (negative) ρ corresponds to a positive (negative) slope in the (x,y) plane.

Another view of the bivariate normal distribution is derived by locating all the pairs of (x,y) values that define a prescribed height for the density.

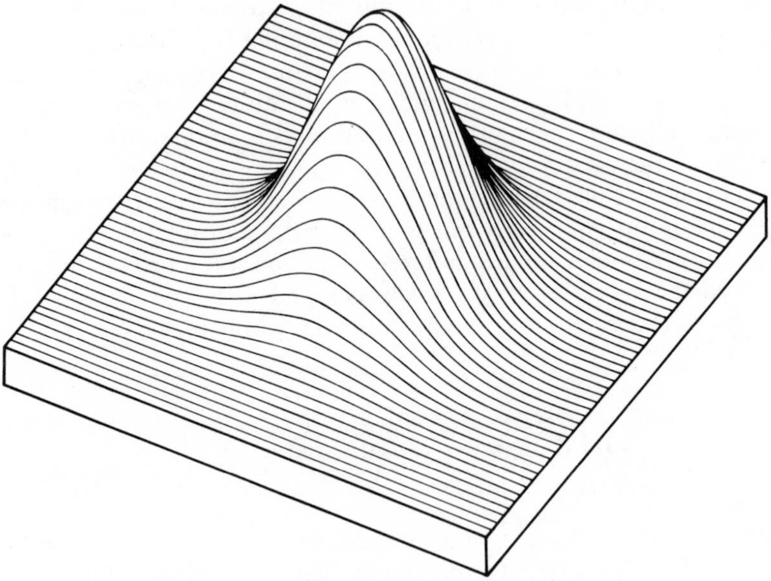

Figure 12.5a Bivariate normal surface with $\rho = 0$.

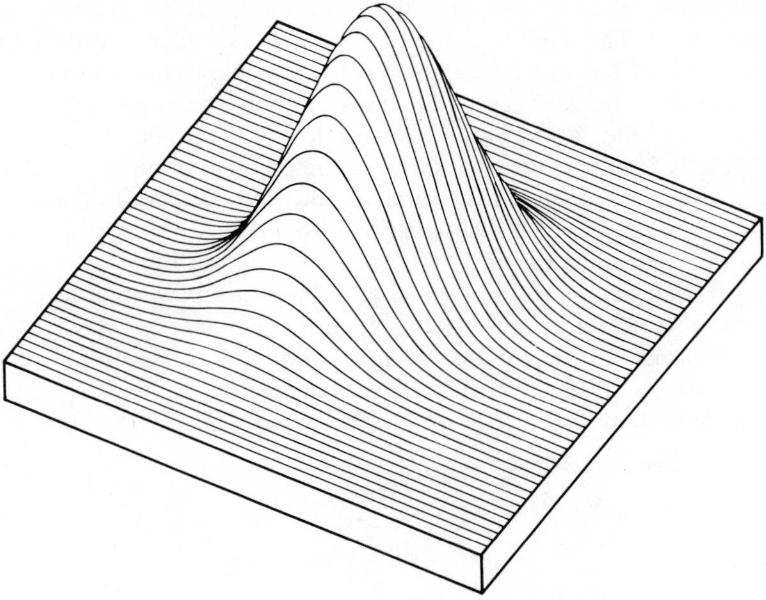

Figure 12.5*b* Bivariate normal surface with $\rho = .4$.

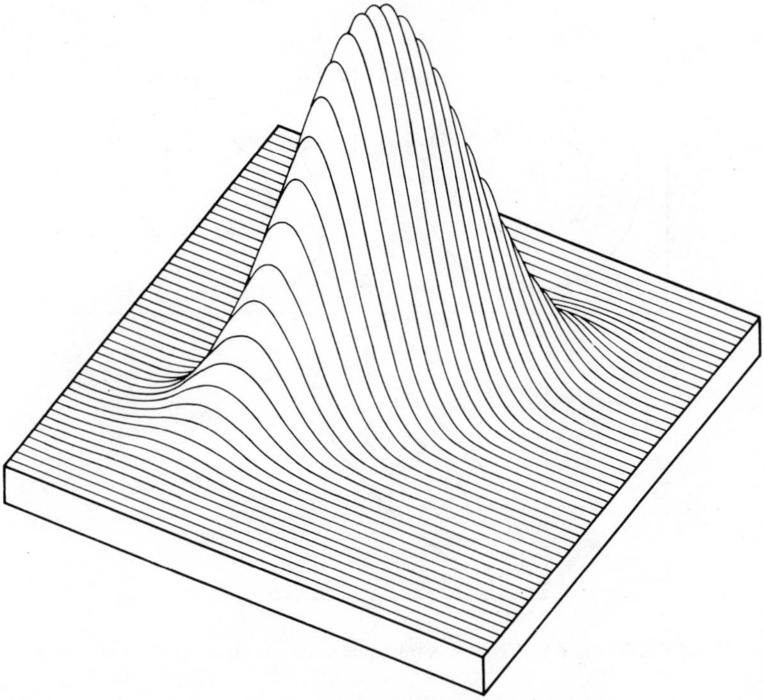

Figure 12.5*c* Bivariate normal surface with $\rho = .75$.

In other words, we find the path in the plane where the hill of density is of constant height. These paths or *contours* are ellipses centered at the pair of means. Figure 12.6 shows the contours corresponding to the densities given in Figure 12.5. In each case the inner contour contains 50% of the probability and the outer contour contains 90% of the probability.

If a bivariate normal distribution were the population model, a random sample would be expected to emulate the population. For instance, roughly 50% and 90% of the observations would be expected to lie within the contours shown in Figure 12.6.

In a bivariate sample an important question to ask is whether or not the two random variables are correlated. When the population is modeled as a bivariate normal, a simple test of the null hypothesis H_0: $\rho=0$ is available. In this model $\rho=0$ is equivalent to the independence of the two variables. The appropriate statistic for testing independence in a bivariate normal model is

$$t=\frac{\sqrt{n-2}\,r}{\sqrt{1-r^2}}$$

which has the Student's t distribution with d.f.$=n-2$. Given a two-sided alternative, the null hypothesis is rejected if the observed value of this test

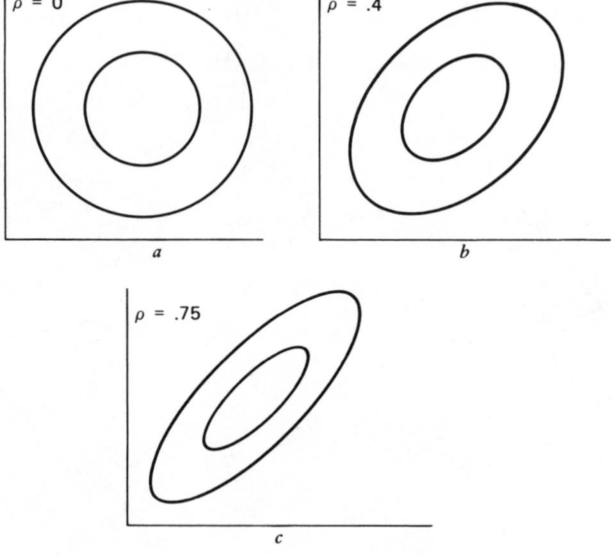

Figure 12.6 The 50% and 90% contours for equal variance bivariate normal distributions.

statistic is greater than $t_{\alpha/2}$ or less than $-t_{\alpha/2}$.

> To test H_0: $\rho=0$ vs. H_1: $\rho\neq 0$ based on n pairs of observations from a bivariate normal population
>
> $$\text{Reject } H_0 \text{ if } \left|\frac{\sqrt{n-2}\ r}{\sqrt{1-r^2}}\right| \geqslant t_{\alpha/2}$$
>
> where d.f. $=n-2$ for t.

Example 12.3 Using the data given in Example 12.1, test the hypothesis that the father's occupational status and the son's occupational status are independent.

We assume that the sample has come from a bivariate normal distribution. The sample correlation coefficient calculated from the data given in Table 12.1 is found to be $r=.412$. The observed value of the test statistic is

$$t=\frac{\sqrt{n-2}\ r}{\sqrt{1-r^2}}=\frac{\sqrt{41}\times .412}{\sqrt{.830}}=2.90, \quad \text{d.f.}=41$$

With $\alpha=.01$, the two-sided tabulated value is $t_{.005}=2.70$. Because the observed value is larger, the null hypothesis H_0: $\rho=0$ is rejected at $\alpha=.01$. ■

To test a more general hypothesis H_0: $\rho=\rho_0$ in a bivariate normal population, a *large-sample test* is based on the fact that

$$Z=\sqrt{n-3}\left(\frac{1}{2}\log_e\frac{1+r}{1-r}-\frac{1}{2}\log_e\frac{1+\rho_0}{1-\rho_0}\right) \qquad \text{is approximately } N(0,1)$$

Given a two-sided alternative and $\alpha=.05$, H_0 is rejected if $|Z|>1.96$.

12.5 SERIAL CORRELATION

When observations about a *single* variable are collected at different points in time, a major concern in statistical analysis is the possibility of dependence among successive observations, otherwise called a *lack of randomness*. In Chapters 8 and 9, which dealt with the basic inferential procedures for one and two populations, we stressed that a serious violation of the assumptions can be caused by the existence of dependence among the observations. This dependence is primarily manifested in observations that are adjacent in time or space. The term *serial correlation* refers to a correlation between observations that are adjacent in time order or any other order of data collection.

Specifically, let $X_1, X_2, \ldots, X_n$ denote the observations made on the variable X at n consecutive points in time. To determine if the consecutive members are correlated, we pair the observations located next to one another, or

$$(X_1, X_2), (X_2, X_3), \ldots, (X_{i-1}, X_i), \ldots, (X_{n-1}, X_n)$$

For the purposes of visual examination, the appropriate scatter diagram consists of a plot of the $n-1$ pairs (X_{i-1}, X_i), taking the first member on the x axis and the second on the y axis. The ordinary correlation coefficient calculated from these $n-1$ pairs is called the *first serial correlation, first-order autocorrelation*, or *lag 1 correlation*. The following slightly modified formula is generally used.

Lag 1 correlation or first-order autocorrelation

$$r_1 = \frac{\sum_{i=2}^{n} (X_{i-1} - \bar{X})(X_i - \bar{X})}{\sum_{i=1}^{n} (X_i - \bar{X})^2}$$

The population analog of the lag 1 correlation statistic is called the *population first-order autocorrelation* and is denoted by ρ_1. A large-sample test of the null hypothesis H_0: $\rho_1 = 0$ is based on the fact that $\sqrt{n}\ r_1$ is approximately distributed as $N(0, 1)$ under H_0. Incidentally, the Durbin–Watson statistic cited in Exercise 11.21 is only a slight generalization of r_1 that is applicable to residuals from a regression model.

Example 12.4 A manufacturer that produces plastic sheets attempts to control the weight near the target value of 125 per unit area. The data for the measurements of deviations from this target recorded at 28 consecutive time points appear in Table 12.4. The scatter diagram plotted to investigate the presence of first-order autocorrelation appears in Figure 12.7. This correlation is found to be $r_1 = .57$.

TABLE 12.4
MEASUREMENTS OF DEVIATIONS FROM TARGET WEIGHT AT 28 CONSECUTIVE TIME POINTS

OBSERVATION NUMBER	MEASUREMENTS									
1–10	6	9	7	2	−1	−6	−8	−2	1	3
11–20	0	1	4	3	3	0	4	−2	−3	−1
21–28	1	2	2	−1	3	6	0	−2		

■

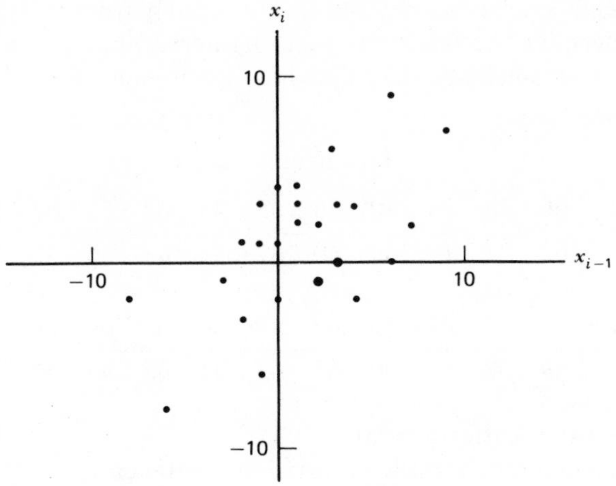

Figure 12.7 Scatter diagram of (x_{i-1}, x_i) for the data given in Table 12.4.

EXERCISES

1. Use both formulas to calculate the sample correlation coefficient for these data:

x	2	4	3	3
y	6	4	6	8

2. The following height X and weight Y measurements were recorded for several recent Miss Americas:

Height (inches)	65	67	66	65.5	65	66.5	66	67	66
Weight (pounds)	114	120	116	118	115	124	124	115	116

Height (inches)	69	67	65.5	68	67	68	69	68
Weight (pounds)	135	125	110	121	118	120	125	119

(*a*) Plot the scatter diagram.
(*b*) Calculate r.

3. It is relatively expensive to measure the dominant tree heights from the ground in a forest plot. An alternative procedure is to estimate the tree heights inexpensively but less accurately from aerial photographs. To determine the feasibility of using aerial photographic methods to estimate tree heights, the following measurements (courtesy of A. Ek) are made:

Tree heights (feet)

Ground x	66	38	20	40	46	55	45	62	55	32	52	58	56
Aerial y	62	23	19	43	40	47	37	44	56	32	55	55	39

x	58	53	44	38	69	40	38	53	54	56	57	58	44
y	59	58	48	40	61	45	41	47	44	51	58	56	47

(*a*) Plot the scatter diagram.
(*b*) Calculate the sample correlation coefficient.

4. As part of a study of the psychobiological correlates of success in athletes, the following measurements (courtesy of W. Morgan) are obtained from members of the U. S. Olympic wrestling team:

Anger x	6	7	5	21	13	5	13	14
Vigor y	28	23	29	22	20	19	28	19

(*a*) Plot the scatter diagram.
(*b*) Calculate r.

5. A director of student counseling is interested in the relationship between the numerical score x and the social science score y on college qualification tests. The following data (courtesy of R. Johnson) are recorded:

x	41	39	53	67	61	67	46	50	55	72	63	59
y	29	19	30	27	28	27	22	29	24	33	25	20

x	53	62	65	48	32	64	59	54	52	64	51	62
y	28	22	27	22	27	28	30	29	21	36	20	29

x	56	38	52	40	65	61	64	64	53	51	58	65
y	34	21	25	24	32	29	27	26	24	25	34	28

(*a*) Plot the scatter diagram.
(*b*) Calculate r.

6. The correlation between the height in inches and the weight in pounds of new recruits is found to be $r = .43$. What is the sample correlation if height and weight are measured in centimeters and grams?
7. Would you expect a positive, negative, or nearly zero correlation for each of the following? Give reasons for your answers.
   (*a*) Government spending and unemployment.
   (*b*) Intelligence scores of husbands and wives.
   (*c*) Number of hours of study the day before an examination and the score on the examination.
   (*d*) Stock prices for IBM and General Motors.
   (*e*) The temperature at a baseball game and beer sales.
8. After examining each of the following situations, state whether you would expect to find a high correlation between the variables. Give reasons why an observed correlation cannot be interpreted as a direct relationship between the variables, and indicate possible lurking variables.
   (*a*) Correlation between the data on the incidence rate X of cancer and per-capita consumption Y of beer, collected from different states.
   (*b*) Correlation between the police budget X and the number of crimes Y recorded during the last 10 years in a city like Chicago.
   (*c*) Correlation between the gross national product X and the number of divorces Y in the country recorded during the last 10 years.
   (*d*) Correlation between the concentration X of air pollutants and the number of riders Y on public transportation facilities when the data are collected from several cities that vary greatly in size.
   (*e*) Correlation between the wholesale price index X and the average speed Y of winning cars in the Indianapolis 500 during the last 10 years.
9. An anthropologist measuring the length and breadth of skulls obtains a correlation of $r = .82$ based on 27 skulls. Assume that the data are

taken from a bivariate normal population. Test H_0: $\rho=0$ vs. H_1: $\rho\neq 0$ with $\alpha=.05$.

10. A food technologist studying hospital meal services finds a correlation of $r=.68$ between preparation time and serving time based on $n=11$ observations. Can the preparation and serving times be considered independent when constructing a model for the uncertainty of the total meal time? Take $\alpha=.05$.
11. Test H_0: $\rho=0$, using the data in Exercise 3. (Take $\alpha=.10$.)
12. Test H_0: $\rho=0$, using the data in Exercise 4. (Take $\alpha=.05$.)
13. Test H_0: $\rho=0$, using the data in Exercise 5. (Take $\alpha=.05$.)
14. A company personnel manager obtains a sample correlation of .57 between applicants' scores on two screening tests. This correlation coefficient is based on $n=103$ observations. Test H_0: $\rho\leqslant .5$ vs. H_1: $\rho>.5$, with $\alpha=.05$. (Assume that the scores are normally distributed.)
15. Test H_0: $\rho=.4$ vs. H_1: $\rho\neq .4$, using the data in Exercise 5. (Take $\alpha=.01$.)
16. Test H_0: $\rho\leqslant .8$ vs. H_1: $\rho>.8$, using the data in Exercise 2. (Take $\alpha=.05$.)

*17. We can also use $Z^*=\frac{1}{2}\log_e[(1+r)/(1-r)]$ to obtain approximate confidence intervals for ρ in *normal* populations. A 95% confidence interval for $\frac{1}{2}\log_e[(1+\rho)/(1-\rho)]$ is given by

$$\frac{1}{2}\log_e\frac{(1+r)}{(1-r)}\pm\frac{1.96}{\sqrt{n-3}}$$

and these end points can be transformed to limits for ρ. The resulting interval is

$$\tanh\left(Z^*-\frac{1.96}{\sqrt{n-3}}\right)<\rho<\tanh\left(Z^*+\frac{1.96}{\sqrt{n-3}}\right)$$

where the tanh is the *hyperbolic tangent function*. Construct a 95% confidence interval for ρ when $r=.7$, based on $n=39$ observations from a normal distribution.

18. Construct an 80% confidence interval for ρ, using the data in Exercise 2.
19. Construct a 90% confidence interval for ρ using the data in Exercise 5.
20. Evaluate the sample lag 1 correlation for the observations $x_1=2$, $x_2=4$, $x_3=3$, $x_4=5$.
21. The following noise measurements were recorded at an intersection in the time order they were observed (courtesy of J. Bollinger): 65, 64, 63, 61, 60, 58, 63, 64, 62, 64, 63, 63, 62, 60, 62, 64, 66, 68, 68, 69.

Investigate the presence of lag 1 dependence by:
(*a*) Plotting the scatter diagram.
(*b*) Calculating the first serial correlation coefficient r_1.
(*c*) Testing the null hypothesis that the population lag 1 correlation ρ_1 is zero.

22. The following consecutive measurements (courtesy of M. Phadke) of the sulfur content of molten metal from a blast furnace were recorded: 2.0, 2.8, 2.0, 1.9, 1.1, 1.4, 1.2, 2.4, 2.5, 2.3, 3.1, 5.2, 2.8, 2.7, 2.6, 2.4, 3.3, 3.9, 3.0, 3.8, 2.9
(*a*) Plot the scatter diagram for the pairs (X_{i-1}, X_i).
(*b*) Calculate the lag 1 correlation coefficient.
(*c*) Test H_0: $\rho_1 = 0$ vs. H_1: $\rho_1 \neq 0$ with $\alpha = .05$.

23. *Correlation coefficient for grouped data*: Often the data for two characteristics X and Y are in the form of a bivariate frequency distribution, as shown in the table below. The values of X are grouped into I cells with midpoints $m_{x1}, \ldots, m_{xI}$ and the values of Y are grouped into J cells with midpoints $m_{y1}, \ldots, m_{yJ}$. The frequency of the (i,j)th cell is denoted by f_{ij}. The structure is similar to that of a joint probability distribution, expect that the cell entries are the frequencies rather than the relative frequencies that are the sample analogs of the probabilities. The marginal totals are $f_{i\cdot} = \sum_{j=1}^{J} f_{ij}$, the frequency of X values in the class with midpoint m_{xi}, and $f_{\cdot j} = \sum_{i=1}^{I} f_{ij}$, the frequency of Y values in the class with midpoint m_{yj}.

Cell midpoints of x	Cell midpoints of y						Total
	m_{y1}	m_{y2}	$\cdots$	m_{yj}	$\cdots$	m_{yJ}	
m_{x1}	f_{11}	f_{12}	$\cdots$	f_{1j}	$\cdots$	f_{1J}	$f_{1\cdot}$
m_{x2}	f_{21}	f_{22}	$\cdots$	f_{2j}	$\cdots$	f_{2J}	$f_{2\cdot}$
$\vdots$	$\vdots$	$\vdots$		$\vdots$		$\vdots$	$\vdots$
m_{xi}	f_{i1}	f_{i2}	$\cdots$	f_{ij}	$\cdots$	f_{iJ}	$f_{i\cdot}$
$\vdots$	$\vdots$	$\vdots$		$\vdots$		$\vdots$	$\vdots$
m_{xI}	f_{I1}	f_{I2}	$\cdots$	f_{Ij}	$\cdots$	f_{IJ}	$f_{I\cdot}$
Total	$f_{\cdot 1}$	$f_{\cdot 2}$	$\cdots$	$f_{\cdot j}$	$\cdots$	$f_{\cdot J}$	n

As in calculating the mean and the variance from grouped data in

Chapter 2, the sample correlation coefficient can be calculated from the bivariate grouped data by using the class midpoints as the values of the observations. Specifically, the sample means are

$$\bar{x} = \frac{\sum_{i=1}^{I} m_{xi} f_{i\cdot}}{n}, \qquad \bar{y} = \frac{\sum_{j=1}^{J} m_{yj} f_{\cdot j}}{n}$$

and the sample variances are

$$s_x^2 = \frac{\sum_{i=1}^{I} m_{xi}^2 f_{i\cdot}}{n} - (\bar{x})^2, \qquad s_y^2 = \frac{\sum_{j=1}^{J} m_{yj}^2 f_{\cdot j}}{n} - (\bar{y})^2$$

Moreover, the *sample covariance* is calculated

$$s_{xy} = \frac{1}{n} \sum_{\text{all cells}} m_{xi} m_{yj} f_{ij} - \bar{x}\bar{y}$$

The sample correlation coefficient for the grouped data is then obtained from

$$r = \frac{s_{xy}}{\sqrt{s_x^2}\sqrt{s_y^2}}$$

Calculate r for the following data, relating smoking to emphysema, derived from autopsy studies (condensed from HEW publication 73-8704). Assume that the midpoint of 2 or more is 3.

Degree of emphysema	Number of packs smoked per day			
	0	0–1	1–2	2 or more
0–2	55	28	26	5
2–5	0	40	188	101
5–9	0	0	7	6

MATHEMATICAL EXERCISES

1. Show that the sample correlation of the transformed observations $ax + b$ and $cy + d$ is the same as the sample correlation of x and y when $a \cdot c > 0$.

2. (*a*) If all points lie on a straight line $y_i = a + bx_i$, $i = 1, \ldots, n$, then show that $r = \pm 1$ where the sign of r is identical to the sign of b.
   (*b*) Conversely, if $r = \pm 1$, then prove that all points must lie on a straight line with slope $b = \pm S_y / S_x$.

   [*Hint for (b)*: Using

$$\left[b(x_i - \bar{x}) - (y_i - \bar{y}) \right]^2 = b^2 (x_i - \bar{x})^2 - 2b(x_i - \bar{x})(y_i - \bar{y}) + (y_i - \bar{y})^2$$

and summing each side over $i = 1, \ldots, n$, show that

$$\sum_{i=1}^{n} \left[b(x_i - \bar{x}) - (y_i - \bar{y}) \right]^2 = b^2 S_x^2 - 2brS_x S_y + S_y^2$$

$$= (bS_x - rS_y)^2 + S_y^2 (1 - r^2)$$

For the case $r = +1$, the right-hand side of this equation is zero when $b = S_y / S_x$. This implies that all points (x_i, y_i) satisfy the linear relation $y_i - \bar{y} = b(x_i - \bar{x})$ with $b = S_y / S_x$. The argument is similar for the case $r = -1$.]

CHAPTER 13

Analysis of Categorized Data

13.1 INTRODUCTION

Frequently, the sample information collected by an investigator will have the primitive structure of *categorized data*. This term refers to observations that are only classified into categories so that the data set can consist of frequency counts for the categories. Such data occur abundantly in almost all fields of quantitative study, particularly in the social sciences. In a study of religious affiliations, people may be classified as Catholic, Protestant, Jewish, or other; in a survey of job compatibility, employed persons may be classified as being satisfied, neutral, or dissatisfied with their jobs; in plant breeding, the offsprings of a cross-fertilization may be grouped into several genotypes; manufactured items may be partitioned into such categories as "free of defects," "slightly blemished," and "rejects." In all of these examples each category is defined by a qualitative trait. Categories can also be defined by arbitrarily specifying ranges of values on an original numerical measurement scale, such as income that is categorized high, medium, or low, and rainfall that is classified heavy, moderate, or light. Developing this idea even further, a frequency distribution of measurements grouped into class intervals can also be visualized as a form of categorized data in which the class intervals define the categories. Although this last form of data is traditionally labeled "numerical data" or "measurement data," certain inferential procedures designed for categorized data can still be applied to it.

The simplest form of categorized data contains only two categories that are defined as possessing or lacking a particular trait. The role binomial distributions play in modeling these dichotomous data has already been examined in Chapter 5, and the associated inference procedures have been treated in subsequent chapters. Our aim here is to present some procedures that can be used to study data that are grouped into multiple categories. A few examples will help to illustrate the nature of inference and to prepare for our forthcoming analysis.

Example 13.1 Consumer preference: A name-brand household washing machine is sold in five different colors, and a market researcher wishes to study the popularity of the various colors. The frequencies given in Table 13.1 are observed from a random sample of 300 recent sales. As an intial step, the researcher may want to test the hypothesis that all five colors are equally popular.

TABLE 13.1
SALES DATA ON FIVE COLORS OF A NAME-BRAND WASHER

AVOCADO	TAN	RED	BLUE	WHITE	TOTAL
88	65	52	40	55	300

■

Example 13.2 Genetic model: The offspring produced by a cross between two given types of plants can be any of the three genotypes denoted by A, B, and C. A theoretical model of gene inheritance suggests that the offspring of types A, B, and C should be in the ratio 1 : 2 : 1. For experimental verification, 90 plants are bred by crossing the two given types. Their genetic classifications are recorded in Table 13.2. Do these data support or contradict the genetic model?

TABLE 13.2
CLASSIFICATION OF CROSSBRED PLANTS

GENOTYPES			
A	B	C	TOTAL
18	44	28	90

■

Example 13.3 Poisson distribution for insurance claims: An actuarial scientist wishes to study the frequency of insurance claims for in-hospital medical care among families with two children in which both the husband and wife are under 50 years of age. The scientist samples 200 such families and records the frequency distribution of the number of these claims over a four-year period. From the data set given in Table 13.3, does a Poisson distribution appear to adequately describe this data?

TABLE 13.3
FREQUENCY DISTRIBUTION OF THE NUMBER OF CLAIMS FOR IN-HOSPITAL MEDICAL CARE

Number of claims	0	1	2	3	4	5	6	7	Total
Frequency	22	53	58	39	20	5	2	1	200

■

13.2 THE MULTINOMIAL MODEL

To construct a probability model that can accommodate the types of data illustrated in the preceding examples, we must extend the model of Bernoulli trials so that outcomes can be classified in a number of categories. For an obvious reason, such trials are called *multinomial trials*.

> *Multinomial trials*
>
> (*a*) The outcome of each trial belongs to one of k mutually exclusive categories or cells, labeled $1, 2, \ldots, k$.
>
> (*b*) The probability that a trial will result in the ith cell is denoted by p_i, $i = 1, \ldots, k$, with $\Sigma_{i=1}^{k} p_i = 1$. These cell probabilities remain the same for all trials.
>
> (*c*) The trials are independent.

For the special case $k = 2$, this formal definition of multinomial trials coincides with the definition of Bernoulli trials. The conditions defining multinomial trials are closely satisfied in many experimental situations in which trials are repeated independently under identical conditions. When random sampling from a population that consists of elements in several categories, the conditions are fulfilled if the sampling is made with replacement and they are nearly satisfied if only a small fraction of the population is sampled without replacement.

For a series of n repetitions of a multinomial trial, the frequencies of the cells $1, \ldots, k$ are denoted by $n_1, \ldots, n_k$, respectively. Before the trials are actually performed, these are random variables that can assume nonnegative integer values, and their sum must satisfy $n_1 + \ldots + n_k = n$. This basic structure is presented in Table 13.4.

TABLE 13.4
STRUCTURE OF MULTINOMIAL DATA

Cells	1	2	...	k	Total
Probabilities	p_1	p_2	...	p_k	1
Frequencies in n trials	n_1	n_2	...	n_k	n

> *Multinomial distribution*
>
> The joint probability distribution of the cell frequencies $n_1, \ldots, n_k$ in n multinomial trials is called a *multinomial distribution with*

parameters $p_1,\ldots,p_k$, which represent the respective cell probabilities. The probability function is

$$f(n_1,\ldots,n_k)=\frac{n!}{n_1!\ldots n_k!}p_1^{n_1}\ldots p_k^{n_k}$$

$$\text{for} \quad \sum_{i=1}^{k} n_i = n$$

$$\text{The parameters satisfy} \quad \sum_{i=1}^{k} p_i = 1$$

The mathematical formula for the multinomial probability function will be of little importance in our exposition. However, a few properties of the distribution deserve attention here. First, the k variables $n_1,\ldots,n_k$ are constrained by $\Sigma n_i = n$, so that once the values of any $k-1$ of these variables are fixed, the value of the remaining variable is automatically determined. This feature can be expressed by stating that the degrees of freedom associated with $n_1,\ldots,n_k$ are $k-1$. The same statement applies to the set of parameters $p_1,\ldots,p_k$, where the constraint is $\Sigma p_i = 1$. Second, if we designate the first cell "success" and lump the remaining cells into a single cell designated "failure," the random variable n_1 can be recognized as the number of successes in n Bernoulli trials whose distribution is $b(n,p_1)$. Consequently, $E(n_1)=np_1$, $\text{Var}(n_1)=np_1(1-p_1)$. The same argument applied to each individual cell gives the results

$$E(n_i)=np_i, \qquad \text{Var}(n_i)=np_i(1-p_i), \quad i=1,\ldots,k$$

The first result

$$\text{Expected cell frequency} = n(\text{cell probability})$$

will play a major role in the development of a test for goodness of fit to be discussed in the next section.

Because the sum of all the n_i is fixed, it is not surprising to find that they have a negative covariance. Its specific form can be shown to be

$$\text{Cov}(n_i,n_j) = -np_ip_j, \quad i \neq j$$

Examples of tests based on the multinomial structure are given in Section 13.3. This structure provides the foundation for all the statistical reasoning presented in this chapter.

13.3 PEARSON'S TEST FOR GOODNESS OF FIT

We now return to our discussion of the type of problems illustrated in Examples 13.1, 13.2, and 13.3, in which the data constitute a realization of the multinomial cell frequencies and the null hypotheses specifies a structure for the unknown cell probabilities. Our primary goal is to test if the model given by the null hypothesis fits the data, and this is appropriately called *testing for goodness of fit*. Insofar as the specification of a model may or may not involve unknown parameters, we distinguish two cases in our discussion:

13.3.1 Case A: Cell Probabilities Completely Specified by H_0

A null hypothesis that completely specifies the cell probabilities is of the form

$$H_0: p_1 = p_{10}, \ldots, p_k = p_{k0}$$

where $p_{10}, \ldots, p_{k0}$ are given numerical values that satisfy $p_{10} + \ldots + p_{k0} = 1$. Referring to Example 13.1, the market researcher's null hypothesis that all five colors are equally popular can be translated into $H_0: p_1 = \frac{1}{5}, \ldots, p_5 = \frac{1}{5}$. In Example 13.2 the null hypothesis assumes the form $H_0: p_1 = \frac{1}{4}$, $p_2 = \frac{2}{4}$, $p_3 = \frac{1}{4}$.

Once the cell probabilities are specified, the expected cell frequencies can be readily computed by *multiplying* these probabilities *by* the sample size n. A goodness of fit test attempts to determine if a conspicuous discrepancy exists between the observed cell frequencies and those expected under H_0. A useful measure for the overall discrepancy is given by

$$\chi^2 = \sum_{i=1}^{k} \frac{(n_i - np_{i0})^2}{np_{i0}} = \sum_{\text{cells}} \frac{(O-E)^2}{E}$$

where O and E symbolize an observed frequency and the corresponding expected frequency. The discrepancy in each cell is measured by the squared difference between the observed and the expected frequencies divided by the expected frequency. The χ^2 measure is the sum of these quantities for all cells.

The χ^2 statistic was originally proposed by Karl Pearson (1857–1936), who found the distribution for large n to be approximately a χ^2 distribution with d.f. $= k-1$. Due to this distribution, the statistic is denoted by χ^2 and is called *Pearson's χ^2 statistic for goodness of fit*. Because a large value of the overall discrepancy indicates a disagreement between the data and the hypothesis, the upper tail of the χ^2 distribution constitutes the rejection region.

Pearson's χ^2 test for goodness of fit

Null hypothesis: $H_0: p_1 = p_{10}, \ldots, p_k = p_{k0}$

Test statistic:
$$\chi^2 = \sum_{i=1}^{k} \frac{(n_i - np_{i0})^2}{np_{i0}} = \sum_{\text{cells}} \frac{(O-E)^2}{E}$$

Distribution: When n is so large that no expected cell frequency is too small, the test statistic has an approximate χ^2 distribution with d.f. $= k-1 =$ No. of cells -1.

Rejection region: $\chi^2 \geqslant \chi^2_\alpha$, where χ^2_α is the upper α point of the χ^2 distribution with d.f. $= k-1$.

It should be remembered that Pearson's χ^2 test for goodness of fit is an approximate test that is valid only for large samples. As a rule of thumb, n should be large enough so that the expected frequency of each cell is at least 5. Some statisticians suggest that the expected frequency may be relaxed to about 1 at the end cell of a unimodal distribution, such as a binomial, Poisson, or normal distribution.

Example 13.4 From the data given in Example 13.1, test the null hypothesis at $\alpha = .05$ that all five colors of washers are equally popular.

Here the null hypothesis is $H_0: p_1 = p_2 = \ldots = p_5 = .2$. The computations for the test statistic appear in Table 13.5, where the expected frequencies are computed by multiplying the cell probabilities by $n = 300$.

The upper 5% point of the χ^2 distribution with d.f. $= 4$ is 9.487. Because the observed χ^2 value is much larger than this, the null hypothesis is rejected at $\alpha = .05$. Consequently, a significant discrepancy appears to exist

TABLE 13.5

χ^2 GOODNESS OF FIT TEST FOR THE DATA IN TABLE 13.1

CELLS	AVOCADO	TAN	RED	BLUE	WHITE	TOTAL
Observed frequency O	88	65	52	40	55	300
Probability under H_0	.2	.2	.2	.2	.2	1.0
Expected frequency E	60	60	60	60	60	300
$\frac{(O-E)^2}{E}$	$\frac{(28)^2}{60}$	$\frac{5^2}{60}$	$\frac{8^2}{60}$	$\frac{(20)^2}{60}$	$\frac{5^2}{60}$	
	$= 13.067$	$= .417$	$= 1.067$	$= 6.667$	$= .417$	$21.635 = \chi^2$ d.f. $= 4$

between the observations and the hypothesis of equal popularity of all colors. In fact, the evidence is strong enough to reject H_0 with $\alpha = .01$.

To further understand the nature of the departure from the model of equal cell probabilities, we consider the contributions of the individual cells to the χ^2 value in the last row of Table 13.5. The greatest contribution to the χ^2 statistic is made by Avocado, for which the difference $O - E$ is positive; the other substantial contribution to this test statistic is made by Blue, whose sales are lower than would be predicted. ■

13.3.2 Case B: Cell Probabilities Not Completely Specified by H_0

Computing the expected cell frequencies often requires a knowledge of some parameters that are not specified by the null hypothesis. Example 13.3 illustrates such a situation; there we must know the value of m to obtain probabilities from the Poisson table. Similarly, in testing if the data are consonant with a normal distribution, the values of μ and σ are required to compute the probabilities. The rule in such cases is first to estimate the unknown parameters from the observed frequency table and then to use the estimated values, as if they were the correct values, to determine the cell probabilities. The expected frequencies and the χ^2 statistic are then computed in the manner discussed in Case A, Section 13.3.1, but the number of degrees of freedom for χ^2 is reduced by the number of estimated parameters.

Before computing the χ^2 statistic, we should examine the table to see if any of the expected cell frequencies are too small. Because the validity of our goodness of fit test precludes this situation, adjacent cells may have to be combined until the expected frequency in each is at least 5. The χ^2 statistic is computed from the appropriately modified cells, and the number of degrees of freedom is given by

$$\text{d.f.} = (\text{No. of cells}) - 1 - (\text{No. of estimated parameters})$$

Example 13.5 Test if a Poisson model adequately describes the data given in Table 13.3.

Here the null hypothesis of fit by a Poisson distribution leaves the parameter m unspecified. Recalling that m represents the mean of a Poisson distribution, we can estimate it by the sample mean computed from Table 13.3, or

$$\bar{x} = \frac{\sum(\text{Value} \times \text{Frequency})}{n}$$

$$= \frac{[(0\times 22) + (1\times 53) + \ldots + (7\times 1)]}{200}$$

$$= \frac{410}{200} = 2.05 \approx 2.0$$

Consulting Appendix Table 3 with the estimated value 2 for m, we obtain the cell probabilities, which we multiply by $n=200$ to obtain the expected frequencies. These values are presented in Table 13.6 with the complete computations for χ^2. Since the sum of the last two expected frequencies is smaller than 5, these cells are combined with the preceding cell. Taking $\alpha=.05$, the upper 5% point of χ^2 with d.f. $=4$ is 9.487, which is larger than the observed value 2.33. Therefore the null hypothesis is not rejected at $\alpha=.05$, and we can conclude that the Poisson model does not seem to contradict the data.* ■

TABLE 13.6

χ^2 GOODNESS OF FIT OF A POISSON MODEL FOR THE DATA IN TABLE 13.3

NO. OF CLAIMS	0	1	2	3	4	5	6	7	TOTAL
Observed frequency O	22	53	58	39	20	5	2	1	200
							8		
Poisson probability with $m=2$	.135	.271	.271	.180	.090	.036	.012	.005	1.000
Expected frequency E	27.0	54.2	54.2	36.0	18.0	7.2	2.4	1.0	200
							10.6		
$\frac{(O-E)^2}{E}$	.926	.027	.266	.250	.222		.638		$2.33=\chi^2$
									d.f. $=6-1-1$ $=4$

In any goodness-of-fit application it is recommended that the parameters be estimated from the grouped data by one of two special methods: the *method of maximum likelihood* or the *method of minimum chi-square*. Estimations made from ungrouped data may alter the accuracy of the χ^2 approximation somewhat. It is also suggested that, insofar as it is possible, the cells be selected to have approximately equal probabilities.

In testing to determine if a data set appears to have come from a normal population, it may be advantageous to consider an alternative procedure devised by Shapiro and Wilk [1].

In conclusion, all tests for goodness of fit suffer from the serious drawback of being unlikely to detect even violent departures from the assumed distribution when the sample size is small. Many different models

*Another approximate test is based on the fact that the variance of the Poisson distribution is equal to the mean. If $s^2/\bar{x}$ differs greatly from 1 by being $<\chi^2_{.975}/(n-1)$ or $>\chi^2_{.025}/(n-1)$, the tentative hypothesis of a Poisson distribution is rejected. This approximate rejection region is based on the χ^2 statistic with d.f. $=n-1$.

will be consonant with the data. On the other hand, small and unimportant departures from the assumed distribution often result in a rejection of the model when the sample size is very large. Agreement with the null hypothesis does *not* imply that the proposed model is correct but merely that it is one of several models consonant with the data.

13.4 CONTINGENCY TABLES

When two or more traits are observed for each sample element, the data can be simultaneously classified with respect to the levels of occurrence of each of these traits. To cite a few examples: a sample of employed persons may be classified according to educational attainment and type of occupation; college students may be classified according to a year in college and attitude toward a dormitory regulation; flowering plants may be classified with respect to both type of foliage and size of flower; a sales record may consist of the brand of product sold and the store location. Frequency data arising from the simultaneous classification of more than one characteristic are called *contingency tables* in statistical language, but they are more popularly known as *cross-classified* or *cross-tabulated data*.

A typical inferential aspect of cross-tabulation is the study of whether particular characteristics appear to be manifested independently or whether certain levels of one characteristic tend to be associated or contingent with some levels of another. For simplicity here, we concentrate on the simultaneous classification of two characteristics, which produces a two-way frequency table.

Example 13.6 A random sample of 500 persons is questioned regarding political affiliation and attitude toward an energy-rationing program. From the observed frequency table given in Table 13.7, we wish to find answers to the following questions.

(*a*) Do the data indicate that the pattern of opinion is independent of political affiliation?
(*b*) Can we quantitatively measure the strength of association between the two characteristics?

TABLE 13.7
CONTINGENCY TABLE FOR POLITICAL AFFILIATION AND OPINION

	FAVOR	INDIFFERENT	OPPOSED	TOTAL
Democrat	138	83	64	285
Republican	64	67	84	215
Total	202	150	148	500

Before developing any formal statistical analysis, we consider the table from a descriptive viewpoint. To do this, we perform the fundamental operation of reducing the counts to proportions. It is instructive to compute the fraction of observations for a given row that fall into each category. Dividing each element in a row by the row total (for instance, $138/285 = .48$) we obtain Table 13.7*a*.

TABLE 13.7*a*
PROPORTIONS BY ROWS

	FAVOR	INDIFFERENT	OPPOSED	TOTAL
Democrat	.48	.29	.22	.99*
Republican	.30	.31	.39	1.00

*Rounding error of .01.

A similar calculation could be performed using the column totals, but the proportions in this direction are not of interest in this particular example. Another set of proportions is created if we divide each cell entry by the total number of observations. This operation produces Table 13.7*b*.

TABLE 13.7*b*
PROPORTION OF OBSERVATIONS IN EACH CELL

	FAVOR	INDIFFERENT	OPPOSED	TOTAL
Democrat	.276	.166	.128	.570
Republican	.128	.134	.168	.430
Total	.404	.300	.296	1

A visual inspection of these tables reveals apparent differences in the distributions across rows or columns. If the distribution seems to vary with rows, we suspect the presence of association. The first row in Table 13.7*a* decreases, whereas the second increases. A purview of the proportion of total observations in each cell should be focused on cells with a heavy concentration of observations and also on any cells with a very small fraction of observations. All the cells in Table 13.7*b* contain moderate proportions, with the Democrat-Favor cell being the largest. Any apparent association should be confirmed by a statistical test. ■

For a general treatment of the test of independence in a contingency table, let the two characteristics to be studied be designated by A and B and suppose there are r categories $A_1,\ldots,A_r$ for A and c categories $B_1,\ldots,B_c$ for B. By arranging the A categories in rows and the B categories in columns, we can create a two-way table in which each cell is the intersection of an A category and a B category. A random sample of n

elements classified into these cells will produce the frequency table given in Table 13.8, where

n_{ij} = frequency of A_iB_j

n_{i0} = ith row total, or frequency of A_i

n_{0j} = jth column total, or frequency of B_j

Table 13.8 is an $r \times c$ contingency table. (A 2×3 contingency table has already been illustrated in Example 13.6.)

TABLE 13.8
AN $r \times c$ CONTINGENCY TABLE

	B_1	B_2	$\cdots$	B_c	ROW TOTAL
A_1	n_{11}	n_{12}	$\cdots$	n_{1c}	n_{10}
A_2	n_{21}	n_{22}	$\cdots$	n_{2c}	n_{20}
$\vdots$	$\vdots$	$\vdots$		$\vdots$	$\vdots$
A_r	n_{r1}	n_{r2}	$\cdots$	n_{rc}	n_{r0}
Column total	n_{01}	n_{02}	$\cdots$	n_{0c}	n

TABLE 13.9
CELL PROBABILITIES

	B_1	B_2	$\cdots$	B_c	ROW TOTAL
A_1	p_{11}	p_{12}	$\cdots$	p_{1c}	p_{10}
A_2	p_{21}	p_{22}	$\cdots$	p_{2c}	p_{20}
$\vdots$	$\vdots$	$\vdots$		$\vdots$	$\vdots$
A_r	p_{r1}	p_{r2}	$\cdots$	p_{rc}	p_{r0}
Column total	p_{01}	p_{02}	$\cdots$	p_{0c}	1

Imagining a classification of the entire population, the unknown population proportions (i.e., the probabilities of the cells) are represented by the entires in Table 13.9, where

$p_{ij} = P(A_iB_j)$, probability of the joint occurrence of A_i and B_j

$p_{i0} = P(A_i)$, total probability in the ith row

$p_{0j} = P(B_j)$, total probability in the jth column

We are concerned with testing the null hypothesis that the A and B classifications are independent. Recall from Chapter 3 that the probability of the intersection of independent events is the product of their probabilities. Therefore, the null hypothesis stating that the events $A_1, \ldots, A_r$ are independent of events $B_1, \ldots, B_c$ can be rephrased $P(A_i B_j) = P(A_i)P(B_j)$ for all $i = 1, \ldots, r$ and $j = 1, \ldots, c$.

The null hypothesis of independence

$$H_0: p_{ij} = p_{i0} p_{0j}, \quad \text{for all cells } (i,j)$$

The problem now becomes testing the goodness of fit for the model of independence, and the concepts outlined in Section 13.3 can be invoked to delineate a test procedure. The model specifies the cell probabilities in terms of the marginal probabilities, which are unknown parameters. Because $p_{i0} = P(A_i)$, a natural estimator is the sample relative frequency of A_i

$$\hat{p}_{i0} = \frac{n_{i0}}{n}$$

Similarly, p_{0j} is estimated by

$$\hat{p}_{0j} = \frac{n_{0j}}{n}$$

Incidentally, these estimates are also obtained by using the method of maximum likelihood and can therefore be employed in the development of the χ^2 test. Using these estimates, the probability in the (i,j)th cell is estimated to be

$$\hat{p}_{ij} = \hat{p}_{i0} \hat{p}_{0j} = \frac{n_{i0} \times n_{0j}}{n^2}$$

Therefore, the estimated expected frequency under the model of independence is

$$E_{ij} = n\hat{p}_{ij} = \frac{n_{i0} n_{0j}}{n}$$

The test statistic then becomes

$$\chi^2 = \sum_{\substack{\text{all} \\ rc \text{ cells}}} \frac{(n_{ij} - E_{ij})^2}{E_{ij}}, \quad E_{ij} = \frac{n_{i0} n_{0j}}{n}$$

which has an approximate χ^2 distribution for large n. The number of degrees of freedom is again determined by d.f. = No. of cells − 1 − No. of estimated parameters. Because the total row probabilities $p_{10},\ldots,p_{r0}$ are bound by the constraint $\Sigma p_{i0} = 1$, there are really $r-1$ parameters among them. Similarly, there are $c-1$ parameters among the set of column probabilities $p_{01},\ldots,p_{0c}$. Therefore, the number of estimated parameters is $(r-1)+(c-1) = r+c-2$, and consequently

$$\begin{aligned}\text{d.f. for } \chi^2 &= rc - 1 - (r+c-2)\\ &= rc - r - c + 1 = (r-1)(c-1)\end{aligned}$$

> The number of degrees of freedom for χ^2 in an $r \times c$ contingency table is
>
> $$(r-1)(c-1)$$

The upper tail of the χ^2 distribution with d.f. $=(r-1)(c-1)$ is used as the rejection region. Once again, it is a large sample test, and its validity seems to require the expected frequency of each cell to be at least 5.

Example 13.7 With the data given in Example 13.6, test if the pattern of attitude is independent of political affiliation.

The contingency table is reproduced in Table 13.10, where the computed expected frequencies are presented in parentheses under the observed frequencies. For instance

$$E_{11} = \frac{n_{10}n_{01}}{n} = \frac{285 \times 202}{500} = 115.14$$

$$E_{23} = \frac{n_{20}n_{03}}{n} = \frac{215 \times 148}{500} = 63.64$$

TABLE 13.10

THE CONTINGENCY TABLE FOR EXAMPLE 13.6

	FAVOR	INDIFFERENT	OPPOSED	TOTAL
Democrat	138 (115.14)	83 (85.50)	64 (84.36)	285
Republican	64 (86.86)	67 (64.50)	84 (63.64)	215
Total	202	150	148	500

The χ^2 statistic has the observed value

$$\chi^2 = \frac{(138-115.14)^2}{115.14} + \frac{(83-85.50)^2}{85.50} + \frac{(64-84.36)^2}{84.36}$$
$$+ \frac{(64-86.86)^2}{86.86} + \frac{(67-64.50)^2}{64.50} + \frac{(84-63.64)^2}{63.64}$$
$$= 4.539 + .073 + 4.914 + 6.016 + .097 + 6.514$$
$$= 22.153$$

with d.f. $=(2-1)(3-1)=2$. Using the level of significance $\alpha = .05$, the tabulated upper 5% point of χ^2 is 5.991. Because the observed χ^2 is larger than the tabulated value, the null hypothesis of independence is rejected at $\alpha = .05$. In fact, it would be rejected even for $\alpha < .005$.

Since the null hypothesis has been rejected, we must continue our investigation. The contributions of individual cells to the χ^2 value and the nature of the differences between the observed and expected frequencies are summarized in Table 13.11.

TABLE 13.11

CALCULATIONS FOR INVESTIGATING A SIGNIFICANT χ^2 VALUE

	$O-E$			$(O-E)^2/E$		
	FAVOR	INDIFFERENT	OPPOSED	FAVOR	INDIFFERENT	OPPOSED
Democrat	22.86	−2.5	−20.36	4.54	.07	4.91
Republican	−22.86	2.5	20.36	6.02	.10	6.51

The large contribution from the corner cells in conjunction with the signs of the corresponding differences substantiates a decreasing trend from left to right for Democrats and an increasing trend for Republicans. ■

Undoubtedly, the tedium of computation rapidly increases with the number of rows and columns in a contingency table. Because standard computer programs are available that can perform the job almost instantaneously, it is important to concentrate on the concepts developed here instead of becoming preoccupied with lengthy calculations.

13.4.1 Spurious Dependence

When the χ^2 test leads to a rejection of the null hypothesis of independence, we conclude that the data provide evidence of a statistical association between the two characteristics. However, we must refrain from making the hasty interpretation that these characteristics are directly related. A claim of causal relationship must be confirmed by common sense, which statistical evidence must not be allowed to supersede.

Two characteristics may appear to be strongly related due to the common influence of a third factor that is not included in the study. In such cases, the dependence is called a *nonsense* or a *spurious dependence*. For instance, if a sample of individuals are classified in a 2×2 contingency table according to whether or not they are heavy drinkers and whether or not they suffer from respiratory trouble, we would probably find a high value for χ^2 and would conclude that a strong association exists between these two characteristics. But the reason for the association may be that most heavy drinkers are also heavy smokers and the smoking habit is a direct cause of respiratory trouble. In this context, we should recall a similar warning given in Chapter 12 regarding the interpretation of a correlation coefficient between two sets of measurements.

13.4.2 Measures of Association in a Contingency Table

We may wish to measure the strength of association on a numerical scale in much the same way that we employ the coefficient of correlation as an index of association between two sets of measurements. Insofar as the χ^2 statistic represents an overall deviation from the model of independence, it is intuitively reasonable to use this statistic to gauge the strength of this association. One difficulty in employing the χ^2 statistic as a numerical index of association is that the number of degrees of freedom attached to this statistic depends on the dimensionality of the contingency table. A χ^2 value of 20.7 in a 2×2 contingency table would reflect a significant association, but this would not be so in a 5×9 table. Several measures have been proposed in the literature that essentially purport to adjust the χ^2 statistic to a common scale that is irrespective of the dimensionality of the contingency table.

In the following formulas large values of a measure indicate a strong association, q represents the number of rows or columns, whichever is *smaller*, and n indicates sample size.

Cramér's contingency coefficient:

$$Q_1=\frac{\chi^2}{n(q-1)}, \quad 0\leqslant Q_1\leqslant 1$$

Pearson's coefficient of mean square contingency:

$$Q_2=\sqrt{\frac{\chi^2}{n+\chi^2}}, \quad 0\leqslant Q_2\leqslant\sqrt{\frac{q-1}{q}}$$

Pearson's phi coefficient in 2×2 *table*:

$$\phi = \frac{(n_{11}n_{22} - n_{12}n_{21})}{\sqrt{n_{10}n_{20}n_{01}n_{02}}}, \qquad -1 \leqslant \phi \leqslant 1$$

These measures are basically variants of one another, and the use of one in preference to another is largely a matter of tradition in individual fields. Large values of these measures are certainly indicative of a strong association, but a general interpretation is lacking for small or intermediate values. This is because a large sample size n tends to make a measure small even though the χ^2 value may be highly significant. A few improved measures have been developed to adjust the χ^2 statistic, but their formulas are more involved. Interested readers are referred to Kruskal [2], who reviews several more advanced methods. The better measures require a sufficient structure to allow the categories for each characteristic to be ordered. Once this order is achieved, a direction can be given to any association. The direction is designated positive if the cells with joint high and joint low categories exhibit high proportions and negative if they exhibit low proportions.

13.5 CONTINGENCY TABLES WITH ONE MARGIN FIXED (TEST OF HOMOGENEITY)

The χ^2 test of independence discussed in Section 13.4 is based on the sampling scheme in which a single random sample of size n is classified with respect to two characteristics simultaneously. In the resulting contingency table, both sets of marginal total frequencies are random variables. An alternative sampling scheme involves a division of the population into subpopulations or *strata* according to the categories of one characteristic. A random sample of a predetermined size is drawn from each stratum and classified into categories of the other characteristic. For instance, in the context of Example 13.6, we may draw independent random samples of sizes 200 and 300 from the populations Democrat and Republican, respectively, and classify these samples according to attitude. Although the resulting frequency table is still called a contingency table, it differs markedly from the contingency table previously discussed in that one set of marginal totals now represents the prespecified sample sizes rather than the outcomes of random sampling. Our current intent is to study the proportions in these individual categories to determine if they are approximately the same for all the different populations; that is, we wish to test whether the populations are homogeneous with respect to the cell probabilities. In discussing procedural details, we will learn that the same

χ^2 statistic with the same number of degrees of freedom presented in Section 13.4 is also useful in testing the null hypothesis of homogeneity.

Suppose that independent random samples of sizes $n_{10},\ldots,n_{r0}$ are taken from r populations $A_1,\ldots,A_r$, respectively. Classifying each sample into the categories $B_1,\ldots,B_c$, we obtain the $r\times c$ contingency table shown in Table 13.12, where the row totals are the fixed sample sizes. The probabilities of the various B categories within each population appear in Table 13.13, where each w denotes a conditional probability.

$$w_{ij} = P(B_j|A_i) = \text{the probability of } B_j \text{ within the population } A_i$$

Note that the total probability in each row of Table 13.13 is 1, but that column totals have little meaning.

TABLE 13.12
AN $r\times c$ TABLE WITH FIXED ROW TOTALS

	B_1	B_2	$\cdots$	B_c	TOTAL
A_1	n_{11}	n_{12}	$\cdots$	n_{1c}	n_{10}
$\vdots$	$\vdots$	$\vdots$		$\vdots$	$\vdots$
A_r	n_{r1}	n_{r2}	$\cdots$	n_{rc}	n_{r0}
Total	n_{01}	n_{02}	$\cdots$	n_{0c}	n

TABLE 13.13
PROBABILITIES OF B CATEGORIES WITHIN EACH POPULATION

	B_1	B_2	$\cdots$	B_c	TOTAL
A_1	w_{11}	w_{12}	$\cdots$	w_{1c}	1
$\vdots$	$\vdots$	$\vdots$		$\vdots$	$\vdots$
A_r	w_{r1}	w_{r2}	$\cdots$	w_{rc}	1

The null hypothesis that the probability of each B category is the same for all r populations can now be formally stated as:

Null hypothesis of homogeneity

$$H_0: w_{1j} = w_{2j} = \ldots = w_{rj} \quad \text{for every } j = 1,\ldots,c.$$

Under H_0, the common probability of the category B_j can be estimated from the pooled samples by observing that out of a total of n elements sampled, n_{0j} possess the category B_j. The estimated probability is

$$\hat{w}_{1j} = \hat{w}_{2j} = \dots = \hat{w}_{rj} = \frac{n_{0j}}{n}$$

and the estimated expected frequency in the (i,j)th cell under H_0 is

$$E_{ij} = (\text{No. of } A_i \text{ sampled}) \times (\text{Estimated probability of } B_j \text{ within } A_i)$$

$$= n_{i0}\hat{w}_{ij} = \frac{n_{i0}n_{0j}}{n}$$

The test statistic is given by

$$\chi^2 = \sum_{\text{cells}} \frac{(n_{ij} - E_{ij})^2}{E_{ij}}$$

Each of the r populations has c cells and contributes $(c-1)$ degrees of freedom to the χ^2 statistic. The total $r(c-1)$ is reduced by the number of estimated parameters. Because we have estimated c column probabilities that sum to 1, the number of estimated parameters is $(c-1)$. Consequently

$$\text{d.f. of } \chi^2 = r(c-1) - (c-1) = (r-1)(c-1)$$

In comparison with the development in Section 13.4, the formulas for χ^2 as well as the degrees of freedom can be seen to be the same. Only the sampling method and the formalization of the null hypothesis are different for the two situations.

Example 13.8 A survey is undertaken to determine the incidence of alcoholism in different professional groups. Random samples of the clergy, educators, executives, and merchants are interviewed, and the data given in Table 13.14 are obtained.

Construct a test to determine if the incidence rate of alcoholism appears to be the same in all four groups.

TABLE 13.14
CONTINGENCY TABLE OF ALCOHOLISM VS. PROFESSION

	ALCOHOLIC	NONALCOHOLIC	SAMPLE SIZE
Clergy	32 (58.25)	268 (241.75)	300
Educators	51 (48.54)	199 (201.46)	250
Executives	67 (58.25)	233 (241.75)	300
Merchants	83 (67.96)	267 (282.04)	350
Total	233	967	1200

Departing from the notations used in our general discussion, we denote the proportions of alcoholics in the populations of the clergy, educators, executives, and merchants by p_1, p_2, p_3, and p_4, respectively. Based on independent random samples from four binomial populations, we want to test the null hypothesis

$$H_0: p_1 = p_2 = p_3 = p_4$$

The expected cell frequencies, shown in parentheses in Table 13.14, are computed by multiplying the row and column totals and dividing this total by 1200. The computed value of the χ^2 statistic is

$$\chi^2 = \frac{(32-58.25)^2}{58.25} + \frac{(268-241.75)^2}{241.75} + \dots + \frac{(267-282.04)^2}{282.04}$$

$$= 20.59$$

$$\text{d.f. of } \chi^2 = (4-1)(2-1) = 3$$

With d.f. = 3, the tabulated upper 5% point of χ^2 is 7.815 so that the null hypothesis is rejected at $\alpha = .05$. It would also be rejected at $\alpha < .005$.

The analysis of a contingency table must *never* conclude with the calculation of the χ^2 value and a rejection of the null hypothesis. The matter needs to be pursued further and the differences between the observed and the expected counts must be examined to determine if any pattern exists that could indicate a systematic departure from the null hypothesis. Although the hand computations required for larger contingency tables are tedious, computers easily supply such calculations as the ones given in Table 13.15 for this example.

TABLE 13.15

CALCULATIONS FOR ASSESSING THE NATURE OF SIGNIFICANT χ^2

	$(O-E)$		$(O-E)^2/E$	
	ALCOHOLIC	NONALCOHOLIC	ALCOHOLIC	NONALCOHOLIC
Clergy	−26.25	26.25	11.83	2.85
Educators	2.46	−2.46	.12	.03
Executives	8.75	−8.75	1.31	.32
Merchants	15.04	−15.04	3.33	.80

An examination of both parts of Table 13.15 shows that the major departure from homogeneity is caused by the relatively small fraction of alcoholics among the clergy.

The signs of the differences that make the largest contribution to the χ^2 value indicate the nature of the departures from homogeneity. ■

A graphic display of the contingency table often helps the analyst to portray the data in a more vivid context than the numbers in a contingency table provide. The data given in Example 13.8, which involves samples from four dichotomous populations, can be readily plotted in a graph. To this end, we compute the sample proportions of alcoholics $\hat{p}$ for each group and construct a large-sample 99% confidence interval for the corresponding population proportion p by the formulas $\hat{p} \pm 2.58\sqrt{\hat{p}(1-\hat{p})/n}$. The results are given in Table 13.16 and are plotted in Figure 13.1, where each $\hat{p}$ is located by a dot on the vertical bar representing the confidence interval.

TABLE 13.16

OBSERVED PROPORTIONS OF ALCOHOLICS AND CONFIDENCE INTERVALS FOR THE POPULATION PROPORTIONS

PROFESSION	SAMPLE SIZE	NUMBER OF ALCOHOLICS	$\hat{p}$	99% CONFIDENCE INTERVAL
Clergy	300	32	.107	(.06, .15)
Educators	250	51	.204	(.14, .27)
Executives	300	67	.223	(.16, .29)
Merchants	350	83	.237	(.18, .30)

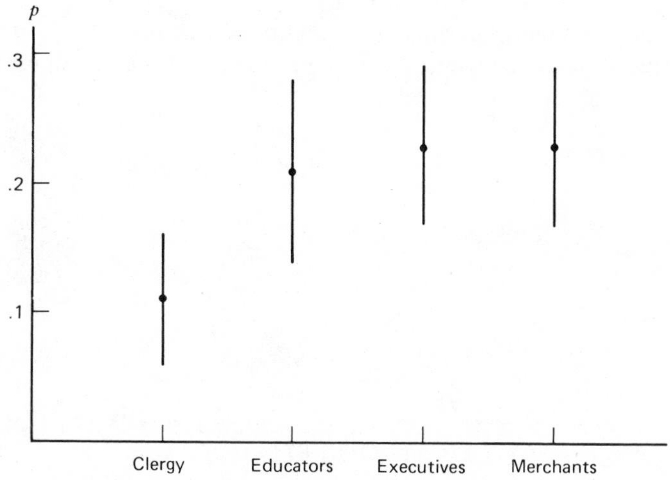

Figure 13.1 A graphic presentation of contingency analysis.

Example 13.9 A 2×2 contingency table: To determine the possible effect of a chemical treatment on the rate of seed germination, 100 chemically treated seeds and 150 untreated seeds are sown. The number of seeds that germinate are recorded in Table 13.17. Do the data provide strong evidence that the rate of germination is different for the treated and untreated seeds?

TABLE 13.17

	GERMINATED	NOT GERMINATED	TOTAL
Treated	84 (86.40)	16 (13.60)	100
Untreated	132 (129.60)	18 (20.40)	150
Total	216	34	250

Letting p_1 and p_2 denote the probabilities of germination for the chemically treated seeds and the untreated seeds, respectively, we wish to test the null hypothesis $H_0: p_1 = p_2$ vs. $H_1: p_1 \neq p_2$. For the χ^2 contingency test, the expected frequencies, computed in the usual way, appear in parentheses in Table 13.17. The computed value of χ^2 is

$$\chi^2 = .067 + .424 + .044 + .282$$

$$= .817$$

$$\text{d.f.} = (2-1)(2-1) = 1$$

Because this value is smaller than 3.841 (the tabulated 5% value of χ^2 with d.f. = 1), the null hypothesis is not rejected.

This problem concerns testing the equality of two binomial proportions and can, alternatively, be solved by using the Z test discussed in Chapter 9. To employ the Z test, we compute

$$\hat{p}_1 = .84, \quad \hat{p}_2 = .88$$

$$\text{combined estimate} \quad \hat{p} = \frac{216}{250} = .864$$

$$Z = \frac{\hat{p}_1 - \hat{p}_2}{\sqrt{\hat{p}(1-\hat{p})}\,\sqrt{(1/n_1) + (1/n_2)}}$$

$$= \frac{-.04}{\sqrt{.864 \times .136}\,\sqrt{(1/100) + (1/150)}} = -.904$$

Because $|z|$ is smaller than $z_{.025} = 1.96$, the null hypothesis is not rejected at $\alpha = .05$.

It can be shown by algebraic calculation that Z^2 is exactly the same as χ^2 for a 2×2 contingency table. This is indeed the case in our example, where $Z^2=(-.904)^2=.817=\chi^2$. Moreover, $(1.96)^2=3.8416$ is the 5% point of χ^2 with d.f. = 1. Thus the two test procedures are equivalent. However, if the alternative hypothesis is one-sided, such as $H_1: p_1 > p_2$, then the χ^2 test is not appropriate. ■

A warning is in order regarding the uncritical combination of entries in contingency tables, especially when the data are believed to be taken from different populations. Pooling contingency tables can produce misleading results (see Exercise 17), and it may be better to combine the results in other ways. Cochran [3] suggests some alternative methods and provides extensive discussions of related topics as well.

13.6 AN EXACT TEST FOR A 2×2 CONTINGENCY TABLE

The χ^2 test for independence in a contingency table is an approximate test that is useful when the sample size is sufficiently large so that no cell frequency is too small. The excessive cost of experimentation or the unavailability of an adequate number of experimental subjects often constrains the sample size, and consequently the χ^2 test may not be applicable. A test procedure that guarantees a specified control of the type 1 error probability, called the *Fisher–Irwin exact test*, was proposed independently by R. A. Fisher and J. O. Irwin. Instead of comparing the observed and the expected cell frequencies, this test is based on an evaluation of the conditional probabilities of the cell frequencies assuming the marginal totals are fixed. Here we outline the principle of the Fisher–Irwin exact test.

Problem. Suppose a study is made to compare the effectiveness of two treatments in curing a rare blood disease. A total of 15 patients afflicted with approximately the same severity of this rare disease are available subjects for the study. Of these, 7 are randomly chosen to receive treatment 1; the remaining 8 are to receive treatment 2. The observed results are given in Table 13.18. From these data, the problem is to test the null hypothesis that the two treatments are equally effective in curing the disease against the one-sided alternative that treatment 1 is more effective.

TABLE 13.18

	CURED	UNCURED	TOTAL
Treatment 1	4	3	7
Treatment 2	1	7	8
Total	5	10	15

If there is no difference between the two treatments, the pooled sample of 5 cures and 10 noncures can be regarded as a random sample from a single population. Viewing the pooled result in itself as a small population, the Fisher–Irwin test addresses itself to the question: Can the two rows of the contingency table be considered homogeneous subsamples from this small population? In making an inference, we reason that strong evidence of lack of homogeneity in the subsamples indicates that the two treatments are not alike.

The model of homogeneous subsamples assumes that the two rows of Table 13.18 result from randomly dividing the 5 cured persons and 10 uncured persons into groups of 7 and 8. The number of ways that 7 persons can be selected from 15 is $\binom{15}{7}$, and each is equally likely to occur according to the concept of random selection. The number of choices of 4 out of 5 cures and 3 out of 10 noncures to complete the first row of the table is $\binom{5}{4}\binom{10}{3}$ by the product rule. Once the selection in the first row has been made, the second row is automatically determined. Therefore, the conditional probability of the observed cell frequencies given the pooled result of 5 cures and 10 noncures is

$$\frac{\binom{5}{4}\binom{10}{3}}{\binom{15}{7}} = \frac{5 \times 120}{6435} = .093$$

With the marginal totals fixed, we begin to search for more extreme configurations of the cell frequencies in the sense that they support the alternative hypothesis more strongly than the observed configuration. Stronger support for the conclusion that treatment 1 is better requires a higher frequency in the upper left-hand corner of the table. The only possible configuration is given in Table 13.19, and its conditional probability is evaluated in the manner just explained.

TABLE 13.19

A MORE EXTREME CONFIGURATION THAN TABLE 13.18

5	2	7
0	8	8
5	10	15

Conditional probability:

$$\frac{\binom{5}{5}\binom{10}{2}}{\binom{15}{7}} = .007$$

For the purposes of illustration, suppose that we have selected a level of significance $\alpha = .15$. To determine if the observed configuration of the cell frequencies contradicts the model of random division into subsamples, we compute the total probability of the observed and the more extreme configurations, or $.093 + .007 = .10$. Because this value is smaller than our chosen level of significance, the hypothesis of randomness is contradicted. Consequently, we reject the null hypothesis of no difference in the treatments at $\alpha = .15$.

Note that with the more traditional choice of the level of significance $\alpha = .05$, the null hypothesis will not be rejected. This may appear somewhat disturbing in view of the rather dramatic difference between the sample proportions of cures: $\frac{4}{7} = .57$ for treatment 1 and $\frac{1}{8} = .125$ for treatment 2. The explanation is that with small sample sizes such as 7 and 8, a large difference between the sample proportions may occur purely by chance, although the population proportions are the same.

In testing the null hypothesis that no difference exists between the two treatments vs. the two-sided alternative, the procedure remains basically the same, except that more extreme configurations must be identified in two directions in anology with the concept of a two-tailed test. A general configuration using the marginal totals in Table 13.18 is represented in Table 13.20.

TABLE 13.20

	CURED	UNCURED	
Treatment 1	x	$7-x$	7
Treatment 2	$5-x$	$3+x$	8
	5	10	15

The difference between the sample proportions in this table is

$$\frac{x}{7} - \frac{5-x}{8}$$

and in the observed table is $\frac{4}{7} - \frac{1}{8}$. The more extreme two-sided configurations can then be identified as all those forms of Table 13.20 where x satisfies

$$\left|\frac{x}{7} - \frac{5-x}{8}\right| > \left|\frac{4}{7} - \frac{1}{8}\right|$$

or equivalently

$$|3x - 7| > 5$$

This criterion is satisfied by Table 13.19 and also by Table 13.21.

TABLE 13.21

0	7	7
5	3	8
5	10	15

Conditional probability:

$$\frac{\binom{5}{0}\binom{10}{7}}{\binom{15}{7}} = .019$$

Hence the significance probability for the two-sided alternative is

$$.093 + .007 + .019 = .119$$

From the viewpoint of the practical application of the Fisher–Irwin exact test, a major difficulty lies in the enumeration of all extreme configurations and the calculation of the corresponding conditional probabilities. However, some tables [4] and computer programs are available to circumvent these tedious enumerations. With large sample sizes, we can use the χ^2 test as an approximation.

REFERENCES

1. Shapiro, S. S., and Wilk, M. B., "An Analysis of Variance Test for Normality (complete samples)," *Biometrika* **52** (1965), 591–611.
2. Kruskal, W. H., "Ordinal Measures of Association," *Journal of American Statistical Association* **53** (1958), 814–61.
3. Cochran, W. G., "Some Methods for Strengthening the Common χ^2 Tests," *Biometrics* **10** (1954), 417–51.
4. Beyer, W., (ed.), *Handbook of Tables for Probability and Statistics*, 2nd Ed., The Chemical Rubber Company, Cleveland, Ohio, 1968.

EXERCISES

1. Test if the data given in Example 13.2 contradict the genetic model that postulates that the offsprings of types A, B, and C occur in the ratio 1 : 2 : 1. (Take $\alpha = .05$.)
2. The following table records the observed number of births at a hospital in four consecutive quarterly periods:

Quarters	Jan.–Mar.	Apr.–Jun.	Jul.–Sep.	Oct.–Dec.
Number of births	110	57	53	80

It is conjectured that twice as many babies are born during the Jan.–Mar. quarter than are born in any of the other three quarters. At $\alpha = .05$, test if these data strongly contradict the stated conjecture.

3. A large-scale nationwide poll is conducted to determine the public's attitude toward the abolition of capital punishment. The percentages in the various response categories are:

Strongly favor	Favor	Indifferent	Oppose	Strongly oppose
20%	30%	20%	20%	10%

From a random sample of 100 law enforcement officers in a metropolitan area, the following frequency distribution is observed:

Strongly favor	Favor	Indifferent	Oppose	Strongly oppose	Total
14	18	18	26	24	100

Do these data provide evidence that the attitude pattern of law enforcement officers differs significantly from the attitude pattern observed in the large-scale national poll?

4. Observations of 80 litters, each containing 3 rabbits, reveal the following frequency distribution of the number of male rabbits per litter:

Number of males in litter	0	1	2	3	Total
Number of litters	19	32	22	7	80

Under the model of Bernoulli trials for the sex of rabbits, the probability distribution of the number of males per litter should be binomial with $n=3$ and $p=$probability of a male birth.

(*a*) From these data, estimate the parameter p as

$$\hat{p}=\frac{\text{Total number of males in 80 litters}}{\text{Total number of rabbits in 80 litters}}$$

(*b*) Perform a test to determine if a binomial distribution fits the data.

5. The number of machine malfunctions per shift at a factory is recorded for 180 shifts, and the following data are obtained:

Number of malfunctions	0	1	2	3	4	5	6	Total
Number of shifts	82	42	31	12	8	3	2	180

(*a*) What is a reasonable probability model for this type of data? (*Hint*: Refer to Chapter 5.)

(*b*) Test if this model adequately describes the data.

6. A medical researcher conjectures that heavy smoking can result in wrinkled skin around the eyes. The smoking habits as well as the presence of prominent wrinkles around the eyes are recorded for a random sample of 500 persons. The following frequency table is obtained:

	Prominent wrinkles	Wrinkles not prominent
Heavy smoker	95	55
Light or nonsmoker	103	247

(*a*) Do these data substantiate an association between wrinkle formation and smoking habit?
(*b*) If the null hypothesis that wrinkle formation is independent of smoking habit is rejected after an analysis of the data, can the researcher readily conclude that heavy smoking causes wrinkles around the eyes? Why or why not?

7. Based on interviews of couples seeking divorces, a social worker compiles the following data related to the period of acquaintanceship before marriage and the duration of marriage:

Acquaintanceship before marriage	Duration of marriage $\leqslant 4$ yrs	>4 yrs	Total
Under $\frac{1}{2}$ yrs	11	8	19
$\frac{1}{2}$–$1\frac{1}{2}$ yrs	28	24	52
Over $1\frac{1}{2}$ yrs	21	19	40
Total	60	51	111

Perform a test to determine if the data substantiate an association between the stability of a marriage and the period of acquaintanceship prior to marriage.

8. By polling a random sample of 350 undergraduate students, a campus press obtains the following frequency counts regarding student attitude toward a proposed change in dormitory regulations:

	Favor	Indifferent	Oppose	Total
Male	93	72	21	186
Female	55	79	30	164
Total	148	151	51	350

Does the proposal seem to appeal differently to male and female students?

9. Calculate Pearson's coefficient of mean square contingency for the data given in Example 13.7.
10. The food supplies department circulates a questionnaire among U.S. armed forces personnel to gather information about preferences for various types of foods. One question is "Do you prefer black olives to green olives on the lunch menu?" A random sample of 435 respondents are classified according to both olive preference and geographical area of home, and the following data are recorded:

Region	Number preferring black olives	Total No. of respondents
West	65	118
East	59	135
South	48	90
Midwest	43	92

Is there a significant variation of preference over the different geographical areas?

11. Many industrial air pollutants adversely affect plants. Sulphur dioxide causes leaf damage in the form of intraveinal bleaching in many sensitive plants. In a study of the effect of a given concentration of sulphur dioxide in the air on three types of garden vegetables, 40 plants of each type are exposed to the pollutant under controlled greenhouse conditions. The frequencies of severe leaf damage are recorded in the following table:

	Leaf damage		
	Severe	Moderate or none	Total
Lettuce	32	8	40
Spinach	28	12	40
Tomato	19	21	40
Total	79	41	120

Analyze these data to determine if the incidence of severe leaf damage is alike for the three types of plants. In particular:

(*a*) Formulate the null hypothesis.

(*b*) Test the null hypothesis with $\alpha = .05$.

(*c*) Construct three individual 98% confidence intervals and plot them on graph paper.

12. Osteoporosis or a loss of bone minerals is a common cause of broken bones in the elderly. A researcher on aging conjectures that bone mineral loss can be reduced by regular physical therapy or by certain kinds of physical activity. A study is conducted on 200 elderly subjects of approximately the same age divided into control, physical therapy, and physical activity groups. After a suitable period of time, the nature of change in bone mineral content is observed.

	Change in bone mineral			
	Appreciable loss	Little change	Appreciable increase	Total
Control	38	15	7	60
Therapy	22	32	16	70
Activity	15	30	25	70
Total	75	77	48	200

Do these data indicate that the change in bone mineral varies for the different groups?

13. When a new product is introduced in a market, it is important for the manufacturer to evaluate its performance during the critical months after its distribution. A study of market penetration, as it is called, involves sampling consumers and assessing their exposure to the product. Suppose that the marketing division of a company selects random samples of 200, 150, and 300 consumers from three cities and obtains the following data from them:

	Never heard of the product	Heard about it but did not buy	Bought it at least once	Total
City 1	36	55	109	200
City 2	45	56	49	150
City 3	54	78	168	300
Total	135	189	326	650

Do these data indicate that the extent of market penetration differs in the three cities?

14. In a study of factors that regulate behavior, three kinds of subjects are identified: overcontrollers, average controllers, and undercontrollers, with the first group being most inhibited. Each subject is given the routine task of filling a box with buttons and all subjects are told they can stop whenever they wish. Whenever a subject indicates he or she wishes to stop, the experimenter asks "Don't you really want to continue?". The number of subjects in each group who stop and the

number who continue are given in the following table:

Controller	Continue	Stop	Total
Over	9	9	18
Average	8	12	20
Under	3	14	17
Total	20	35	55

Do the three groups differ in their responses to the experimenter's influence?

15. An investigation is conducted to determine if the exposure of women to atomic fallout influences the rate of birth defects. A sample of 500 children born to mothers who were exposed to the atomic explosion in Hiroshima is to be studied. A sample of 400 children from a Japanese island far removed from the site of the atomic explosion forms the control group. Suppose that the following data for the incidence of birth defects are obtained:

	Birth defect		
	Present	Absent	Total
Mother exposed	84	416	500
Mother not exposed	43	357	400
Total	127	773	900

(*a*) Do these data indicate that there is a different rate of incidence of birth defects for the two groups of mothers? Use the χ^2 test for homogeneity in a contingency table.

(*b*) Use the normal test for testing the equality of two population proportions with a two-sided alternative. Verify the relation $\chi^2 = z^2$ by comparing their numerical values.

(*c*) If the alternative hypothesis is that the incidence rate is higher for exposed mothers, which of the two tests should be used?

16. Random samples of 250 persons in the 30–40 years age group and 250 persons in the 60–70 year age group are asked about the average number of hours they sleep per night, and the following summary data are recorded:

	Hours of sleep		
Age	≤ 8	>8	Total
30–40	172	78	250
60–70	120	130	250
Total	292	208	500

(*a*) Analyze the data to determine if the sleep needs are different for the two age groups.

(*b*) Denoting by p_1 and p_2 the population proportions in the two groups who have $\leqslant 8$ hours of sleep per night, construct a 95% confidence interval for $p_1 - p_2$. (*Hint:* Refer to Chapter 9.)

17. *Pooling contingency tables can produce spurious association*: A large organization is being investigated to determine if their recruitment is sex biased. Table (*i*) and Table (*ii*), respectively, show the classification of applicants for secretarial and for sales positions according to sex and result of interview. Table (*iii*) is an aggregation of the corresponding entries of Table (*i*) and Table (*ii*).

Table (*i*)
Secretarial positions

	Offered	Denied	Total
Male	25	50	75
Female	75	150	225
Total	100	200	300

Table (*ii*)
Sales positions

	Offered	Denied	Total
Male	150	50	200
Female	75	25	100
Total	225	75	300

Table (*iii*)
Secretarial and sales positions

	Offered	Denied	Total
Male	175	100	275
Female	150	175	325
Total	325	275	600

(*a*) Verify that the χ^2 statistic for testing independence is zero for each of the data sets given in Table (*i*) and Table (*ii*).

(*b*) For the pooled data given in Table (*iii*), compute the value of the χ^2 statistic and test the null hypothesis of independence.

(*c*) Explain the paradoxical result that there is no sex bias in any job category but the combined data indicate sex discrimination.

18. A chemical engineer suspects that the property of brittleness in

polyethylene bars may depend on the duration of heat treatment at a particular phase of the manufacturing process. For experimental verification, the engineer makes 10 bars with 45 seconds of heat treatment and 9 bars with 60 seconds of heat treatment. Each bar is then tested and classified as brittle or tough. The results are:

	Brittle	Tough	Total
45 sec	1	9	10
60 sec	6	3	9
Total	7	12	19

Use the Fisher–Irwin exact test to test the null hypothesis that the brittleness of polyethylene bars does not vary with the two heat treatment settings.

19. In a genetic study of chromosome structures (courtesy of K. Patau), 28 individuals are classified according to the type of structural chromosome aberration and carriers in their parents. The following data are obtained:

Type of Aberration	Carrier: One parent	Carrier: Neither parent	Total
Presumably innocuous	7	1	8
Substantially unbalanced	5	15	20
Total	12	16	28

(*a*) Use the Fisher–Irwin exact test to test the null hypothesis that type of aberration is independent of parental carrier.

(*b*) Use the χ^2 test as an approximation and compare the results with the results in (*a*).

MATHEMATICAL EXERCISES

1. Letting $n_1, n_2, \ldots, n_k$ denote the cell frequencies in n multinomial trials and $p_1, p_2, \ldots, p_k$ denote the corresponding cell probabilities, prove that

$$\sum_{i=1}^{k} \frac{(n_i - np_i)^2}{np_i} = \sum_{i=1}^{k} \frac{n_i^2}{np_i} - n$$

This shows that an alternative formula for Pearson's χ^2 statistic is

$$\chi^2 = \left(\sum_{\text{cells}} \frac{O^2}{E} \right) - n$$

2. Consider the 2×2 contingency table with fixed row totals

	B_1	B_2	
A_1	n_{11}	n_{12}	n_{10}
A_2	n_{21}	n_{22}	n_{20}
	n_{01}	n_{02}	n

Let p_1 and p_2 denote $P(B_1|A_1)$ and $P(B_1|A_2)$, respectively. To test $H_0: p_1 = p_2$, we can employ the normal test with the test statistic

$$Z = \frac{\hat{p}_1 - \hat{p}_2}{\sqrt{\hat{p}(1-\hat{p})} \sqrt{(1/n_{10}) + (1/n_{20})}}$$

where $\hat{p}_1 = n_{11}/n_{10}$, $\hat{p}_2 = n_{21}/n_{20}$, and $\hat{p} = n_{01}/n$. Prove that Z^2 is exactly the same as the χ^2 statistic.

3. Referring to Exercise 2, prove that the formula for the χ^2 statistic for a 2×2 contingency table can also be written

$$\chi^2 = \frac{n(n_{11}n_{22} - n_{12}n_{21})^2}{n_{10}n_{20}n_{01}n_{02}}$$

CHAPTER 14

Design of Experiments and Analysis of Variance

14.1 INTRODUCTION AND BASIC TERMINOLOGY

Careful planning of an experiment is a prerequisite for success when employing statistical principles to draw conclusions from the data. In fact, a well-designed experiment helps to achieve clear interpretations and to avoid complicated analyses. On the other hand, no amount of sophistication in statistical analysis can salvage the confusing or conflicting information produced by a poorly planned experiment. The subject of this chapter, the design of experiments, is therefore of paramount importance in all quantitative fields, especially when several factors are studied simultaneously to identify and compare their effects. The two fundamental principles of a good design, randomization and sampling in matched pairs, have already been discussed in Chapter 9 in the context of comparing two treatments. These ideas are further expanded here to encompass even more practically useful circumstances that involve the comparative study of several treatments when one or more factors are manipulated in the composition of a treatment. A few examples should help to delineate our objective and to facilitate an understanding of the basic terminology of experimental design.

Example 14.1 A civil engineer wishes to compare four different mixing processes for the treatment of sewage in terms of the density of suspended solids in the treated sewage. For this purpose, density measurements are collected by experimenting with each mixing process on a number of sewage specimens. ■

Example 14.2 Members of aircraft crews are the experimental subjects in a psychological experiment. Each individual is subjected to a warning followed by a stimulus, and the reaction time to the stimulus is then recorded. The experiment is designed to study the reaction times using two types of stimuli, auditory and visual, and three different time settings for the interval between warning and stimulus. ■

Example 14.3 An agronomist plans to study the yield rates of four hybrid types of wheat at three arid geographical regions representing different drought conditions. The locations are to be chosen according to the levels of average seasonal rainfall, and the yield per acre of each variety of wheat is to be recorded as the response. ■

A common feature of these three examples is that an individual response measurement originates from a source in which one or more conditions are manipulated by the experimenter. With this in mind, we introduce a few general terms.

> The basic units for which response measurements are collected are called *experimental units* or *subjects*.
>
> Distinct types of conditions that are manipulated on the experimental units are called *factors*.
>
> The different modes of presence of a factor are called *factor levels*.
>
> Each specific combination of the levels of different factors is called a *treatment*.
>
> The number of experimental units on which a particular treatment is applied is called the number of *replications* of that treatment.

If the levels of a factor correspond to intensities measured on a scale, it is called a *quantitative* factor. When the levels differ only in some qualitative characteristic, the factor is said to be *qualitative*. Example 14.1 involves a single qualitative factor, the process of mixing, which has 4 levels corresponding to the four processes being studied. Each sewage specimen is an experimental unit that is subjected to one of four treatments. Example 14.2 concerns a two-factor experiment in which the factors are "type of stimulus" and "time interval between warning and stimulus." The first is a qualitative factor with 2 levels, auditory and visual; the second is a quantitative factor with 3 levels. The number of treatments in this experiment is then $2\times3=6$. Similarly, Example 14.3 illustrates a two-factor experiment involving qualitative factors: "type of plant" with 4 levels, and "geographical location" with 3 levels. Because locations are chosen with respect to average rainfall, the second factor may be regarded as quantitative if the remaining conditions do not vary substantially. The experimental units in Example 14.3 are the geographical plots of predetermined size, and the response is the measurement of crop yield per plot.

The main aspects of designing an experiment can now be summarized as follows:

(*a*) Choose the factors to be studied in an experiment and the levels of each factor that are relevant to the investigation. This determines the treatments to be used.
(*b*) Consider the intended scope of inference and choose the type of experimental units on which treatments are to be applied.
(*c*) From the perspectives of cost and desired precision of inference, decide how many units are to be included in the experiment.
(*d*) Finally, and most importantly, determine the manner in which treatments are to be applied to the experimental units.

A comprehensive description of the many aspects of experimental design, including the pitfalls to avoid, cannot be condensed into a single chapter of an introductory text. Instead, we take the opportunity here to present the reasoning behind an important method of analysis, called the *analysis of variance*, in the context of a few common experimental designs.

The *analysis of variance*, more appropriately *the analysis of variation about means*, consists of partitioning the total variation present in a data set into components. Each component is attributed to an identifiable cause or source of variation; in addition, one component represents the variation due to uncontrolled factors and random errors associated with the response measurements. Specifically, if the data set consists of n measurements $y_1,\ldots,y_n$ and their mean is denoted by $\bar{y}$, the total variation about the mean is embodied in the sum of squared deviations $\Sigma_{i=1}^{n}(y_i-\bar{y})^2$, which is called the *total sum of squares*. The technique of analysis of variance decomposes this total sum of squares into the parts schematically depicted in Figure 14.1 for a case in which three identifiable sources of variation are present in addition to the error component. The number of identifiable sources of variation and the formulas for the component sums

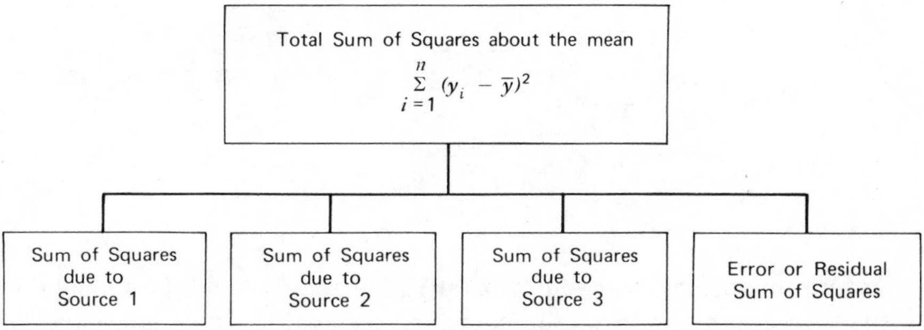

Figure 14.1 The partitioning scheme for the analysis of variance.

of squares are intrinsically connected to the particular experimental design employed in the collection of data and to the statistical model deemed appropriate for the analysis.

14.2 COMPARISON OF SEVERAL TREATMENTS: THE COMPLETELY RANDOMIZED DESIGN

It is usually more expedient both in terms of time and expense to simultaneously compare several treatments than it is to conduct several comparative trials two at a time. The term *completely randomized design* is synonymous with independent random sampling from several populations when each population is identified as the population of responses under a particular treatment. Suppose that out of k treatments to be studied in an experiment, treatment 1 is to be applied to n_1 experimental units, treatment 2 to n_2 units,..., treatment k to n_k units. In a completely randomized design, n_1 experimental units selected at random from the available collection of $n = n_1 + n_2 + \ldots + n_k$ units are to receive treatment 1, n_2 units randomly selected from the remaining lot are to receive treatment 2, and proceeding in this manner, treatment k is to be applied to the remaining n_k units. The special case of this design for a comparison of $k=2$ treatments has already been discussed in Section 9.2. The data structure for the response measurements can be represented by the format shown in Table 14.1 where y_{ij} is the ith observation on treatment j. The summary statistics appear in the last two rows.

TABLE 14.1
DATA STRUCTURE FOR THE COMPLETELY RANDOMIZED DESIGN WITH k TREATMENTS

	TREATMENT 1	TREATMENT 2	...	TREATMENT k
	y_{11}	y_{12}	...	y_{1k}
	y_{21}	y_{22}	...	y_{2k}
	⋮	⋮		⋮
	$y_{n_1 1}$	$y_{n_2 2}$	...	$y_{n_k k}$
Means	$\bar{y}_1$	$\bar{y}_2$	...	$\bar{y}_k$
Sums of Squares of deviations about means	$\sum_{i=1}^{n_1} (y_{i1}-\bar{y}_1)^2$	$\sum_{i=1}^{n_2} (y_{i2}-\bar{y}_2)^2$	...	$\sum_{i=1}^{n_k} (y_{ik}-\bar{y}_k)^2$

Before proceeding with the general case of k treatments, it would be instructive to explain the reasoning behind the analysis of variance and the associated calculations in terms of a numerical example.

Example 14.4 In an effort to improve the quality of recording tapes, the effects of four kinds of coatings A, B, C, D on the reproducing quality of sound are compared. Suppose that the measurements of sound distortion given in Table 14.2 are obtained from tapes treated with the four coatings. (The data are artificial to simplify the calculations.)

TABLE 14.2

SOUND DISTORTIONS OBTAINED WITH FOUR TYPES OF COATING

	A	B	C	D	
	10	14	17	12	
	15	18	16	15	
	8	21	14	17	
	12	15	15	15	
	15		17	16	
			15	15	
			18		Grand mean
Means	$\bar{y}_1 = 12$	$\bar{y}_2 = 17$	$\bar{y}_3 = 16$	$\bar{y}_4 = 15$	$\bar{y} = 15$
Sums of Squares	$\sum_{i=1}^{5} (y_{i1} - \bar{y}_1)^2 = 38$	$\sum_{i=1}^{4} (y_{i2} - \bar{y}_2)^2 = 30$	$\sum_{i=1}^{7} (y_{i3} - \bar{y}_3)^2 = 12$	$\sum_{i=1}^{6} (y_{i4} - \bar{y}_4)^2 = 14$	

Two questions immediately come to mind. Does any significant difference exist among the mean distortions obtained using the four coatings? Can we establish confidence intervals for the mean differences between individual pairs? ■

An analysis of the results essentially consists of decomposing the observations into contributions from different sources. Reasoning that the deviation of an individual observation from the grand mean, $y_{ij} - \bar{y}$, is partly due to differences among the mean qualities of the coatings and partly due to random variations in measurement within the same group, the following decomposition is suggested:

$$\begin{array}{ccccccc} & & \text{Grand} & & \text{Deviation due} & & \\ \text{Observation} & = & \text{mean} & + & \text{to treatment} & + & \text{Residual} \\ y_{ij} & = & \bar{y} & + & (\bar{y}_j - \bar{y}) & + & (y_{ij} - \bar{y}_j) \end{array}$$

For the data given in Table 14.2, the decomposition of all the observations can be presented in the form of the following arrays:

$$\begin{array}{c}\text{Observations}\\ y_{ij}\end{array}$$

$$\begin{bmatrix} 10 & 14 & 17 & 12 \\ 15 & 18 & 16 & 15 \\ 8 & 21 & 14 & 17 \\ 12 & 15 & 15 & 15 \\ 15 & & 17 & 16 \\ & & 15 & 15 \\ & & 18 & \end{bmatrix} =$$

$$\underset{\bar{y}}{\text{Grand mean}} \qquad \underset{(\bar{y}_j - \bar{y})}{\text{Treatment effects}} \qquad \underset{(y_{ij} - \bar{y}_j)}{\text{Residuals}}$$

$$\begin{bmatrix} 15 & 15 & 15 & 15 \\ 15 & 15 & 15 & 15 \\ 15 & 15 & 15 & 15 \\ 15 & 15 & 15 & 15 \\ 15 & & 15 & 15 \\ & & 15 & 15 \\ & & 15 & \end{bmatrix} + \begin{bmatrix} -3 & 2 & 1 & 0 \\ -3 & 2 & 1 & 0 \\ -3 & 2 & 1 & 0 \\ -3 & 2 & 1 & 0 \\ -3 & & 1 & 0 \\ & & 1 & 0 \\ & & 1 & \end{bmatrix} + \begin{bmatrix} -2 & -3 & 1 & -3 \\ 3 & 1 & 0 & 0 \\ -4 & 4 & -2 & 2 \\ 0 & -2 & -1 & 0 \\ 3 & & 1 & 1 \\ & & -1 & 0 \\ & & 2 & \end{bmatrix}$$

For instance, the upper left-hand entries of the arrays show that

$$\begin{aligned} 10 &= 15 + \quad (-3) + \quad (-2) \\ y_{11} &= \bar{y} + (\bar{y}_1 - \bar{y}) + (y_{11} - \bar{y}_1) \end{aligned}$$

If there is really no difference in the mean distortions obtained using the four tape coatings, we can expect the entries of the second array on the right-hand side of the equation, whose terms are $(\bar{y}_j - \bar{y})$, to be close to zero. As an overall measure of the amount of variation due to differences in the treatment means, we calculate the sums of squares of all the entries in this array, or

$$\underbrace{(-3)^2 + \ldots + (-3)^2}_{n_1=5} + \underbrace{2^2 + \ldots + 2^2}_{n_2=4} + \underbrace{1^2 + \ldots + 1^2}_{n_3=7} + \underbrace{0^2 + \ldots + 0^2}_{n_4=6}$$

$$= 5(-3)^2 + 4(2)^2 + 7(1)^2 + 6(0)^2$$

$$= 68$$

Thus the sum of squares due to differences in the treatment means, also called the *treatment sum of squares*, is given by

$$\text{Treatment Sum of Squares} = \sum_{j=1}^{4} n_j(\bar{y}_j - \bar{y})^2 = 68$$

The last array consists of the entries $(y_{ij} - \bar{y}_j)$, which are the deviations of individual observations from the corresponding treatment mean. These deviations reflect inherent variabilities in the experimental material and the measuring device and are called the *residuals*. The overall variation due to random errors is measured by the sum of squares of all these residuals

$$(-2)^2 + 3^2 + (-4)^2 + \ldots + 1^2 + 0^2 = 94$$

Thus we obtain

$$\text{Residual Sum of Squares} = \sum_{j=1}^{4} \sum_{i=1}^{n_j} (y_{ij} - \bar{y}_j)^2 = 94$$

The double summation indicates that the elements are summed within each column and then over different columns. Alternatively, referring to the last row in Table 14.2, we obtain

$$\begin{aligned}\text{Residual Sum of Squares} &= \sum_{i=1}^{5} (y_{i1} - \bar{y}_1)^2 + \sum_{i=1}^{4} (y_{i2} - \bar{y}_2)^2 \\ &\quad + \sum_{i=1}^{7} (y_{i3} - \bar{y}_3)^2 + \sum_{i=1}^{6} (y_{i4} - \bar{y}_4)^2 \\ &= 38 + 30 + 12 + 14 = 94\end{aligned}$$

Finally, the deviations of individual observations from the grand mean $y_{ij} - \bar{y}$, are given by the array

$$\begin{bmatrix} -5 & -1 & 2 & -3 \\ 0 & 3 & 1 & 0 \\ -7 & 6 & -1 & 2 \\ -3 & 0 & 0 & 0 \\ 0 & & 2 & 1 \\ & & 0 & 0 \\ & & 3 & \end{bmatrix}$$

The total variation present in the data is measured by the sum of squares of all these deviations

$$\text{Total Sum of Squares} = \sum_{j=1}^{4} \sum_{i=1}^{n_j} (y_{ij} - \bar{y})^2$$

$$= (-5)^2 + 0^2 + (-7)^2 + \ldots + 0^2$$

$$= 162$$

Note that the total sum of squares is the sum of the treatment sum of squares and the residual sum of squares, a mathematical fact that we prove later in this section.

It is time to turn our attention to another property of this decomposition, the degrees of freedom associated with the sums of squares. In general terms:

$$\left\{\begin{array}{l}\text{Degrees of freedom}\\\text{associated with}\\\text{a sum of squares}\end{array}\right\} = \left\{\begin{array}{l}\text{Number of elements}\\\text{whose squares}\\\text{are summed}\end{array}\right\} - \left\{\begin{array}{l}\text{Number of linear}\\\text{constraints satisfied}\\\text{by the elements}\end{array}\right\}$$

In our present example, the treatment sum of squares is the sum of four terms $n_1(\bar{y}_1 - \bar{y})^2 + n_2(\bar{y}_2 - \bar{y})^2 + n_3(\bar{y}_3 - \bar{y})^2 + n_4(\bar{y}_4 - \bar{y})^2$, where the elements satisfy the single constraint

$$n_1(\bar{y}_1 - \bar{y}) + n_2(\bar{y}_2 - \bar{y}) + n_3(\bar{y}_3 - \bar{y}) + n_4(\bar{y}_4 - \bar{y}) = 0$$

This equality holds because the grand mean $\bar{y}$ is a weighted average of the treatment means, or

$$\bar{y} = \frac{n_1\bar{y}_1 + n_2\bar{y}_2 + n_3\bar{y}_3 + n_4\bar{y}_4}{n_1 + n_2 + n_3 + n_4}$$

Consequently, the number of degrees of freedom associated with the treatment sum of squares is $4 - 1 = 3$. To determine the degrees of freedom for the residual sum of squares, we note that the entries $(y_{ij} - \bar{y}_j)$ in each column of the residual array sum to zero and that there are 4 columns. The number of degrees of freedom for the residual sum of squares is then $(n_1 + n_2 + n_3 + n_4) - 4 = 22 - 4 = 18$. Finally, the number of degrees of freedom for the total sum of squares is $(n_1 + n_2 + n_3 + n_4) - 1 = 22 - 1 = 21$, because the 22 entries $(y_{ij}-\bar{y})$ whose squares are summed satisfy the single constraint that their total is zero.

Guided by this numerical example, we now present the general formulas for the analysis of variance for a comparison of k treatments, using the data structure given in Table 14.1. Beginning with the basic decomposition

$$(y_{ij}-\bar{y})=(\bar{y}_j-\bar{y})+(y_{ij}-\bar{y}_j)$$

and squaring each side of the equation, we obtain

$$(y_{ij}-\bar{y})^2=(\bar{y}_j-\bar{y})^2+(y_{ij}-\bar{y}_j)^2+2(\bar{y}_j-\bar{y})(y_{ij}-\bar{y}_j)$$

When summed over $i=1,\ldots,n_j$, the last term on the right-hand side of this equation reduces to zero due to the relation $\Sigma_{i=1}^{n_j}(y_{ij}-\bar{y}_j)=0$. Therefore, summing each side of the preceding relation over $i=1,\ldots,n_j$ and $j=1,\ldots,k$ provides the decomposition

$$\underbrace{\sum_{j=1}^{k}\sum_{i=1}^{n_j}(y_{ij}-\bar{y})^2}_{\text{Total SS}}=\underbrace{\sum_{j=1}^{k}n_j(\bar{y}_j-\bar{y})^2}_{\text{Treatment SS}}+\underbrace{\sum_{j=1}^{k}\sum_{i=1}^{n_j}(y_{ij}-\bar{y}_j)^2}_{\substack{\text{Residual SS}\\ \text{or error SS}}}$$

$$\text{d.f.}=\sum_{j=1}^{k}n_j-1 \qquad \text{d.f.}=k-1 \qquad \text{d.f.}=\sum_{j=1}^{k}n_j-k$$

It is customary to present the decomposition of the sum of squares and the degrees of freedom in a tabular form called the *analysis of variance table* or, more conveniently, the *ANOVA table*. This table contains the additional column for the *mean square* associated with a component, which is defined

$$\text{Mean Square}=\frac{\text{Sum of Squares}}{\text{d.f.}}$$

The ANOVA table for comparing k treatments in our present example appears in Table 14.3.

TABLE 14.3
ANOVA TABLE FOR COMPARING k TREATMENTS

SOURCE	SUM OF SQUARES	d.f.	MEAN SQUARE
Treatments	$SS_T = \sum_{j=1}^{k} n_j(\bar{y}_j - \bar{y})^2$	$k-1$	$MS_T = \frac{SS_T}{k-1}$
Error	$SSE = \sum_{j=1}^{k} \sum_{i=1}^{n_j} (y_{ij} - \bar{y}_j)^2$	$\sum_{j=1}^{k} n_j - k$	$MSE = \frac{SSE}{\sum_{j=1}^{k} n_j - k}$
Total	$\sum_{j=1}^{k} \sum_{i=1}^{n_j} (y_{ij} - \bar{y})^2$	$\sum_{j=1}^{k} n_j - 1$	

14.3 POPULATION MODEL AND INFERENCES FOR A COMPLETELY RANDOMIZED DESIGN

To implement a formal statistical test for no difference among treatment effects we need to have a population model for the experiment. To this end, we assume that the response measurements with the jth treatment constitute a random sample from a normal population with a mean of μ_j and a common variance of σ^2. The samples are assumed to be mutually independent. Introducing the new parameters

$$\mu = \frac{\sum_{j=1}^{k} \mu_j}{k} \quad \text{(the average of the population means)}$$

$$\beta_j = \mu_j - \mu \quad \text{(the effect of the } j\text{th treatment)}, \quad j = 1, \ldots, k$$

another equivalent representation of the model is:

Population model for comparing k treatments

$$Y_{ij} = \mu + \beta_j + e_{ij}, \quad i = 1, \ldots, n_j, \quad j = 1, \ldots, k$$

where μ = overall mean

β_j = the jth treatment effect, $\sum_{j=1}^{k} \beta_j = 0$

and e_{ij} are all independently distributed as $N(0, \sigma)$.

The null hypothesis that no difference exists among the k population means can now be phrased

$$H_0: \beta_1 = \beta_2 = \ldots = \beta_k = 0$$

and the alternative hypothesis is that some of the β_j values differ from zero. Seeking a criterion to test the null hypothesis, we observe that when the population means are all equal, the treatment mean square $\Sigma n_j(\bar{y}_j - \bar{y})^2/k-1$ is expected to be small and that it is likely to be large when the means differ markedly. The residual mean square, which provides an unbiased estimate of σ^2, can be used as a yardstick for determining how large a treatment mean square should be before it indicates significant differences. Statistical distribution theory tells us that under H_0 the ratio

$$F = \frac{\text{Treatment Mean Square}}{\text{Residual Mean Square}} = \frac{\text{Treatment SS}/(k-1)}{\text{Residual SS}\Big/\left(\sum_{j=1}^{k} n_j - k\right)}$$

has an F distribution with d.f. $=(k-1, n-k)$, where $n = \Sigma_{j=1}^{k} n_j$.

> Reject $H_0: \beta_1 = \beta_2 = \ldots = \beta_k = 0$ if
>
> $$\frac{\text{Treatment SS}/(k-1)}{\text{Residual SS}/(n-k)} > F_\alpha(k-1, n-k)$$
>
> where $n = \Sigma_{j=1}^{k} n_j$ and $F_\alpha(k-1, n-k)$ is the upper α point of the F distribution with d.f. $=(k-1, n-k)$.

The computed value of the F ratio is usually presented in the last column of the ANOVA table.

Example 14.5 Construct the ANOVA table for the data given in Example 14.4 concerning a comparison of four tape coatings. Test the null hypothesis that the means are all equal. Use $\alpha = .05$.

Using our earlier calculations for the component sums of squares, we construct the ANOVA table that appears in Table 14.4.

A test of the null hypothesis $H_0: \beta_1 = \beta_2 = \beta_3 = \beta_4 = 0$ is performed by comparing the observed F value 4.34 with the tabulated value of F with d.f. $=(3, 18)$. At a .05 level of significance, the tabulated value is found to

TABLE 14.4
ANOVA TABLE FOR THE DATA GIVEN IN EXAMPLE 14.4

SOURCE	SUM OF SQUARES	d.f.	MEAN SQUARE	F RATIO
Treatments	68	3	22.67	$\frac{22.67}{5.22}=4.34$
Error	94	18	5.22	
Total	162	21		

be 3.16, which is exceeded by the observed value. We therefore conclude that a significant difference in the four tape qualities is supported by the data. ■

14.3.1 Simultaneous Confidence Intervals

The ANOVA F test is only the initial step in our analysis that determines if significant differences exist among the treatment means. Our goal clearly should be more than to merely conclude that treatment differences are indicated by the data. Rather we must detect likenesses and differences among the treatments. Thus the problem of estimating differences in treatment means is of even greater importance than the overall F test.

Referring to the comparison of k treatments using the data structure given in Table 14.1, let us examine how a confidence interval can be established for $\mu_1-\mu_2$, the mean difference between treatment 1 and treatment 2. Incidentally, note that given our reparameterization $\beta_j=\mu_j-\mu$, the difference $\mu_1-\mu_2$ is equal to $\beta_1-\beta_2$. The estimator $\overline{Y}_1-\overline{Y}_2$ is normally distributed with a mean of $\mu_1-\mu_2$ and a variance of

$$\sigma^2\left(\frac{1}{n_1}+\frac{1}{n_2}\right)$$

Moreover, the quantity SSE/σ^2 has a χ^2 distribution with d.f.$=n-k$, and it is independent of the means $\bar{y}_j$. Therefore, the ratio

$$t=\frac{(\overline{Y}_1-\overline{Y}_2)-(\mu_1-\mu_2)}{\sqrt{\dfrac{\mathrm{SSE}}{n-k}}\sqrt{\dfrac{1}{n_1}+\dfrac{1}{n_2}}}$$

has a t distribution with d.f.$=n-k$, and this can be employed to construct a confidence interval for $\mu_1-\mu_2$. More generally:

Confidence interval for a single difference

A $100(1-\alpha)\%$ confidence interval for $(\mu_j - \mu_{j'})$, the mean difference between the jth and j'th treatments, is given by

$$(\bar{y}_j - \bar{y}_{j'}) \pm t_{\alpha/2}\, s \sqrt{\frac{1}{n_j} + \frac{1}{n_{j'}}}$$

where

$$s = \sqrt{\text{MSE}} = \sqrt{\frac{\text{SSE}}{n-k}}$$

and $t_{\alpha/2}$ is the upper $\alpha/2$ point of t with d.f. $= n-k$.

If the F test *first* shows a significant difference in means, then some statisticians feel that it is reasonable to compare means pairwise according to the preceding intervals. However, many statisticians prefer a more conservative procedure based on the following reasoning.

Without the proviso that the F test is significant, the preceding method provides *individual* confidence intervals for pairwise differences. For instance, with $k=4$ treatments there are $\binom{4}{2}=6$ pairwise differences $(\mu_j - \mu_{j'})$, and this procedure applied to all pairs yields six confidence statements, each having a $100(1-\alpha)\%$ level of confidence. It is difficult to determine what level of confidence will be achieved for claiming that *all* six of these statements are correct. To overcome this dilemma, procedures have been developed for several confidence intervals to be constructed in such a manner that the joint probability that all the statements are true is guaranteed not to fall below a predetermined level. Such intervals are called *multiple confidence intervals* or *simultaneous confidence intervals*. Numerous methods proposed in the statistical literature have achieved varying degrees of success. We present one that can be used simply and conveniently in general applications.

The procedure, called the *Multiple-t confidence intervals*, consists of setting confidence intervals for the differences $(\mu_j - \mu_{j'})$ in much the same way we just did for the individual differences, except that a different percentage point is read from the t table.

Multiple-t Confidence Intervals

A set of $100(1-\alpha)\%$ simultaneous confidence intervals for m number of pairwise differences $(\mu_j - \mu_{j'})$ is given by

$$(\bar{y}_j - \bar{y}_{j'}) \pm t_{\alpha/2m}\, s \sqrt{\frac{1}{n_j} + \frac{1}{n_{j'}}}$$

where $s = \sqrt{\text{MSE}}$, $m =$ the number of confidence statements, and $t_{\alpha/2m} =$ the upper $\alpha/2m$ point of t with d.f. $= n-k$. Using this procedure, the probability of all the m statements being correct is at least $(1-\alpha)$.

Operationally, the construction of these confidence intervals does not require any new concepts or calculations, but it usually involves some nonstandard percentage point of t. For example, with $k=3$ and $(1-\alpha) = .95$, if we want to set simultaneous intervals for all $m = \binom{k}{2} = 3$ pairwise differences, we require the upper $\alpha/2m = .05/6 = .0083$ point of a t distribution. An extended set of percentage points for the t distributions are given in Appendix Table 8. With a larger number m of confidence statements, the confidence intervals constructed using the present method become too wide. Consequently, the simultaneous t intervals should be used for a subset of all pairwise differences, namely those that are of primary interest to the experimenter.

Example 14.6 An experiment is conducted to determine the soil moisture deficit resulting from varying amounts of residual timber left after cutting trees in a forest. The three treatments are treatment 1: no timber left; treatment 2: 2000 bd ft left; treatment 3: 8000 bd ft left. (Board feet is a particular unit of measurement of timber volume.) The measurements of moisture deficit are given in Table 14.5. Perform the ANOVA test and construct confidence intervals for the treatment differences.

Our analysis employs convenient alternative forms of the expressions for sums of squares involving totals.

The total number of observations $n = 5+6+6 = 17$.

$$\text{Total SS} = \sum_{j=1}^{3} \sum_{i=1}^{n_j} (y_{ij} - \bar{y})^2 = \sum_{j=1}^{3} \sum_{i=1}^{n_j} y_{ij}^2 - \frac{T^2}{n} = 71.3047 - 64.2531 = 7.0516$$

TABLE 14.5
MOISTURE DEFICIT IN SOIL

TREATMENT 1	TREATMENT 2	TREATMENT 3	
1.52	1.63	2.56	
1.38	1.82	3.32	
1.29	1.35	2.76	
1.48	1.03	2.63	
1.63	2.30	2.12	
	1.45	2.78	
Total $T_1=7.30$	$T_2=9.58$	$T_3=16.17$	Grand total $T=33.05$
Mean $\bar{y}_1=1.460$	$\bar{y}_2=1.597$	$\bar{y}_3=2.695$	Grand mean $\bar{y}=1.944$

$$\text{Treatment SS} = \sum_{j=1}^{3} n_j(\bar{y}_j-\bar{y})^2 = \sum_{j=1}^{3}\frac{T_j^2}{n_j}-\frac{T^2}{n} = 69.5322-64.2531=5.2791$$

$$\text{Residual SS} = \text{Total SS} - \text{Treatment SS} = 1.7725$$

The ANOVA table appears in Table 14.6.

TABLE 14.6
ANOVA TABLE FOR COMPARISON OF MOISTURE DEFICIT

SOURCE	SUM OF SQUARES	d.f.	MEAN SQUARE	F RATIO
Treatments	5.2791	2	2.640	20.79
Error	1.7725	14	.127	
Total	7.0516	16		

Because the observed value of F is larger than the tabulated value $F_{.05}(2, 14)=3.74$, the null hypothesis of no difference in the treatment effects is rejected at $\alpha=.05$. In fact, this would be true at almost any significance level. In constructing a set of 95% multiple-t confidence intervals for pairwise differences, note that there are $\binom{3}{2}=3$ pairs:

$$\frac{\alpha}{2m}=\frac{.05}{(2\times 3)}=.0083$$

From Appendix Table 5, the upper .0083 point of t with d.f. = 14 is 2.742.

The confidence intervals are then obtained

$$\mu_2 - \mu_1: \qquad (1.597 - 1.460) \pm 2.742 \times .356 \times \sqrt{\frac{1}{6} + \frac{1}{5}}$$
$$= .137 \pm .591 = (-.45, .73)$$

$$\mu_3 - \mu_2: \qquad (2.695 - 1.597) \pm 2.742 \times .356 \times \sqrt{\frac{1}{6} + \frac{1}{6}}$$
$$= 1.098 \pm .564 = (.54, 1.66)$$

$$\mu_3 - \mu_1: \qquad (2.695 - 1.460) \pm 2.742 \times .356 \times \sqrt{\frac{1}{6} + \frac{1}{5}}$$
$$= 1.235 \pm .591 = (.64, 1.83)$$

These confidence intervals indicate that treatments 1 and 2 do not differ appreciably but that the mean of treatment 3 is considerably different from the means of 1 and 2. ■

In addition to testing hypotheses and setting confidence intervals, an analysis of data must include a critical examination of the assumptions involved in a model. As in regression analysis, of which analysis of variance is a special case, the residuals must be examined for evidence of serious violations of the assumptions. This aspect of the analysis is ignored in the ANOVA table summary.

Example 14.7 Determine the residuals for the moisture data given in Example 14.6 (see Table 14.7) and graphically examine them for possible violations of the assumptions.

TABLE 14.7
RESIDUALS $y_{ij} - \bar{y}_j$ FOR THE DATA GIVEN IN TABLE 14.5

TREATMENT 1	TREATMENT 2	TREATMENT 3
.06	.03	−.14
−.08	.22	.62
−.17	−.25	.06
.02	−.57	−.07
.17	.70	−.58
	−.15	.08

The residual plots of these data are shown in Figure 14.2, where the combined dot diagram is presented in 14.2*a* and the dot diagrams of residuals corresponding to individual treatments appear in 14.2*b*.

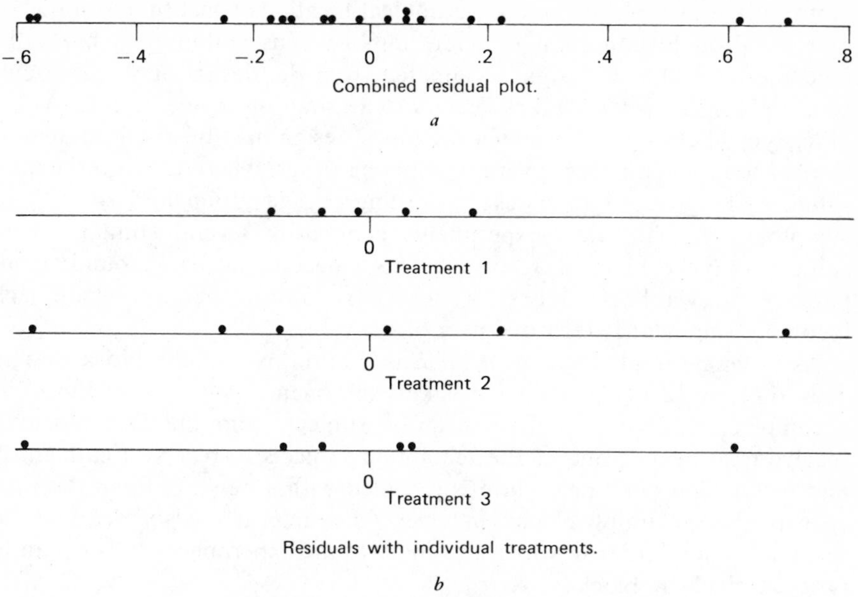

Figure 14.2 Residual plots for the data given in Example 14.6.

From an examination of the dot diagrams, the variability in the points for treatment 1 appears to be somewhat smaller than the variabilities in the points for treatments 2 and 3. However, given so few observations it is difficult to determine if this has occurred by chance or if treatment 1 actually has a smaller variance. A few more observations are usually necessary to obtain a meaningful pattern for the individual treatment plots.

Fortunately, the testing procedure is *robust* in the sense that small or moderate departures from normality and constant variance do not seriously affect its performance. ■

14.4 RANDOMIZED BLOCK EXPERIMENTS FOR COMPARING k TREATMENTS

Just as we can pair like subjects or experimental units to improve the procedure of taking independent samples, we can also arrange or *block* subjects into homogeneous groups of size k when comparing k treatments. Then if each treatment is applied to exactly one unit in the block and if comparisons are only drawn between treatment responses from the same block, extraneous variability should be greatly reduced. This is the concept underlying the *randomized* block design. In Example 14.3 the agricultural plots chosen at each of three geographical locations constitute a block, so that the drought condition is made homogeneous within each block. In fact, the term "block design" originated from the design of such agricultural field trials, where "block" refers to a group of adjacent plots. A few other typical examples for which the block design may be appropriate are: clinical trials to compare several competing drugs, where the experimental subjects are grouped in blocks according to age group and severity of symptoms; psychological experiments comparing several stimuli, where subjects may be blocked according to socioeconomic background; comparison of several techniques for storing fruits or vegetables, where each incoming shipment is regarded as a block.

As its name implies, *randomization* is a basic part of the block design. This time, once the grouping of like experimental subjects in blocks is accomplished, we randomly select one subject from the first block to receive treatment 1, one of the remaining subjects to receive treatment 2, and so on. The same procedure is repeated with a new randomization for each of the remaining blocks. Whenever possible, it is also preferable to select a random order in which to run all of the experiments rather than to proceed block by block.

Once the data are obtained, they can be arranged in rows according to the blocks and in columns according to the treatments. Designating the measurement corresponding to block i and treatment j by y_{ij}, the data structure of a randomized block design with b blocks and k treatments is shown in Table 14.8. The row and column means are denoted by

$$i\text{th block (row) mean} \quad \bar{y}_{i\cdot} = \frac{1}{k} \sum_{j=1}^{k} y_{ij}$$

$$j\text{th treatment (column) mean} \quad \bar{y}_{\cdot j} = \frac{1}{b} \sum_{i=1}^{b} y_{ij}$$

TABLE 14.8

DATA STRUCTURE OF A RANDOMIZED BLOCK DESIGN WITH b BLOCKS AND k TREATMENTS

	TREATMENT 1	TREATMENT 2	...	TREATMENT k	BLOCK MEANS
Block 1	y_{11}	y_{12}	...	y_{1k}	$\bar{y}_{1\cdot}$
Block 2	y_{21}	y_{22}	...	y_{2k}	$\bar{y}_{2\cdot}$
.	.	.	...	.	.
.	.	.	...	.	.
.	.	.	...	.	.
Block b	y_{b1}	y_{b2}	...	y_{bk}	$\bar{y}_{b\cdot}$
Treatment means	$\bar{y}_{\cdot 1}$	$\bar{y}_{\cdot 2}$	...	$\bar{y}_{\cdot k}$	$\bar{y}_{\cdot\cdot}$

These means are shown in the margins of the table. Here an overbar on y indicates an average and a dot in the subscript denotes that the average is taken over the subscript appearing in that place.

We now discuss the analysis of variance for a randomized block design with illustrative calculations based on the data given in Example 14.8.

Example 14.8 The cutting speeds of four types of tools are being compared in an experiment. Five materials of varying degrees of hardness are to be used as experimental blocks. The data pertaining to measurements of cutting time in seconds appear in Table 14.9.

TABLE 14.9
MEASUREMENTS OF CUTTING TIME ACCORDING TO TYPES OF TOOL (TREATMENTS) AND HARDNESS OF MATERIAL (BLOCKS)

	TREATMENT				BLOCK
BLOCK	1	2	3	4	MEANS
1	12	20	13	11	$\bar{y}_{1\cdot}=14$
2	2	14	7	5	$\bar{y}_{2\cdot}=7$
3	8	17	13	10	$\bar{y}_{3\cdot}=12$
4	1	12	8	3	$\bar{y}_{4\cdot}=6$
5	7	17	14	6	$\bar{y}_{5\cdot}=11$
Treatment means	$\bar{y}_{\cdot 1}=6$	$\bar{y}_{\cdot 2}=16$	$\bar{y}_{\cdot 3}=11$	$\bar{y}_{\cdot 4}=7$	$\bar{y}_{\cdot\cdot}=\frac{200}{20}=10$

■

The observations form a two-way table, and their decomposition includes both a term for column (treatment) deviations and for row (block) deviations.

Decomposition of observations

$$\text{Observation} = \begin{matrix}\text{Grand}\\ \text{mean}\end{matrix} + \begin{matrix}\text{Deviation}\\ \text{due to}\\ \text{treatment}\end{matrix} + \begin{matrix}\text{Deviation}\\ \text{due to}\\ \text{block}\end{matrix} + \text{Residual}$$

$$y_{ij} = \bar{y}_{\cdot\cdot} + (\bar{y}_{\cdot j}-\bar{y}_{\cdot\cdot}) + (\bar{y}_{i\cdot}-\bar{y}_{\cdot\cdot}) + (y_{ij}-\bar{y}_{i\cdot}-\bar{y}_{\cdot j}+\bar{y}_{\cdot\cdot})$$

Taking the observation $y_{11}=12$ in Example 14.8, we obtain

$$\begin{aligned}12 &= 10+(6-10)+(14-10)+(12-14-6+10)\\ &= 10+(-4)+(4)+(2)\end{aligned}$$

Tables 14.10 and 14.11 contain the results of the decomposition of all the observations in two different formats.

TABLE 14.10

DECOMPOSITION OF OBSERVATIONS FOR THE RANDOMIZED BLOCK EXPERIMENT IN TABLE 14.9

OBSERVATION y_{ij} = GRAND MEAN $\bar{y}_{..}$ + BLOCK EFFECT $\bar{y}_{i\cdot}-\bar{y}_{..}$ +

$$\begin{bmatrix} 12 & 20 & 13 & 11 \\ 2 & 14 & 7 & 5 \\ 8 & 17 & 13 & 10 \\ 1 & 12 & 8 & 3 \\ 7 & 17 & 14 & 6 \end{bmatrix} = \begin{bmatrix} 10 & 10 & 10 & 10 \\ 10 & 10 & 10 & 10 \\ 10 & 10 & 10 & 10 \\ 10 & 10 & 10 & 10 \\ 10 & 10 & 10 & 10 \end{bmatrix} + \begin{bmatrix} 4 & 4 & 4 & 4 \\ -3 & -3 & -3 & -3 \\ 2 & 2 & 2 & 2 \\ -4 & -4 & -4 & -4 \\ 1 & 1 & 1 & 1 \end{bmatrix} +$$

TREATMENT EFFECT $\bar{y}_{\cdot j}-\bar{y}_{..}$ + RESIDUAL $y_{ij}-\bar{y}_{i\cdot}-\bar{y}_{\cdot j}+\bar{y}_{..}$

$$\begin{bmatrix} -4 & 6 & 1 & -3 \\ -4 & 6 & 1 & -3 \\ -4 & 6 & 1 & -3 \\ -4 & 6 & 1 & -3 \\ -4 & 6 & 1 & -3 \end{bmatrix} + \begin{bmatrix} 2 & 0 & -2 & 0 \\ -1 & 1 & -1 & 1 \\ 0 & -1 & 0 & 1 \\ -1 & 0 & 1 & 0 \\ 0 & 0 & 2 & -2 \end{bmatrix}$$

TABLE 14.11

AN ALTERNATIVE FORMAT OF THE DECOMPOSITION TABLE FOR A RANDOMIZED BLOCK EXPERIMENT

BLOCK	TREATMENT 1 2 ... j ... k	DEVIATION OF BLOCK MEAN FROM GRAND MEAN
1	⋮	$(\bar{y}_{1\cdot}-\bar{y}_{..})$ ⋮
2 ⋮		
i	... $(\bar{y}_{ij}-\bar{y}_{i\cdot}-\bar{y}_{\cdot j}+\bar{y}_{..})$ ...	$(\bar{y}_{i\cdot}-\bar{y}_{..})$
⋮	⋮	⋮
b		$(\bar{y}_{b\cdot}-\bar{y}_{..})$
Deviation of treatment mean from grand mean	$(\bar{y}_{\cdot 1}-\bar{y}_{..})$ ... $(\bar{y}_{\cdot j}-\bar{y}_{..})$ ... $(\bar{y}_{\cdot k}-\bar{y}_{..})$	Grand mean $\bar{y}_{..}$

The number of distinct entries in the treatment deviations array is k, and the single constraint is that they must sum to zero. Thus $k-1$ degrees of freedom are associated with treatment sum of squares.

Sum of Squares due to Treatment

$$SS_T = \sum_{i=1}^{b} \sum_{j=1}^{k} (\bar{y}_{\cdot j} - \bar{y}_{\cdot\cdot})^2$$

$$= b \sum_{j=1}^{k} (\bar{y}_{\cdot j} - \bar{y}_{\cdot\cdot})^2$$

with d.f. $= k-1$.

In the array of treatment effects in Table 14.10, each entry appears $b=5$ times, once in each block. The treatment sum of squares for this example is then

$$SS_T = 5(-4)^2 + 5(6)^2 + 5(1)^2 + 5(-3)^2 = 310$$

with d.f. $= (4-1) = 3$.

In a similar manner:

Sum of Squares due to Block

$$SS_B = \sum_{i=1}^{b} \sum_{j=1}^{k} (\bar{y}_{i\cdot} - \bar{y}_{\cdot\cdot})^2$$

$$= k \sum_{i=1}^{b} (\bar{y}_{i\cdot} - \bar{y}_{\cdot\cdot})^2$$

with d.f. $= b-1$.

Referring to the array of block effects in Table 14.10, the block sum of squares for our example is found to be

$$SS_B = 4(4)^2 + 4(-3)^2 + 4(2)^2 + 4(-4)^2 + 4(1)^2 = 184$$

with d.f. $= 5-1 = 4$.

The number of degrees of freedom associated with the residual array is $(b-1)(k-1)$. To understand why this is so, note that among the $b \times k$ residuals the following constraints are satisfied. One constraint is that the sum of all entries be zero. The fact that all row sums are zero introduces $b-1$ additional constraints. This is so because having fixed any $b-1$ row totals and the grand total, the remaining row total is automatically fixed. By the same reasoning, $k-1$ additional constraints arise from the fact that all column totals are zero. Consequently, the number of degrees of freedom for the residuals is

$$bk-1-(b-1)-(k-1)=(b-1)(k-1)$$

Residual Sum of Squares

$$\text{SSE}=\sum_{i=1}^{b}\sum_{j=1}^{k}(y_{ij}-\bar{y}_{i\cdot}-\bar{y}_{\cdot j}+\bar{y}_{\cdot\cdot})^2$$

with d.f. $=(b-1)(k-1)$.

In our example

$$\text{SSE}=2^2+(0)^2+(-2)^2+0^2+\ldots+(-2)^2=24$$

with d.f. $=(5-1)(4-1)=12$.

Finally, the total sum of squares equals the sum of squares of each observation about the grand mean, or

$$\text{Total Sum of Squares}=\sum_{i=1}^{b}\sum_{j=1}^{k}(y_{ij}-\bar{y}_{\cdot\cdot})^2=\sum_{i=1}^{b}\sum_{j=1}^{k}y_{ij}^2-bk\bar{y}_{\cdot\cdot}^2$$

with d.f. $= bk-1$. In our example, we sum the square of each entry in the array for y_{ij} and subtract the sum of squares of entries in the $\bar{y}_{\cdot\cdot}$ array.

$$\text{Total Sum of Squares}=(12)^2+(2)^2+\ldots+(6)^2-\left[(10)^2+(10)^2+\ldots(10)^2\right]$$

$$=518$$

This provides a check on our previous calculations, because those sums of squares and degrees of freedom must sum to these totals.

These calculations are conveniently summarized in the ANOVA tables shown in Table 14.12 for the general case and in Table 14.13 for the data in our numerical example. The last column of F ratios will be explained after we discuss the population model.

TABLE 14.12
ANOVA TABLE FOR A RANDOMIZED BLOCK DESIGN

SOURCE	SUM OF SQUARES	d.f.	MEAN SQUARE	F RATIO
Treatments	$SS_T = b \sum_{j=1}^{k} (\bar{y}_{\cdot j} - \bar{y}_{\cdot\cdot})^2$	$k-1$	$MS_T = \frac{SS_T}{k-1}$	$\frac{MS_T}{MSE}$
Blocks	$SS_B = k \sum_{i=1}^{b} (\bar{y}_{i\cdot} - \bar{y}_{\cdot\cdot})^2$	$b-1$	$MS_B = \frac{SS_B}{b-1}$	$\frac{MS_B}{MSE}$
Residual	$SSE = \sum_{i=1}^{b} \sum_{j=1}^{k} (y_{ij} - \bar{y}_{i\cdot} - \bar{y}_{\cdot j} + \bar{y}_{\cdot\cdot})^2$	$(b-1)(k-1)$	$MSE = \frac{SSE}{(b-1)(k-1)}$	
Total	$\sum_{i=1}^{b} \sum_{j=1}^{k} (y_{ij} - \bar{y}_{\cdot\cdot})^2$	$bk-1$		

TABLE 14.13
ANOVA TABLE FOR THE DATA GIVEN IN EXAMPLE 14.8

SOURCE	SUM OF SQUARES	d.f.	MEAN SQUARE	F RATIO
Treatments	310	3	103.3	51.7
Blocks	184	4	46	23
Residual	24	12	2	
Total	518	19		

Again, a statistical test of treatment differences is based on an underlying population model.

Population model for a randomized block experiment

$$\text{Observation} = \begin{matrix}\text{Overall}\\ \text{mean}\end{matrix} + \begin{matrix}\text{Block}\\ \text{effect}\end{matrix} + \begin{matrix}\text{Treatment}\\ \text{effect}\end{matrix} + \text{Error}$$

$$Y_{ij} = \mu + \alpha_i + \beta_j + e_{ij}$$

$$i = 1, \ldots, b, \qquad j = 1, \ldots, k$$

where the parameters satisfy

$$\sum_{i=1}^{b} \alpha_i = 0, \qquad \sum_{j=1}^{k} \beta_j = 0$$

and the e_{ij} are random errors independently distributed as $N(0, \sigma)$.

Tests for the absence of treatment differences or the absence of differences in block effects can now be performed by comparing the corresponding mean square with the yardstick of the residual mean square by using an F test.

> Reject $H_0: \beta_1 = \ldots = \beta_k = 0$ (no treatment differences) if
>
> $$\frac{\text{MS}_T}{\text{MSE}} > F_\alpha(k-1, (b-1)(k-1))$$
>
> Reject $H_0: \alpha_1 = \ldots = \alpha_b = 0$ (no block differences) if
>
> $$\frac{\text{MS}_B}{\text{MSE}} > F_\alpha(b-1), (b-1)(k-1))$$

To test the hypothesis of no treatment differences for the analysis of variance in Table 14.13, we find that the tabulated .05 point of $F(3, 12)$ is 3.49, a value far exceeded by the observed F ratio for treatment effect. We therefore conclude that a highly significant treatment difference is indicated by the data. The block effects are also highly significant, because the observed F value of 24 is much larger than the tabulated value $F_{.05}(4, 12) = 3.26$.

Again we stress that a serious violation of the model assumptions is likely to jeopardize the conclusions drawn from the preceding analyses and that a careful examination of the residuals should be an integral part of the analysis. In addition to plotting the whole set of residuals in a graph, separate plots for individual treatments and individual blocks should also be studied. When observations are collected over time, a plot of the residuals vs. the time order is also important.

In addition to normality and constant variance, a somewhat restrictive assumption of the present model is that both the treatment and the block effects are additive. In other words, the population mean response of the jth treatment applied to a unit in the ith block is modeled $\mu + \alpha_i + \beta_j$, which aside from the overall mean, is the sum of a component purely due to treatment effect and a component purely due to block effect. When the relative performance of two treatments varies appreciably from one block to another, we say that there is a *block–treatment interaction*. The additive model is then inappropriate. If the different types of cutting tools in Example 14.8 perform best on different kinds of materials, then the additive model is vitiated.

Confidence Intervals for Treatment Differences

In addition to performing the overall F test for detecting treatment differences, the experimenter typically establishes confidence intervals to compare specific pairs of treatments. This is particularly important when the F test leads to a rejection of the null hypothesis, thus signifying the presence of treatment differences.

The difference $\beta_j - \beta_{j'}$ of the mean responses of treatments j and j' is estimated by $\bar{Y}_{.j} - \bar{Y}_{.j'}$, which is normally distributed with a mean of $\beta_j - \beta_{j'}$ and a variance of

$$\sigma^2\left(\frac{1}{b} + \frac{1}{b}\right) = \sigma^2\left(\frac{2}{b}\right)$$

The ratio

$$t = \frac{\left(\bar{Y}_{.j} - \bar{Y}_{.j'}\right) - (\beta_j - \beta_{j'})}{\sqrt{\text{MSE}}\sqrt{(2/b)}}$$

has a t distribution with d.f. $= (b-1)(k-1)$, which can be used to construct a confidence interval for an individual difference $\beta_j - \beta_{j'}$.

When several such pairwise comparisons are to be integrated into a combined confidence statement, the concept of simultaneous confidence intervals, discussed in Section 14.3, is again applied. We have already discussed the procedure called multiple-t confidence intervals. The present design has an additional feature in that all treatments have an equal number of replications. A design having this property is said to be *balanced*. More efficient methods of multiple comparisons are available for balanced designs.

14.5 FACTORIAL EXPERIMENTS

A major conceptual advancement in experimental design is exemplified by factorial designs. More often than not, scientific investigations are conducted to study the effects of several factors on the response. By varying the factors simultaneously during the course of an experiment, information can be gathered not only about the effects of individual factors but also about the manner in which these factors interact. This is in direct contrast to the old-fashioned concept of conducting separate experiments in which one factor is varied at a time. A major drawback of these one-at-a-time experiments is that they do not provide any information about the possible interaction between factors. Moreover, the set of measurements, collected by varying one factor, is useful in making inferences about that factor alone. In factorial designs, an assessment of each individual factor effect is based on the whole set of measurements, so that a more efficient utilization of experimental resources is achieved in these designs.

For an introduction to the basic ideas of a factorial design, let us suppose that two factors A and B are to be studied in an experiment, where

Factor A has p levels.
Factor B has q levels.

Each combination of a level of A with a level of B defines a treatment, so that there are pq number of treatments in this experiment. The design is then called a $p \times q$ factorial design. A single replication consists of assigning pq experimental units at random, one to each treatment. The resulting data structure can be presented in a two-way table, as shown in Table 14.14. If this experiment is repeated r times using r sets of pq experimental units, we will have a $p \times q$ factorial experiment with r replicates. Example 14.2 illustrated a 2×3 factorial experiment with factors of stimulus type and warning–stimulus time interval.

TABLE 14.14
DATA STRUCTURE FOR A $p \times q$ FACTORIAL DESIGN WITH A SINGLE REPLICATE

		LEVELS OF FACTOR B			
		B_1	B_2	...	B_q
Levels of	A_1	y_{11}	y_{12}	...	y_{1q}
Factor A	A_2	y_{21}	y_{22}	...	y_{2q}
	.	.	.	...	.
	.	.	.	...	.
	.	.	.	...	.
	A_p	y_{p1}	y_{p2}	...	y_{pq}

We can now note that the data structure in Table 14.14 is identical to the randomized block design in Table 14.8. In the randomized block design, the factor "Block" serves to reduce the error variance and thereby to increase the precision of inferences about the primary factor "treatment." The intent of that design is to eliminate the block effect, not to study it. In the factorial design, inferences are desired for each individual factor effect and, if possible, about their interactions as well. However, the decomposition of observations and the ANOVA table are formally identical to those for a randomized block design, with blocks playing the role of factor A and treatments playing the role of factor B. A single replication of the design does not readily permit us to make inferences about possible interactions between the factors. The analysis is based on the assumption of an additive model.

Additive model for a two factor experiment

$$Y_{ij}=\mu+\alpha_i+\beta_j+e_{ij}$$

where μ = overall mean

α_i = effect of the ith level of A, $\sum_{i=1}^{p} \alpha_i=0$

β_j = effect of the jth level of B, $\sum_{j=1}^{q} \beta_j=0$

and e_{ij} are random errors independently distributed as $N(0,\sigma)$.

Decomposition of observations

Observation = Grand mean + Deviation due to Factor A + Deviation due to Factor B + Residual

$$y_{ij} = \bar{y}_{..} + (\bar{y}_{i.}-\bar{y}_{..}) + (\bar{y}_{.j}-\bar{y}_{..}) + (y_{ij}-\bar{y}_{i.}-\bar{y}_{.j}+\bar{y}_{..})$$

The decomposition of the sum of squares appears in the ANOVA table given in Table 14.15, where the mean square for each factor is compared with the residual mean square by using an F test.

TABLE 14.15
ANOVA TABLE FOR A $p\times q$ FACTORIAL DESIGN WITH ONE REPLICATE

SOURCE	SUM OF SQUARES	d.f.	MEAN SQUARE	F RATIO
Factor A	$SS_A=q\sum_{i=1}^{p}(\bar{y}_{i.}-\bar{y}_{..})^2$	$p-1$	MS_A	$\frac{MS_A}{MSE}$
Factor B	$SS_B=p\sum_{j=1}^{q}(\bar{y}_{.j}-\bar{y}_{..})^2$	$q-1$	MS_B	$\frac{MS_B}{MSE}$
Residual	$SSE=\sum_{i=1}^{p}\sum_{j=1}^{q}(y_{ij}-\bar{y}_{i.}-\bar{y}_{.j}+\bar{y}_{..})^2$	$(p-1)(q-1)$	MSE	
Total	$\sum_{i=1}^{p}\sum_{j=1}^{q}(y_{ij}-\bar{y}_{..})^2$	$pq-1$		

14.5.1 Interaction and Its Interpretation

In the models described thus far, we have assumed that

$$E(Y_{ij}) = E(\mu + \alpha_i + \beta_j + e_{ij}) = \mu + \alpha_i + \beta_j$$

Now we consider a comparison of the population means for levels 1 and 3 of factor A at a fixed level j of factor B. The expected values are compared by

$$\begin{pmatrix}\text{Mean of level}\\ 1 \text{ of } A\end{pmatrix} - \begin{pmatrix}\text{Mean of level}\\ 3 \text{ of } A\end{pmatrix} = \mu + \alpha_1 + \beta_j - (\mu + \alpha_3 + \beta_j)$$
$$= \alpha_1 - \alpha_3$$

for all levels j of factor B. The difference is the same whatever the level of factor B. This model is called an *additive model*, because the effects due to each of the factors are added. This property is graphically illustrated in Figure 14.3, where the difference between any two curves is a constant for all levels of factor B.

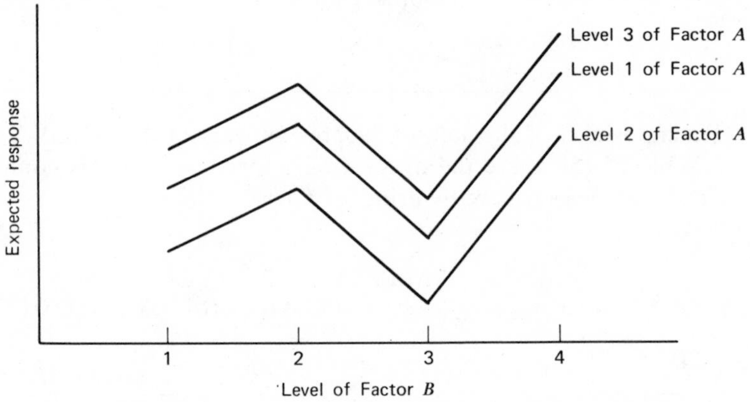

Figure 14.3 Curves for expected response without interaction.

A contrasting situation is graphed in Figure 14.4, where the difference in mean response between levels 1 and 3 of factor A is sometimes positive and sometimes negative. Because the behavior here depends on the level of factor B, factors A and B are said to *interact*. In such cases the additive model does not hold.

Interactions frequently occur in practice, and it is important to recognize their presence. If they are not properly accounted for, they can obscure the

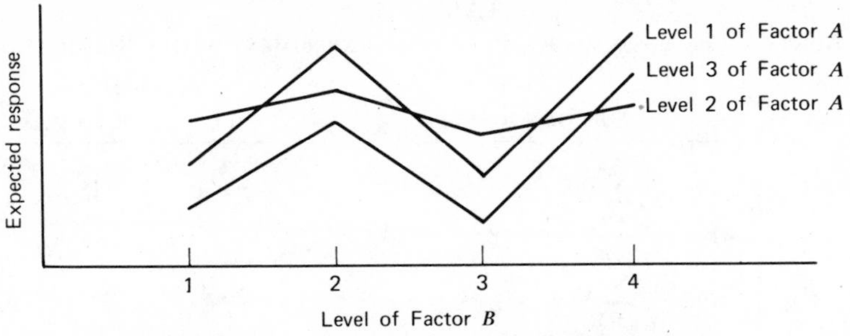

Figure 14.4 Curves for expected response with interaction.

main effects of an experiment. Our model can be extended to include an interaction term γ_{ij}, which depends on the level i of factor A and the level j of factor B, or

$$\text{Observation} = \mu + \alpha_i + \beta_j + \gamma_{ij} + \text{Error}$$

However, as we mentioned earlier, a single replication of the $p \times q$ experiment does not permit us to make inferences about both the main effects and the interactions. If we attempt to do so, the component we previously termed "residual" must now be used to estimate the interaction parameter γ_{ij}, and no term remains with which to estimate the error variance. When the interaction component is included in the model, more than one replication of the factorial experiment is necessary. With r number of replications of a $p \times q$ factorial experiment, we denote by y_{ijk} the kth observation with treatment (A_i, B_j). Again using dots in the subscripts to designate averages, the decomposition is then

Observation	=	Grand mean	+	Deviation due to Factor A	+	Deviation due to Factor B
y_{ijk}	=	$\bar{y}_{\dots}$	+	$(\bar{y}_{i\cdot\cdot} - \bar{y}_{\dots})$	+	$(\bar{y}_{\cdot j\cdot} - \bar{y}_{\dots})$
			+	Interaction of A_i with B_j	+	Residual
			+	$(\bar{y}_{ij\cdot} - \bar{y}_{i\cdot\cdot} - \bar{y}_{\cdot j\cdot} + \bar{y}_{\dots})$	+	$(y_{ijk} - \bar{y}_{ij\cdot})$

The analysis of variance appears in Table 14.16.

An example of a 2×2 factorial design with two replicates should help to clarify these computations. A more vivid interpretation of these effects and interactions is given in geometric terms in Section 14.6.

TABLE 14.16

ANOVA Table for a $p \times q$ Factorial Experiment with r Replicates

SOURCE	SUM OF SQUARES	d.f.	MEAN SQUARE	F RATIO
Factor A	$SS_A = qr \sum_{i=1}^{p} (\bar{y}_{i\cdot\cdot} - \bar{y}_{\dots})^2$	$p-1$	MS_A	$\frac{MS_A}{MSE}$
Factor B	$SS_B = pr \sum_{j=1}^{q} (\bar{y}_{\cdot j\cdot} - \bar{y}_{\dots})^2$	$q-1$	MS_B	$\frac{MS_B}{MSE}$
Interaction $A \times B$	$SS_{AB} = r \sum_{i=1}^{p} \sum_{j=1}^{q} (\bar{y}_{ij\cdot} - \bar{y}_{i\cdot\cdot} - \bar{y}_{\cdot j\cdot} + \bar{y}_{\dots})^2$	$(p-1)(q-1)$	MS_{AB}	$\frac{MS_{AB}}{MSE}$
Residual	$SSE = \sum_{i=1}^{p} \sum_{j=1}^{q} \sum_{k=1}^{r} (y_{ijk} - \bar{y}_{ij\cdot})^2$	$pq(r-1)$	MSE	
Total	$\sum_{i=1}^{p} \sum_{j=1}^{q} \sum_{k=1}^{r} (y_{ijk} - \bar{y}_{\dots})^2$	$pqr-1$		

Example 14.9 A photographer who wishes to improve the clarity of developed pictures adds two amounts of Metol (1.5 and 2.5 grams) and two amounts of hydroquinone (4 and 6 grams) to a liter of negative developer. The resulting clarity readings appear in Table 14.17.

TABLE 14.17

Clarity Readings

		HYDROQUINONE 4	6	
Metol	1.5	28, 30	33, 33	$\bar{y}_{1\cdot\cdot} = 31$
	2.5	42, 38	40, 42	$\bar{y}_{2\cdot\cdot} = 40.5$
		$\bar{y}_{\cdot 1\cdot} = 34.5$	$\bar{y}_{\cdot 2\cdot} = 37$	$\bar{y}_{\dots} = 35.75$

Construct the ANOVA table for these data.

Identifying Metol as factor A and hydroquinone as factor B, the notation is $2 = q = p, \quad r = 2$. Thus

$$SS_A = 2 \times 2\left[(31 - 35.75)^2 + (40.5 - 35.75)^2\right] = 180.5$$

$$SS_B = 2 \times 2\left[(34.5 - 35.75)^2 + (37 - 35.75)^2\right] = 12.5$$

$$SS_{AB} = 2\left[(29 - 31 - 34.5 + 35.75)^2 + \cdots + (41 - 40.5 - 37 + 35.75)^2\right] = 4.5$$

$$\text{SSE} = (28-29)^2 + (30-29)^2 + (33-33)^2 + (33-33)^2 + (42-40)^2$$

$$+ (38-40)^2 + (40-41)^2 + (42-41)^2$$

$$= 12.0$$

$$\text{Total SS} = (28-35.75)^2 + (30-35.75)^2 + \cdots + (42-35.75)^2 = 209.5$$

The ANOVA table for these data is given in Table 14.18.

TABLE 14.18
ANOVA TABLE FOR CLARITY READINGS

SOURCE	SUM OF SQUARES	d.f.	MEAN SQUARE	F RATIO
Metol	180.5	1	180.5	60.2
Hydroquinone	12.5	1	12.5	4.2
Interaction	4.5	1	4.5	1.5
Residual	12.0	4	3	
Total	209.5	7		

The upper 5% point of $F(1,4)$ is 7.71, so that over the range of values in the experiment only the amount of Metol has a significant effect on photographic clarity. ■

Even when replicates are unavailable, the array of residuals may reveal a systematic pattern that indicates the presence of an interaction. In such cases, this pattern can be further investigated by conducting repeated experimental trials.

14.6 TWO-LEVEL FACTORIAL DESIGNS

The full force of the power of factorial design is dramatically illustrated by the *two-level factorial design*, which derives its name from the fact that each factor in the design is tried at two settings or in two versions. All combinations of the input variables are included in the two designs discussed here.

Historically, scientific workers tend to overemphasize the experimental technique of varying the input variables *one at a time*. According to this method, the experimenter must first isolate the elementary factors that may

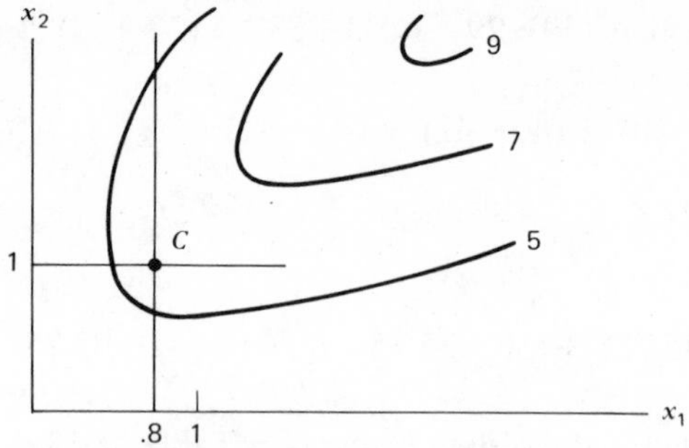

Figure 14.5 False location of the maximum.

be causal variables and then vary one factor while holding all the others fixed. Although this experimental design gives the appearance of simplicity, this one-at-a-time procedure can result in seriously misleading conclusions and its use is rather limited in current scientific research.

To see why this is true, suppose that factor levels x_1 and x_2 are varied to obtain an optimum response. Figure 14.5 illustrates a situation in which the response increases from 5 to 7 to 9 as (x_1, x_2) move toward the upper right-hand corner of the graph. If we first vary x_1 and hold x_2 fixed at 1, we conclude that point C yields the best response. If we then hold x_1 fixed at .8 and vary x_2, we *wrongly* conclude that condition C gives the maximum response.

Factorial designs are particularly well suited to situations in which a large number of factors can influence the response. These designs provide an efficient method of studying not only the individual (main) effects but also the effects caused by interactions among the factors.

14.6.1 2^2 Factorial Design

When each of two factors are tried in two versions, there are $2 \times 2 = 2^2$ treatments or experimental conditions. This fact gives rise to the name *2^2 factorial design*.

TABLE 14.19
2^2 FACTORIAL DESIGN IN STANDARD ORDER

DESIGN	
x_1	x_2
-1	-1
$+1$	-1
-1	$+1$
$+1$	$+1$

It is convenient to code the factors so that the low level is -1 and the high level is $+1$. If the two levels of a factor are nonquantitative (say, two different brands of soap), we simply designate one level -1 and the other $+1$. Let us reconsider Example 14.9 from this viewpoint. The two factors are quantitative and can be coded by the formulas

	Low	High	
Metol:	1.5	2.5	coded $x_1 = \dfrac{M-(1.5+2.5)/2}{(2.5-1.5)/2} = \dfrac{M-2}{.5}$
Hydroquinone:	4	6	coded $x_2 = \dfrac{H-(4+6)/2}{(6-4)/2} = H-5$

The four possible experimental conditions shown in Table 14.19 in standard order are then $(-1,-1)$ $(+1,-1)$, $(-1,+1)$, and $(+1,+1)$, where, for instance $(-1,+1)$ is the experimental run using 1.5 grams of Metol and 6 grams of hydroquinone. Geometrically, these four sets of conditions are positioned at the corners of a square, as shown in Figure 14.6.

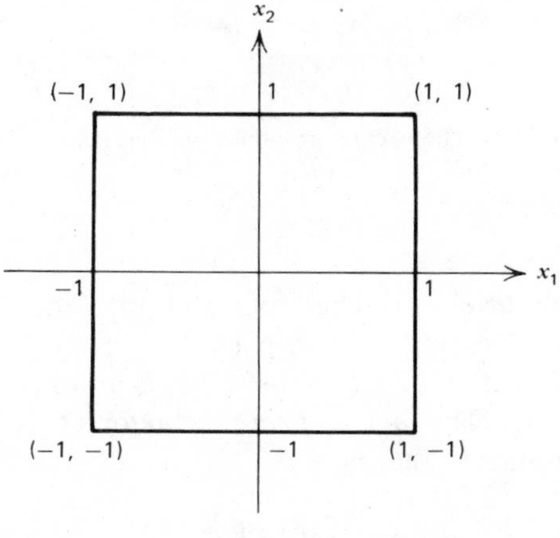

Figure 14.6 The coded 2^2 design.

The complete experiment is conducted twice, so that eight observations, two for each set of experimental conditions, are available. Dropping the 1s and retaining only the signs, we can write the design and observations shown in Table 14.20.

We now describe a method for estimating the effects of the two factors and their interaction. The averages of the two responses at fixed experimental conditions are used to estimate the effects. The difference between

TABLE 14.20
OBSERVATIONS ON PHOTOGRAPHIC CLARITY

RUN	DESIGN x_1	x_2	OBSERVATIONS REPLICATE 1	REPLICATE 2	AVERAGE
1	−	−	28	30	$\bar{y}_1=29$
2	+	−	42	38	$\bar{y}_2=40$
3	−	+	33	33	$\bar{y}_3=33$
4	+	+	40	42	$\bar{y}_4=41$

the two responses then contributes to an estimate of the error variance.

A major but simple step toward understanding the data is recording the average responses on the square, as shown in Figure 14.7.

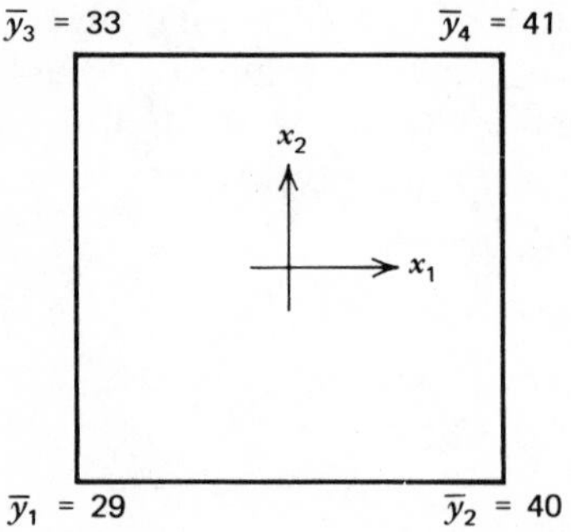

Figure 14.7 The average response on the square at each condition.

14.6.2 Estimates of the Main Effects

Runs 1 and 2 both have x_2 at the low level -1. Therefore, the difference $(40-29)$ is a measure of the effect of increasing x_1 from -1 to $+1$. Runs 3 and 4 also have the same x_2 value. Thus the difference $(41-33)$ is another estimate of the effect of changing x_1. The *main effect* of factor A is estimated by averaging these two estimates. That is,

$$\text{estimate of main effect of Factor } A = \frac{(\bar{y}_2-\bar{y}_1)+(\bar{y}_4-\bar{y}_3)}{2}$$

$$= \frac{(40-29)+(41-33)}{2} = 9.5$$

Similarly,

$$\text{estimate of main effect of Factor } B = \frac{(\bar{y}_3-\bar{y}_1)+(\bar{y}_4-\bar{y}_2)}{2}$$

$$= \frac{(33-29)+(41-40)}{2} = 2.5$$

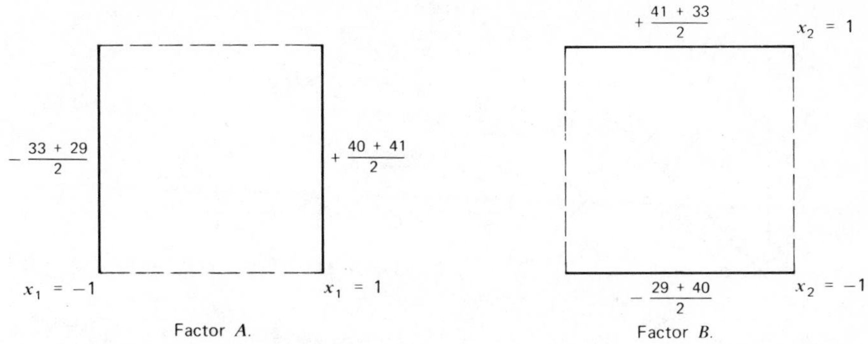

Figure 14.8 Estimates of the main effects of factors A and B in the 2^2 design.

These estimates can also be interpreted as the average response on the face of the square with level $+1$ minus the average response on the face of the square with level -1. This view is shown in Figure 14.8.

14.6.3 Interaction Effect

We can also define an estimate of interaction by taking the difference between the increase in yield at the high level $(41-33)$ and at the low level $(40-29)$ of factor B. If the two ingredients Metol and hydroquinone do not interact, these increases should be approximately the same and their difference should be nearly zero. Therefore, the

$$\text{estimate of interaction effect} = \frac{(\bar{y}_4-\bar{y}_3)-(\bar{y}_2-\bar{y}_1)}{2}$$

$$= \frac{(41-33)-(40-29)}{2} = -1.5$$

Note that the same answer is obtained if we subtract the differences for the increases due to the second factor, or $[(41-40)-(33-29)]/2 = -1.5$. A second interpretation in terms of the average response on diagonals is illustrated in Figure 14.9.

In developing an estimate of the variance, we first note that each estimate of an effect is of the form

$$\frac{1}{2}\left[\pm\bar{y}_1 \pm\bar{y}_2 \pm\bar{y}_3 \pm\bar{y}_4\right]$$

where the signs correspond to the signs of x_1 for the Metol effect, x_2 for hydroquinone effect, and the product $x_1 \cdot x_2$ for the interaction. Before

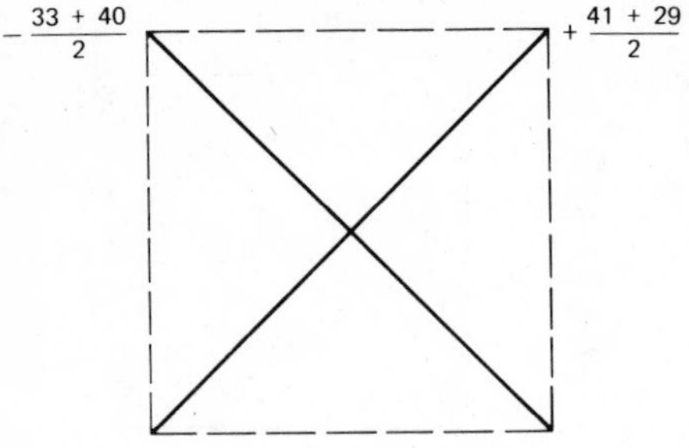

Figure 14.9 Estimate of interaction.

these observations are made, the means are random variables. If r replications are available, the variance of each sample mean is σ^2/r. Because these means are independent, the variances can be summed. Thus

$$\text{Var(estimator)} = \frac{1}{4}\left[\frac{\sigma^2}{r} + \frac{\sigma^2}{r} + \frac{\sigma^2}{r} + \frac{\sigma^2}{r}\right] = \frac{\sigma^2}{r}$$

where r represents the number of replications.

We have already calculated an estimate of σ^2 as the MSE entry shown in Table 14.16. Another approach to the same calculation is to extend the concept of pooling employed with the two-sample t statistic. Letting $y_{11}, y_{12}, \ldots, y_{1r}$ denote r observations at $x_1 = -1$, $x_2 = -1$, we find $(r-1)s_1^2 = \sum_{j=1}^{r}(y_{1j} - \bar{y}_1)^2$ and, similarly, $(r-1)s_2^2$, $(r-1)s_3^2$, and $(r-1)s_4^2$ from observations at the other (x_1, x_2) settings. An estimate of σ^2 can then be derived by pooling or adding all these contributions and dividing by the number of degrees of freedom $(r-1)+(r-1)+(r-1)+(r-1) = 4(r-1)$.

$$s^2_{\text{pooled}} = \frac{1}{4}\left(s_1^2 + s_2^2 + s_3^2 + s_4^2\right)$$

In our photography example, $r=2$ and $(2-1)s_1^2 = (28-29)^2 + (30-29)^2 = 2$. Moreover, $s_2^2 = (42-40)^2 + (38-40)^2 = 8$, $s_3^2 = 0$, and $s_4^2 = 2$, so that $s^2_{\text{pooled}} = (2+8+2)/4 = 3$. Of course, this is equal to the value previously calculated for MSE in Table 14.18.

14.6.4 Confidence Intervals for the Effects

Each estimated effect is known to have a variance of σ^2/r, which can be estimated by s^2_{pooled}/r.

A 95% confidence interval for an effect, based on r replications of a 2^2 factorial design, is

$$\text{Estimated effect} \pm \sqrt{\frac{s^2_{\text{pooled}}}{r}}\; t_{.025}$$

where d.f. $= 4(r-1)$ for $t_{.025}$.

We are now in a position to summarize the conclusions of our photography experiment. The number of replicates is $r=2$, and consulting the t table for d.f. $=4$, we find that $t_{.025}=2.776$. Then, $\sqrt{s^2_{\text{pooled}}/2}\; t_{.025} = \sqrt{3/2}\; 2.776 = 3.4$, so that

Metol effect	9.5 ± 3.4	or	$(6.1, 12.9)$
Hydroquinone effect	2.5 ± 3.4	or	$(-.9, 5.9)$
Interaction	-1.5 ± 3.4	or	$(-4.9, 1.9)$

Our conclusion is that only the main effect of Metol seems to be real. Moreover, we have now constructed a confidence interval for the size of the effect.

A slight caution should be given here, because several confidence statements are being made simultaneously using a procedure that is designed for single statements. We could use the $t_{.025/3}$ value to be on the safe side.

If the interaction effect is ever judged to be significant, we *cannot* conclude that the main effect is not significant. Instead, in the presence of interaction, the factors must be considered jointly. The best summary is then given by a two-way table of means, as shown in Table 14.21.

TABLE 14.21

A SUMMARY OF A 2^2 FACTORIAL DESIGN WHEN INTERACTION IS PRESENT

	FACTOR B	
FACTOR A	LOW	HIGH
Low	$\bar{y}_1$	$\bar{y}_3$
High	$\bar{y}_2$	$\bar{y}_4$

14.6.5 2^3 Factorial Designs

By considering one additional factor or a total of three factors, we arrive at the 2^3 factorial design. Considering three factors, each at two levels, provides a total of $2 \times 2 \times 2 = 2^3$ or 8 combinations. Coding the low level -1 and the high level $+1$ for each variable, the design can be written in *standard form*, as shown in Table 14.22. There, the first column x_1 contains $-+-+-+-+$, the second x_2 contains $--++--++$, and the third column x_3 contains $----++++$.

TABLE 14.22

A 2^3 Factorial Design in Standard Order

RUN	x_1	x_2	x_3
1	−	−	−
2	+	−	−
3	−	+	−
4	+	+	−
5	−	−	+
6	+	−	+
7	−	+	+
8	+	+	+

We develop our examination of 2^3 factorial designs on the basis of the following problem.

A designer in charge of modifying carburetors wishes to know if using a new spring and slightly changing two critical dimensions will increase the running time achieved with one gallon of gas. The following experiment is conducted:

	Factor	Low	High	Coding
1	Spring	old	new	$x_1 = -1$ old spring; $+1$ new spring
2	Dimension A	50 mm	55 mm	$x_2 = \dfrac{A - 52.5}{2.5}$
3	Dimension B	20 mm	25 mm	$x_3 = \dfrac{B - 22.5}{2.5}$

The design and the observations are shown in Table 14.23.

Estimates of the main effects and interactions derived below will all be of the form

$$\frac{1}{2^{3-1}}\left[\pm y_1 \pm y_2 \ldots \pm y_8\right]$$

TABLE 14.23
A 2^3 FACTORIAL DESIGN

DESIGN			MULTIPLICATION OF DESIGN COLUMNS				OBSERVATION TIME
x_1	x_2	x_3	x_1x_2	x_1x_3	x_2x_3	$x_1x_2x_3$	(IN MINUTES)
−	−	−	+	+	+	−	$y_1=68$
+	−	−	−	−	+	+	$y_2=65$
−	+	−	−	+	−	+	$y_3=72$
+	+	−	+	−	−	−	$y_4=70$
−	−	+	+	−	−	+	$y_5=48$
+	−	+	−	+	−	−	$y_6=50$
−	+	+	−	−	+	−	$y_7=78$
+	+	+	+	+	+	+	$y_8=80$

where the signs are obtained from the corresponding x columns or their products. For example, the main effect of factor 2 is estimated by $[-y_1-y_2+y_3+y_4-y_5-y_6+y_7+y_8]/2^{3-1}$. Although the experimental runs are *always* performed in random order, the advantage of arranging the results in the standard order should now be apparent.

A plot of these data, given in parentheses at the corners $(\pm 1, \pm 1, \pm 1)$ of the cube in Figure 14.10, already suggest the improved design conditions.

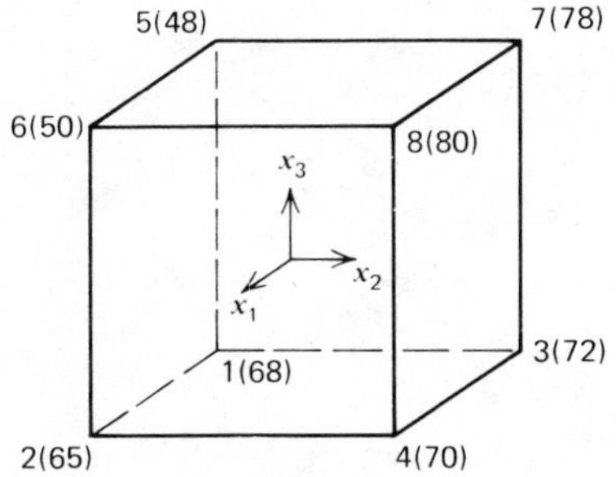

Figure 14.10 A visual presentation of the responses in the 2^3 design.

14.6.6 Estimates of the Main Effects

The difference (y_2-y_1) measures the effect of increasing factor 1 from -1 to $+1$, because both x_2 and x_3 are at their low levels. Similarly, (y_4-y_3), (y_8-y_7), and (y_6-y_5) also measure the increase in response when x_1 changes from -1 to $+1$. We define the main effect as the average of these four differences. Thus the

$$\begin{aligned}\text{estimate of the main effect of factor 1} &= \frac{1}{2^{3-1}}\left[(y_2-y_1)+(y_4-y_3)+(y_6-y_5)+(y_8-y_7)\right]\\ &= \frac{1}{4}\left[-y_1+y_2-y_3+y_4-y_5+y_6-y_7+y_8\right]\\ &= \frac{1}{4}\left[-68+65-72+70-48+50-78+80\right]\\ &= \frac{1}{4}\left[-1\right]=-.25\end{aligned}$$

A second interpretation of the estimate can also be made. The four positive responses are all positioned on the face of the cube with $x_1=1$, so that the estimate of the main effect of factor 1 is the difference between two averages. Both of these calculation methods are illustrated geometrically in Figure 14.11.

Using Figure 14.11(b), we can alternatively calculate the estimated main effect of factor 1 as

$$\frac{1}{4}(65+70+50+80)-\frac{1}{4}(68+72+48+78)=-.25$$

Similiarly, we find

$$\begin{aligned}\text{estimate of the main effect of factor 2} &= \frac{1}{2^{3-1}}\left[-y_1-y_2+y_3+y_4-y_5-y_6+y_7+y_8\right]\\ &= \frac{1}{4}\left[-68-65+72+70-48-50+78+80\right]\\ &= \frac{69}{4}=17.25\end{aligned}$$

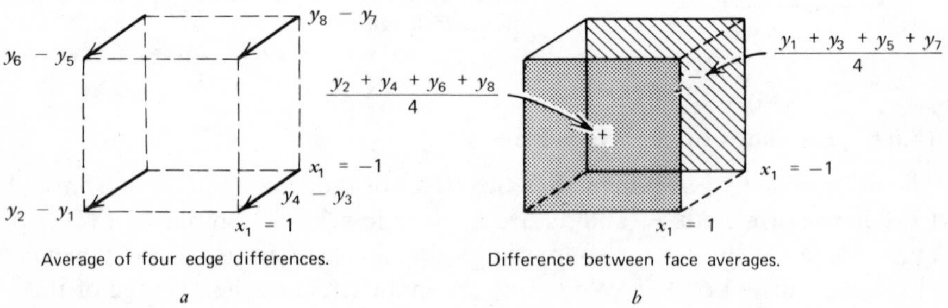

Figure 14.11 Geometric views of the estimate of the main effect of factor 1 in the 2^3 design.

$$\begin{aligned}\text{estimate of the main effect of factor 3} &= \frac{1}{2^{3-1}}[-y_1-y_2-y_3-y_4+y_5+y_6+y_7+y_8]\\ &= \frac{1}{4}[-68-65-72-70+48+50+78+80]\\ &= \frac{-19}{4} = -4.75\end{aligned}$$

14.6.7 Estimates of Two-factor Interactions

Focusing again on the calculation for the 2^2 factorial design, we can see that the bottom of the cube provides the value $(y_1+y_4)/2-(y_2+y_3)/2$ as an estimate of the interaction between factor 1 and factor 2 when factor 3 is held fixed at the low level. Similarly, the top of the cube provides the value $(y_5+y_8)/2-(y_6+y_7)/2$ at the high level of factor 3. An average of these two values, then yields

$$\begin{aligned}\text{Estimate of 1–2 interaction} &= \frac{1}{2^{3-1}}[y_1-y_2-y_3+y_4+y_5-y_6-y_7+y_8]\\ &= \frac{1}{4}[1] = .25\end{aligned}$$

$$\begin{aligned}\text{Estimate of 2–3 interaction} &= \frac{1}{2^{3-1}}[y_1+y_2-y_3-y_4-y_5-y_6+y_7+y_8]\\ &= \frac{1}{4}[51] = 12.75\end{aligned}$$

$$\begin{aligned}\text{Estimate of 1–3 interaction} &= \frac{1}{2^{3-1}}[y_1-y_2+y_3-y_4-y_5+y_6-y_7+y_8]\\ &= \frac{1}{4}[9] = 2.25\end{aligned}$$

The geometric views of the 1–2 interaction calculation appear in Figure 14.12.

14.6.8 Estimates of Three-factor Interaction

Referring to our examination of the 1–2 interaction, we note that if factor 3 did not influence the first factors 1 and 2, the calculations of interaction on the top and bottom faces of the cube would be approximately equal. The average difference of these two interactions is then considered to be

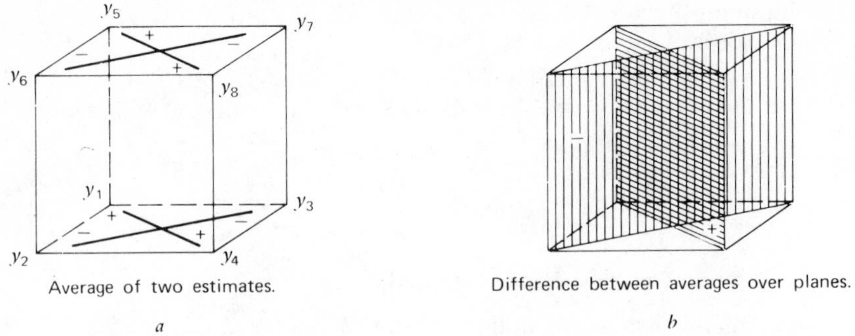

Figure 14.12 Calculation of the 1–2 interaction effect in the 2^3 design.

the estimate of the 1–2–3 interaction:

$$\begin{array}{l}\text{Estimate of} \\ \text{1–2–3 interaction}\end{array} = \frac{1}{2^{3-1}}\left[-y_1+y_2+y_3-y_4+y_5-y_6-y_7+y_8\right]$$

$$= \frac{1}{4}[-1] = -.25$$

It may prove interesting to verify that the result would be the same if we worked with any two-factor interaction. The geometric view of the 1–2–3 interaction calculation is presented in Figure 14.13. Referring back to Table 14.23, we can see that the signs attached to the ys are the same as the signs in the $x_1x_2x_3$ column.

If the experiment is not repeated and if no external estimate of σ^2 is available, the absolute value of the three-factor interaction estimate can be used if it appears small. Three-factor interactions are more rare than two-factor interactions, and if indeed it is absent its estimate is likely to be estimating zero. Accordingly, we can attach the approximate error bounds $\pm 2|.25| = \pm .5$ to the estimates in our carburetor study. Given this yardstick, it appears that only factors 1 and 2 do not interact. However, this

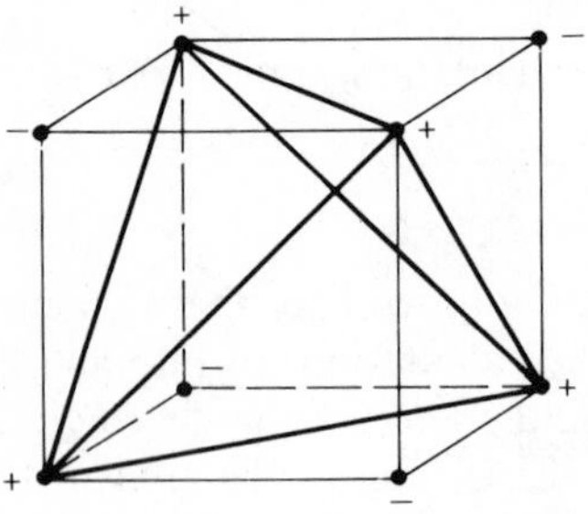

Figure 14.13 The three-factor interaction.

procedure is not entirely satisfactory, and it is better to estimate the variance on the basis of repeated runs.

When the complete design is replicated r times, the averages are used as entries in the preceding calculations.

> Estimated effects in the 2^3 factorial design
>
> $$\frac{1}{2^{3-1}}[\pm\bar{y}_1 \pm \bar{y}_2 \ldots \pm \bar{y}_8]$$
>
> where the signs correspond to the columns in Table 14.23.

Because the variance of each average, when considered to be a random variable is σ^2/r,

$$\text{Var(estimator)} = \text{Var}\left(\frac{1}{4}\left[\pm \bar{Y}_1 \pm \bar{Y}_2 \ldots \pm \bar{Y}_8\right]\right)$$

$$= \frac{1}{16}\left(\frac{\sigma^2}{r} + \frac{\sigma^2}{r} + \ldots + \frac{\sigma^2}{r}\right) = \frac{\sigma^2}{2r}$$

Again, we can pool the estimates $(r-1)s_1^2 = \Sigma_{j=1}^{r}(y_{1j} - \bar{y}_1)^2, \ldots, (r-1)s_8^2$ to obtain

$$s_{\text{pooled}}^2 = \frac{(r-1)s_1^2 + \ldots + (r-1)s_8^2}{8(r-1)} = \frac{1}{8}\left(s_1^2 + \ldots + s_8^2\right)$$

as an estimate of σ^2, based on d.f. $=(r-1)+(r-1)+\ldots+(r-1)=8(r-1)$.

> *A 95% confidence interval for an effect*, based on r replications of a 2^3 factorial design, is
>
> $$\text{Estimated effect} \pm \sqrt{\frac{s_{\text{pooled}}^2}{2r}}\; t_{.025}$$
>
> where d.f. $=8(r-1)$ for t.

For a more detailed treatment of factorial and other experimental designs, the interested student is referred to [1] and [2].

REFERENCES

1. Cochran, W. G., and Cox, G. M., *Experimental Design*, 2nd Ed., John Wiley & Sons, New York, 1957.
2. Cox, D. R., *Planning of Experiments*, John Wiley & Sons, New York, 1958.

EXERCISES

1. (*a*) Provide a decomposition for the following observations from a completely randomized design with three treatments:

Treatment 1	Treatment 2	Treatment 3
19	16	13
18	11	16
21	13	18
18	14	11
	11	15
		11

(*b*) Verify that the sum of all entries in the array $(\bar{y}_j - \bar{y})$ is zero, or that $\Sigma_{j=1}^{k} n_j(\bar{y}_j - \bar{y}) = 0$. Another way of expressing this is to state that $\Sigma_{j=1}^{k}\Sigma_{i=1}^{n_j}(\bar{y}_j - \bar{y}) = 0$, so that the sum over all entries of a $\bar{y}$ value *multiplied by* the value $(\bar{y}_j - \bar{y})$ from the second array is zero. (DEFINITION: Two arrays are said to be *orthogonal* if the sum of the products of the corresponding entries of the two arrays is zero.) The preceding result shows that the $\bar{y}$ array and the $(\bar{y}_j - \bar{y})$ array are orthogonal.

(*c*) Verify that the $(\bar{y}_j - \bar{y})$ array and the array of residuals are orthogonal.

2. Compute the sums of squares and construct the ANOVA table for the data given in Exercise 1.
3. As part of a multilab study, four fabrics are tested for flammability at the National Bureau of Standards. The following burn times in seconds are recorded after a paper tab is ignited on the hem of a dress made of each fabric:

Fabric 1	Fabric 2	Fabric 3	Fabric 4
17.8	11.2	11.8	14.9
16.2	11.4	11.0	10.8
17.5	15.8	10.0	12.8
17.4	10.0	9.2	10.7
15.0	10.4	9.2	10.7

(*a*) State the statistical model and present the ANOVA table. With $\alpha = .05$, test the null hypothesis of no difference in degree of flammability for the four fabrics.
(*b*) If the null hypothesis is rejected, construct confidence intervals to determine the fabric(s) with the lowest mean burn time.
(*c*) Plot the residuals and comment on the plausibility of the assumptions.
(*d*) If the tests had been conducted one at a time on a single mannequin, how would you have randomized the fabrics in this experiment?

4. Subjects in a study of the impact of child abuse on IQ [Sandgrund et. al., *American Journal of Mental Deficiency* **79(3)** (1974), 327–30] are selected from families receiving public assistance. The criterion for "abuse" is that an ongoing situation be confirmed by investigation, the "neglect" category is based on legal findings with regard to lack of adequate care, and the "nonabuse control" group is selected from out-patients at a clinic. The summary statistics for boys are:

	Verbal IQ	
Abuse	Neglect	Nonabuse Control
$\bar{y}_1 = 81.06$	$\bar{y}_2 = 78.56$	$\bar{y}_3 = 87.81$
$s_1 = 17.05$	$s_2 = 15.43$	$s_3 = 14.36$
$n_1 = 32$	$n_2 = 16$	$n_3 = 16$

where $s_j^2 = \Sigma_{i=1}^{n_j}(y_{ij} - \bar{y}_j)^2/(n_j - 1)$.
(*a*) Present the ANOVA table for these data.
(*b*) Calculate simultaneous confidence intervals for the differences.
(*c*) Because randomization of treatments is impossible, what conclusion regarding abuse and low IQ is still unanswered?

5. (*a*) Provide a decomposition for the following observations from a randomized block experiment.
(*b*) Find the sum of squares for each array.
(*c*) Determine the degrees of freedom by checking the constraints for each array.

	Treatment		
Block	1	2	3
1	35	19	21
2	24	14	16
3	28	14	21
4	21	13	14

6. Referring to Exercise 5, verify that the arrays of treatment effects, block effects, and residuals are mutually orthogonal. [If necessary, refer to Exercise 1(*b*) for definition.]
7. Again referring to Exercise 5, present the ANOVA table. What conclusions can you draw from the two F tests?
8. Three loaves of bread, each made according to a different recipe, are baked in one oven at the same time. Because of possible uncontrolled variations in oven performance, each baking is treated as a block. This procedure is repeated five times and the following measurements of density are obtained:

	Recipe		
Block	1	2	3
1	.95	.71	.69
2	.86	.85	.68
3	.71	.62	.51
4	.72	.72	.73
5	.74	.64	.44

   (*a*) How should the three oven positions of the three loaves have been selected for each trial?
   (*b*) Perform an analysis of variance for these data.
9. Referring to Exercise 8:
   (*a*) Obtain simultaneous confidence intervals for the pairwise differences in mean density for the three recipes.
   (*b*) Calculate the residuals and plot them on normal probability paper.
10. As part of a cooperative study on the nutritional quality of oats, 6 varieties of oat kernels with their hulls removed are subjected to a mineral analysis. The plants are grown according to a randomized block design, and the following measurements of protein by percent of dry weight are recorded:

	Variety					
Block	1	2	3	4	5	6
1	19.09	16.28	16.31	17.50	16.25	21.09
2	20.29	17.88	18.17	18.05	16.92	21.37
3	20.31	16.88	17.38	17.59	15.88	21.38
4	19.60	17.57	17.53	17.64	14.78	20.52
5	18.62	16.72	16.34	17.38	15.97	21.09
6	20.10	17.32	17.88	18.04	16.66	21.58

   (Data courtesy of D. Peterson, L. Schrader, and V. Youngs.)
   (*a*) Perform an analysis of variance for these data.

(*b*) Calculate and plot the residuals. Does the model appear to be adequate?

11. Referring to Exercise 10, suppose that the variety 6 is of special interest. Construct simultaneous 95% confidence intervals for the differences between the mean of variety 6 and each of the other means.
12. A garden insecticide can be manufactured using one of three different base powders and one of two different ingredients. The response is the measure of strength after spreading. The following data are obtained:

	Powder		
Ingredient	A_1	A_2	A_3
B_1	81	103	118
B_2	123	142	180

(*a*) Present the ANOVA table for these data.
(*b*) Calculate the residuals.
(*c*) Construct confidence intervals to show the plausible values of any important differences.

13. Referring to Exercise 12, suppose that a second replicate to the experiment provides the following measurements:

	Powder		
Ingredient	A_1	A_2	A_3
B_1	70	110	138
B_2	131	143	189

(*a*) State the statistical model to be used if the two replicates are to be analyzed together.
(*b*) Present the ANOVA table for these data.
(*c*) Do the data indicate a significant interaction between base powder and ingredient?

14. An experiment is designed to study the progress of pregnancy in hamsters by artificially administering two important hormones normally produced in the ovaries. A sample of 18 Golden hamsters are included in the experiment. The hamsters' ovaries are surgically removed on the eleventh day of pregnancy, and the hormones are administered in the following dosages:

Hormone A (Progesterone): 0, 0.1 mg/day, 10 mg/day
Hormone B (Estradiol): 0, 0.5 mg/day

Weight gain during the eleventh to sixteenth days of pregnancy is taken as the indicator of maintained pregnancy. The data obtained

Weight gain (in grams)

Hormone B	Hormone A: 0	0.1 mg/day	10 mg/day
0	−19	8	7
	−11	−18	23
	−14	−9	23
0.5 mg/day	−10	−3	32
	−19	−10	29
	−28	−4	18

(courtesy of Richard Marsh.)

(*a*) State the statistical model and perform an analysis of variance for these data. Test for the main effects and also for interaction.

(*b*) Calculate and plot the residuals.

(*c*) Construct confidence intervals for the pairwise differences between the levels of hormone *A*.

(*d*) How would you randomize the treatments in this experiment?

15. Subjects in a learning experiment are given either general directions or a list of goals. They then read one of two forms of a 2400-word passage describing life on an imaginary planet. In one form of the text, all sentences relevant to a goal occur consecutively; in the other form of the text, information relating to the goals is dispersed. Knowledge is to be demonstrated by a fill-in-the-blank test. E. Gagné, in a 1974 Ph.D. thesis at the University of Wisconsin, reported the following summary based on 38 subjects in each category. The measurements are based on the number of correctly recalled goal-relevant items.

Text	Directions: Goals	General
Nondispersed	$\bar{y}_{11\cdot}=14.45$ $s_{11}=\text{sd}=5.82$	$\bar{y}_{12\cdot}=11.03$ $s_{12}=\text{sd}=4.12$
Dispersed	$\bar{y}_{21\cdot}=11.74$ $s_{21}=4.82$	$\bar{y}_{22\cdot}=10.53$ $s_{22}=4.24$

Present the ANOVA table for this factorial design. [*Hint*: $\text{MSE}=s^2_{\text{pooled}}=(s^2_{11}+s^2_{12}+s^2_{21}+s^2_{22})/4$.]

16. Referring to Exercise 15 and using the coding:

	Factor	Low (−1)	High (+1)
1	Text	Nondispersed	Dispersed
2	Directions	goals	general

(*a*) Estimate the two main effects and the interaction.
(*b*) Construct confidence intervals for the effects and summarize your conclusions.
(*c*) Verify that $SS_{\text{factor}} = r$ (estimated effect)^2.
(See Mathematical Exercise 3.)

17. Three factors, temperature, pH, and concentration, may influence the yield of a medicine from a manufacturing process. A 2^3 factorial design is employed, and the following results are obtained.

Design: Temperature x_1	pH x_2	Concentration x_3	Yield
−	−	−	6
+	−	−	12
−	+	−	2
+	+	−	12
−	−	+	8
+	−	+	14
−	+	+	12
+	+	+	30

(*a*) Attach the yield numbers to the corners of a cube.
(*b*) Estimate the main effects and the two-factor interactions.
(*c*) Estimate the three-factor interaction. (NOTE: With a single replicate, no degree of freedom remains for the residual if all the two-factor interactions and the three-factor interaction are incorporated in the model in addition to the main effects. For the purposes of analysis, the three-factor interaction is assumed to be absent.

18. Referring to Exercise 17, construct the ANOVA table (see Mathematical Exercise 3).
19. Considerable effort has been expended to reduce the loss of potato crops due to soft rot. In an experiment each potato tuber is inoculated with bacteria, and the maximum diameter of rotted tissue is recorded after five days. Three factors are used in an experiment, each at two

settings. As part of a larger study, the following results are obtained for Red Nugget potatoes:

Design			Diameter of rot Replicate				
x_1	x_2	x_3	1	2	3	4	5
−	−	−	13	11	3	20	10
+	−	−	26	19	24	26	27
−	+	−	15	2	7	6	5
+	+	−	20	24	8	18	11
−	−	+	13	10	11	5	15
+	−	+	17	21	17	18	17
−	+	+	10	8	6	10	16
+	+	+	18	15	9	8	18

(courtesy of Arthur Kelman.)

Factor	Low	High
Temperature x_1	10°C	16°C
Oxygen x_2	2%	10%
Tuber treatment x_3	control	Rindite

(*a*) Estimate the main effects and all interactions.
(*b*) Using the confidence interval approach, what can you conclude about the factors?

20. In a study of the influence of sex x_1 and the level of aspiration x_2 on a measure of perceptual discrimination y, the following summary statistics based on 40 subjects divided into groups of 10 are obtained:

x_1	x_2	Sample mean
M	low	20.8
F	low	17.4
M	high	17.1
F	high	35.1

Mean square error = 24.1
d.f. = 9 × 4 = 36

[P. Peretti, *Social Behavior and Personality* **2(1)** (1974), 4–9 Copyright by the Society For Personality Research Incorporated. Reprinted by permission] Present the ANOVA table and a summary of the statistically significant results for these data.

21. Brain function in Rhesus monkeys is studied by inflicting a bilateral

inferior temporal lesion on the part of the brain associated with vision. The factors in the study are the effects of the operation and the age at which it is performed. The following data (courtesy of H. Harlow, U.W. Primate Laboratory) reflect performance on a visual acuity test that requires discrimination between a vertical bar and a broken bar, the top half of which is displaced by 8 mm. The measurements are based on the number of correct responses out 100.

	Age at operation (days)		
	100	300	730
Control	93, 90	79, 94	89, 91
	91, 88	50, 70	86, 84
Operated	86, 71	76, 91	60, 72
	68, 70	51, 96	93, 68

(*a*) Present the ANOVA table and interpret the results for these data.
(*b*) Calculate and plot the residuals. Comment on your graph.

MATHEMATICAL EXERCISES

1. Suppose we have r replicates of a $p \times q$ factorial experiment. Consider the decomposition of observations according to

$$y_{ijk} = \bar{y}_{\dots} + (\bar{y}_{i\cdot\cdot} - \bar{y}_{\dots}) + (\bar{y}_{\cdot j\cdot} - \bar{y}_{\dots}) + (\bar{y}_{ij\cdot} - \bar{y}_{i\cdot\cdot} - \bar{y}_{\cdot j\cdot} + \bar{y}_{\dots}) + (y_{ijk} - \bar{y}_{ij\cdot})$$

(*a*) Show that the array of values for any term on the right-hand side of this equation is orthogonal to every other array. For example, the sum over all $p \times q \times r$ entries $(\bar{y}_{i\cdot\cdot} - \bar{y}_{\dots})(\bar{y}_{ij\cdot} - \bar{y}_{i\cdot\cdot} - \bar{y}_{\cdot j\cdot} + \bar{y}_{\dots})$ is zero.
(*b*) Using your answer to (*a*), conclude that the sum of squares can be added; that is

$$\text{Total SS} = \text{SS}_A + \text{SS}_B + \text{SS}_{AB} + \text{SSE}$$

2. The model for a two-way analysis of variance with interaction is

$$Y_{ijk} = \mu + \alpha_i + \beta_j + \gamma_{ij} + e_{ijk}, \quad E(e_{ijk}) = 0$$

and

$$0 = \sum_{i=1}^{p} \alpha_i = \sum_{j=1}^{q} \beta_j = \sum_{i=1}^{p} \gamma_{ij} = \sum_{j=1}^{q} \gamma_{ij}$$

Show that:
(*a*) $E(\overline{Y}_{i..} - \overline{Y}_{...}) = \alpha_i$, $E(\overline{Y}_{.j.} - \overline{Y}_{...}) = \beta_j$
(*b*) $E(\overline{Y}_{ij.} - \overline{Y}_{i..} - \overline{Y}_{.j.} + \overline{Y}_{...}) = \gamma_{ij}$.

3. (*a*) Given a 2^2 factorial with r replicates, show that

$$SS_{\text{factor}} = r(\text{estimated effect})^2$$

(*b*) Show that a similar result holds for a 2^3 factorial.
[*Hint*:

$$2(\bar{y}_{1..} - \bar{y}_{...}) = 2\left(\bar{y}_{1..} - \frac{\bar{y}_{1..} + \bar{y}_{2..}}{2}\right) = (\bar{y}_{1..} - \bar{y}_{2..})$$

$$= \left(\frac{\bar{y}_{11.} + \bar{y}_{12.}}{2} - \frac{\bar{y}_{21.} + \bar{y}_{22.}}{2}\right)$$

is the estimate of the Factor 1 effect. Also $(\bar{y}_{1..} - \bar{y}_{...}) = -(\bar{y}_{2..} - \bar{y}_{...})$.]

4. Verify the alternative calculation formulas for the one-way analysis given in Example 14.6.

CHAPTER 15

Nonparametric Inference

15.1 INTRODUCTION

With the exception of our analysis of categorized data in Chapter 13, most of the models for inference procedures that we have discussed thus far tentatively assume a specific structure for the population distribution, including the treatment of tests for the binomial and Poisson parameters in Chapter 6. The Student's t test for inferences about a population mean and the comparison of two means, the χ^2 and F tests for inferences about variances, the inferences presented for regression models, and the analysis of variance are all based on the assumption that the response measurements constitute samples from normal populations. These procedures are designed to make inferences about the values of the parameters μ and σ that appear in the prescription for the mathematical curve of the normal population. Collectively, they are called *normal-theory parametric inference procedures*.

Nonparametric statistics is a body of inference procedures that is valid under a much wider range of shapes for the population distribution. The term *nonparametric inference* is derived from the fact that the usefulness of these procedures does not require modeling a population in terms of a specific parametric form of density curves, such as normal distributions. In testing hypotheses, nonparametric test statistics typically utilize some simple aspects of the sample data, such as the signs of the measurements, order relationships, or category frequencies. These general features do not require the existence of a meaningful numerical scale for measurements. More importantly, stretching or compressing the scale does not alter them. As a consequence, the null distribution of a nonparametric test statistic can be determined without regard to the shape of the underlying population distribution. For this reason, these procedures are also called *distribution-free tests*. This distribution-free property is their strongest advantage, and many statisticians actually prefer to refer to these procedures as distribution-free rather than nonparametric.

What major difficulties are associated with parametric inference procedures based on the t, χ^2, F, or similar distributions, and how do we overcome these difficulties by using a nonparametric approach? First, the distributions of these parametric model statistics and their tabulated percentage points are valid if the underlying population distribution is approximately normal. Although the normal distribution does approximate many real-life situations, exceptions are numerous, so that it is unwise to presume that normality is a fact of life. Drastic departures from normality often occur in the forms of conspicuous asymmetry, sharp peaks, or heavy tails. The intended strength of parametric inference procedures can be seriously affected when such departures occur, especially when the sample size is small or moderate. For instance, a t test with an intended level of significance of $\alpha = .05$ may have an actual type I error probability far in excess of this value. Similarly, the strength of confidence statements can be seriously distorted. These effects are most pronounced when small or moderate sample sizes are being considered, because it is more difficult to assess the shape of the population in such cases. The selection of a parametric procedure leaves the data analyst with the question: Does my normality assumption make sense in the present situation? To avoid this risk, a nonparametric method could be used in which inferences rest on the safer ground of distribution-free properties.

Second, parametric procedures require that observations be recorded on an unambiguous scale of measurements and then make explicit use of specific numerical aspects of the data, such as the sample mean, the standard deviation, and other sums of squares. In a great many situations, particularly in the social and behaviorial sciences, responses are difficult or impossible to measure on a specific and meaningful numerical scale. Characteristics like degree of apathy, taste preference, and surface gloss cannot be evaluated on an objective numerical scale, and an assignment of numbers is, therefore, bound to be arbitrary. Also, when people are asked to express their views on a 5-point rating scale, where 1 represents strongly disagree and 5 represents strongly agree, the numbers have little physical meaning beyond the fact that higher scores indicate greater agreement. Data of this type are called *ordinal data*, because only the order of the numbers is meaningful and the distance between two numbers does not lend itself to practical interpretation. Any controversy surrounding a scale of measurement for ordinal data is readily transmitted to parametric statistical procedures through the use of sample means and standard deviations. Nonparametric procedures that utilize information only on order or rank are therefore particularly suited to measurements on an ordinal scale.

When the data constitute measurements on a meaningful numerical scale and the assumption of normality holds, parametric procedures are certainly more efficient in the sense that tests have higher power and confidence intervals are generally shorter than their nonparametric counterparts. This brings to mind the old adage "You get what you pay for." A willingness to assume more about the population form leads to improved inference procedures. However, trying to get too much for your money by assuming more about the population than is reasonable can lead to the "purchase" of invalid conclusions. A choice between the parametric and nonparametric approach should be guided by a consideration of loss of efficiency and the degree of protection desired against possible violations of the assumptions.

15.2 THE WILCOXON RANK-SUM TEST FOR COMPARING TWO TREATMENTS

The problem of comparing two populations based on independent random samples has already been discussed in Section 9.2 under the assumptions of normality and equal variances for the population distributions. There the parametric procedure was based on Student's t statistic. Here we describe a useful nonparametric procedure originally proposed by F. Wilcoxon (1945). An equivalent alternative version was independently proposed by H. Mann and D. Whitney (1947).

For a comparative study of two treatments A and B, a set of $n = n_A + n_B$ experimental units are randomly divided into two groups of sizes n_A and n_B, respectively. Treatment A is applied to the n_A units, and treatment B is applied to the n_B units. The response measurements, recorded in a slightly different notation than before, are

Treatment A	X_{11}	X_{12}	$\cdots$	X_{1n_A}
Treatment B	X_{21}	X_{22}	$\cdots$	X_{2n_B}

These two treatments constitute independent random samples from two populations. Assuming that larger responses indicate a better treatment, we wish to test the null hypothesis that there is no difference between the two treatment effects vs. the one-sided alternative that treatment A is more effective than treatment B. In the present nonparametric setting, the distributions are assumed to be continuous.

Model: Both distributions are continuous.

Hypotheses:

H_0: The two population distributions are identical.

H_1: The distribution of population A is shifted to the right of the distribution of population B.

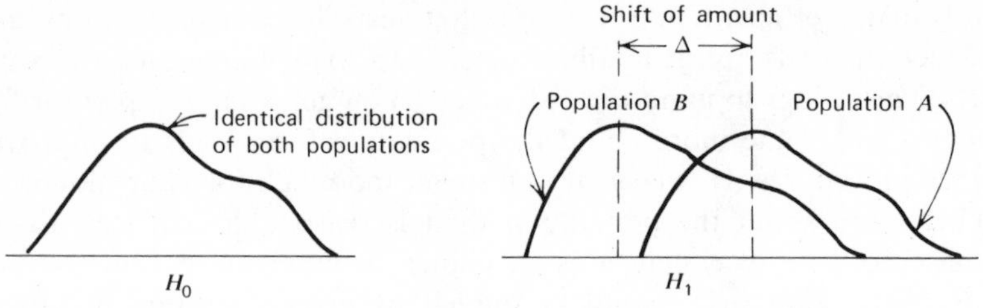

Figure 15.1 Representation of H_0 and H_1 in terms of the amount of shift Δ

Note that no assumption is made regarding the *shape* of the population distribution. This is in sharp contrast to our parametric procedure in Chapter 9, where we assumed that the population distributions were normal with equal variances. Figure 15.1 illustrates the above hypotheses H_0 and H_1.

The basic concept underlying the rank-sum test can now be explained by the following intuitive line of reasoning. Suppose that the two sets of observations are plotted on the same diagram, using different markings (dots and crosses) to identify their sources. Under H_0, the samples come from the same population, so that the two sets of points should be well mixed. However, if the larger observations are more often associated with the first sample, for example, we can infer that population A is possibly shifted to the right of population B. These two situations are diagrammed in Figure 15.2, where the combined set of points in each case is serially numbered from left to right. These numbers are called the *combined sample ranks*. In Figure 15.2*a* large as well as small ranks are associated with each sample, whereas in Figure 15.2*b* most of the larger ranks are associated with the first sample. Therefore, considering the sum of the ranks associated with the first sample as a test statistic, a large value of this statistic should reflect that the first population is located to the right of the second.

To establish a rejection region with a specified level of significance, we must consider the distribution of the rank-sum statistic under the null

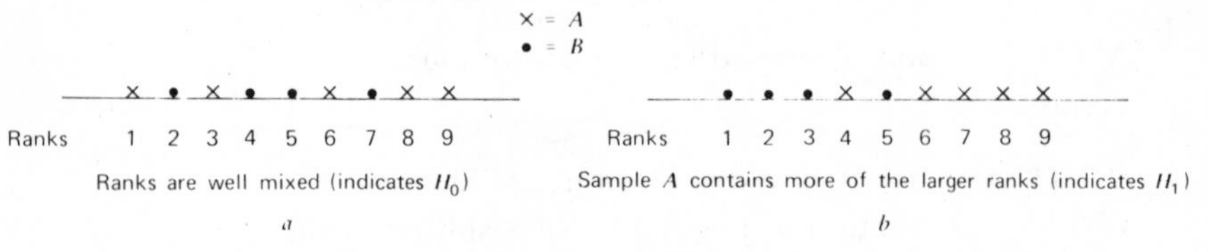

Figure 15.2 Combined plot of the two samples and the combined sample ranks.

hypothesis. This concept is explored in Example 15.1, where small sample sizes are investigated to avoid an overlong series of enumerations.

Example 15.1 To determine if a new hybrid seedling produces a bushier flowering plant than a currently popular variety, a horticulturist plants 2 new hybrid seedlings and 3 currently popular seedlings in a garden plot. After the plants mature, the following measurements of shrub girth in inches are recorded:

	SHRUB GIRTH (IN INCHES)		
New hybrid A	31.8	39.1	
Current variety B	35.5	27.6	21.3

Do these data strongly indicate that the new hybrid produces larger shrubs than the current variety?

We wish to test the null hypothesis

H_0: A and B populations are identical.

vs. the alternative hypothesis

H_1: Population A is shifted from B toward larger values.

In the rank-sum test, the two samples are placed together and ranked from smallest to largest:

Combined sample ordered observations	21.3	27.6	31.8	35.5	39.1
Ranks	1	2	3	4	5
Variety	B	B	A	B	A

Rank sum for A: $W_A = 3+5 = 8$
Rank sum for B: $W_B = 1+2+4 = 7$

Because larger measurements and therefore higher ranks for variety A tend to support H_1, the rejection region of our test should consist of large values for W_A:

$$\text{Reject } H_0 \quad \text{if} \quad W_A \geqslant c$$

To determine the critical value c so that the type I error probability is controlled at a specified level α, we evaluate the probability distribution of W_A under H_0. When the two samples come from the same population, every pair of integers out of $\{1,2,3,4,5\}$ is equally likely to be the ranks for the two A measurements. There are $\binom{5}{2} = 10$ potential pairs, so that each collection of possible ranks has a probability of $\frac{1}{10} = .1$ under H_0. These rank collections are listed in Table 15.1 with their corresponding W_A values.

TABLE 15.1
RANK COLLECTIONS FOR TREATMENT A WITH SAMPLE SIZES
$n_A=2,\ n_B=3$

RANKS OF A	RANK SUM W_A	PROBABILITY
1,2	3	.1
1,3	4	.1
1,4	5	.1
1,5	6	.1
2,3	5	.1
2,4	6	.1
2,5	7	.1
3,4	7	.1
3,5	8	.1
4,5	9	.1
	Total	1.0

The null distribution of W_A can be obtained immediately from Table 15.1 by collecting the probabilities of identical values (see Table 15.2).

TABLE 15.2
DISTRIBUTION OF THE RANK SUM W_A FOR SAMPLE SIZES
$n_A=2,\ n_B=3$

Values of W_A	3	4	5	6	7	8	9
Probability	.1	.1	.2	.2	.2	.1	.1

The observed value $W_A=8$ has the significance probability $P_{H_0}(W_A \geqslant 8) = .1+.1=.2$. In other words, we must tolerate a type I error probability of .2 for the observed value $W_A=8$ to result in the rejection of H_0. The rank-sum test leads us to conclude that the evidence is not sufficiently strong to reject H_0. Note that even if the A measurements did receive the highest ranks of 4 and 5, a significance level of $\alpha=.1$ would be required to reject H_0. ■

Guided by Example 15.1, we now state the rank-sum test procedure in a general setting.

Wilcoxon rank-sum test

Let $X_{11},\ldots,X_{1n_A}$ and $X_{21},\ldots,X_{2n_B}$ be independent random samples from continuous populations A and B, respectively. To test H_0: The populations are identical:

(*a*) Rank the combined sample of $n=n_A+n_B$ observations in increasing order of magnitude.

(*b*) Find the rank sum W_A of the first sample.

(*c*) (*i*) For H_1: Population A is shifted to the right of population B; set the rejection region at the upper tail of W_A.

(*ii*) For H_1: Population A is shifted to the left of population B; set the rejection region at the lower tail of W_A.

(*iii*) For H_1: Populations are different; set the rejection region at both tails of W_A having equal probabilities.

A determination of the null distribution of the rank-sum statistic by direct enumeration becomes more tedious as the sample sizes increase. However, tables for the null distribution of this statistic have been prepared for small samples, and an approximation is available for large samples. To explain the use of Appendix Table 9, first we note some features of the rank sums W_A and W_B.

The total of the two rank sums $W_A + W_B$ is a constant, which is the sum of the integers $1, 2, \ldots, n$, where n is the combined sample size. For instance, in Example 15.1

$$W_A + W_B = (3+5) + (1+2+4)$$

$$= 1+2+3+4+5 = 15$$

Therefore, a test that rejects H_0 for large values of W_A is equivalent to a test that rejects H_0 for small values of W_B. We can just as easily designate W_B the test statistic and set the rejection region at the lower tail. Consequently, we can always concentrate on the rank sum of the smaller sample and set the rejection region at the lower (or upper) tail, depending on whether the alternative hypothesis states that the corresponding population distribution is shifted to the left (or right).

Second, the distribution of each of the rank-sum statistics W_A and W_B is symmetric. In fact, W_A is symmetric about $n_A(n_A + n_B + 1)/2$ and W_B is symmetric about $n_B(n_A + n_B + 1)/2$. Table 15.2 illustrates the symmetry of the W_A distribution for the case $n_A = 2$, $n_B = 3$. The interested student may wish to determine the W_B distribution for this case and to verify its symmetry. These properties of the test statistic greatly simplify the construction of the distribution tables.

15.2.1 The Use of Appendix Table 9

The Wilcoxon rank-sum test statistic is taken as

W_s = sum of ranks of the smaller sample in the combined sample ranking.

When the sample sizes are equal, take the sum of ranks for either of the samples. The Appendix Table 9 gives the upper- as well as the lower-tail probabilities:

$$P[W_s \geqslant x]: \quad \text{Upper-tail probability}$$

$$P[W_s \leqslant x^*]: \quad \text{Lower-tail probability}$$

By the symmetry of the distribution, these probabilities are equal when x and x^* are at equal distances from the center. The table includes the x^* values corresponding to the x's at the upper tail. For instance, with sample sizes 3 and 7, the table shows $P[W_s \geqslant 25] = .033 = P[W_s \leqslant 8]$. The steps to follow when using Appendix Table 9 in performing a rank-sum test are:

Use the rank-sum of the smaller sample W_s as the test statistic. If the sample sizes are equal, take either rank sum as W_s. If H_1 states that the population corresponding to W_s is:

(*a*) Shifted to the right of the other population; set a rejection region of the form $W_s \geqslant c$ and take c as the smallest x value for which P is $\leqslant \alpha$.
(*b*) Shifted to the left; set a rejection region of the form $W_s \leqslant c$ and take c as the largest x^* value for which $P \leqslant \alpha$.
(*c*) Shifted in either direction; set a rejection region of the form $[W_s \leqslant c_1$ or $W_s \geqslant c_2]$ and read c_1 from the x^* column and c_2 from the x column, so that $P \leqslant \alpha/2$.

Example 15.2 Two geological formations are compared with respect to richness of mineral content. The mineral contents of 7 specimens of ore collected from formation 1 and 5 specimens collected from formation 2 are measured by chemical analysis. The following data are obtained:

	MINERAL CONTENT						
Formation 1	7.6	11.1	6.8	9.8	4.9	6.1	15.1
Formation 2	4.7	6.4	4.1	3.7	3.9		

Do the data provide strong evidence that formation 1 has a higher mineral content than formation 2? Test with α near .05.

To use the rank-sum test, first we rank the combined sample and determine the sum of ranks for the second sample, which has the smaller size. The observations from the second sample and their ranks are underlined here for quick identification:

Combined ordered values	$\underline{3.7}$	$\underline{3.9}$	$\underline{4.1}$	$\underline{4.7}$	4.9	6.1	$\underline{6.4}$	6.8	7.6	9.8	11.1	15.1
Ranks	$\underline{1}$	$\underline{2}$	$\underline{3}$	$\underline{4}$	5	6	$\underline{7}$	8	9	10	11	12

The observed value of the rank-sum statistic is

$$W_s = 1+2+3+4+7 = 17$$

We wish to test the null hypothesis that the two population distributions are identical vs. the alternative hypothesis that the second population, corresponding to W_s, lies to the left of the first. The rejection region is therefore at the lower tail of W_s.

Reading Appendix Table 9 with smaller sample size = 5 and larger sample size = 7, we find $P[W_s \leqslant 21] = .037$ and $P[W_s \leqslant 22] = .053$. Hence the rejection region with $\alpha = .053$ is established as $W_s \leqslant 22$. Because the observed value falls in this region, the null hypothesis is rejected at $\alpha = .053$. In fact, it would be rejected if α were as low as $P[W_s \leqslant 17] = .005$. ■

Example 15.3 Flame-retardant materials are tested by igniting a paper tab on the hem of a dress worn by a mannequin. One response is the vertical length of damage to the fabric measured in inches. The following data (courtesy B. Joiner) for 5 samples, each taken from two fabrics, are obtained by researchers at the National Bureau of Standards as part of a larger cooperative study:

Fabric *A*	5.7	7.3	7.6	6.0	6.5
Fabric *B*	4.9	7.4	5.3	4.6	6.2

Do the data provide strong evidence that a difference in flammability exists between the two fabrics? Test with α near .05.

The sample sizes are equal, so that we can take the rank sum of either sample as the test statistic. We compute the rank sum for the second sample:

Ordered values	<u>4.6</u>	<u>4.9</u>	<u>5.3</u>	5.7	6.0	<u>6.2</u>	6.5	7.3	<u>7.4</u>	7.6
Ranks	<u>1</u>	<u>2</u>	<u>3</u>	4	5	<u>6</u>	7	8	<u>9</u>	10

$$W_s = 1+2+3+6+9 = 21$$

Because the alternative hypothesis is two-sided, the rejection region includes both tails of W_s. From Appendix Table 9, we find that

$$P[W_s \geqslant 37] = .028 = P[W_s \leqslant 18]$$

Thus with $\alpha = .056$, the rejection region is $W_s \geqslant 37$ or $W_s \leqslant 18$. The observed value does not fall in the rejection region, and the null hypothesis is not rejected at $\alpha = .056$. ■

15.2.2 Large-Sample Approximation

When the sample sizes are large, the null distribution of the rank-sum statistic is approximately normal and the test can therefore be performed using the normal table. Specifically, with W_A denoting the rank sum of the sample of size n_A, suppose that both n_A and n_B are large. Then W_A is approximately normally distributed.

> Under H_0
>
> $$\text{mean of } W_A = \frac{n_A(n_A+n_B+1)}{2}$$
>
> $$\text{variance of } W_A = \frac{n_A n_B(n_A+n_B+1)}{12}$$

> *Large sample approximation to the rank-sum statistic:*
>
> $$Z = \frac{W_A - n_A(n_A+n_B+1)/2}{\sqrt{n_A n_B(n_A+n_B+1)/12}}$$
>
> is approximately $N(0,1)$ when H_0 is true.

The rejection region for the Z statistic can be determined by using the standard normal table.

To illustrate the amount of error involved in this approximation, we consider the case $n_A=9$, $n_B=10$. With $\alpha=.05$, the approximate one-sided rejection region is

$$R:\ \frac{W_A - 9(20)/2}{\sqrt{9\times10\times20/12}} = \frac{W_A-90}{12.247} \geqslant 1.64$$

which simplifies to R: $W_A \geqslant 110.1$. From Appendix Table 9 we find $P[W_s \geqslant 110] = .056$ and $P[W_s \geqslant 111] = .047$, which are quite close to $\alpha = .05$. The error decreases with increasing sample sizes.

15.2.3 Handling Tied Observations

In the preceding examples observations in the combined sample are all distinct and therefore the ranks are determined without any ambiguity.

Often, however, due to imprecision in the measuring scale or a basic discreteness of the scale, such as 5-point preference rating scale, observed values may be repeated in one or both samples. For example, consider the data sets

Sample 1	20	24	22	24	26
Sample 2	26	28	26	30	18

The ordered combined sample is

$$18 \quad 20 \quad 22 \quad \underbrace{24 \quad 24}_{\text{tie}} \quad \underbrace{26 \quad 26 \quad 26}_{\text{tie}} \quad 28 \quad 30$$

Here two ties are present; the first has 2 elements, and the second has 3. The two positions occupied by 24 are eligible for the ranks 4 and 5, and we assign the average rank $(4+5)/2=4.5$ to each of these observations. Similarly, the three tied observations 26, eligible for the ranks 6, 7, and 8, are each assigned the average rank $(6+7+8)/3=7$. After assigning average ranks to the tied observations and usual ranks to the other observations, the rank-sum statistic can then be calculated. When ties are present in small samples, the distribution in Appendix Table 9 no longer holds exactly. Strictly speaking, given a tie structure, it is best to calculate the null distribution of W_s by direct enumeration. This can be conveniently performed with the aid of an electronic computer, but instead of giving the details of this procedure here, we present a modified form of the standardized statistic for use in large samples.

For large sample sizes, a normal approximation to the rank-sum statistic again applies. The mean is given by the same formula that is used in the untied case, but now

$$\text{variance of } W_A = \frac{n_A n_B (n+1)}{12}\left[1 - \frac{\sum_{j=1}^{l} q_j(q_j^2-1)}{n(n^2-1)}\right]$$

where

$$n = n_A + n_B$$

$$l = \text{number of ties}$$

$$q_j = \text{number of elements in the } j\text{th tie}, \quad j=1,\ldots,l$$

When there are relatively few ties, this corrected variance should even improve the normal approximation for moderate sample sizes.

15.3 PAIRED COMPARISONS

In the presence of extensive heterogeneity in the experimental units, two treatments can be compared more efficiently if internally homogeneous pairs of units are selected and then the two treatments are applied one to each member of the pair. We were first exposed to this concept in Section 9.4, where the t test was based on the paired differences. In small samples this test assumes that these paired differences constitute a random sample from a normal population. In this section we discuss two nonparametric tests, the *sign test* and the *Wilcoxon signed-rank test*, that can be safely applied to paired differences when the assumption of normality is suspect. The data structure of a paired-comparison experiment is given in Table 15.3, where the observations on the ith pair are denoted by (X_{1i}, X_{2i}). The null hypothesis of primary interest is that there is no difference, or

$$H_0\text{: no difference in the treatment effects.}$$

TABLE 15.3
DATA STRUCTURE OF PAIRED SAMPLING

PAIR	1	2	...	n
Treatment A	X_{11}	X_{12}	...	X_{1n}
Treatment B	X_{21}	X_{22}	...	X_{2n}
Difference $(A-B)$	D_1	D_2	...	D_n

15.3.1 The Sign Test

This nonparametric test is notable for its intuitive appeal and ease of application. As its name suggests, the *sign test* is based on the signs of the response differences D_i. The test statistic is

$$S = \text{Number of pairs in which treatment } A \text{ has a higher response than treatment } B.$$
$$= \text{Number of positive signs among the differences } D_1, \ldots, D_n.$$

When the two treatment effects are actually alike, the response difference D_i in each pair is as likely to be positive as it is to be negative. Moreover, if measurements are made on a continuous scale, the occurrence of identical responses in a pair can be neglected probabilistically. The null hypothesis

$$H_0\text{: } P[+] = .5 = P[-]$$

can therefore be formulated. Identifying a plus sign as a success, the test statistic S is simply the number of successes in n trials and therefore has the binomial distribution $b(n, .5)$ under H_0. If the alternative hypothesis states that treatment A has higher responses than treatment B, which is translated $P[+] > .5$, then large values of S should be in the rejection region. For two-sided alternatives H_1: $P[+] \neq .5$, a two-tailed test should be employed.

Example 15.4 Mileage tests are conducted to compare an innovative vs. a conventional spark plug. A sample of 12 cars ranging from subcompacts to large station wagons are included in the study. The gasoline mileage for each car is recorded, once with the conventional plug and once with the new plug. The results are given in Table 15.4.

Looking at the differences $(A - B)$, we can see that there are 8 plus signs in the sample of size $n = 12$. Thus the observed value of the sign-test statistic is $S = 8$. In testing

H_0: No difference between A and B, or $P[+] = .5$

vs.

H_1: A is better than B, or $P[+] > .5$

we will reject H_0 for large values of S. Consulting the binomial table for $n = 12$ and $p = .5$, we find $P[S \geqslant 9] = .073$ and $P[S \geqslant 10] = .019$. If we wish to control α below .05, the rejection region should be established at $S \geqslant 10$. The observed value $S = 8$ is too low to be in the rejection region, so that at the level of significance $\alpha = .019$ the data do not sustain the claim of mileage improvement. The significance probability of the observed value is $P[S \geqslant 8] = .194$. ■

An application of the sign test does not require the numerical values of the differences to be calculated. The number of positive signs can be obtained by glancing at the data. Aside from the merit of requiring minimal numerical work, the sign test has a broad scope of applications. Even when a response cannot be measured on a well-defined numerical scale, we can often determine which of the two responses in a pair is better visually or by some other method of comparison. This is the only information that is required to conduct the sign test.

For large samples, the sign test can be performed by using the normal approximation to the binomial distribution. With large n, the binomial distribution $b(n, .5)$ is close to the normal distribution with a mean of $n/2$ and a standard deviation of $\sqrt{n/4}$.

Large sample approximation to the sign test statistic:

Under H_0,
$$Z = \frac{S - n/2}{\sqrt{n/4}}$$
is approximately distributed as $N(0, 1)$.

TABLE 15.4

Car number	1	2	3	4	5	6	7	8	9	10	11	12
New *A*	26.4	10.3	15.8	16.5	32.5	8.3	22.1	30.1	12.9	12.6	27.3	9.4
Conventional *B*	24.3	9.8	16.9	17.2	30.5	7.9	22.4	28.6	13.1	11.6	25.5	8.6
Difference $(A-B)$	+2.1	+.5	−1.1	−.7	+2.0	+.4	−.3	+1.5	−.2	+1.0	+1.8	+.8

When the two responses in a pair are exactly equal, we say that there is a *tie*. Because a tied pair has zero difference, it does not have a positive or a negative sign. In the presence of ties, the sign test is performed by discarding the tied pairs, thereby reducing the sample size. For instance, when a sample of $n=20$ pairs has 10 plus signs, 6 minus signs, and 4 ties, the sign test is performed with the effective sample size $n'=20-4=16$ and with $S=10$.

15.3.2 The Wilcoxon Signed-Rank Test

We have already noted that the sign test extends to ordinal data for which the response in a pair can be compared without being measured on a numerical scale. However, when numerical measurements are available, the sign test may result in a considerable loss of information because it includes only the signs of the differences and disregards their magnitudes. Compare the two sets of paired differences plotted in the dot diagrams in Figure 15.3. In both cases there are $n=6$ data points with 4 positive signs, so that the sign test will lead to identical conclusions. However, the plot in Figure 15.3*b* exhibits more of a shift toward the positive side, because the positive differences are farther away from zero than the negative differences. Instead of attaching equal weights to all the positive signs, as is done in the sign test, we should attach larger weights to the plus signs that are farther away from zero. This is precisely the concept underlying the signed-rank test.

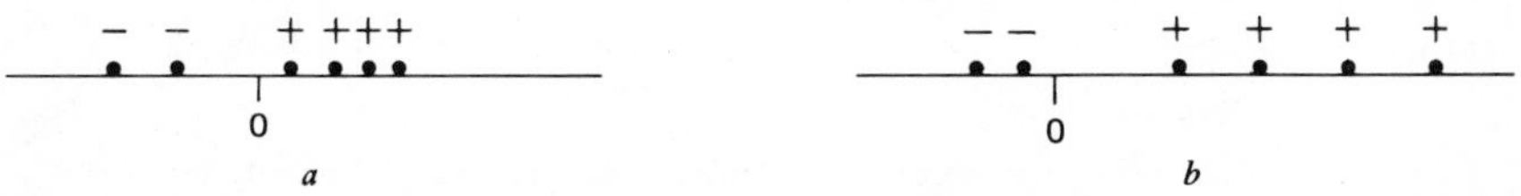

Figure 15.3 Two plots of paired differences with the same number of + signs but with different locations for the distributions.

In the signed-rank test, the paired differences are ordered according to their numerical values without regard to signs, and then the ranks associated with the positive observations are added to form the test statistic. To illustrate, we refer to the mileage data given in Example 15.4, where the paired differences appear in the last row of Table 15.4. We attach ranks by arranging these differences in increasing order of their *absolute* values and record the corresponding signs:

The signed-rank statistic T^+ is then calculated as

$$\begin{aligned} T^+ &= \text{Sum of the ranks associated with positive observations} \\ &= 3+4+6+7+9+10+11+12 \\ &= 62 \end{aligned}$$

Paired differences	2.1	.5	−1.1	−.7	2.0	.4	−.3	1.5	−.2	1.0	1.8	.8
Ordered absolute values	.2	.3	.4	.5	.7	.8	1.0	1.1	1.5	1.8	2.0	2.1
Ranks	1	2	3	4	5	6	7	8	9	10	11	12
Signs	−	−	+	+	−	+	+	−	+	+	+	+

If the null hypothesis of no difference in treatment effects is true, then the paired differences $D_1, D_2, \ldots, D_n$ constitute a random sample from a population that is symmetric about zero. On the other hand, the alternative hypothesis that treatment A is better asserts that the distribution is shifted from zero toward positive values. Under H_1, not only are more plus signs anticipated, but the positive signs are also likely to be associated with larger ranks. Consequently, T^+ is expected to be large under the one-sided alternative, and we select a rejection region in the upper tail of T^+.

Steps in the signed-rank test

(*a*) Calculate the differences $D_i = X_{1i} - X_{2i}$, $i = 1, \ldots, n$.

(*b*) Assign ranks by arranging the absolute values of the D_i in increasing order; also record the corresponding signs.

(*c*) Calculate the signed-rank statistic $T^+ =$ Sum of ranks of positive differences D_i.

(*d*) Set the rejection region at the upper or lower tail or at both tails of T^+, according to whether treatment A is stated to have a higher, lower, or different response than treatment B under the alternative hypothesis.

Selected tail probabilities of the null distribution of T^+ are given in Appendix Table 10 for $n=3$ to $n=15$. By symmetry of the distribution around $n(n+1)/4$, we obtain

$$P[T^+ \geqslant x] = P[T^+ \leqslant x^*]$$

when $x^* = n(n+1)/4 - x$. The x and x^* values in the Appendix Table 10 satisfy this relation. To illustrate the use of this table, we refer once again to the mileage data given in Example 15.4. There $n=12$ and the observed value of T^+ is found to be 62. From the table, we find $P[T^+ \geqslant 61] = .046$. Thus the null hypothesis is rejected at the level of significance $\alpha = .046$, and a significant mileage improvement using the new type of spark plug is indicated.

With increasing sample size n, the null distribution of T^+ is closely approximated by a normal curve, with mean $= n(n+1)/4$ and variance $= n(n+1)(2n+1)/24$.

Large sample approximation to signed-rank statistic:

Under the null hypothesis

$$Z = \frac{T^+ - n(n+1)/4}{\sqrt{n(n+1)(2n+1)/24}}$$

is approximately distributed as $N(0,1)$.

This result can be used to perform the signed-rank test with large samples.

In computing the signed-rank statistic, ties may occur in two ways: Some of the differences D_i may be zero, or some nonzero differences D_i may have the same absolute value. The first type of tie is handled by discarding the zero values and simultaneously modifying the sample size downward to

$$n' = n - \text{No. of zeros}$$

The second type of tie is handled by assigning the average rank to each observation in a group of tied observations with nonzero differences D_i. With a large sample, a normal approximation is again employed, with

$$\text{mean of } T^+ = \frac{n'(n'+1)}{4}$$

$$\text{and variance of } T^+ = \frac{n'(n'+1)(2n'+1)}{24} - \frac{1}{48}\sum_{j=1}^{l} q_j\left(q_j^2 - 1\right)$$

where l = number of ties

q_j = number of elements in the j-th tie, $j = 1, \ldots, l$

Example 15.5 A diet research program selects 20 volunteers to rate the taste of each of two breakfast drinks on a preference scale of 0–100. Table 15.5 lists these preference ratings and their differences. Use the signed-rank test to determine if there is a significant difference between taste preferences for the two drinks.

TABLE 15.5

RATINGS OF TWO BREAKFAST DRINKS BY 20 PERSONS

Drink A	70	85	73	75	65	50	80	71	80	51
Drink B	65	41	45	80	84	50	71	52	42	78
Difference $(A-B)$	5	44	28	−5	−19	0	9	19	38	−27

Drink A	72	76	79	65	59	72	84	90	56	57
Drink B	62	38	80	65	54	67	87	90	38	43
Difference $(A-B)$	10	38	−1	0	5	5	−3	0	18	14

The ordered numerical values of the nonzero differences, their ranks, and the corresponding signs are:

Ordered absolute values with signs	−1	−3	5	−5	5	5	9	10	14	18
			tie	tie	tie					
Ranks	1	2	4.5	4.5	4.5	4.5	7	8	9	10

Ordered absolute values with signs	−19	19	−27	28	38	38	44
	tie	tie			tie	tie	
Ranks	11.5	11.5	13	14	15.5	15.5	17

The number of nonzero differences is $n'=17$, and 3 ties are present among their absolute values. Assigning the average rank to each member of a group of tied observations, as above, we can calculate the signed rank statistic

$$T^{+}=4.5+4.5+4.5+7+8+9+10+11.5+14+15.5+15.5+17$$
$$=121$$

For the large-sample normal approximation, the mean, variance, and standard deviation are

$$\text{Mean}=17\times\frac{18}{4}=76.5$$

$$\text{Variance}=\frac{17\times18\times35}{24}-\frac{1}{48}\left[4\times15+2\times3+2\times3\right]$$

$$=446.25-1.5=444.75$$

$$\text{sd}=\sqrt{444.75}=21.09$$

The standardized value of the test statistic is $z=(121-76.5)/21.09=2.11$. Because this value is larger than the two-sided standard normal critical value 1.96 corresponding to $\alpha=.05$, the null hypothesis of no difference in quality is rejected by the signed-rank test at $\alpha=.05$. ■

15.4 CONFIDENCE INTERVALS BASED ON RANK TESTS

It is important to emphasize here that nonparametric procedures do not apply only to tests of hypotheses. The correspondence between confidence intervals and the simultaneous testing of many null hypotheses has already been clarified in Section 8.6. This relationship can also be employed to determine confidence intervals that correspond to the nonparametric tests presented in this chapter. The two-sided or two-tailed tests give rise to the intervals we discuss here.

15.4.1 Confidence Intervals Based on Paired Differences

First we treat observations originating from subjects or experimental units that are paired according to the structure described in Section 15.3. Under the null hypothesis of equal treatments, the paired differences $D_1, D_2, \ldots, D_n$ are assumed to have a distribution that is *symmetric* about zero. When there is a difference in treatments, the distribution is still assumed to be symmetric but to have a center of symmetry M. Because this center divides the population in half, it is also the population median. These distributions are illustrated in Figure 15.4.

The Wilcoxon signed-rank test, detailed in Section 15.3, is a test of H_0: $M=0$. To test the more general null hypothesis H_0: $M=M_0$, only a simple modification is required. Under this second null hypothesis, the quantities $(D_1-M_0), \ldots, (D_n-M_0)$ are symmetrically distributed about zero. The signed rank test performed with these modified differences $D_i'=D_i-M_0$ then provides a test for H_0: $M=M_0$. A $100(1-\alpha)\%$ confidence interval for M can now be conceptualized as the range of values of M_0 such that the

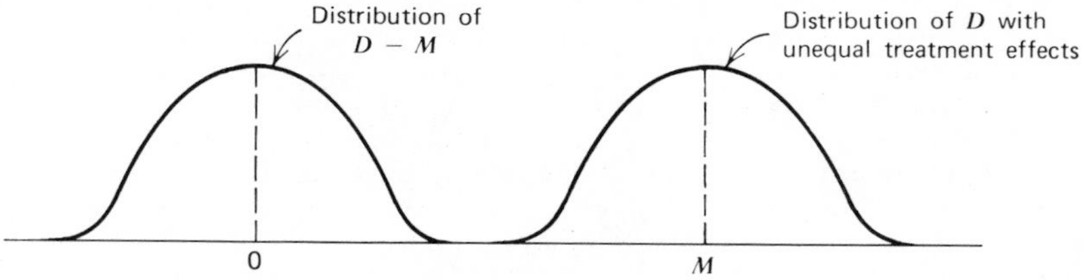

Figure 15.4 Distributions for a difference.

null hypothesis H_0: $M = M_0$ is not rejected in favor of H_1: $M \neq M_0$ at the level of significance α. Fortunately, instead of performing the two-sided signed-rank test for all possible values of M_0, there is an alternative way to arrive at the same confidence interval:

(*a*) Calculate the average for all pairs of differences, including the averages $(D_1 + D_1)/2, \ldots, (D_n + D_n)/2$, or

$$\frac{D_1 + D_1}{2}, \frac{D_1 + D_2}{2}, \ldots, \frac{D_1 + D_n}{2}, \frac{D_2 + D_2}{2}, \frac{D_2 + D_3}{2}, \ldots, \frac{D_n + D_n}{2}$$

(*b*) Order these $\binom{n}{2} + n = n(n+1)/2$ averages from smallest to largest.

(*c*) A confidence interval for the center of symmetry M is given by the interval

$$d\text{th smallest to } d\text{th largest}$$

including end points. (The values for d and the confidence coefficients are given in Appendix Table 11.)

Actually, there is no need to calculate all the $[n(n+1)/2]$ averages. With some experience, it is possible to systematically determine the extreme averages after arranging the original differences in increasing order.

Example 15.6 Construct a confidence interval for M based on the mileage data given in Example 15.4.

Appendix Table 11 provides the value $d = 14$ for a 95.8% confidence interval with a sample of size $n = 12$. Thus only the 14 smallest and the 14 largest averages for the differences in Table 15.4 are needed. The differences arranged in increasing order are

$$-1.1 \quad -.7 \quad -.3 \quad -.2 \quad .4 \quad .5 \quad .8 \quad 1.0 \quad 1.5 \quad 1.8 \quad 2.0 \quad 2.1$$

The smallest average is $[(-1.1)+(-1.1)]/2 = -1.1$, and the next higher average is $[(-1.1)+(-.7)]/2 = -.9$. Proceeding in this manner, we record

Smallest	−1.1	−.9	−.7	−.7	−.65	−.5	−.45	−.35
Largest	··· 1.4	1.45	1.5	1.5	1.55	1.65	1.75	1.8

Smallest	−.3	−.3	−.25	−.2	−.15	−.15	···
Largest	1.8	1.9	1.95	2.0	2.05	2.1	

A 95.8% confidence interval for M is then given by

$$[-.15, 1.4]$$

Note that the value $M = 0$ is included, indicating that no difference in treatments is a plausible conclusion. ■

In performing these calculations, zero differences should not be ignored and the sample size should not be reduced, as in testing. When ties are present, the intervals have at least the stated confidence when the end points are included in the interval. If a dot diagram reveals a serious lack of symmetry, the interval is inappropriate, and a similar result based on the sign test may be used to determine an interval of values for the median (See Exercise 24).

15.4.2 Confidence Intervals for Shift: Independent Samples

Rank comparisons based on two independent samples also produce confidence statements. Referring again to Figure 15.1, we can see that the alternative hypothesis specifies that the distribution of population A is shifted to the right of population B by an amount Δ. The crucial property to be considered now is that if Δ is subtracted from each observation from population A, the modified observations $X_{1i}-\Delta$ will have the same distribution as the observations from population B. This suggests that we can subtract a possible value of Δ and then apply the two-sided Wilcoxon rank-sum test. A confidence interval can be obtained which consists of all these values of Δ for which the null hypothesis of no treatment difference is *not* rejected. Again, an equivalent but somewhat simpler calculation is available:

(*a*) For all possible $n_A \cdot n_B$ pairs of observations, one from sample A and one from sample B, determine the difference $X_{1i}-X_{2j}$. (Each is an estimate of Δ.)
(*b*) Order these differences from smallest to largest.
(*c*) A confidence interval for the amount of shift Δ is given by the interval

$$d\text{th smallest to } d\text{th largest}$$

including the end points. The values for d and the confidence coefficients are given in Appendix Table 12.

When ties are present, the intervals have at least the stated confidence.

Example 15.7 Construct a confidence interval for the amount of shift between the distributions of richness of mineral content for the data given in Example 15.2.

Only the smallest and largest few differences must be calculated. For example, $x_{17}-x_{25}=15.1-3.9=11.2$. Proceeding in this manner, we record:

Smallest	-1.5	$-.3$	.2	.4	.8	1.0	1.2	1.2	⋯
Largest	⋯	7.2	7.4	8.7	10.4	11	11.2	11.4	

Consulting Appendix Table 12 for the smaller sample size 5 and for the larger sample size 7, we find $d=6$ for $1-\alpha=.952$. Thus

$$[1.0, 7.4]$$

is a 95.2% confidence interval for Δ. Because this interval does not include zero, we can conclude that the two formations have significantly different mineral content. ■

If a dot diagram reveals that the two distributions appear to differ in some aspect other than shift, the confidence interval for shift may not be appropriate. This is true if one sample exhibits considerably more spread than the other sample in addition to a difference in location.

Both of the procedures discussed in this section become quite tedious for even moderately large sample sizes. Fortunately, the procedures we examined in Section 9.2, which are based on the normal population, are reliable for most populations when the sample sizes are moderate.

15.5 MEASURES OF CORRELATION BASED ON RANKS

Ranks may also be employed to determine the degree of association between two random variables. These two variables could be mathematical ability and musical aptitude or the aggressiveness scores of first- and second-born sons on a psychological test. We encountered this same general problem in Chapter 12, where we introduced Pearson's product moment correlation coefficient

$$r=\frac{\sum_{i=1}^{n}(X_i-\bar{X})(Y_i-\bar{Y})}{\sqrt{\sum_{i=1}^{n}(X_i-\bar{X})^2}\sqrt{\sum_{i=1}^{n}(Y_i-\bar{Y})^2}}$$

as a measure of association between X and Y. Serving as a descriptive statistic, r provides a numerical value for the amount of linear dependence between X and Y. However, the sampling distribution given for r only holds under the quite restrictive assumption that the joint distribution for X and Y is normal. Rank correlation methods surmount this limitation and also demonstrate the stronger attribute of measuring certain relationships that are not linear.

Structure of the observations

The n pairs $(X_1, Y_1), (X_2, Y_2), \ldots, (X_n, Y_n)$ are independent, and each pair has ther same continuous bivariate distribution. The $X_1, \ldots, X_n$ are then ranked among *themselves*, and the $Y_1, \ldots, Y_n$ are ranked among themselves:

Pair no.	1	2	...	n
Ranks of X_i	R_1	R_2	...	R_n
Ranks of Y_i	S_1	S_2	...	S_n

Before we present a measure of association, we note a few simplifying properties. Each of the ranks $1, 2, \ldots, n$ must occur exactly once in the set $R_1, R_2, \ldots, R_n$, so that the average rank $\bar{R} = (1+2+\cdots+n)/n = n(n+1)/2n = (n+1)/2$, whatever the outcome.* Similarly, $\bar{S}$ is always equal to $(n+1)/2$. For the same reason

$$\sum_{i=1}^{n} (R_i - \bar{R})^2 = \sum_{i=1}^{n} R_i^2 - n\bar{R}^2 = (1^2 + 2^2 + \cdots + n^2) - \frac{n(n+1)^2}{4}$$

$$= \frac{n(n+1)(2n+1)}{6} - \frac{n(n+1)^2}{4}$$

$$= \frac{n(n^2-1)}{12}$$

and

$$\sum_{i=1}^{n} (S_i - \bar{S})^2 = \frac{n(n^2-1)}{12}$$

for all possible outcomes.

A measure of correlation is defined by C. Spearman that is analogous to Pearson's correlation, r, except that Spearman replaces the observations

* *Mathematical facts*: $\sum_{i=1}^{n} i = \frac{n(n+1)}{2}$ and $\sum_{i=1}^{n} i^2 = \frac{n(n+1)(2n+1)}{6}$

with their ranks. Spearman's rank correlation r_{Sp} is defined by

$$r_{\text{Sp}} = \frac{\sum_{i=1}^{n} (R_i - \bar{R})(S_i - \bar{S})}{\sqrt{\sum_{i=1}^{n} (R_i - \bar{R})^2} \sqrt{\sum_{i=1}^{n} (S_i - \bar{S})^2}} = \frac{\sum_{i=1}^{n} \left(R_i - \frac{n+1}{2}\right)\left(S_i - \frac{n+1}{2}\right)}{n(n^2-1)/12}$$

This rank correlation shares the properties of r that $-1 \leqslant r_{\text{Sp}} \leqslant 1$ and that values near $+1$ indicate a tendency for the larger values of X to be paired with the larger values of Y. However, the rank correlation is more meaningful, because its interpretation does not require the relationship to be linear.

Spearman's rank correlation

$$r_{\text{Sp}} = \frac{\sum_{i=1}^{n} \left(R_i - \frac{n+1}{2}\right)\left(S_i - \frac{n+1}{2}\right)}{n(n^2-1)/12}$$

(*a*) $-1 \leqslant r_{\text{Sp}} \leqslant 1$

(*b*) r_{Sp} near $+1$ indicates a tendency for the larger values of X to be associated with the larger values of Y. Values near -1 indicate the opposite relationship.

(*c*) The association need not be linear; only an increasing/decreasing relationship is required.

Example 15.8 An interviewer in charge of hiring large numbers of typists wishes to determine the strength of the relationship between ranks given on the basis of an interview and scores on an aptitude test. The data for 6 applicants are

Interview rank	5	2	3	1	6	4
Aptitude score	47	32	29	28	56	38

Calculate r_{Sp}.

There are 6 ranks, so that $\bar{R} = (n+1)/2 = 7/2 = 3.5$ and $n(n^2-1)/12 =$

$35/2 = 17.5$. Ranking the aptitude scores, we obtain

Interview R_i	5	2	3	1	6	4
Aptitude S_i	5	3	2	1	6	4

Thus

$$\sum_{i=1}^{n}\left(R_i - \frac{n+1}{2}\right)\left(S_i - \frac{n+1}{2}\right)$$

$$= (5-3.5)(5-3.5) + (2-3.5)(3-3.5) + \cdots + (4-3.5)(4-3.5)$$

$$= 1.5(1.5) + (-1.5)(.5) + \cdots + (.5)(.5)$$

$$= 16.5$$

and

$$r_{\text{Sp}} = \frac{16.5}{17.5} = .943$$

The relationship between interview rank and aptitude score appears to be quite strong. ■

Figure 15.5 helps to stress the point that r_{Sp} is a measure of any monotone relationship, not merely a linear relation.

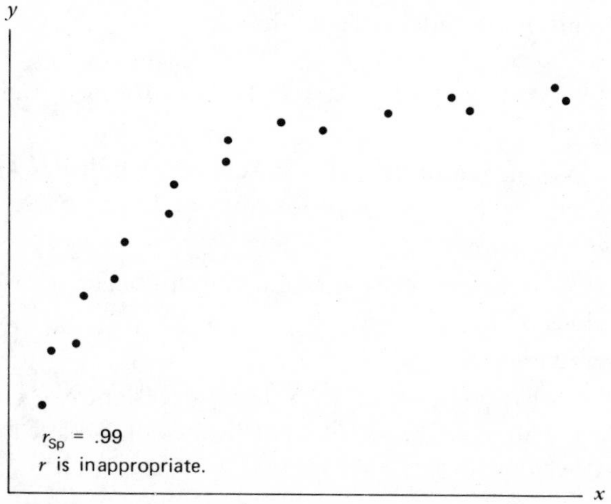

Figure 15.5 r_{Sp} is a measure of any monotone relationship.

Without introducing any more restrictive assumptions about the population, we can employ rank correlation to test the independence of X and Y vs. alternatives of increasing (decreasing) relationship. Before we proceed, however, it is convenient at this point to simplify the calculation of this statistic further. Because

$$\sum_{i=1}^{n}\left(R_i-\frac{n+1}{2}\right)\left(S_i-\frac{n+1}{2}\right)$$

$$=\sum_{i=1}^{n} R_iS_i-\frac{(n+1)}{2}\sum_{i=1}^{n} R_i-\frac{(n+1)}{2}\sum_{i=1}^{n} S_i+n\left(\frac{n+1}{2}\right)^2$$

$$=\sum_{i=1}^{n} R_iS_i-n\left(\frac{n+1}{2}\right)^2$$

the only random part of r_{Sp} is $\sum_{i=1}^{n} R_iS_i$. Consequently, we can equivalently base our test on this quantity and tabulate its distribution, at least for small n.

Hypotheses for a Test of Independence

$(X_1, Y_1), \ldots, (X_n, Y_n)$ is a random sample from a continuous bivariate distribution.

H_0: X and Y are independent; that is

$$P[X \leqslant x, Y \leqslant y]=P[X \leqslant x]P[Y \leqslant y] \quad \text{for all } x \text{ and } y.$$

As a consequence of H_0, all the sets of $n!$ pairings (R_i, S_i) are equally likely, each having a probability of $1/n!$.

One-sided alternative:

H_1: Large values of X have a tendency to occur with large values of Y.

Two-sided alternative:

H_1: Either (a) large values of X have a tendency to occur with large values of Y, or (b) small values of X have a tendency to occur with large values of Y.

The test based on $\sum_{i=1}^{n} R_i S_i$ or, equivalently, on r_{Sp} is implemented by using the distribution of $\sum_{i=1}^{n} R_i S_i$ tabulated in Appendix Table 13 for $n \leqslant 10$.

Test of independence based on r_{Sp}

One-sided alternative:

Reject H_0 in favor of H_1 if

$$T = \sum_{i=1}^{n} R_i S_i \geqslant x$$

where

$$P\left[\sum_{i=1}^{n} R_i S_i \geqslant x\right] \leqslant \alpha$$

Two-sided alternative:

Reject H_0 in favor of H_1 if

$$\sum_{i=1}^{n} R_i S_i \geqslant x \quad \text{or} \quad \leqslant x^*$$

where

$$\frac{\alpha}{2} \geqslant P\left[\sum_{i=1}^{n} R_i S_i \geqslant x\right] = P\left[\sum_{i=1}^{n} R_i S_i \leqslant x^*\right]$$

In large samples for which n is beyond the range of Appendix Table 13, we can employ the result:

Under H_0, $\sqrt{n-1}\, r_{Sp}$ is approximately distributed as $N(0, 1)$.

In other words, reject H_0 in favor of a one-sided H_1 in large samples if $\sqrt{n-1}\, r_{Sp} > z_\alpha$, the upper α point of the standard normal distribution.

Example 15.9 Referring to Example 15.8, test the independence of interview rank and aptitude test score.

We consider the alternative to be two-sided, with either positive or negative correlation. The observed value of the test statistic is

$$\sum R_i S_i = 5\times5+2\times3+3\times2+1\times1+6\times6+4\times4$$
$$=90$$

From Appendix Table 13, we see that the upper-tail critical value of 89 has $P\left[\sum_{i=1}^{6} R_i S_i \geqslant 89\right]=.017$, so that the observed value is in the critical region for $\alpha=.05$ and $\alpha/2=.025$ for each tail. The hypothesis of independence is therefore rejected. ∎

When ties are present, the large-sample normal approximation becomes

$$\frac{\sqrt{n-1}\, r_{\text{Sp}}}{\sqrt{(1-c_x)(1-c_y)}} \quad \text{approximately } N(0,1)$$

where

$$c_x = \sum_{j=1}^{l} \frac{q_j^3 - q_j}{n^3 - n}$$

q_j = number of elements in the jth tie of the X sample

and c_y is the corresponding correction for the Y sample.

15.5.1 Kendall's Tau

Another frequently used measure of rank correlation, named after its proposer, is *Kendall's Tau*

$$r_{\text{tau}} = \sum_{\substack{\text{all pairs}\\(i,j)\, i<j}} \frac{\operatorname{sign}(S_i - S_j)\operatorname{sign}(R_i - R_j)}{\frac{1}{2}n(n-1)}$$

where $\operatorname{sign}(R_i - R_j) = +1$ if $R_i > R_j$ and -1 if $R_i < R_j$.

Kendall's Tau has the same properties as r_{Sp}, and in many ways the two correlations are similar. Null distribution tables and information about the large-sample distribution for Kendall's Tau are readily available (see [1]).

Two pairs (X_1, Y_1), (X_2, Y_2) are said to be *concordant* if $(X_1 - X_2)$ and $(Y_1 - Y_2)$ have the same sign; otherwise these pairs are said to be *discordant*. The correlation r_{tau} is an unbiased estimator of

$$P(\text{concordant}) - P(\text{discordant}) = 2P(\text{concordant}) - 1$$

$$= 2P[(X_1 - X_2)(Y_1 - Y_2) > 0] - 1$$

Under the null hypothesis of independence, the expectation of r_{tau} is zero.

15.6 THE KRUSKAL–WALLIS TEST FOR COMPARING k TREATMENTS

In the completely randomized design for comparing k treatments originally described in Section 14.2, a set of $n = n_1 + n_2 + \cdots + n_k$ experimental units is randomly divided into groups of $n_1, n_2, \ldots, n_k$ units, and these groups receive treatments $1, 2, \ldots, k$, respectively. Under the model of a normal distribution for the responses, the analysis of variance F test is an efficient method to employ in testing the equality of treatment effects. If a serious violation of the assumption of normality is anticipated, a rank test procedure proposed by W. Kruskal and W. Wallis can be employed.

The data structure of the completely randomized design is given in Table 15.6, where the k columns constitute independent random samples from k populations. The problem concerns testing:

> H_0: All k continuous population distributions are identical.
> H_1: Not all k distributions are identical.

TABLE 15.6
DATA STRUCTURE OF A COMPLETELY RANDOMIZED DESIGN

TREATMENT 1	TREATMENT 2	$\cdots$	TREATMENT k
X_{11}	X_{21}	$\cdots$	X_{k1}
X_{12}	X_{22}	$\cdots$	X_{k2}
$\vdots$	$\vdots$		$\vdots$
X_{1n_1}	X_{2n_2}	$\cdots$	X_{kn_k}

Total sample size $n = \sum_{i=1}^{k} n_i$

For the Kruskal–Wallis rank test, we pool all the k samples together and rank the $n = n_1 + n_2 + \cdots + n_k$ observations in increasing order. Having determined the ranks in the combined sample, we replace the observations in Table 15.6 with their corresponding ranks. The k sets of ranks thus obtained are presented in Table 15.7.

TABLE 15.7

RANKS OF OBSERVATIONS IN THE POOLED SAMPLE

	TREATMENT 1	TREATMENT 2	$\cdots$	TREATMENT k
	R_{11}	R_{21}	$\cdots$	R_{k1}
	R_{12}	R_{22}	$\cdots$	R_{k2}
	$\vdots$	$\vdots$		$\vdots$
	R_{1n_1}	R_{2n_2}	$\cdots$	R_{kn_k}
Rank sum	W_1	W_2	$\cdots$	W_k
Average rank	$\bar{R}_1 = \dfrac{W_1}{n_1}$	$\bar{R}_2 = \dfrac{W_2}{n_2}$	$\cdots$	$\bar{R}_k = \dfrac{W_k}{n_k}$

The average ranks for individual samples

$$\bar{R}_i = \frac{1}{n_i} \sum_{j=1}^{n_i} R_{ij} = \frac{W_i}{n_i}$$

appear in the last row of Table 15.7. Because the ranks in the pooled sample consist of the set of integers $\{1, 2, \ldots, n\}$, the pooled-sample average rank is

$$\bar{R} = \frac{1 + 2 + \cdots + n}{n} = \frac{n+1}{2}$$

Under the null hypothesis that the k populations are identical, the sample average ranks should all be close to the pooled average $(n+1)/2$. The differences

$$\left(\bar{R}_1 - \frac{n+1}{2}\right), \ldots, \left(\bar{R}_k - \frac{n+1}{2}\right)$$

reflect the deviations of the individual sample average ranks from the grand mean. The *Kruskal–Wallis statistic* H is an overall measure of heterogeneity among the samples, and it is given by

$$H=\frac{12}{n(n+1)}\left[n_1\left(\bar{R}_1-\frac{n+1}{2}\right)^2+n_2\left(\bar{R}_2-\frac{n+1}{2}\right)^2+\cdots+n_k\left(\bar{R}_k-\frac{n+1}{2}\right)^2\right]$$

$$=\frac{12}{n(n+1)}\sum_{i=1}^{k}n_i\left(\bar{R}_i-\frac{n+1}{2}\right)^2$$

The alternative form

$$H=\frac{12}{n(n+1)}\left[\frac{W_1^2}{n_1}+\cdots+\frac{W_k^2}{n_k}\right]-3(n+1)$$

is more conveneint for use in calculations. The null hypothesis is rejected for large values of H. Tabulation of exact percentage points for H is awkward, because a table is required for each k as well as for each combination of sample sizes $n_1,\ldots,n_k$. Some sets of limited tables are available (the interested student is again referred to [1]). For larger sample sizes, an approximation to the distribution of H is provided by the χ^2 distribution with d.f. $=k-1$. The large-sample testing procedure is then

> Reject H_0 at level α if $H \geqslant \chi^2_\alpha$
>
> where
>
> $$H=\frac{12}{n(n+1)}\sum_{i=1}^{k}n_i\left(\bar{R}_i-\frac{n+1}{2}\right)^2$$
>
> and χ^2_α is the upper α point of χ^2 with d.f. $=k-1$.

When ties are present, average ranks are assigned to the sets of tied observations, and H is calculated using the preceding formula. However, to approximate the χ^2 distribution, an adjustment in the test statistic is required just as it is in the Wilcoxon test. The modified statistic

$$\frac{H}{1-\sum_{j=1}^{l} q_j(q_j^2-1)/n(n^2-1)} \text{ is approximately } \chi^2_{k-1}$$

where

$$l = \text{number of ties}$$

$$q_j = \text{number of elements in the } j\text{th tie}$$

Example 15.10 An endocrinologist conducts an experiment to study the effect of diet on the activity of an enzyme that consumes fat in the body. These three diets are regularly given to groups of rats of an identical breed:

Diet 1: Protein-rich food; free feeding.
Diet 2: Protein-rich food; controlled feeding.
Diet 3: Carbohydrate-rich food; controlled feeding.

Measurements of fatty deposits found on the kidneys and intestines of the rats given each diet are recorded in Table 15.8.

TABLE 15.8
MEASUREMENTS OF FATTY DEPOSITS WITH THREE DIETS

	DIET 1	DIET 2	DIET 3
	120 (22)	96 (17)	98 (19)
	93 (12.5)	62 (2)	92 (11)
	95 (15)	84 (8)	81 (7)
	96 (17)	86 (9)	93 (12.5)
	105 (20)	69 (3)	75 (5)
	96 (17)	74 (4)	61 (1)
	110 (21)	78 (6)	94 (14)
			87 (10)
Rank sums	$W_1 = 124.5$	$W_2 = 49.0$	$W_3 = 79.5$

Use the Kruskal–Wallis statistic to test the hypothesis that there is no difference in fat buildup for the three diets.

The ranks (shown in parentheses in Table 15.8) for the combined sample of $n = 7 + 7 + 8 = 22$ observations are obtained by arranging the observations in increasing order. Note that there are two ties: Two observations of 93 are elegible for ranks 12 and 13, and three observations of 96 are eligible for ranks 16, 17, and 18. The respective average ranks of 12.5 and 17 are assigned to these two groups. The rank sums are calculated in the last row of Table 15.8.

The unadjusted Kruskal–Wallis statistic has the observed value

$$H = \frac{12}{22 \times 23}\left[\frac{(124.5)^2}{7} + \frac{(49.0)^2}{7} + \frac{(79.5)^2}{8}\right] - 3 \times 23$$

$$= 79.38 - 69 = 10.38$$

To adjust for ties, we calculate

$$\frac{10.38}{1-\left[3(3^2-1)+2(2^2-1)\right]/22(22^2-1)}=\frac{10.38}{.997}=10.41$$

With d.f. = 2, $\chi^2_{.05}=5.99$. Because the observed value of 10.41 is larger than 5.99, the null hypothesis is rejected at $\alpha=.05$. It would also be rejected at $\alpha=.01$. ■

15.7 CONCLUDING REMARKS

In the preceding sections we have introduced a few basic nonparametric tests that are popular for reasons of their simplicity, distribution-free properties, and satisfactory performance in detecting population differences. Nonparametric tests that employ other scores in place of ranks have been proposed with the objective of improved performance for specified classes of distributions. Rank tests have also been developed to test hypotheses concerning the equality of spread for two populations, parameters of the straight-line regression model, and randomized block designs, to name only a few. Hollander and Wolfe [1] contains concise descriptions of many of these procedures as well as pertinent tables and references.

We remarked earlier that nonparametric tests are particularly useful when responses cannot be recorded on a specific numerical scale but when data about their order relationships are available. When numerical measurements are available, however, nonparametric procedures are often criticized, because they can overlook valuable information that is contained in the numbers. To some degree, this is true when measurements are discarded and only their ranks and possibly signs are retained. In contrast, when the two populations under comparison are normal, the t statistic explicitly employs all the information in the numerical measurements, which is transmitted in turn by the sample means and the standard deviations. The price we pay for not using the t statistic in such situations is partial loss of information.

Then what can we expect to gain by employing only primitive aspects of the data, such as ranks, and abandoning numerical details? The answer to this question lies in the fact that numerical aspects of the data that are valuable under one model may be misleading under another. The sample mean and the standard deviation for the t statistic are excellent respective measures of center and spread for normal samples, but they can be very unreliable in the presence of heavily tailed or skewed population distributions. Uncertainty concerning the form of the population model motivates the use of nonparametric procedures. When the form of the model is difficult to specify, meaningful numerical aspects of the data may not be apparent. This case is particularly acute for small sample sizes. Nonparametric methods circumvent this dilemma by selecting aspects such as ranks that may not be the best characteristics for a specific model but that provide good protection against a wide spectrum of population shapes.

Tests are judged by two criteria: control of the type I error probability, and the power to detect alternatives. Nonparametric tests guarantee the desired control of the type I error probability α, whatever the form of the distribution. However, a parametric test established at $\alpha=.05$ for a normal distribution may suffer a much larger α when a departure from normality occurs, particularly in small sample sizes. To achieve universal protection, nonparametric tests, quite expectedly, must forfeit some power to detect alternatives when normality actually prevails. As plausible as this argument sounds, it is rather surprising that the loss in power is often marginal with such simple procedures as the Wilcoxon rank-sum test and the signed-rank test. To illustrate this point, Table 15.9 gives the powers of the two-sample t test and of the Wilcoxon rank-sum test for sample sizes $n_1=7=n_2$ from normal populations. Table 15.9 is taken from Milton [2], where other comparisons are also made.

TABLE 15.9

POWER OF TWO-SAMPLE TESTS FOR SAMPLING FROM NORMAL POPULATIONS ($n_1=7$; $n_2=7$; $\alpha=.05$; $\delta=(\mu_1-\mu_2)/\sigma$; ONE-SIDED TESTS)

Alternative δ	.2	.4	.6	.8	1.0	1.5	2.0	3.0
Power of t test	.098	.174	.279	.408	.547	.841	.970	.9999
Power of Wilcoxon test	.096	.169	.269	.391	.525	.818	.959	.9996

We can see that the differences in power between the two tests are quite small. Another viewpoint can be gained by considering the length of confidence intervals. Typically, the t-based confidence intervals are somewhat shorter than intervals based on the Wilcoxon test when the normal model holds. The interested student can confirm this behavior by undertaking the class project at the end of the chapter.

In relatively simple problems, such as a comparison of two treatments using independent or matched-pair samples, nonparametric tests compete well with the optimal parametric tests. However, in more complex testing situations when several factors are to be considered, the loss of power of the nonparametric tests can be quite substantial. When no serious violation of the normal assumption is suspected, the use of parametric procedures is expected to provide more efficient inferences. Even if a moderate or a large sample reveals noticeable departures from normality, normalizing transformations often improve the situation to the point where parametric procedures are more effective than nonparametric methods.

Finally, the presence of dependence among the observations affects the usefulness of nonparametric and parametric methods in much the same manner. Using either method, the level of the test may seriously differ from the nominal value selected by the analyst. In such cases parametric methods can often be rectified by incorporating a structure for dependence, but nonparametric methods are not so easily corrected.

> Caution:
> When successive observations are dependent, nonparametric test procedures lose their distribution-free property, and conclusions drawn from them can be seriously misleading.

REFERENCES

1. Hollander, M., and Wolfe, D. A., *Nonparametric Statistical Methods*, John Wiley & Sons, New York, 1973.
2. Milton, R., *Rank Order Probabilities*, John Wiley & Sons, New York, 1970.

EXERCISES

1. Independent random samples of sizes $n_A = 4$ and $n_B = 2$ are taken from two continuous populations.
   (*a*) Enumerate all possible collections of ranks associated with the smaller sample in the combined sample ranking. Attach probabilities to these rank collections under the null hypothesis that the populations are identical.
   (*b*) Obtain the null distribution of $W_s =$ sum of ranks of the smaller sample. Verify that the tail probabilities agree with the tabulated values.
2. Using Appendix Table 9, find:
   (*a*) $P[W_s \geqslant 42]$ when $n_1 = 5$, $n_2 = 7$.
   (*b*) $P[W_s \leqslant 25]$ when $n_1 = 6$, $n_2 = 6$.
   (*c*) $P[W_s \geqslant 81$ or $W_s \leqslant 45]$ when $n_1 = 10$, $n_2 = 7$.
   (*d*) The point c such that $P[W_s \geqslant c] = .036$ when $n_1 = 8$, $n_2 = 4$.
   (*e*) The points c_1 and c_2 such that $P[W_s \geqslant c_2] = P[W_s \leqslant c_1] = .05$ when $n_1 = 3$, $n_2 = 9$.

3. A mixture of compounds called phenolics occurs in wood waste products. It has been found that when phenolics are present in large quantities, the waste becomes unsuitable for use as a livestock feed. To compare two species of wood, a dairy scientist measures the percentage content of phenolics from 6 batches of waste of species A and 7 batches of waste of species B. The following data are obtained:

Percentage of phenolics

Species A	2.38	4.19	1.39	3.73	2.86	1.21	
Species B	4.67	5.38	3.89	4.67	3.58	4.96	3.98

Use the Wilcoxon rank-sum test to determine if the phenolics content of species B is significantly higher than that of species A. Use α close to .05.

4. In a study of the cognitive capacities of nonhuman primates, 19 monkeys of the same age are randomly divided into two groups of 10 and 9. The groups are trained by two different teaching methods to recollect an acoustic stimulus. The monkeys' scores on a subsequent test are:

Memory scores

Method 1	167	149	137	178	179	155	164	104	151	150
Method 2	98	127	140	103	116	105	100	95	131	

Do the data strongly indicate a difference in the recollection abilities of monkeys trained by the two methods? Use the Wilcoxon rank-sum test with α close to .10.

5. A project (courtesy of Howard Garber) is constructed to prevent the decline of intellectual performance in children who have a high risk of the most common type of mental retardation, called cultural-familial. It is believed that this can be accomplished by a comprehensive family intervention program. Seventeen children in the high-risk category are chosen in early childhood and given special schooling until the age of $4\frac{1}{2}$. Another 17 children in the same high-risk category form the control group. Measurements of the psycholinguistic quotient (PLQ) are recorded for the control and for the experimental groups at the age of $4\frac{1}{2}$ years (data on p. 541).
Do the data strongly indicate improved PLQs for the children who received special schooling? Use the Wilcoxon rank-sum test with a large-sample approximation; use $\alpha = .05$.

6. The following data pertain to the serum calcium measurements in units of IU/L and the serum alkaline phosphate measurements in units of

	PLQ at age $4\frac{1}{2}$ years								
Experimental group	105.4	118.1	127.2	110.9	109.3	121.8	112.7	120.3	
Control group	79.6	87.3	79.6	76.8	79.6	98.2	88.9	70.9	

Experimental group	110.9	120.0	100.0	122.8	121.8	112.9	107.0	113.7	103.6
Control group	87.0	77.0	96.4	100.0	103.7	61.2	91.1	87.0	76.4

μg/ml for two breeds of pigs, Chester White and Hampshire:

Chester White

Calcium	116	112	82	63	117	69	79	87
Phosphate	47	48	57	75	65	99	97	110

Hampshire

Calcium	62	59	80	105	60	71	103	100
Phosphate	230	182	162	78	220	172	79	58

Using the Wilcoxon rank-sum procedure, test if the serum calcium level is different for the two breeds.

7. Referring to the data in Exercise 6, is there strong evidence of a difference in the serum phosphate level between the two breeds?
8. *The two-sample median test*: Like the Wilcoxon rank-sum test, the *median test* is a nonparametric procedure for testing the equality of two population distributions based on independent random samples. Suppose that random samples of sizes n_A and n_B are taken from continuous populations A and B. By ordering all the $n = n_A + n_B$ observations, determine the combined sample median X^*. Then count the number of first-sample observations below the combined sample median X^* and use T to denote this number. Under the null hypothesis, we would expect to find close to half of the first sample and half of the second sample on either side of X^*. A shift of population A to the right of B is indicated by small values of T.
When the sample sizes are large, the test statistic T is approximately normally distributed under H_0, with

$$\text{Mean} = \frac{n_A}{2}, \qquad \text{Variance} = \frac{n_A n_B}{4(n_A + n_B)}$$

This fact can be employed in performing the test based on large samples with one-sided or two-sided alternatives.
Alternatively, we can construct the following 2×2 table (for simplicity, we consider only the case when n is even):

	Number of observations Below X^*	Above X^*	Total
First sample	T	$n_A - T$	n_A
Second sample	$\frac{n}{2} - T$	$\frac{n}{2} - n_A + T$	n_B
Total	$\frac{n}{2}$	$\frac{n}{2}$	n

The structural similarity to a 2×2 contingency table is noticeable. For large samples and two-sided alternatives, the median test can therefore be performed as a χ^2 test in a 2×2 contigency table (see Chapter 13).

For an application of the median test, suppose that a driving-skills test is given to a group of regular marijuana smokers and to a group of nonsmokers. All the subjects are male college students of approximately the same age who have a minimum of three years' driving experience. Their scores on individual test items such as alertness, speed, and driving habits are combined, and the following composite scores are obtained, (where a high score indicates good driving) skills:

	Composite scores on a driving-skills test													
Smokers	28	53	39	27	41	68	27	28	45	48	65	78		
Nonsmokers	32	35	61	43	82	44	78	38	85	63	46	30	47	57

Using the median test, determine if a significant difference in driving skills between marijuana smokers and nonsmokers is demonstrated by the data.

9. Referring to Exercise 22 in Chapter 9 (page 329), use the sign test to determine if the claim of increased growth with the chemical additive is supported by the data.
10. A social researcher interviews 25 newly married couples. Each husband and wife are independently asked the question: "How many children would you like to have?" The following data are obtained:

	Answer of			Answer of	
Couple	Husband	Wife	Couple	Husband	Wife
1	3	2	14	2	1
2	1	1	15	3	2
3	2	1	16	2	2
4	2	3	17	0	0
5	5	1	18	1	2
6	0	1	19	2	1
7	0	2	20	3	2
8	1	3	21	4	3
9	2	2	22	3	1
10	3	1	23	0	0
11	4	2	24	1	2
12	1	2	25	1	1
13	3	3			

Do the data show a significant difference of opinion between husbands and wives regarding an ideal family? Use the sign test with α close to .05.

11. The null distribution of the Wilcoxon signed-rank statistic T^+ is determined from the fact that under the null hypothesis of a symmetric distribution about zero, each of the ranks $1, 2, \ldots, n$ is equally likely to be associated with a positive sign or a negative sign. Moreover, the signs are independent of the ranks.
   (*a*) Considering the case $n=3$, identify all $2^3=8$ possible associations of signs with the ranks 1, 2, and 3 and determine the value of T^+ for each association.
   (*b*) Assigning the equal probability of $\frac{1}{8}$ to each case, obtain the distribution of T^+ and verify that the tail probabilities agree with the tabulated values.
12. Using Appendix Table 10, find:
   (*a*) $P[T^+ \geqslant 65]$ when $n=12$.
   (*b*) $P[T^+ \leqslant 10]$ when $n=10$.
   (*c*) The value c such that $P[T^+ \geqslant c]=.039$ when $n=8$.
   (*d*) The values c_1 and c_2 such that $P[T^+ \leqslant c_1]=P[T^+ \geqslant c_2]=.027$ when $n=11$.
13. Charles Darwin performed an experiment to determine if self-fertilized and cross-fertilized plants have different growth rates. Pairs of Zea mays plants, one self- and the other cross-fertilized, were planted in pots, and their heights were measured after a specified period of time. The data Darwin obtained were:

Pair	Plant height (in $\frac{1}{8}$ inches) Cross-	Self-	Pair	Plant height (in $\frac{1}{8}$ inches) Cross-	Self-
1	188	139	9	146	132
2	96	163	10	173	144
3	168	160	11	186	130
4	176	160	12	168	144
5	153	147	13	177	102
6	172	149	14	184	124
7	177	149	15	96	144
8	163	122			

(SOURCE: Darwin, C., "The Effects of Cross and Self Fertilization in the Vegetable Kingdom," D. Appleton and Co., New York, 1902.)
   (*a*) Calculate the paired differences and plot a dot diagram for the data. Does the assumption of normality seem plausible?

(*b*) Perform the Wilcoxon signed-rank test to determine if crossed plants have a higher growth rate than self-fertilized plants.

14. Refer to Exercise 3. With a confidence level close to 95%, construct a confidence interval for the shift between the distributions of the phenolics contents of species *A* and species *B*.
15. Refer to Exercise 4. With a confidence level close to 90%, establish a confidence interval for the shift between the distributions of scores using method 1 and method 2.
16. Referring to Exercise 13, calculate a 90.5% confidence interval for the median of the difference in height between the cross- and self-fertilized plants.
17. Five finalists in a figure-skating contest are rated by two judges on a 10-point scale as follows:

Contestants	*A*	*B*	*C*	*D*	*E*
Judge 1	6	9	2	8	5
Judge 2	8	10	4	7	3

Calculate the Spearman's rank correlation r_{Sp} between the two ratings.

18. The following scores are obtained on a test of dexterity and aggression administered to a random sample of 10 high-school seniors:

Student	1	2	3	4	5	6	7	8	9	10
Dexterity	23	29	45	36	49	41	30	15	42	38
Aggression	45	48	16	28	38	21	36	18	31	37

Using Spearman's statistic, test the null hypothesis that the manifestations of dexterity and aggression are independent.

19. Referring to Exercise 5 in Chapter 12 (page 415), use the large-sample approximation to the distribution of the Spearman statistic to test if numerical score and social science score are independent.
20. Referring to Exercise 6:
    (*a*) Plot the scatter diagram of the serum calcium and serum phosphate measurements for Chester White pigs.
    (*b*) Calculate Spearman's rank correlation.
    (*c*) Test if the levels of calcium and phosphate are independent.
21. Calculate Kendall's Tau for the data given in Exercise 17.
22. The possible synergetic effect of insecticides and herbicides is a matter of concern to many environmentalists. It is feared that farmers who apply both herbicides and insecticides to a crop may enhance the toxicity of the insecticide beyond the desired level. An experiment is

conducted with a particular insecticide and herbicide to determine the toxicity of three treatments:

Treatment 1: A concentration of .25 μg per gram of soil of insecticide with no herbicide.

Treatment 2: Same dosage of insecticide used in treatment 1 plus 100 μg of herbicide per gram of soil.

Treatment 3: Same dosage of insecticide used in treatment 1 plus 400 μg of herbicide per gram of soil.

Several batches of fruit flies are exposed to each treatment, and the mortality percent is recorded as a measure of toxicity. The following data are obtained:

Treatment 1	Treatment 2	Treatment 3
40	38	68
28	49	51
31	56	45
38	25	75
43	37	75
46	30	69
29	41	
18		

Use the Kruskal–Wallis statistic to determine if the data strongly indicate different toxicity levels among the three treatments.

23. Morphologic measurements of a particular type of fossil excavated from four different geologic sites provide the following data:

Site			
1	2	3	4
1.38	1.49	3.12	1.31
1.42	1.32	2.19	1.46
1.59	2.01	2.76	1.86
1.36	1.59	3.96	1.58
1.91	1.76	2.23	1.64

Do the data strongly indicate that fossils at the four sites differ with respect to the particular morphology measured?

24. *Confidence interval for median using the sign test*. Let $X_1,\ldots,X_n$ be a random sample from a continuous population whose median is denoted by M. For testing H_0: $M=M_0$ we can use the sign test statistic $S=$No. of $X_i>M_0$, $i=1,\ldots,n$. H_0 is rejected at level α in favor of H_1:

$M \neq M_0$ if $S \leqslant r$ or $S \geqslant n-r+1$ where $\Sigma_{x=0}^{r} b(x;n,.5) = \alpha/2$. Repeating this test procedure for all possible values of M_0, a $100(1-\alpha)\%$ confidence interval for M is then the range of values M_0 so that S is in the acceptance region. Ordering the observations from smallest to largest, verify that this confidence interval becomes

$$(r+1)st \text{ smallest to } (r+1)st \text{ largest.}$$

(*a*) Refer to Example 15.4. Using the sign test, construct a confidence interval for the population median of the differences $(A-B)$, with level of confidence close to 95%.

(*b*) Refer to Exercise 5. Using the sign test, construct a confidence interval for the median PLQ of the experimental group, with level of confidence close to 95%.

CLASS PROJECTS

Compare the lengths of confidence intervals obtained by distribution-free methods with those based on the t distribution.

(*A*)(*i*) Select a sample of size $n=10$ from the standard normal distribution with the help of a table or a computer. Calculate a confidence interval for the mean using (1), $\bar{x} \pm t_{.025} s/\sqrt{n}$, and (2) the signed-rank test with $\alpha = .05$.

(*ii*) Combine the results in (*i*) for the whole class. For each approach:

- (*a*) Draw a dot diagram of the lengths.
- (*b*) Determine the number of times the interval covers the true mean.
- (*c*) Calculate the sample mean of the lengths.
- (*d*) Compute the sample standard deviation of the lengths.

(*B*)(*i*) Select a sample of size 6 from a $N(20,4)$ distribution and a sample size 9 from a $N(23,4)$ distribution. Calculate confidence intervals for the difference of means using (1) the two-sample t method, and (2) the Wilcoxon statistic. (Take $\alpha = .05$.)

(*ii*) Calculate the measures in $A(ii)$ for the results in $B(i)$.

What can you conclude from this study?

MATHEMATICAL EXERCISES

1. *The Mann–Whitney version of the Wilcoxon rank-sum test*:
   Let $X_{11}, \ldots, X_{1n_A}$ and $X_{21}, \ldots, X_{2n_B}$ be independent random samples from

two continuous populations. For every observation X_{1i} of the first sample and X_{2j} of the second sample, record

$$U_{ij} = \begin{cases} 1 & \text{if } X_{1i} > X_{2j} \\ 0 & \text{otherwise} \end{cases}$$

When testing the equality of two population distributions, the Mann–Whitney statistic is

$$U = \sum_{i=1}^{n_A} \sum_{j=1}^{n_B} U_{ij} = \begin{array}{l}\text{total number of times an observation from the} \\ \text{first sample exceeds an observation} \\ \text{from the second sample.}\end{array}$$

Prove that U is related to the rank sum W_A of the first sample by

$$U = W_A - \frac{n_A(n_A+1)}{2}$$

[*Hint*: Let $R_{(1)} < R_{(2)} < \cdots < R_{(n_A)}$ be the ordered ranks of the first sample in combined sample ranking. The observation having the rank $R_{(1)}$ exceeds $(R_{(1)} - 1)$ observations of the second sample; the observation having the rank $R_{(2)}$ exceeds $(R_{(2)} - 2)$ observations of the second sample; and so on. Hence, $U = \sum_{i=1}^{n_A} [R_{(i)} - i]$.]

2. Suppose that $n = (n_A + n_B)$ identical cards bearing the numbers $v_1, v_2, \ldots, v_n$ are shuffled and that n_A of the cards are drawn at random without replacement. If $V_1, V_2, \ldots, V_{n_A}$ denote the numbers observed on the sampled cards, show that

$$E\left(\sum_{i=1}^{n_A} V_i\right) = n_A \bar{v} \quad \text{where } \bar{v} = \frac{1}{n} \sum_{a=1}^{n} v_a$$

$$\operatorname{Var}\left(\sum_{i=1}^{n_A} V_i\right) = \frac{n_A n_B}{n-1} \sigma_v^2 \quad \text{where } \sigma_v^2 = \frac{1}{n} \sum_{a=1}^{n} (v_a - \bar{v})^2$$

[*Hint*: $E(V_i) = \frac{1}{n} \sum_{a=1}^{n} v_a, \quad E(V_i^2) = \frac{1}{n} \sum_{a=1}^{n} v_a^2$

$$E(V_i V_j) = \frac{1}{n(n-1)} \sum_{a=1}^{n} \sum_{\substack{b=1 \\ a \neq b}}^{n} v_a v_b = \frac{1}{n(n-1)} \left[\left(\sum_{a=1}^{n} v_a \right)^2 - \sum_{a=1}^{n} v_a^2 \right]]$$

3. Use your results in Mathematical Exercise 2 to prove the following: Under the null hypothesis that the two population distributions are identical, the mean and the variance of the Wilcoxon rank-sum statistic are

$$E(W_A)=n_A\frac{(n+1)}{2}, \qquad \text{Var}(W_A)=\frac{n_An_B(n+1)}{12}$$

4. The Wilcoxon signed-rank statistic is of the form

$$T^+=1\cdot Z_1+2\cdot Z_2+\cdots+n\cdot Z_n$$

where

$$Z_i=\begin{cases} +1 & \text{if rank } i \text{ is associated with a } + \text{ sign.} \\ 0 & \text{if rank } i \text{ is associated with a } - \text{ sign.} \end{cases}$$

Under the null hypothesis of a symmetric distribution about zero, the positive and negative signs occur with equal probabilities of $\frac{1}{2}$ and are independent of the ranks. Use these facts to prove that

$$E(T^+)=\frac{n(n+1)}{4}, \qquad \text{Var}(T^+)=\frac{n(n+1)(2n+1)}{24}$$

5. *An alternative form of* r_{Sp}: When no ties are present, show that Spearman's rank correlation r_{Sp} can also be written

$$r_{\text{Sp}}=1-\frac{6\sum_{i=1}^{n}(R_i-S_i)^2}{n(n^2-1)}$$

[*Hint*: $\sum_{i=1}^{n}(R_i-S_i)^2=\sum_{i=1}^{n}(R_i^2+S_i^2-2R_iS_i)$

$$=2\sum_{i=1}^{n}i^2-2\sum_{i=1}^{n}R_iS_i$$]

CHAPTER 16

Sample Surveys

16.1 WHY TAKE ONLY A SAMPLE WHEN THE POPULATION IS FINITE?

Sample surveys are employed by the ecologist who wishes to obtain information to construct maps of plant communities, by the forester who wishes to know the timber yield of a forest, and by the manager of a rating service who wishes to ascertain the popularity of TV programs among viewers. Surveys are also used to forecast crop harvests, to identify prevailing social and economic conditions, such as unemployment, health care, and inflation, and to examine people's attitudes toward proposed legislation. A barometer of public opinions, so important to a democracy, is readily provided by sample-based estimates of public reaction to the effects of such events as a new wage regulation, a major change in trade policy, or the actions of world leaders.

Whatever population characteristic is of interest, a *census* or complete evaluation of all members of the population can conceivably provide all of the desired information. However, circumstances often prevent such an extensive evaluation. Both cost considerations and the lack of qualified personnel and, if required, highly specialized equipment can severely limit the size of a proposed survey. For these reasons and/or when fairly accurate information must be obtained quickly, it is prudent to forego a census and to survey a *"representative" sample* from the population in question.

The primary purpose of this chapter is to introduce the reader to procedures for collecting such a sample and to methods for analyzing the sample data. When properly planned and executed, samples numbering a few thousand can provide accurate information about populations numbering in the hundreds of thousands.

The inference techniques examined in this chapter are not unlike the procedures discussed in previous chapters, except that the latter have been based primarily on the framework of an infinite population where a random sample consisted of independent and identically distributed random variables. The sample survey problems treated here involve finite (although often quite large) populations, and except for the simple case of sampling with replacement, the observations in survey samples cannot strictly speaking, be considered to be independent.

To provide some insight into the nature of survey sampling, which is common to many fields of application, we begin with a few definitions and then discuss the idea of bias and the method of selecting a random sample. We subsequently introduce the concept of *stratification*, which leads us to a slightly more complex but very useful method for obtaining a representative sample from a population.

16.2 SPECIFICATION OF THE POPULATION AND THE CHARACTERISTIC OF INTEREST

Once the decision to obtain information by means of a sample survey has been made, we are immediately faced with two tasks: carefully defining the population we wish to study, and selecting the characteristic or characteristics to be recorded.

> The *target population* is the population about which we wish to draw inferences on the basis of a sample.

Although specifying a population may seem to be a straightforward procedure, some borderline cases can present difficulties in even the simplest survey. To conduct a survey of the leisure activities of college students, for example, we would have to decide whether or not to include part-time students and students who are carrying less than the specified minimum number of credits because they dropped courses during the semester.

The population to be sampled should coincide with the target population. When the sampled population differs substantially by being more restrictive, any conclusions reached should be considered to apply only to the sampled population. A major practical difficulty encountered at this stage can be the construction of a list of all the members of the population to be sampled. Individual members of the population are called *sampling units* or *units*, and a list of all the members in a population is called a *frame*. Constructing a frame is a basic part of any objective process of sample selection. Can you imagine the difficulties of constructing a frame for cats living in a city or for people who drink excessively? Nevertheless, it is usually possible to develop a reasonably good frame by expending some care and thought about the structure of the target population.

> The *characteristic* is the basic information of interest concerning the sampling units.

The characteristic can be a person's opinion about welfare programs or the dollar amount spent on charity. Our discussion in Section 16.3 illustrates the underlying concepts of survey sampling in terms of a single characteristic, although several characteristics are studied simultaneously in most large-scale surveys.

16.3 PROBABILITY SAMPLING

Once the target population and the characteristic have been specified, attention is focused on choosing a method to obtain a sample that will be representative of the entire population as far as the particular characteristic is concerned. To be able to correctly employ statistical methods to draw inferences about a population from a sample, it is essential that randomness enter the selection process in an explicit manner. Specifically, before the selection is made from the frame, the method of selection should specify the probability that any particular member or group of members will be included in the sample. All of the sampling methods that satisfy this criterion are called *probability samples*. The known probabilities of the units to be included in the sample allow us to determine interval and point estimates for the value of a population quantity. The two most basic forms of probability sampling, *simple random sampling* and *stratified sampling*, are examined in this chapter.

A sample that is not at least approximately a probability sample is called a *nonprobability sample*. Nonprobability sampling methods have the serious drawback that no assessment of variance or uncertainty of the estimate can be given. Nonprobability selection methods should be avoided whenever possible.

16.4 BIAS AND ITS SOURCES

In probability sampling, *bias* is defined as the difference between the expected value of the estimator and the population quantity being estimated.

$$\text{Bias} = E(\text{estimator}) - (\text{Target population value})$$

When this difference is zero, the estimator is said to be *unbiased*. In choosing an estimator, care should be taken to ensure that it does not systematically underestimate or overestimate the population quantity. The criterion of unbiasedness is intended to safeguard against this undesirable feature.

In the literature of sample surveys, the word "bias" has much wider implications than are embodied in the distributional property of an estimator. Any source or cause that tends to make a sample estimate differ systematically from the target population quantity is called a *source of bias*. Of course, choosing the wrong formula for an estimator can be a source of bias, but this is far from the most important source of bias. Due to a faulty measuring device, the sample observations themselves can differ from what they are intended to measure. Even though the formula for an unbiased estimator is used, the sample can still produce biased estimates. In estimating the average weight of a population of children, the sample mean weight can be shown to be an unbiased estimator under simple random sampling. But if the zero setting of the weighing scale is in error, each measurement will be affected by this constant error and the estimate will be biased. Poorly worded questionnaires can also be the source of distorted observations, because respondents may frequently answer questions incorrectly. Questions that are too technical may generate many responses that are sheer guesses.

Aside from bias introduced by a faulty measuring device, a major source of bias is often the existence of a substantial difference between the sampled population and the target population. One of the most dramatic situations in which this problem surfaced was the failure of the *Literary Digest's* poll to predict a winner in the national election of 1936 between the presidential candidates F. D. Roosevelt and A. Landon. Although a large-scale survey was conducted, these pollsters drew their sample from such sources as lists of telephone and car owners. In those days, such amenities were much more common among upper-income groups, and the sample consequently failed to adequately represent low-income groups. Because support for the Republican candidate happened to be strongest in upper-income classes, the poll wrongly predicted a defeat for Roosevelt. Moreover, the sampling was nonprobabilistic; no error limits could be placed on the estimated percentage of votes, even for the sampled population. The blunder was capped by ignoring the 75% who did not respond.

Another primary source of bias emerges when there are a large number of nonrespondents to a survey. Nonrespondents typically differ from respondents regarding the characteristic being surveyed, making the population actually sampled quite different from the target population. A follow-up study of nonrespondents is often conducted to rectify this possible source of bias. Also, substituting units that are conveniently available for nonresponding units can introduce bias. An interviewer who finds no one home at a designated residence may decide to interview the neighbors, who may have an entirely different lifestyle from the intended subjects.

16.5 USING A RANDOM NUMBER TABLE

We are now ready to examine the technical aspects of drawing a random sample. Given a list of population members, we can conceivably number them from 1 to N and also number a set of small balls from 1 to N. These balls can then be placed in a barrel, mixed, and drawn one at a time until we have selected n balls when n is the desired sample size. The population members corresponding to the numbers on the sampled balls can then be included in the sample, and the characteristics of these sampled units can be measured.

As we illustrate in Section 16.6, we prefer to sample by *not replacing* a ball before the next ball is drawn. For the present, however, we recall the two main types of sampling originally examined in Chapter 5:

> *Random sampling with replacement*: The balls are replaced after each individual draw.
>
> *Random sampling without replacement*: The balls are *not* replaced after each individual draw.

If the population is quite large, this mechanical method of random selection may be difficult or practically impossible to implement. This leads us to a consideration of the random number table. Appendix Table 14 contains 5000 digits. Ideally, these numbers are generated by a mechanism such that each digit is the outcome of a trial that consists of drawing one number out of $0, 1, \ldots, 9$ with an equal probability of $\frac{1}{10}$; the digits in different positions are the results of independent repetitions of such trials. As a first conceptually simple procedure, suppose that 10 identical balls numbered $0, 1, \ldots, 9$ are placed in a urn. After mixing the balls, one is drawn blindly and the digit on it is recorded. The ball is returned to the urn, and the operation is repeated. Due to practical considerations, however, the tables of random numbers are generated by a computer that closely simulates this procedure, and the resulting sets of numbers are then carefully checked for conformity with the requirement of independence and equal probability.

The model underlying a random number table ensures that all single digits have the same probability of $\frac{1}{10}$ of occurrence, that all pairs of digits $00, 01, \ldots, 99$ have an equal probability of $\frac{1}{100}$ of occurrence, and so on. How does such a table help us to draw a random sample from a specific finite population? To illustrate the use of the random number table, suppose that we have 40 boxes of dehydrated camping suppers and that we wish to take a sample of size $n=4$ to study their condition. Our first step is to number the boxes from 1 to 40 or to pile them in some order so that they can be identified. In Appendix Table 14, the digits must be chosen two at a time because the population size $N=40$ is a two-digit number. We begin by arbitrarily selecting a page, a row, and a column of the table. Suppose that our selection is row 60, and column 4. We read the pairs of digits in columns 4 and 5,

13 02 18 74 59 13 74 33

We ignore the numbers greater than 40 and also any repeated number when it appears a second time, as 13 does here. We continue reading pairs of digits until four different units

13 2 18 33

are selected. We then test the corresponding packages of camping suppers.

For large-scale samplings or frequent applications, *A Million Random Digits*, published by the Rand Corporation [1], or a well-tested random number generator on a convenient computer are recommended.

16.6 SIMPLE RANDOM SAMPLING

According to the well-established terminology of survey sampling, sampling without replacement is known as *simple random sampling*. To see why this method is always preferred to sampling with replacement, suppose that we have $N=4$ units u_1, u_2, u_3, and u_4 in the population and that the corresponding measurements of the characteristic are

$$\boxed{x_1^*=5} \quad \boxed{x_2^*=3} \quad \boxed{x_3^*=1} \quad \boxed{x_4^*=2}$$

For the purposes of discussion, this characteristic could be the number of people living in each of the four dwelling units that constitute a population. A comparison is to be drawn between sampling with and without replacement for a sample of size $n=2$. First we list all possible unordered samples of size $n=2$, according to the values of the characteristic:

WITH REPLACEMENT				WITHOUT REPLACEMENT		
{5,5}	{3,3}	{1,1}	{2,2}	{5,3}	{3,1}	{1,2}
{5,3}	{3,1}	{1,2}		{5,1}	{3,2}	
{5,1}	{3,2}			{5,2}		
{5,2}						

From this listing, we see that any sample that can occur when sampling without replacement is also possible when sampling with replacement. However, the samples that contain repeated units are not possible when sampling without replacement. Because measuring a unit more than once does not contribute new information, it is apparent that sampling without replacement tends to assemble more information about the population than can be obtained by sampling with replacement.

Continuing with our example, we might ask how closely the distribution of $\bar{X}=(X_1+X_2)/2$ is clustered about the population mean $(5+3+1+2)/4=2.75$ in each of the two cases. Although we know the population mean for purposes of this example, in applications it cannot be obtained without a complete census.

Sampling with Replacement

The unordered sample {5,3} consists of the union of [5 first, 3 second] and [3 first, 5 second]. The probability for each of these last two events is $\frac{1}{16}$, because each of the four units are equally likely to occur in every trial. Therefore, for this sample, $\bar{x}=(5+3)/2=4$ with an associated probability of $\frac{2}{16}$. Proceeding in this manner, the complete distribution of $\bar{X}$ is obtained; from this distribution, the expectation and the variance are then calculated:

Distribution of $\bar{X}=\dfrac{X_1+X_2}{2}$

Value of $\bar{x}$	1	1.5	2	2.5	3	3.5	4	5
Probability	$\frac{1}{16}$	$\frac{2}{16}$	$\frac{3}{16}$	$\frac{2}{16}$	$\frac{3}{16}$	$\frac{2}{16}$	$\frac{2}{16}$	$\frac{1}{16}$

$$E(\bar{X})=1\times\frac{1}{16}+1.5\times\frac{2}{16}+\cdots+5\times\frac{1}{16}=2.75$$

$$E(\bar{X}^2)=1^2\times\frac{1}{16}+(1.5)^2\times\frac{2}{16}+\cdots+5^2\times\frac{1}{16}=8.656$$

$$\operatorname{Var}(\bar{X})=8.656-(2.75)^2=1.094$$

Sampling without Replacement

Each of the six samples is equally likely when sampling without replacement.

Distribution of $\bar{X}=\dfrac{X_1+X_2}{2}$

Value of $\bar{x}$	1.5	2	2.5	3	3.5	4
Probability	$\frac{1}{6}$	$\frac{1}{6}$	$\frac{1}{6}$	$\frac{1}{6}$	$\frac{1}{6}$	$\frac{1}{6}$

$$E(\bar{X})=1.5\times\frac{1}{6}+2\times\frac{1}{6}+\cdots+4\times\frac{1}{6}=2.75$$

$$E(\bar{X}^2)=(1.5)^2\times\frac{1}{6}+2^2\times\frac{1}{6}+\cdots+4^2\times\frac{1}{6}=8.292$$

$$\text{Var}(\bar{X})=8.292-(2.75)^2=0.729$$

Using either method of sampling, the sample mean $\bar{X}$ has an expected value that is equal to the population mean. However, the variance of $\bar{X}$ is smaller when sampling without replacement, so that the distribution of $\bar{X}$ is more concentrated about the population mean. These conclusions, which can be shown to hold regardless of population or sample size, support the method of sampling without replacement or simple random sampling.

Before studying estimators, let us define the basic population quantities within a general conceptual framework. The population consists of N units where the characteristic has a value of x_1^* on unit u_1, x_2^* on unit $u_2,\ldots,x_N^*$ on unit u_N. The population mean is then the average of the characteristic over all units:

$$\text{Population mean}=\mu=\frac{\sum_{i=1}^{N} x_i^*}{N}$$

When defining variability for a finite population, it proves convenient to use the divisor $N-1$ in analogy with the formula for a sample variance. Strictly speaking, the term *variance* should be reserved for the expression with divisor N, but we take this liberty to avoid the introduction of the latter quantity.

$$\text{Population variance}=\sigma^2=\frac{\sum_{i=1}^{N}(x_i^*-\mu)^2}{N-1}$$

A primary purpose of sampling is to learn about μ. Inferences about the population mean are naturally based on the sample mean $\bar{X}$, calculated from the n units selected by simple random sampling. Moreover, the unknown population variance can be estimated by employing the sample variance s^2.

Simple random sample: $X_1, X_2, \ldots, X_n$

Sample mean: $\bar{X} = \frac{1}{n} \sum_{i=1}^{n} X_i$

Sample variance: $s^2 = \frac{\sum_{i=1}^{n} (X_i - \bar{X})^2}{n-1}$

We state without proof that $E(\bar{X}) = \mu$ (the population mean), so that $\bar{X}$ is unbiased. Also, $E(s^2) = \sigma^2$, so that the sample variance is an unbiased estimator of σ^2. The variance of $\bar{X}$ is given by

$$\text{Var}(\bar{X}) = \sigma^2\left(\frac{1}{n} - \frac{1}{N}\right) = \frac{\sigma^2}{n}(1-f)$$

where $f = n/N$ is the fraction of the population included in the sample. It is important to note that the finiteness of the population reduces the variance of $\bar{X}$ from the value for infinite populations σ^2/n to $\sigma^2(1-f)/n$. The factor $(1-f)$ is called the *finite population correction*. When the sampling fraction f is less than .1, this correction can usually be ignored.

Properties of $\bar{X}$ and s^2 with simple random sampling

$$E(\bar{X}) = \mu$$

$$\text{Var}(\bar{X}) = \frac{\sigma^2}{n}(1-f), \quad f = \frac{n}{N}$$

$$E(s^2) = \sigma^2$$

$$\text{Estimated sd. }(\bar{X}) = \frac{s}{\sqrt{n}}\sqrt{(1-f)}$$

To those who are unfamiliar with the subject, it may seem startling that information sampled from a small percentage of the population can quite accurately determine a population value. Inspection of the expression for $\text{Var}(\bar{X})$ provides an explanation, because it shows that the standard deviation of $\bar{X}$ essentially decreases as $1/\sqrt{n}$. Consequently a sample of a few thousand typically produces a small value for the standard deviation of the estimator $\bar{X}$, whatever the population size.

Inference about μ under simple random sampling

Point estimator: $\bar{X}$

Approximate 95% error bound: $\pm 2\frac{s}{\sqrt{n}}\sqrt{(1-f)}\ , \quad f=\frac{n}{N}$

where $\bar{X}=\frac{X_1+\cdots+X_n}{n}$ and $s^2=\frac{\sum_1^n (X_i-\bar{X})^2}{(n-1)}$

The error bound $\pm 2s\sqrt{1-f}\,/\sqrt{n}$ is approximate, but this approximation is quite good when both the sample size n and $N-n$ are large. Under these circumstances, the distribution of $\bar{X}$ is nearly normal and $\bar{X}\pm 2s\sqrt{1-f}\,/\sqrt{n}$ can be treated as a 95% confidence interval for μ.

Example 16.1 Someone who is interested in how elementary-school principals spend their time takes a simple random sample of 12 out of 30 schools in a particular district. The 12 principals are then asked how much time they require to handle discipline problems each week. The responses yield

$$\bar{x}=9.1 \text{ hours} \qquad s^2=22.3$$

Obtain an approximate 95% error bound for your estimate of $\mu=$mean number of hours for the district.

The estimate is $\bar{x}=9.1$, and the fraction of the population sampled is $f=n/N=12/30=.4$. The value of f is too large to be ignored, and the approximate error bound is

$$\pm 2\frac{s}{\sqrt{n}}\sqrt{1-f} \quad \text{or} \quad \pm 2\frac{\sqrt{22.3}}{\sqrt{12}}\sqrt{.6} = \pm 2.11$$

■

Example 16.2 Interest is expressed in the amount of money students spend each month for housing. A simple random sample of 160 students from a university population of 32,400 provides the statistics

$$\bar{x} = \$105.30 \qquad s^2 = 453.6$$

Find an approximate 95% confidence interval for the population mean amount.

The approximate confidence interval is

$$\bar{x} \pm 2\frac{s}{\sqrt{n}}\sqrt{1-f} \quad \text{or} \quad 105.30 \pm 2\frac{\sqrt{453.6}}{\sqrt{160}}\sqrt{1-\frac{160}{32{,}400}}$$

$$\text{or} \quad (101.94,\ 108.66)$$

The correction factor can be ignored because it is extremely small. Dropping this factor, the confidence interval becomes $105.30 \pm 2\sqrt{453.6}/\sqrt{160}$, or (101.93, 108.67), which is almost identical to the interval just calculated. ■

A complete summary of the sample information requires more than the calculation of $\bar{X}$ and the error bound. The discussion of descriptive methods in Chapter 2 applies equally here. A histogram should be constructed to provide some insight into the manner in which the characteristic is distributed over the population. It may also be advisable to record any extreme values and to calculate other descriptive measures if the distribution does not appear to be symmetric.

16.7 SAMPLING TO DETERMINE A PROPORTION

Sometimes we wish to estimate the proportion of units in the population that have a certain attribute. This may be the proportion of unemployed persons or the proportion of voters who support a particular issue. The value 1 is assigned to units possessing the attribute of interest, and the value 0 is assigned to the remaining units. In this way, the population is divided into two groups or types, according to the value of the numerical characteristic. Under simple random sampling, referred to as "sampling without replacement" in Chapter 5, we know that the hypergeometric distribution governs the number of units X in the sample that possess the attribute in question.

The following population and sample compositions were originally presented in Section 5.6:

QUANTITY	POPULATION	SAMPLE
Total number of units	N	n
Number of units with the given attribute	M	X
Proportion of units with the given attribute	$p=\dfrac{M}{N}$	$\hat{p}=\dfrac{X}{n}$

The sample proportion is an unbiased estimator

$$E(\hat{p})=E\left(\frac{X}{n}\right)=p$$

and the variance is given by

$$\text{Var}(\hat{p})=\frac{p(1-p)}{n}\left(\frac{N-n}{N-1}\right)$$

Of course, we could estimate this variance by replacing p with X/n. However, the *unbiased* estimator

$$\hat{p}(1-\hat{p})\frac{(N-n)}{N(n-1)}$$

is typically employed. The inference procedures for proportions and totals can be summarized:

Simple random sampling: Inference on proportions

Point estimator of p: $\quad \hat{p}=\dfrac{X}{n}$

Approximate 95% error bound for p: $\quad \pm 2\sqrt{\hat{p}(1-\hat{p})\dfrac{(N-n)}{N(n-1)}}$

Point estimator of the population total $M=Np$: $\quad \hat{M}=N\dfrac{X}{n}=N\hat{p}$

Approximate 95% error bound for M: $\quad \pm 2\sqrt{N\hat{p}(1-\hat{p})\left(\dfrac{N-n}{n-1}\right)}$

The results for the population total M are derived from

$$E\left(N\frac{X}{n}\right)=NE\left(\frac{X}{n}\right)=p$$

and

$$\operatorname{Var}\left(N\frac{X}{n}\right)=N^2\operatorname{Var}\left(\frac{X}{n}\right)$$

In Chapter 5, we concluded that sampling without replacement is almost the same as sampling with replacement when the population size N is large and the sampling fraction n/N is small. For sampling with replacement the binomial distribution applies, and the variance of X/n is $p(1-p)/n$, which differs from the preceding expression by the absence of the factor $(N-n)/(N-1)$. This factor makes the variance smaller for simple random sampling. In any case, with increasing sample size n, the variance of the estimator X/n decreases at the rate $1/n$. One conclusion is that the variance of the proportion of supportive votes based on a sample size of 5000 is about the same for populations of 100,000 and 1 million. This result, which is not intuitive, partially explains why modern election predictions based on only a small fraction of the electorate are often so successful in forecasting election results.

Example 16.3 Suppose that we have a population of tents u_1, u_2, u_3, u_4, u_5, u_6, where u_2 and u_4 are not watertight. Let us consider estimating the proportion of defective tents in the population on the basis of a simple random sample size of 2.

Each of the following samples are equally likely; the values for $X=$ number of defective tents in the sample are given in parentheses:

u_1, u_2 (1)	u_2, u_3 (1)	u_3, u_4 (1)	u_4, u_5 (1)	u_5, u_6 (0)
u_1, u_3 (0)	u_2, u_4 (2)	u_3, u_5 (0)	u_4, u_6 (1)	
u_1, u_4 (1)	u_2, u_5 (1)	u_3, u_6 (0)		
u_1, u_5 (0)	u_2, u_6 (1)			
u_1, u_6 (0)				

The distribution of X and the mean for this population are:

x	0	1	2
$P[X=x]$	$\frac{6}{15}$	$\frac{8}{15}$	$\frac{1}{15}$

$$E(X)=0\times\frac{6}{15}+1\times\frac{8}{15}+2\times\frac{1}{15}=\frac{2}{3}$$

$$E\left(\frac{X}{n}\right)=\frac{1}{2}\times\frac{2}{3}=\frac{1}{3}$$

Note that $E(X/n)=\frac{1}{3}$, which is the population proportion. The interested student should also verify that $P[X=x]$ is given by the hypergeometric distribution $\binom{2}{x}\binom{4}{2-x}/\binom{6}{2}$.

If our sample consists of u_2 and u_5, then our estimate of the population proportion is $\frac{1}{2}$, and the estimated variance of the sample proportion is $\frac{1}{2}(1-\frac{1}{2})\dfrac{(6-2)}{6(2-1)}=\frac{1}{6}$. ■

Example 16.4 To investigate voting irregularities, a simple random sample of size 60 is taken from the list of 1024 registered voters in a particular ward. It is found that 12 in the sample are persons registered at nonexistent addresses. What can we estimate for the ward?

Here $N=1024$, $n=60$, and $x=12$, so that

$$\hat{p}=\frac{x}{n}=\frac{12}{60}=.2$$

is an estimate of the population proportion. The variance of $\hat{p}$ is estimated by

$$(.2)(.8)\frac{(1024-60)}{1024(60-1)}=.0026$$

The 95% error bound is $\pm 2\sqrt{.0026}=\pm .10$. Similarly, $(.2)(1024)=204.8$ is an estimate for the total number registered at nonexistent addresses. ■

16.8 STRATIFIED RANDOM SAMPLING

The principal objective of a sampling design is to make efficient use of the budget allocated to a study by obtaining as precise an estimate of a population quantity as possible. Simple random sampling is the most basic sampling technique that not only ensures a representative sample but also yields an estimate of a population quantity and a statement of precision. Many ramifications have evolved from this central concept of simple random sampling that permit more precise inferences to be attained for different types of populations. One of the most practically useful designs, called *stratified random sampling*, first divides the population into homogeneous segments and then draws independent simple random samples from these individual subpopulations.

At first, it may seem surprising that the technique of simple random sampling can be improved. To clarify this point, we consider a city in which the northern districts are predominantly high-income areas and the southern districts are primarily low-income areas. To estimate the average cost of housing for the whole city, it is intuitively apparent that relatively small simple random samples taken separately from the northern and southern districts are likely to provide more accurate information than a single sample taken from the whole city. The essence of stratification is that it capitalizes on the known homogeneity of the subpopulations, so that only relatively small samples are required to estimate the characteristic for each subpopulation. These individual estimates can then be easily combined to produce an estimate for the whole population. In addition to savings in sample size, a valuable by-product of the stratified sampling scheme is that estimates obtained for different parts of the population can be subsequently used to draw comparisons.

For a general description of stratified random sampling and the methods of inference associated with this procedure, we suppose that the population is divided into h *subpopulations* or *strata* of known sizes $N_1, N_2, \ldots, N_h$ such that the units in each stratum are homogeneous with respect to the characteristic in question. The unknown mean and the variance for the ith stratum are denoted by μ_i and σ_i^2, respectively.

Structure of population

	Strata			
	1	2	...	h
Size (known)	N_1	N_2	...	N_h
Mean	μ_1	μ_2	...	μ_h
Variance	σ_1^2	σ_2^2	...	σ_h^2

Size of whole population: $N = \sum_{i=1}^{h} N_i$

Mean of whole population: $\mu = \frac{1}{N} \sum_{i=1}^{h} N_i \mu_i$

Stratified random sampling consists of taking independent simple random samples of predetermined sizes $n_1, n_2, \ldots, n_h$ from the strata $1, 2, \ldots, h$, respectively, and measuring the characteristic for each sampled unit. Denoting the jth observation in the sample from the ith stratum by X_{ij}, we can record the summary statistics:

Summary statistics

	Strata			
	1	2	$\cdots$	h
Sample size	n_1	n_2	$\cdots$	n_h
Sample mean	$\bar{X}_1$	$\bar{X}_2$	$\cdots$	$\bar{X}_h$
Sample variance	s_1^2	s_2^2	$\cdots$	s_h^2

where $\bar{X}_i = \frac{1}{n_i} \sum_{j=1}^{n_i} X_{ij}$, $\quad s_i^2 = \frac{1}{n_i - 1} \sum_{j=1}^{n_i} (X_{ij} - \bar{X}_i)^2$

Applying the property of simple random sampling to individual subpopulations, the sample mean $\bar{X}_i$ is an unbiased estimator of μ_i and its variance is

$$\text{Var}(\bar{X}_i) = \frac{\sigma_i^2}{n_i}\left(1 - \frac{n_i}{N_i}\right)$$

Because the overall population mean μ is the weighted average

$$\mu = \frac{N_1}{N}\mu_1 + \frac{N_2}{N}\mu_2 + \cdots + \frac{N_h}{N}\mu_h$$

where each N_i is a known subpopulation size, an unbiased estimator of μ is obtained as

$$\bar{X}_{\text{st}} = \frac{N_1}{N}\bar{X}_1 + \frac{N_2}{N}\bar{X}_2 + \cdots + \frac{N_h}{N}\bar{X}_h$$

The suffix st emphasizes the fact that the estimator is constructed from stratified samples. Also, due to the independence of the samples, the variance of $\bar{X}_{\text{st}}$ is the sum of the variances of the components, where

$$\text{Var}\left(\frac{N_i}{N}\bar{X}_i\right) = \frac{N_i^2}{N^2}\text{Var}(\bar{X}_i) = \frac{N_i^2}{N^2}\frac{\sigma_i^2}{n_i}\left(1 - \frac{n_i}{N_i}\right)$$

$$= \frac{N_i}{N^2}\frac{\sigma_i^2}{n_i}(N_i - n_i)$$

In establishing error bounds and confidence intervals, the unknown σ_i^2 can be estimated by the corresponding sample variance s_i^2.

Estimation of μ by stratified sampling

Point estimator: $$\bar{X}_{\text{st}} = \frac{N_1}{N}\bar{X}_1 + \frac{N_2}{N}\bar{X}_2 + \cdots + \frac{N_h}{N}\bar{X}_h = \frac{1}{N}\sum_{i=1}^{h} N_i\bar{X}_i$$

$$E(\bar{X}_{\text{st}}) = \mu$$

$$\text{Var}(\bar{X}_{\text{st}}) = \frac{1}{N^2}\left[N_1(N_1 - n_1)\frac{\sigma_1^2}{n_1} + N_2(N_2 - n_2)\frac{\sigma_2^2}{n_2} + \cdots + N_h(N_h - n_h)\frac{\sigma_h^2}{n_h}\right]$$

$$= \frac{1}{N^2}\sum_{i=1}^{h} N_i(N_i - n_i)\frac{\sigma_i^2}{n_i}$$

Approximate 95% error bound for μ: $$\bar{X}_{\text{st}} \pm \frac{2}{N}\sqrt{\sum_{i=1}^{h} N_i(N_i - n_i)\frac{s_i^2}{n_i}}$$

Incidentally, it should be noted that the unbiased estimator $\bar{X}_{\text{st}}$ is generally different from the combined sample mean

$$\bar{X} = \frac{n_1}{n}\bar{X}_1 + \frac{n_2}{n}\bar{X}_2 + \cdots + \frac{n_h}{n}\bar{X}_h, \quad \text{where } n = \sum_{i=1}^{h} n_i$$

However, $\bar{X}$ and $\bar{X}_{\text{st}}$ coincide when the strata sample sizes satisfy

$$\frac{n_1}{n} = \frac{N_1}{N}, \frac{n_2}{n} = \frac{N_2}{N}, \ldots, \frac{n_h}{n} = \frac{N_h}{N}$$

This situation is called *proportional allocation* due to the fact that the total sample size n is allocated to different strata in proportion to strata size.

Example 16.5 Each tree in the apple orchard shown in Figure 16.1 is labeled with its yield in bushels. The last three rows of trees are much younger than the others. Estimate the population mean yield per tree, based on stratified samples of size 6 from the subpopulation of the first five rows and of size 2 from the subpopulation of the last three rows.

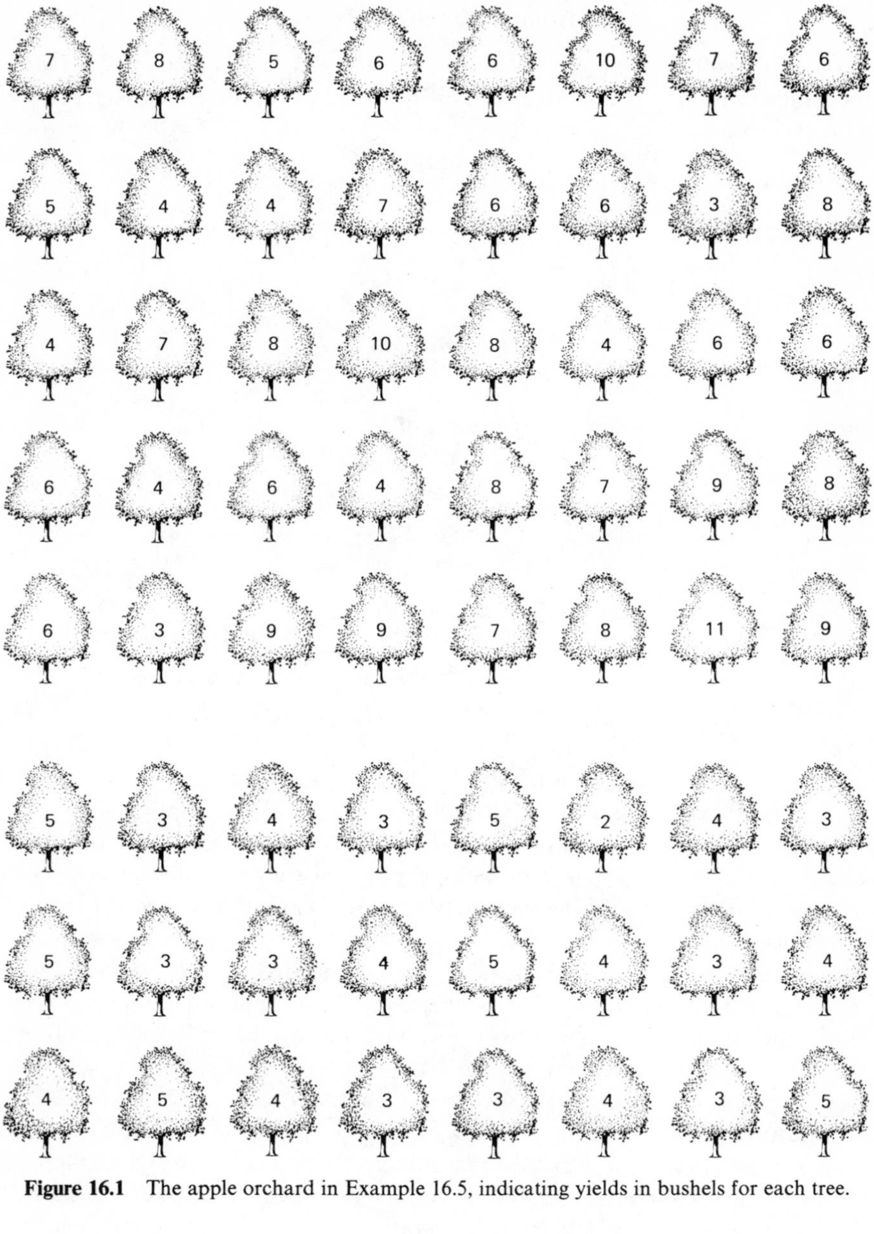

Figure 16.1 The apple orchard in Example 16.5, indicating yields in bushels for each tree.

Here $N_1=40$, $N_2=24$, $n_1=6$, and $n_2=2$. Reading pairs of numbers from the random number table, one for row and one for column, we select trees with the values

Stratum 1: 6, 4, 7, 6, 7, 9 $\quad \bar{X}_1=6.5, \quad s_1^2=2.7$

Stratum 2: 5, 4 $\quad \bar{X}_2=4.5, \quad s_2^2=.5$

Our point estimate of μ is

$$\bar{X}_{st}=\frac{N_1}{N}\bar{X}_1+\frac{N_2}{N}\bar{X}_2=\frac{40}{64}(6.5)+\frac{24}{64}(4.5)=5.75$$

with an approximate 95% error bound of

$$\pm\frac{2}{N}\left[N_1(N_1-n_1)\frac{s_1^2}{n_1}+N_2(N_2-n_2)\frac{s_2^2}{n_2}\right]^{\frac{1}{2}}$$

$$=\pm\frac{2}{64}\left[40(40-6)\frac{2.7}{6}+24(24-2)\frac{(.5)}{2}\right]^{\frac{1}{2}}$$

$$=\pm .85$$

■

We began this discussion by stating that more precise inferences, regarding different populations, can be made using stratified sampling than can be made using simple random sampling. We now consider an ideal situation to see why this is true. Suppose that we wish to obtain information on salaries and that the payroll office tells us that $N_1=20$ workers are at one salary level and $N_2=4$ are at another. Clearly, given this stratification, a sample of size 1 from each stratum would provide the correct mean salary for the population of 24 workers. If simple random sampling were used, the sample mean of two observations would still be unbiased. However, it is possible that both of these observations could be selected from the same stratum. If the two strata salary levels are \$100 and \$400, for example, then the sample mean would be \$100, \$250, or \$400. This estimate will vary in repeated samplings, whereas stratified sampling provides the exact result. A similar comparison extends to more practical situations when the population is divided into internally homogeneous strata.

Example 16.6 Compare the variance of $\bar{X}$, based on a simple random sample of size 8, with the variance of $\bar{X}_{st}$ for the apple orchard in Figure 16.1.

Computations involving the yield records of $N=64$ trees in the entire population give us

$$\text{Population mean:} \qquad \mu=5.5625$$

$$\text{Population variance:} \qquad \sigma^2=4.66$$

and separate calculations involving the first stratum of 40 trees and the second stratum of 24 trees give us

	Mean	Variance
Stratum 1:	$\mu_1=6.625$	$\sigma_1^2=3.984$
Stratum 2:	$\mu_2=3.792$	$\sigma_2^2=0.781$

Consequently, with

SIMPLE RANDOM SAMPLING	STRATIFIED RANDOM SAMPLING
$\text{Var}(\bar{X})=\frac{\sigma^2}{n}(1-\frac{n}{N})$	$\text{Var}(\bar{X}_{st})=\frac{1}{N^2}\left[N_1(N_1-n_1)\frac{\sigma_1^2}{n_1}+N_2(N_2-n_2)\frac{\sigma_2^2}{n_2}\right]$
$=\frac{4.66}{8}(1-\frac{8}{64})$	$=\frac{1}{(64)^2}\left[40(40-6)\frac{3.98}{6}+24(24-2)\frac{(0.781)}{2}\right]$
$=0.51$	$=0.27$

and the estimate based on stratified sampling has a smaller variance. ■

16.9 ALLOCATION OF SAMPLE SIZES

Although the total sample size n is generally limited by the budget available for a survey, the allocation of the sample size to the individual strata remains at the discretion of the sampler. Intuitively, the most plausible choice may be the *proportional allocation* that appropriates the sample sizes to strata in proportion to the sizes of the subpopulations.

> *Proportional allocation*: $n_i=n\left(\frac{N_i}{N}\right), \quad i=1,\ldots,h$

Proportional allocation is motivated by the consideration of a representative sample: if a stratum comprises a large portion of the whole population, it should be allowed to contribute a large share to the sample.

Because our major goal is to improve the precision of our estimate (i.e., to reduce its variance), a more important criterion of allocating the sample sizes should be the minimization of $\mathrm{Var}(\bar{X}_{st})$. This is feasible when the stratum variances σ_i^2 are known or at least when some estimates of σ_i^2 are available from a pilot survey. Specifically, the allocation $n_1, \ldots, n_h$ with $(n_1 + \cdots + n_h) = n$ fixed that minimizes $\mathrm{Var}(\bar{X}_{st})$ is given by:

$$\text{Optimal allocation:} \quad n_i = n \frac{N_i \sigma_i}{\sum_{j=1}^{h} N_j \sigma_j}$$

This requires that the sample size be proportional to the product of the stratum size and the stratum standard deviation. When all the stratum standard deviations are equal, the optimal allocation coincides with the proportional allocation.

Example 16.7 The sizes of three small towns comprising a population are $N_1 = 40{,}000$, $N_2 = 20{,}000$, and $N_3 = 30{,}000$. A stratified random sample is to be taken with a total sample size of $n = 400$. Determine the sample size to be taken from each town individually using (*a*) proportional allocation, and (*b*) optimal allocation when it is roughly known from a previous survey that $\sigma_1 = 20$, $\sigma_2 = 12$, and $\sigma_3 = 14$.

(*a*) *Proportional allocation*:

$$n_1 = n\left(\frac{N_1}{N}\right) = 400\left(\frac{4}{9}\right) = 178$$

$$n_2 = n\left(\frac{N_2}{N}\right) = 400\left(\frac{2}{9}\right) = 89$$

$$n_3 = n\left(\frac{N_3}{N}\right) = 400\left(\frac{3}{9}\right) = 133$$

(*b*) *Optimal allocation*:

$$N_1\sigma_1 = 800{,}000$$
$$N_2\sigma_2 = 240{,}000$$
$$N_3\sigma_3 = 420{,}000$$
$$\text{Total} = 1{,}460{,}000$$

$$n_1 = n \frac{N_1 \sigma_1}{\sum N_i \sigma_i} = 400\left(\frac{800}{1460}\right) = 219$$

$$n_2 = n \frac{N_2 \sigma_2}{\sum N_i \sigma_i} = 400\left(\frac{240}{1460}\right) = 66$$

$$n_3 = n \frac{N_3 \sigma_3}{\sum N_i \sigma_i} = 400\left(\frac{420}{1460}\right) = 115$$

■

We conclude this discussion by delineating the situations for which stratification is a profitable sampling technique. First, stratification usually produces a reduction in the variance of the estimator of a population characteristic. This reduction can be substantial if each stratum is homogeneous but differs from the other strata with respect to the characteristic. Second, if estimates are required for certain subdivisions of the population, it may be helpful to treat the subdivisions as strata to obtain these estimates. For example, we may wish to estimate income among members of a certain minority group while conducting a survey of the income of an urban population.

16.10 STRATIFIED SAMPLING TO DETERMINE A PROPORTION

Stratification techniques also provide improved estimates of the population proportion. The notation and the structure are

		STRATA 1	2	...	h	COMBINED
Population	Size	N_1	N_2	...	N_h	$N = \sum_{i=1}^{h} N_i$
	Proportion	p_1	p_2	...	p_h	$p = \sum_{i=1}^{h} \frac{N_i p_i}{N}$
Sample	Size	n_1	n_2	...	n_h	$n = \sum_{i=1}^{h} n_i$
	Count	X_1	X_2	...	X_h	
	Proportion	$\hat{p}_1 = \frac{X_1}{n_1}$	$\hat{p}_2 = \frac{X_2}{n_2}$	...	$\hat{p}_h = \frac{X_h}{n_h}$	

The formulas for the unbiased estimator of the population proportion and its standard error are obtained in exactly the same manner as they are for the estimator of a population mean.

Stratified sampling: Inference about a proportion

Point estimator: $\hat{p}_{\text{st}} = \frac{N_1}{N}\hat{p}_1 + \frac{N_2}{N}\hat{p}_2 + \cdots + \frac{N_h}{N}\hat{p}_h = \frac{1}{N}\sum_{i=1}^{h} N_i \hat{p}_i$

Approximate 95% error bound: $\pm \frac{2}{N}\sqrt{\sum_{i=1}^{h} \frac{N_i(N_i - n_i)}{(n_i - 1)} \hat{p}_i(1-\hat{p}_i)}$

Regarding the allocation of sample sizes, the proportional allocation $n_i = n(N_i/N)$ can be conveniently implemented. The *optimal allocation* that minimizes the variance of $\hat{p}_{\text{st}}$ requires n_i to be proportional to $N_i\sqrt{p_i(1-p_i)}$, or

$$\text{Optimal allocation:} \qquad n_i = n\frac{N_i\sqrt{p_i(1-p_i)}}{\sum_{j=1}^{h} N_j\sqrt{p_j(1-p_j)}}$$

Obviously, an implementation of the optimal allocation requires some prior knowledge about the approximate value of each strata proportion p_i.

Many surveys, particularly those that consist of mailing a questionnaire to a select number of people, are often jeopardized because there are a large number of nonrespondents. If a survey requests current salary information about a 1975 graduating class, persons who earn low salaries may be less likely to respond than persons who earn high salaries. The sample estimate will then be seriously biased toward the high-income side. To remedy this, we could regard the nonrespondents as a stratum and initiate some type of follow-up interview to obtain the necessary information from at least a few nonrespondents. We could then combine this information with the data obtained earlier to arrive at estimates for the whole population.

Example 16.8 Those students attending a large midwestern university who live in apartments are surveyed by telephone. One of the survey questions is whether or not a security deposit was required on the apartment. Originally, 200 students are selected from the campus directory, ignoring students who live in dorms. Security deposits are required for 108 of the 160 students responding in three survey attempts. The 40 nonrespondents are then treated as a stratum, and four are selected at random to be subjects in a follow-up survey. Of the four, one student was required to make a security deposit.

Estimate the proportion of students living in apartments who were required to make a security deposit.

We can approximate that the population is divided into two strata in the same proportions as those in the observed sample. In other words, we assume that N_1/N is the same as $160/200 = .8$ and that N_2/N is the same as $40/200 = .2$. The estimate then becomes

$$\hat{p}_{\text{st}} = \frac{N_1}{N}\hat{p}_1 + \frac{N_2}{N}\hat{p}_2 = .8\left(\frac{108}{160}\right) + .2\left(\frac{1}{4}\right) = .59$$

Taking the factors $1 - n_1/N_1$ and $1 - n_2/N_2$ as 1, the approximate 95%

error bound can then be written

$$\pm 2\left[\frac{(.8)^2\left(\frac{108}{160}\right)\left(\frac{52}{160}\right)}{159}+\frac{(.2)^2\left(\frac{1}{4}\right)\left(\frac{3}{4}\right)}{3}\right]^{\frac{1}{2}}=\pm .116 \qquad \blacksquare$$

16.11 OTHER METHODS OF SAMPLING

Randomization and stratification constitute the core concepts of survey sampling. However, many other sampling designs have been developed either to exploit specific population structures or for administrative convenience. Here we describe a few of these additional methods and briefly discuss their advantages and disadvantages.

16.11.1 Systematic Sampling

As the name suggests, *systematic sampling* involves selecting the units in a systematic and therefore nonrandom manner. The purpose of this type of sampling technique is usually to spread these units evenly over the frame. Specifically, if $N/n=k$, where N is the population size and n is the desired sample size, this method takes one unit from the first k units of the frame and every kth unit thereafter. An element of randomness is often introduced by letting the first unit be randomly selected. However, the selection of the first unit determines the remainder of the sample units.

The major advantage of systematic sampling is its operational convenience, especially when a sample is to be selected from a list, such as a class roster, a telephone directory, or a stack of file cards. Systematic sampling may be regarded as an approximation to simple random sampling if the list order is not relevant to the characteristic in question, such as recording "age of student" when the list is alphabetized according to last name. Also, if the consecutive sets of k units are viewed as strata, a systematic sample will have the appearance of a stratified sample. However, the performance of a systematic sample may be greatly inferior to a properly executed stratified random sample, because here the strata are arbitrarily constructed without considering internal homogeneity.

As appealing as it may appear, systematic sampling often produces a seriously nonrepresentative sample when the list contains hidden periodicities. For instance, a combined team-by-team listing of baseball players may have a catcher's name in every second place; a systematic sample of every second place would therefore contain only catcher's names.

16.11.2 Cluster Sampling

In many situations, a substantial cost saving can be achieved by conducting a survey with randomly selected groups or *clusters* of the sampling units instead of by taking a simple random sample from the population. Suppose that a sample is to be selected from the population of all fifth-grade students in a particular state. We can view each school in the state as a cluster of the basic sampling units, the fifth-grade students. In *cluster sampling*, first we choose a simple random sample of a few schools in the state and then interview *all* of the fifth graders in those schools.

Collecting a simple random sample of comparable size may be prohibitively expensive for two reasons. First, to construct a frame, we must list every fifth-grade student in all the schools in the state. Second, the units selected in a simple random sample will be typically scattered throughout the state, and the interviewers' travel expenses will consequently be quite high. Cluster sampling avoids the necessity of constructing a frame for the entire population, which is often an exhausting and expensive job in itself. In addition, because the units in a cluster are adjacent and therefore easy to approach, the sampling process is remarkably expedient. However, a price is paid in that the accuracy of the estimates is reduced. A simple random sample of 400 students usually better represents the entire population—and therefore provides better information about that population—than a group of 100 students studied in each of four specified schools. Thus the choice between the two sampling methods should be guided by cost considerations and by the degree of precision desired in the estimates. ■

Most of the sample survey results such as the Gallup poll, the consumer price index, and unemployment figures that regularly appear in print and other mass media not only employ stratification and sampling without replacement but also use strata within strata or even a combination of stratification and cluster sampling. More sophisticated methods, such as double sampling and sampling with probability proportional to size, are also used to obtain representative samples and precise estimators. The ideas presented here offer no more than an introduction to some of the basic concepts that underlie the planning of a survey sampling. Cochran [2] is an excellent reference for interested students who wish to become more familiar with advanced sampling techniques.

When a nonprobability sampling method is employed, extra precaution must be taken to avoid bias. If the interviewer is given too much latitude, only subjects who are easy to contact may be included in the survey. The opinions of people selected from the streets of New York City at noon may differ from those of people who stay inside at noon; an estimate of the proportion of students who do not have summer jobs should clearly not be based on a sample collected at a city beach.

16.12 PLANNING A SURVEY

In previous sections we introduced a few basic sampling methods and the inferential procedures associated with both simple and stratified random sampling. Planning an elaborate survey is usually a far more intricate process that requires careful reflection on the complexities involved in a population structure, on the practical feasibility of the sampling methods, on the coordination and supervision of the field work, and, finally, on the processing and analyses of the data. To briefly introduce these issues, the principal steps involved in planning and executing a survey are examined here. Because different populations as well as the facilities and personnel available for sampling can present diverse difficulties in conducting a survey, our treatment is intended to be illustrative rather than exhaustive.

Purpose of the Study

The necessity for a clear statement of the *purpose* of the study cannot be overemphasized. Without establishing the goal of an investigation, including what we hope to learn from the data, any deliberation as to the choice among alternative sampling methods will be meaningless. If you don't know what you are looking for, you cannot know where to search. The reward, for the care and thought expended in initially defining the purpose of a survey as specifically as possible, is that vital information is unlikely to be overlooked when the units are sampled.

Target Population

The population from which inferences are to be drawn, called the *target population*, must be defined as clearly as possible. In the course of conducting the survey, care must be taken to ensure that the sampled population does not deviate drastically from the target population. When the sampled population is restricted for practical convenience, we must be cautious in extending inferences to the target population.

What Data Should Be Collected?

Guided by the statement of purpose, we should determine the *nature of the data* that are to be collected from the sampled units. Care should be taken to include all the essential data and, at the same time, to avoid collecting data that are irrelevant to the purpose of the survey. In sampling human populations, the primary vehicle of gathering data is the questionnaire, and a well-designed questionnaire is crucial to the success of a survey. Frequently, there is a tendency to prepare an overly long questionnaire in the mistaken belief that more questions will provide more information. Conversely, an unusually large number of questions can bore the respondents and may erode the quality of the data gathered on vital issues. The intended manner of conducting the survey should be kept in mind while designing the questionnaire, and the questions should be carefully worded to avoid guiding responses in any particular direction. A question should not indicate the desired answer. (Avoid such questions as "Don't you think that···?".) On the other hand, a question should not be stated in an ambiguous manner. The question "How many unemployed persons are there in your family?" does not indicate whether children, students, and the retired should be counted. Above all, questions should be limited to relevant issues, and brevity should be a primary consideration. Each question must pass this test: what pertinent information will the answer provide?

What Sampling Method Should Be Used?

Determining the *sampling design* to be used and choosing the *sample sizes* are key issues in planning a survey. The selection of an appropriate sampling method is based on such factors as the structure of the population, the type of information sought, and the administrative facilities and personnel available to conduct the survey. In conjunction with choosing the appropriate sampling method, the required sample size is determined by a specification of the degree of precision desired in the estimates. It should also then be checked if this sample size is feasible within the budget allocated for the survey.

Pilot Study

Frequently, it is advantageous to expend a portion of the budget to conduct a small-scale trial survey called a *pilot survey* or a *pretest*. A pilot survey provides an opportunity to field test the questionnaire to detect and correct any serious irregularities or inadequacies. A pilot survey can also provide information or suggestions that improve the sampling design. In stratified sampling, for example, information about the strata variances obtained in the pilot survey can be used to achieve a nearly optimal allocation of sample sizes in the full-scale survey.

Analyzing the Data

Once the survey has been completed, the full force of graphic and numerical techniques can be employed to interpret the results. Histograms and joint frequency tables help to show the correlation between responses.

Ingenuity in creating plots can suggest interesting relationships and conclusions. In reporting the estimates of the population quantities, a statement of uncertainty should be made in terms of probabilistic error bounds or confidence intervals.

REFERENCES

1. *A Million Random Digits with* 100,000 *Normal Deviates*, The RAND Corporation, The Free Press, New York, 1955.
2. Cochran, W. G., *Sampling Techniques*, 2nd Ed., John Wiley & Sons, New York, 1963.

EXERCISES

1. Discuss the appropriate choices of the sampling units and the frame in each of the following surveys:
   (*a*) A state justice department wishes to estimate the average duration of pretrial detention for persons in a particular metropolitan area who are arrested on charges of committing a major criminal offense.
   (*b*) The marketing division of a pharmaceutical company wishes to ascertain the percentage of hospitals in the country that use their brand of disinfecting solution to sterilize surgical equipment.
   (*c*) A university governing body is interested in studying student opinion about a change in the academic calendar recently proposed by a faculty committee.
   (*d*) An elected judge wishes to determine how her constituency feels about a proposed piece of antiobscenity legislation.
   (*e*) A state public health agency is undertaking a project to assess the quality of health care in the state. Part of the project is designed to estimate the average annual expenditure for dental care incurred by families.
2. For each survey in Exercise 1, discuss:
   (*a*) The construction of the frame and any difficulties that might be encountered in the process.
   (*b*) Suitable methods of conducting the sampling: by telephone, mailing a questionnaire, or other relevant methods.
   (*c*) The advantages and possible disadvantages of using your suggested method.
3. Identify the major sources of bias in each of the following situations:
   (*a*) A survey is conducted to study the extent of the use of convenience foods (such as TV dinners and precooked canned foods)

by households in a community. A random sample of households is selected, and the data are collected by telephone interviews made during working hours (8 A.M. to 5 P.M.). The nonrespondents are ignored.

(*b*) To study the participation of residents in a particular city in outdoor sporting activities, data are collected by interviewers who visit the sampled households, usually during weekend afternoons. If the residents of a household are not at home, a neighboring house is visited instead.

(*c*) An agency decides to use a convenient and inexpensive sampling method to conduct a public opinion poll. Interviewers stationed at major supermarkets during weekday afternoons collect data from shoppers every three minutes as they prepare to enter the stores.

(*d*) A radio station conducts a public opinion poll on a political issue by broadcasting a request to its listeners to call the station and state their opinions.

4. A resort area has 32 motels, which are collectively taken as a population. The characteristic to be studied is the charge per day for double-occupancy rooms. The population values are the following rates per day in dollars: 25, 20, 35, 21, 22, 22, 24, 25, 30, 28, 24, 20, 20, 25, 20, 19, 25, 23, 20, 24, 28, 24, 24, 22, 28, 26, 23, 25, 22, 27, 25, 23. Using the random numbers in Appendix Table 14, draw a simple random sample of size 10 from this population.
5. Referring to Exercise 4, use the simple random sample you drew to calculate the sample mean. Also estimate the variance of the mean of a simple random sample of size 10.
6. The variance of the population in Exercise 4 is $\sigma^2 = 13.97$. What is the variance of $\bar{X}$ based on a simple random sample of size 10? Of size 20? Of size 30?
7. Suppose that 588 farms located in a particular area constitute a population and that their last year's capital expenditure on farming machinery and equipment is the characteristic to be studied. The complete record for the whole population appears in Table 16-1, divided into three data sets by a farm size classification (courtesy of Harlan Hughes). From the whole population of 588 farms, draw a simple random sample of size 60 using the random numbers in Appendix Table 14. Record the expenditures of the sampled units and retain your data to use in Exercise 8.
8. Referring to Exercise 7, suppose that the population values are *not* known and that you have only the data provided by the sample of size 60. Estimate the population mean expenditure per farm and establish a 95% error bound for your estimate.

TABLE 16-1

CAPITAL EXPENDITURES ON MACHINERY AND EQUIPMENT (IN THOUSANDS OF DOLLARS)

SMALL FARMS

17.	38.	9.	7.	11.	14.	17.	10.	31.	24.	22.	21.	9.	41.	19.
9.	13.	26.	36.	18.	8.	11.	23.	19.	16.	14.	14.	17.	20.	20.
9.	18.	6.	19.	52.	14.	5.	27.	14.	14.	28.	17.	9.	11.	12.
25.	19.	28.	15.	18.	24.	23.	27.	24.	20.	21.	27.	21.	34.	26.
21.	9.	29.	22.	10.	18.	45.	24.	16.	95.	40.	42.	11.	17.	17.
13.	14.	23.	17.	27.	18.	34.	18.	16.	17.	20.	23.	18.	42.	22.
18.	23.	16.	26.	11.	37.	23.	32.	24.	16.	24.	34.	37.	31.	29.
15.	41.	38.	21.	34.	23.	24.	27.	34.	5.	34.	29.	22.	26.	30.
26.	27.	39.	30.	31.	28.	39.	28.	34.	28.	24.	44.	22.	23.	40.
16.	5.	19.	36.	36.	17.	21.	43.	21.	19.	14.	14.	31.	27.	39.
30.	41.	28.	19.	32.	18.	19.	33.	27.	28.	26.	23.	32.	36.	21.
24.	32.	19.	18.	31.	25.	26.	21.	18.	36.	29.	47.	26.	31.	26.
32.	27.	43.	45.	45.	25.	17.	30.	27.	28.	16.	44.	20.	15.	31.
21.	42.	27.	32.	33.	21.	35.	44.	24.	26.	38.	57.	54.	24.	37.
21.	33.	19.	20.	32.										

MEDIUM FARMS

37.	30.	41.	17.	38.	29.	32.	21.	39.	41.	28.	33.	35.	24.	36.
28.	20.	23.	27.	34.	33.	36.	25.	28.	39.	36.	22.	25.	54.	53.
36.	14.	22.	32.	21.	35.	35.	39.	32.	40.	24.	48.	41.	30.	42.
20.	38.	23.	17.	38.	16.	23.	28.	32.	18.	60.	28.	47.	61.	25.
22.	25.	48.	53.	35.	25.	23.	44.	18.	56.	42.	55.	39.	24.	38.
42.	27.	30.	34.	43.	29.	35.	43.	62.	25.	15.	66.	34.	25.	11.
45.	28.	40.	32.	38.	33.	48.	46.	54.	45.	35.	31.	30.	42.	22.
23.	46.	14.	42.	33.	31.	75.	50.	44.	33.	41.	32.	45.	44.	51.
39.	35.	22.	44.	35.	24.	29.	23.	32.	30.	35.	50.	28.	21.	21.
12.	30.	28.	60.	35.	49.	33.	22.	58.	25.	23.	39.	40.	44.	41.
14.	37.	32.	22.	27.	23.	37.	59.	50.	46.	40.	47.	41.	38.	48.
40.	32.	31.	22.	24.	25.	33.	54.	36.	52.	39.	61.	46.	36.	16.
37.	38.	51.	25.	35.	49.	9.	46.	35.	53.	43.	59.	41.	52.	51.
47.	72.	46.	29.	25.	42.	42.	43.	46.	43.	29.	58.	47.	85.	52.
48.	23.	39.	40.	43.	52.	36.	35.	27.	56.	47.	39.	51.	48.	48.
23.	24.	39.	30.	59.	35.	39.	32.	51.	18.	27.	38.	36.	41.	11.
42.	42.	65.	27.	34.	72.	49.	39.	44.	57.	64.	51.	53.	55.	63.
39.	31.	48.												

LARGE FARMS

53.	63.	44.	66.	40.	42.	48.	44.	27.	56.	37.	39.	37.	40.	66.
49.	39.	54.	30.	68.	36.	42.	28.	29.	41.	57.	30.	39.	28.	80.
79.	61.	81.	53.	57.	54.	29.	94.	77.	52.	61.	49.	52.	67.	36.
35.	57.	63.	32.	48.	57.	50.	62.	51.	52.	59.	55.	22.	18.	84.
57.	86.	50.	54.	96.	45.	28.	59.	64.	42.	41.	77.	76.	83.	36.
42.	39.	72.	84.	34.	55.	51.	66.	96.	63.	88.	87.	63.	91.	117.
107.	48.	56.	71.	54.	64.	45.	61.	59.	68.	50.	74.	100.	144.	80.
64.	101.	105.	77.	85.	60.	63.	66.	36.	95.					

9. A simple random sample of size 350 is taken from a population of 4000 migrant farm workers in a particular state, and the data on their hourly wages are recorded. The sample mean and the standard deviation are found to be \$3.45 and \$1.07, respectively. Construct a 90% confidence interval for the mean hourly wage of this population.
10. Errors present in the accounts receivable are of vital concern to auditors. An auditor who works for a shipping company wishes to estimate the proportion of instances in which customers receive faulty bills. During a given period of time, suppose that the customer clearances of 2325 accounts receivable are filed at the auditor's office. A simple random sample of 500 of these accounts is taken, and 48 of them are found to involve faulty billing. Construct a 95% confidence interval for the proportion of accounts receivable that involved faulty billing to customers.
11. A forester wishes to estimate the proportion of trees in a lowland forest that have wetwood infections. These infections can be detected by the presence of a particular bacteria in the juice extracted from bore holes. It is known that there are about 5000 trees in the forest. Borehole tests performed on a simple random sample of 400 trees indicate that 139 trees are affected. Construct a 95% confidence interval for the population proportion of trees having wetwood infections.
12. A nutrition survey is conducted to determine the quality of elementary school children's food intake. From a total of 1500 children in elementary schools in a particular city, a simple random sample of 80 is selected, and the parents are asked to keep a record of the food eaten by their children for a period of one week. The average caloric intake per day is then determined for each child. From the data obtained for the sample of 80 children, the sample mean and the standard deviation of caloric intake are found to be 752 and 138, respectively. Based on this information, construct a 95% confidence interval for the mean daily caloric intake of the population.
13. Referring to Exercise 7, suppose that the three groups of farms are to be treated as three strata. Using the random numbers in Appendix Table 14, draw independent simple random samples of sizes 22, 26, and 12 from the strata of small, medium, and large farms, respectively, and retain your data to use in Exercise 14.
14. Using your sample information from Exercise 13 and the information on the strata sizes, estimate the population mean capital expenditure per farm. Also construct a 95% confidence interval for the population mean.
15. A city transportation department is conducting a survey to determine the gasoline usages of its residents. Stratified random sampling is

used, and the four city wards are treated as the strata. The amount of gasoline purchased in the last week is recorded for each household sampled. The strata sizes and the summary information obtained from the sample are:

	Strata			
	I	II	III	IV
Stratum size	3750	3272	1387	2475
Sample size	50	45	30	30
Sample mean (in gallons)	12.6	14.5	18.6	13.8
Sample variance	2.8	2.9	4.8	3.2

Estimate the mean weekly gasoline usage per household for the city population and construct a 95% error bound for your estimate.

16. The publisher of a news magazine is conducting a survey to determine subscribers' views regarding the magazine's coverage of international affairs. Because subscriber records covering different parts of the country are maintained at three regional headquarters, it is convenient to draw a stratified random sample, treating the three regions as the strata. The subscribers sampled from each stratum are asked: "Are you satisfied with the magazine's coverage of international affairs?". The summary data obtained are:

	Strata		
	I	II	III
Stratum size	2 million	5 million	3 million
Sample size	500	600	600
Number responding "yes"	225	336	286

Estimate the population proportion of subscribers who are satisfied with the magazine's coverage of international affairs and construct a 90% error bound for your estimate.

17. Referring to Exercise 15, suppose that we wish to select a stratified random sample from a total sample size of 1000 households. Determine the sample sizes for the individual strata according to proportional allocation.

18. Referring to Exercise 15, suppose that the sampling is actually a pilot study designed to estimate the strata variances for the purpose of determining the optimal allocation of a full-scale survey to be conducted with a total sample size of 1000. Using the information on strata sizes and the estimates of strata variances from Exercise 15, determine the optimal allocation for the total sample size of 1000.

*19. *Optimal Allocation with Cost (Neyman Allocation):* In stratified sampling, the cost of sampling a unit often varies from one stratum to another. For instance, the travel expenses for interviewers may be substantially less to units located in major cities than to units located in remote rural areas. If the sampling cost varies from stratum to stratum, this should be considered when determining the optimal allocation of sample sizes.

Let C_i denote the cost of sampling a unit from the ith stratum $i=1,\ldots,h$, where C represents the total budget for the survey, and let C_0 be a fixed cost irrespective of sample size. The optimal allocation, the one that minimizes $\operatorname{Var}(\overline{X}_{\text{st}})$ for the total budget outlay C, is then

$$n_i = n\frac{N_i\sigma_i/\sqrt{C_i}}{\sum_{j=1}^{h} N_j\sigma_j/\sqrt{C_j}}$$

where the total sample size n is given by

$$n=(C-C_0)\frac{\sum_{j=1}^{h} N_j\sigma_j/\sqrt{C_j}}{\sum_{j=1}^{h} N_j\sigma_j\sqrt{C_j}}$$

Now suppose that it is known that the population is divided into three strata

Size:	$N_1=1000$	$N_2=5000$	$N_3=2000$
Sd.:	$\sigma_1=20$	$\sigma_2=32$	$\sigma_3=40$
Cost:	$C_1=1$	$C_2=.25$	$C_3=4$

and that the fixed cost is $C_0=10$.

(*a*) Determine the approximate optimal allocation when the total budget outlay is 200.

(*b*) What is $\operatorname{Var}(\overline{X}_{\text{st}})$ given your allocation?

20. Referring to Exercise 19, the same allocation will also minimize the total cost for a specified value v_0^2 for $\operatorname{Var}(\overline{X}_{\text{st}})$. In this case the total sample size n is given by

$$n=\frac{\left(\sum_{j=1}^{h}\frac{N_j}{N}\sigma_j\sqrt{C_j}\right)\left(\sum_{j=1}^{h}\frac{N_j}{N}\sigma_j/\sqrt{C_j}\right)}{v_0^2+\frac{1}{N}\sum_{j=1}^{h}\frac{N_j}{N}\sigma_j^2}$$

(*a*) Using the values given in Exercise 19, determine the optimal allocation when $v_0^2 = 4$.

(*b*) What is the total cost of this survey? (NOTE: The total cost is $C = C_0 + C_1 n_1 + \ldots + C_h n_h$.)

CLASS PROJECTS

1. Organize a survey using simple random sampling. (Suggestions: a student rent survey or an attitude survey; if it is election time, you may wish to predict the outcome.) Choose a survey problem that requires both the estimation of a population proportion and the estimation of a population mean.
2. In the context of your chosen survey (Project 1), discuss the feasibility and the expected advantages of stratification vs. other types of sampling.

MATHEMATICAL EXERCISES

1. Using simple random sampling, show that $E(\overline{X}) = \mu$, so that the sample mean is an unbiased estimator of the population mean.
   [*Hint*: Let the population values be denoted by $x_1^*, x_2^*, \ldots, x_N^*$ and the observations from a simple random sample of size n be denoted by $X_1, \ldots, X_n$. An individual X_i can assume any of the values $x_1^*, x_2^*, \ldots, x_N^*$ with an equal probability of $1/N$, so that $E(X_i) = \frac{1}{N} \sum_{k=1}^{N} x_k^* = \mu$.]
2. Using simple random sampling, show that

$$\operatorname{Var}(\overline{X}) = \frac{N-n}{N} \frac{\sigma^2}{n} \qquad \text{where } \sigma^2 = \frac{\sum_{k=1}^{N} (x_k^* - \mu)^2}{N-1}$$

$\Big[$ *Hint*: $\operatorname{Var}(\overline{X}) = E(\overline{X}^2) - \mu^2$

$$\overline{X}^2 = \frac{1}{n^2} \left[\sum_{i=1}^{n} X_i^2 + \sum_{i \neq j} \sum X_i X_j \right]$$

$$E(X_i^2) = \frac{1}{N} \sum_{k=1}^{N} x_k^{*2} = \frac{N-1}{N} \sigma^2 + \mu^2$$

Each pair (X_i, X_j) can assume all pairs of population values (x_k^*, x_l^*) with an equal probability of $1/N(N-1)$, so that

$$E(X_iX_j)=\frac{1}{N(N-1)}\sum_{k\neq l}^{N}\sum^{N}x_k^*x_l^*=\frac{1}{N(N-1)}\left[\left(\sum_{k=1}^{N}x_k^*\right)^2-\sum_{k=1}^{N}x_k^{*2}\right]$$

$$=\frac{1}{N(N-1)}\left[N^2\mu^2-(N-1)\sigma^2-N\mu^2\right]$$

3. The Cauchy–Schwarz inequality states that for any two sets of numbers $\{a_1,a_2,\ldots,a_h\}$, $\{b_1,b_2,\ldots,b_h\}$

$$\left(\sum_{i=1}^{h}a_i^2\right)\left(\sum_{i=1}^{h}b_i^2\right)\geqslant\left(\sum_{i=1}^{h}a_ib_i\right)^2$$

and that equality is attained only when

$$\frac{a_1}{b_1}=\frac{a_2}{b_2}=\cdots=\frac{a_h}{b_h}$$

(see Mathematical Exercise 8.1). Using this inequality, show that optimum allocation in stratified sampling without consideration of cost is given by

$$n_i=n\frac{N_i\sigma_i}{\sum_{j=1}^{h}N_j\sigma_j}$$

$$\left[\textit{Hint}:\quad \operatorname{Var}(\bar{X}_{\text{st}})=\sum_{i=1}^{h}\frac{W_i\sigma_i^2}{n_i}-\frac{1}{N}\sum_{i=1}^{h}W_i\sigma_i^2\quad\text{where } W_i=\frac{N_i}{N}\right.$$

The second term on the right-hand side of this equation is a constant, so that the problem becomes choosing $n_1,\ldots,n_h$ such that $n=n_1+\cdots+n_h$ is fixed and $\sum_{i=1}^{h}W_i\sigma_i^2/n_i$ is minimized.

Apply the Cauchy–Schwarz inequality, with $a_i=W_i\sigma_i/\sqrt{n_i}$ and

$b_i = \sqrt{n_i}$, $i = 1, \ldots, h$. Then

$$\left(\sum_{i=1}^{h} \frac{W_i^2 \sigma_i^2}{n_i} \right) \left(\sum_{i=1}^{h} n_i \right) \geqslant \left(\sum_{i=1}^{h} W_i \sigma_i \right)^2$$

so that the minimum possible value of $\sum_{i=1}^{h} W_i^2 \sigma_i^2 / n_i$ is $\frac{1}{n} (\sum_{i=1}^{h} W_i \sigma_i)^2$. The minimum is achieved by taking $n_1, \ldots, n_h$ such that

$$\frac{a_1}{b_1} = \frac{a_2}{b_2} = \cdots = \frac{a_h}{b_h}$$

]

APPENDIX

TABLE 1
THE NUMBER OF COMBINATIONS $\binom{n}{r}$

n \ r	2	3	4	5	6	7	8	9	10
2	1								
3	3	1							
4	6	4	1						
5	10	10	5	1					
6	15	20	15	6	1				
7	21	35	35	21	7	1			
8	28	56	70	56	28	8	1		
9	36	84	126	126	84	36	9	1	
10	45	120	210	252	210	120	45	10	1
11	55	165	330	462	462	330	165	55	11
12	66	220	495	792	924	792	495	220	66
13	78	286	715	1,287	1,716	1,716	1,287	715	286
14	91	364	1,001	2,002	3,003	3,432	3,003	2,002	1,001
15	105	455	1,365	3,003	5,005	6,435	6,435	5,005	3,003
16	120	560	1,820	4,368	8,008	11,440	12,870	11,440	8,008
17	136	680	2,380	6,188	12,376	19,448	24,310	24,310	19,448
18	153	816	3,060	8,568	18,564	31,824	43,758	48,620	43,758
19	171	969	3,876	11,628	27,132	50,388	75,582	92,378	92,378
20	190	1,140	4,845	15,504	38,760	77,520	125,970	167,960	184,756

SOURCE: Hoel, Paul G. *Elementary Statistics*, John Wiley & Sons, New York, (1971).

TABLE 2

CUMULATIVE BINOMIAL PROBABILITIES $P[X \leqslant c] = \sum_{x=0}^{c} \binom{n}{x} p^x (1-p)^{n-x}$

	c	.05	.10	.20	.30	.40	*p* .50	.60	.70	.80	.90	.95
$n=1$	0	.950	.900	.800	.700	.600	.500	.400	.300	.200	.100	.050
	1	1.000	1.000	1.000	1.000	1.000	1.000	1.000	1.000	1.000	1.000	1.000
$n=2$	0	.902	.810	.640	.490	.360	.250	.160	.090	.040	.010	.002
	1	.997	.990	.960	.910	.840	.750	.640	.510	.360	.190	.097
	2	1.000	1.000	1.000	1.000	1.000	1.000	1.000	1.000	1.000	1.000	1.000
$n=3$	0	.857	.729	.512	.343	.216	.125	.064	.027	.008	.001	.000
	1	.993	.972	.896	.784	.648	.500	.352	.216	.104	.028	.007
	2	1.000	.999	.992	.973	.936	.875	.784	.657	.488	.271	.143
	3	1.000	1.000	1.000	1.000	1.000	1.000	1.000	1.000	1.000	1.000	1.000
$n=4$	0	.815	.656	.410	.240	.130	.063	.026	.008	.002	.000	.000
	1	.986	.948	.819	.652	.475	.313	.179	.084	.027	.004	.000
	2	1.000	.996	.973	.916	.821	.688	.525	.348	.181	.052	.014
	3	1.000	1.000	.998	.992	.974	.938	.870	.760	.590	.344	.185
	4	1.000	1.000	1.000	1.000	1.000	1.000	1.000	1.000	1.000	1.000	1.000
$n=5$	0	.774	.590	.328	.168	.078	.031	.010	.002	.000	.000	.000
	1	.977	.919	.737	.528	.337	.188	.087	.031	.007	.000	.000
	2	.999	.991	.942	.837	.683	.500	.317	.163	.058	.009	.001
	3	1.000	1.000	.993	.969	.913	.813	.663	.472	.263	.081	.023
	4	1.000	1.000	1.000	.998	.990	.969	.922	.832	.672	.410	.226
	5	1.000	1.000	1.000	1.000	1.000	1.000	1.000	1.000	1.000	1.000	1.000
$n=6$	0	.735	.531	.262	.118	.047	.016	.004	.001	.000	.000	.000
	1	.967	.886	.655	.420	.233	.109	.041	.011	.002	.000	.000
	2	.998	.984	.901	.744	.544	.344	.179	.070	.017	.001	.000
	3	1.000	.999	.983	.930	.821	.656	.456	.256	.099	.016	.002
	4	1.000	1.000	.998	.989	.959	.891	.767	.580	.345	.114	.033
	5	1.000	1.000	1.000	.999	.996	.984	.953	.882	.738	.469	.265
	6	1.000	1.000	1.000	1.000	1.000	1.000	1.000	1.000	1.000	1.000	1.000
$n=7$	0	.698	.478	.210	.082	.028	.008	.002	.000	.000	.000	.000
	1	.956	.850	.577	.329	.159	.063	.019	.004	.000	.000	.000
	2	.996	.974	.852	.647	.420	.227	.096	.029	.005	.000	.000
	3	1.000	.997	.967	.874	.710	.500	.290	.126	.033	.003	.000
	4	1.000	1.000	.995	.971	.904	.773	.580	.353	.148	.026	.004

TABLE 2 (*Continued*)

	c	p = .05	.10	.20	.30	.40	.50	.60	.70	.80	.90	.95
	5	1.000	1.000	1.000	.996	.981	.938	.841	.671	.423	.150	.044
	6	1.000	1.000	1.000	1.000	.998	.992	.972	.918	.790	.522	.302
	7	1.000	1.000	1.000	1.000	1.000	1.000	1.000	1.000	1.000	1.000	1.000
$n=8$	0	.663	.430	.168	.058	.017	.004	.001	.000	.000	.000	.000
	1	.943	.813	.503	.255	.106	.035	.009	.001	.000	.000	.000
	2	.994	.962	.797	.552	.315	.145	.050	.011	.001	.000	.000
	3	1.000	.995	.944	.806	.594	.363	.174	.058	.010	.000	.000
	4	1.000	1.000	.990	.942	.826	.637	.406	.194	.056	.005	.000
	5	1.000	1.000	.999	.989	.950	.855	.685	.448	.203	.038	.006
	6	1.000	1.000	1.000	.999	.991	.965	.894	.745	.497	.187	.057
	7	1.000	1.000	1.000	1.000	.999	.996	.983	.942	.832	.570	.337
	8	1.000	1.000	1.000	1.000	1.000	1.000	1.000	1.000	1.000	1.000	1.000
$n=9$	0	.630	.387	.134	.040	.010	.002	.000	.000	.000	.000	.000
	1	.929	.775	.436	.196	.071	.020	.004	.000	.000	.000	.000
	2	.992	.947	.738	.463	.232	.090	.025	.004	.000	.000	.000
	3	.999	.992	.914	.730	.483	.254	.099	.025	.003	.000	.000
	4	1.000	.999	.980	.901	.733	.500	.267	.099	.020	.001	.000
	5	1.000	1.000	.997	.975	.901	.746	.517	.270	.086	.008	.001
	6	1.000	1.000	1.000	.996	.975	.910	.768	.537	.262	.053	.008
	7	1.000	1.000	1.000	1.000	.996	.980	.929	.804	.564	.225	.071
	8	1.000	1.000	1.000	1.000	1.000	.998	.990	.960	.866	.613	.370
	9	1.000	1.000	1.000	1.000	1.000	1.000	1.000	1.000	1.000	1.000	1.000
$n=10$	0	.599	.349	.107	.028	.006	.001	.000	.000	.000	.000	.000
	1	.914	.736	.376	.149	.046	.011	.002	.000	.000	.000	.000
	2	.988	.930	.678	.383	.167	.055	.012	.002	.000	.000	.000
	3	.999	.987	.879	.650	.382	.172	.055	.011	.001	.000	.000
	4	1.000	.998	.967	.850	.633	.377	.166	.047	.006	.000	.000
	5	1.000	1.000	.994	.953	.834	.623	.367	.150	.033	.002	.000
	6	1.000	1.000	.999	.989	.945	.828	.618	.350	.121	.013	.001
	7	1.000	1.000	1.000	.998	.988	.945	.833	.617	.322	.070	.012
	8	1.000	1.000	1.000	1.000	.998	.989	.954	.851	.624	.264	.086
	9	1.000	1.000	1.000	1.000	1.000	.999	.994	.972	.893	.651	.401
	10	1.000	1.000	1.000	1.000	1.000	1.000	1.000	1.000	1.000	1.000	1.000
$n=11$	0	.569	.314	.086	.020	.004	.000	.000	.000	.000	.000	.000
	1	.898	.697	.322	.113	.030	.006	.001	.000	.000	.000	.000
	2	.985	.910	.617	.313	.119	.033	.006	.001	.000	.000	.000
	3	.998	.981	.839	.570	.296	.113	.029	.004	.000	.000	.000

TABLE 2 (*Continued*)

		p										
	c	.05	.10	.20	.30	.40	.50	.60	.70	.80	.90	.95
	4	1.000	.997	.950	.790	.533	.274	.099	.022	.002	.000	.000
	5	1.000	1.000	.988	.922	.753	.500	.247	.078	.012	.000	.000
	6	1.000	1.000	.998	.978	.901	.726	.467	.210	.050	.003	.000
	7	1.000	1.000	1.000	.996	.971	.887	.704	.430	.161	.019	.002
	8	1.000	1.000	1.000	.999	.994	.967	.881	.687	.383	.090	.015
	9	1.000	1.000	1.000	1.000	.999	.994	.970	.887	.678	.303	.102
	10	1.000	1.000	1.000	1.000	1.000	1.000	.996	.980	.914	.686	.431
	11	1.000	1.000	1.000	1.000	1.000	1.000	1.000	1.000	1.000	1.000	1.000
$n=12$	0	.540	.282	.069	.014	.002	.000	.000	.000	.000	.000	.000
	1	.882	.659	.275	.085	.020	.003	.000	.000	.000	.000	.000
	2	.980	.889	.558	.253	.083	.019	.003	.000	.000	.000	.000
	3	.998	.974	.795	.493	.225	.073	.015	.002	.000	.000	.000
	4	1.000	.996	.927	.724	.438	.194	.057	.009	.001	.000	.000
	5	1.000	.999	.981	.882	.665	.387	.158	.039	.004	.000	.000
	6	1.000	1.000	.996	.961	.842	.613	.335	.118	.019	.001	.000
	7	1.000	1.000	.999	.991	.943	.806	.562	.276	.073	.004	.000
	8	1.000	1.000	1.000	.998	.985	.927	.775	.507	.205	.026	.002
	9	1.000	1.000	1.000	1.000	.997	.981	.917	.747	.442	.111	.020
	10	1.000	1.000	1.000	1.000	1.000	.997	.980	.915	.725	.341	.118
	11	1.000	1.000	1.000	1.000	1.000	1.000	.998	.986	.931	.718	.460
	12	1.000	1.000	1.000	1.000	1.000	1.000	1.000	1.000	1.000	1.000	1.000
$n=13$	0	.513	.254	.055	.010	.001	.000	.000	.000	.000	.000	.000
	1	.865	.621	.234	.064	.013	.002	.000	.000	.000	.000	.000
	2	.975	.866	.502	.202	.058	.011	.001	.000	.000	.000	.000
	3	.997	.966	.747	.421	.169	.046	.008	.001	.000	.000	.000
	4	1.000	.994	.901	.654	.353	.133	.032	.004	.000	.000	.000
	5	1.000	.999	.970	.835	.574	.291	.098	.018	.001	.000	.000
	6	1.000	1.000	.993	.938	.771	.500	.229	.062	.007	.000	.000
	7	1.000	1.000	.999	.982	.902	.709	.426	.165	.030	.001	.000
	8	1.000	1.000	1.000	.996	.968	.867	.647	.346	.099	.006	.000
	9	1.000	1.000	1.000	.999	.992	.954	.831	.579	.253	.034	.003
	10	1.000	1.000	1.000	1.000	.999	.989	.942	.798	.498	.134	.025
	11	1.000	1.000	1.000	1.000	1.000	.998	.987	.936	.766	.379	.135
	12	1.000	1.000	1.000	1.000	1.000	1.000	.999	.990	.945	.746	.487
	13	1.000	1.000	1.000	1.000	1.000	1.000	1.000	1.000	1.000	1.000	1.000
$n=14$	0	.488	.229	.044	.007	.001	.000	.000	.000	.000	.000	.000
	1	.847	.585	.198	.047	.008	.001	.000	.000	.000	.000	.000
	2	.970	.842	.448	.161	.040	.006	.001	.000	.000	.000	.000

TABLE 2 (*Continued*)

							p					
	c	.05	.10	.20	.30	.40	.50	.60	.70	.80	.90	.95
	3	.996	.956	.698	.355	.124	.029	.004	.000	.000	.000	.000
	4	1.000	.991	.870	.584	.279	.090	.018	.002	.000	.000	.000
	5	1.000	.999	.956	.781	.486	.212	.058	.008	.000	.000	.000
	6	1.000	1.000	.988	.907	.692	.395	.150	.031	.002	.000	.000
	7	1.000	1.000	.998	.969	.850	.605	.308	.093	.012	.000	.000
	8	1.000	1.000	1.000	.992	.942	.788	.514	.219	.044	.001	.000
	9	1.000	1.000	1.000	.998	.982	.910	.721	.416	.130	.009	.000
	10	1.000	1.000	1.000	1.000	.996	.971	.876	.645	.302	.044	.004
	11	1.000	1.000	1.000	1.000	.999	.994	.960	.839	.552	.158	.030
	12	1.000	1.000	1.000	1.000	1.000	.999	.992	.953	.802	.415	.153
	13	1.000	1.000	1.000	1.000	1.000	1.000	.999	.993	.956	.771	.512
	14	1.000	1.000	1.000	1.000	1.000	1.000	1.000	1.000	1.000	1.000	1.000
$n=15$	0	.463	.206	.035	.005	.000	.000	.000	.000	.000	.000	.000
	1	.829	.549	.167	.035	.005	.000	.000	.000	.000	.000	.000
	2	.964	.816	.398	.127	.027	.004	.000	.000	.000	.000	.000
	3	.995	.944	.648	.297	.091	.018	.002	.000	.000	.000	.000
	4	.999	.987	.836	.515	.217	.059	.009	.001	.000	.000	.000
	5	1.000	.998	.939	.722	.403	.151	.034	.004	.000	.000	.000
	6	1.000	1.000	.982	.869	.610	.304	.095	.015	.001	.000	.000
	7	1.000	1.000	.996	.950	.787	.500	.213	.050	.004	.000	.000
	8	1.000	1.000	.999	.985	.905	.696	.390	.131	.018	.000	.000
	9	1.000	1.000	1.000	.996	.966	.849	.597	.278	.061	.002	.000
	10	1.000	1.000	1.000	.999	.991	.941	.783	.485	.164	.013	.001
	11	1.000	1.000	1.000	1.000	.998	.982	.909	.703	.352	.056	.005
	12	1.000	1.000	1.000	1.000	1.000	.996	.973	.873	.602	.184	.036
	13	1.000	1.000	1.000	1.000	1.000	1.000	.995	.965	.833	.451	.171
	14	1.000	1.000	1.000	1.000	1.000	1.000	1.000	.995	.965	.794	.537
	15	1.000	1.000	1.000	1.000	1.000	1.000	1.000	1.000	1.000	1.000	1.000
$n=16$	0	.440	.185	.028	.003	.000	.000	.000	.000	.000	.000	.000
	1	.811	.515	.141	.026	.003	.000	.000	.000	.000	.000	.000
	2	.957	.789	.352	.099	.018	.002	.000	.000	.000	.000	.000
	3	.993	.932	.598	.246	.065	.011	.001	.000	.000	.000	.000
	4	.999	.983	.798	.450	.167	.038	.005	.000	.000	.000	.000
	5	1.000	.997	.918	.660	.329	.105	.019	.002	.000	.000	.000
	6	1.000	.999	.973	.825	.527	.227	.058	.007	.000	.000	.000
	7	1.000	1.000	.993	.926	.716	.402	.142	.026	.001	.000	.000
	8	1.000	1.000	.999	.974	.858	.598	.284	.074	.007	.000	.000
	9	1.000	1.000	1.000	.993	.942	.773	.473	.175	.027	.001	.000
	10	1.000	1.000	1.000	.998	.981	.895	.671	.340	.082	.003	.000

TABLE 2 (*Continued*)

						p					
	c	.05	.10	.20	.30	.40	.50	.60	.70	.80	.90	.95
	11	1.000	1.000	1.000	1.000	.995	.962	.833	.550	.202	.017	.001
	12	1.000	1.000	1.000	1.000	.999	.989	.935	.754	.402	.068	.007
	13	1.000	1.000	1.000	1.000	1.000	.998	.982	.901	.648	.211	.043
	14	1.000	1.000	1.000	1.000	1.000	1.000	.997	.974	.859	.485	.189
	15	1.000	1.000	1.000	1.000	1.000	1.000	1.000	.997	.972	.815	.560
	16	1.000	1.000	1.000	1.000	1.000	1.000	1.000	1.000	1.000	1.000	1.000
$n=17$	0	.418	.167	.023	.002	.000	.000	.000	.000	.000	.000	.000
	1	.792	.482	.118	.019	.002	.000	.000	.000	.000	.000	.000
	2	.950	.762	.310	.077	.012	.001	.000	.000	.000	.000	.000
	3	.991	.917	.549	.202	.046	.006	.000	.000	.000	.000	.000
	4	.999	.978	.758	.389	.126	.025	.003	.000	.000	.000	.000
	5	1.000	.995	.894	.597	.264	.072	.011	.001	.000	.000	.000
	6	1.000	.999	.962	.775	.448	.166	.035	.003	.000	.000	.000
	7	1.000	1.000	.989	.895	.641	.315	.092	.013	.000	.000	.000
	8	1.000	1.000	.997	.960	.801	.500	.199	.040	.003	.000	.000
	9	1.000	1.000	1.000	.987	.908	.685	.359	.105	.011	.000	.000
	10	1.000	1.000	1.000	.997	.965	.834	.552	.225	.038	.001	.000
	11	1.000	1.000	1.000	.999	.989	.928	.736	.403	.106	.005	.000
	12	1.000	1.000	1.000	1.000	.997	.975	.874	.611	.242	.022	.001
	13	1.000	1.000	1.000	1.000	1.000	.994	.954	.798	.451	.083	.009
	14	1.000	1.000	1.000	1.000	1.000	.999	.988	.923	.690	.238	.050
	15	1.000	1.000	1.000	1.000	1.000	1.000	.998	.981	.882	.518	.208
	16	1.000	1.000	1.000	1.000	1.000	1.000	1.000	.998	.977	.833	.582
	17	1.000	1.000	1.000	1.000	1.000	1.000	1.000	1.000	1.000	1.000	1.000
$n=18$	0	.397	.150	.018	.002	.000	.000	.000	.000	.000	.000	.000
	1	.774	.450	.099	.014	.001	.000	.000	.000	.000	.000	.000
	2	.942	.734	.271	.060	.008	.001	.000	.000	.000	.000	.000
	3	.989	.902	.501	.165	.033	.004	.000	.000	.000	.000	.000
	4	.998	.972	.716	.333	.094	.015	.001	.000	.000	.000	.000
	5	1.000	.994	.867	.534	.209	.048	.006	.000	.000	.000	.000
	6	1.000	.999	.949	.722	.374	.119	.020	.001	.000	.000	.000
	7	1.000	1.000	.984	.859	.563	.240	.058	.006	.000	.000	.000
	8	1.000	1.000	.996	.940	.737	.407	.135	.021	.001	.000	.000
	9	1.000	1.000	.999	.979	.865	.593	.263	.060	.004	.000	.000
	10	1.000	1.000	1.000	.994	.942	.760	.437	.141	.016	.000	.000
	11	1.000	1.000	1.000	.999	.980	.881	.626	.278	.051	.001	.000
	12	1.000	1.000	1.000	1.000	.994	.952	.791	.466	.133	.006	.000
	13	1.000	1.000	1.000	1.000	.999	.985	.906	.667	.284	.028	.002
	14	1.000	1.000	1.000	1.000	1.000	.996	.967	.835	.499	.098	.011

TABLE 2 (*Continued*)

n	c	.05	.10	.20	.30	.40	.50	.60	.70	.80	.90	.95
	15	1.000	1.000	1.000	1.000	1.000	.999	.992	.940	.729	.266	.058
	16	1.000	1.000	1.000	1.000	1.000	1.000	.999	.986	.901	.550	.226
	17	1.000	1.000	1.000	1.000	1.000	1.000	1.000	.998	.982	.850	.603
	18	1.000	1.000	1.000	1.000	1.000	1.000	1.000	1.000	1.000	1.000	1.000
$n=19$	0	.377	.135	.014	.001	.000	.000	.000	.000	.000	.000	.000
	1	.755	.420	.083	.010	.001	.000	.000	.000	.000	.000	.000
	2	.933	.705	.237	.046	.005	.000	.000	.000	.000	.000	.000
	3	.987	.885	.455	.133	.023	.002	.000	.000	.000	.000	.000
	4	.998	.965	.673	.282	.070	.010	.001	.000	.000	.000	.000
	5	1.000	.991	.837	.474	.163	.032	.003	.000	.000	.000	.000
	6	1.000	.998	.932	.666	.308	.084	.012	.001	.000	.000	.000
	7	1.000	1.000	.977	.818	.488	.180	.035	.003	.000	.000	.000
	8	1.000	1.000	.993	.916	.667	.324	.088	.011	.000	.000	.000
	9	1.000	1.000	.998	.967	.814	.500	.186	.033	.002	.000	.000
	10	1.000	1.000	1.000	.989	.912	.676	.333	.084	.007	.000	.000
	11	1.000	1.000	1.000	.997	.965	.820	.512	.182	.023	.000	.000
	12	1.000	1.000	1.000	.999	.988	.916	.692	.334	.068	.002	.000
	13	1.000	1.000	1.000	1.000	.997	.968	.837	.526	.163	.009	.000
	14	1.000	1.000	1.000	1.000	.999	.990	.930	.718	.327	.035	.002
	15	1.000	1.000	1.000	1.000	1.000	.998	.977	.867	.545	.115	.013
	16	1.000	1.000	1.000	1.000	1.000	1.000	.995	.954	.763	.295	.067
	17	1.000	1.000	1.000	1.000	1.000	1.000	.999	.990	.917	.580	.245
	18	1.000	1.000	1.000	1.000	1.000	1.000	1.000	.999	.986	.865	.623
	19	1.000	1.000	1.000	1.000	1.000	1.000	1.000	1.000	1.000	1.000	1.000
$n=20$	0	.358	.122	.012	.001	.000	.000	.000	.000	.000	.000	.000
	1	.736	.392	.069	.008	.001	.000	.000	.000	.000	.000	.000
	2	.925	.677	.206	.035	.004	.000	.000	.000	.000	.000	.000
	3	.984	.867	.411	.107	.016	.001	.000	.000	.000	.000	.000
	4	.997	.957	.630	.238	.051	.006	.000	.000	.000	.000	.000
	5	1.000	.989	.804	.416	.126	.021	.002	.000	.000	.000	.000
	6	1.000	.998	.913	.608	.250	.058	.006	.000	.000	.000	.000
	7	1.000	1.000	.968	.772	.416	.132	.021	.001	.000	.000	.000
	8	1.000	1.000	.990	.887	.596	.252	.057	.005	.000	.000	.000
	9	1.000	1.000	.997	.952	.755	.412	.128	.017	.001	.000	.000
	10	1.000	1.000	.999	.983	.872	.588	.245	.048	.003	.000	.000
	11	1.000	1.000	1.000	.995	.943	.748	.404	.113	.010	.000	.000
	12	1.000	1.000	1.000	.999	.979	.868	.584	.228	.032	.000	.000
	13	1.000	1.000	1.000	1.000	.994	.942	.750	.392	.087	.002	.000
	14	1.000	1.000	1.000	1.000	.998	.979	.874	.584	.196	.011	.000

TABLE 2   (*Continued*)

	.05	.10	.20	.30	.40	p .50	.60	.70	.80	.90	.95
c											
15	1.000	1.000	1.000	1.000	1.000	.994	.949	.762	.370	.043	.003
16	1.000	1.000	1.000	1.000	1.000	.999	.984	.893	.589	.133	.016
17	1.000	1.000	1.000	1.000	1.000	1.000	.996	.965	.794	.323	.075
18	1.000	1.000	1.000	1.000	1.000	1.000	.999	.992	.931	.608	.264
19	1.000	1.000	1.000	1.000	1.000	1.000	1.000	.999	.988	.878	.642
20	1.000	1.000	1.000	1.000	1.000	1.000	1.000	1.000	1.000	1.000	1.000
$n=25$ 0	.277	.072	.004	.000	.000	.000	.000	.000	.000	.000	.000
1	.642	.271	.027	.002	.000	.000	.000	.000	.000	.000	.000
2	.873	.537	.098	.009	.000	.000	.000	.000	.000	.000	.000
3	.966	.764	.234	.033	.002	.000	.000	.000	.000	.000	.000
4	.993	.902	.421	.090	.009	.000	.000	.000	.000	.000	.000
5	.999	.967	.617	.193	.029	.002	.000	.000	.000	.000	.000
6	1.000	.991	.780	.341	.074	.007	.000	.000	.000	.000	.000
7	1.000	.998	.891	.512	.154	.022	.001	.000	.000	.000	.000
8	1.000	1.000	.953	.677	.274	.054	.004	.000	.000	.000	.000
9	1.000	1.000	.983	.811	.425	.115	.013	.000	.000	.000	.000
10	1.000	1.000	.994	.902	.586	.212	.034	.002	.000	.000	.000
11	1.000	1.000	.998	.956	.732	.345	.078	.006	.000	.000	.000
12	1.000	1.000	1.000	.983	.846	.500	.154	.017	.000	.000	.000
13	1.000	1.000	1.000	.994	.922	.655	.268	.044	.002	.000	.000
14	1.000	1.000	1.000	.998	.966	.788	.414	.098	.006	.000	.000
15	1.000	1.000	1.000	1.000	.987	.885	.575	.189	.017	.000	.000
16	1.000	1.000	1.000	1.000	.996	.946	.726	.323	.047	.000	.000
17	1.000	1.000	1.000	1.000	.999	.978	.846	.488	.109	.002	.000
18	1.000	1.000	1.000	1.000	1.000	.993	.926	.659	.220	.009	.000
19	1.000	1.000	1.000	1.000	1.000	.998	.971	.807	.383	.033	.001
20	1.000	1.000	1.000	1.000	1.000	1.000	.991	.910	.579	.098	.007
21	1.000	1.000	1.000	1.000	1.000	1.000	.998	.967	.766	.236	.034
22	1.000	1.000	1.000	1.000	1.000	1.000	1.000	.991	.902	.463	.127
23	1.000	1.000	1.000	1.000	1.000	1.000	1.000	.998	.973	.729	.358
24	1.000	1.000	1.000	1.000	1.000	1.000	1.000	1.000	.996	.928	.723
25	1.000	1.000	1.000	1.000	1.000	1.000	1.000	1.000	1.000	1.000	1.000

TABLE 3

CUMULATIVE POISSON PROBABILITIES $P[X \leqslant c] = \sum_{x=0}^{c} \frac{e^{-m}m^x}{x!}$

c	.10	.20	.30	.40	.50	.60	.70	.80	.90	1.00
0	.905	.819	.741	.670	.607	.549	.497	.449	.407	.368
1	.995	.982	.963	.938	.910	.878	.844	.809	.772	.736
2	1.000	.999	.996	.992	.986	.977	.966	.953	.937	.920
3	1.000	1.000	1.000	.999	.998	.997	.994	.991	.987	.981
4	1.000	1.000	1.000	1.000	1.000	1.000	.999	.999	.998	.996
5	1.000	1.000	1.000	1.000	1.000	1.000	1.000	1.000	1.000	.999
6	1.000	1.000	1.000	1.000	1.000	1.000	1.000	1.000	1.000	1.000
7	1.000	1.000	1.000	1.000	1.000	1.000	1.000	1.000	1.000	1.000

c	1.10	1.20	1.30	1.40	1.50	1.60	1.70	1.80	1.90	2.00
0	.333	.301	.273	.247	.223	.202	.183	.165	.150	.135
1	.699	.663	.627	.592	.558	.525	.493	.463	.434	.406
2	.900	.879	.857	.833	.809	.783	.757	.731	.704	.677
3	.974	.966	.957	.946	.934	.921	.907	.891	.875	.857
4	.995	.992	.989	.986	.981	.976	.970	.964	.956	.947
5	.999	.998	.998	.997	.996	.994	.992	.990	.987	.983
6	1.000	1.000	1.000	.999	.999	.999	.998	.997	.997	.995
7	1.000	1.000	1.000	1.000	1.000	1.000	1.000	.999	.999	.999
8	1.000	1.000	1.000	1.000	1.000	1.000	1.000	1.000	1.000	1.000
9	1.000	1.000	1.000	1.000	1.000	1.000	1.000	1.000	1.000	1.000

c	2.10	2.20	2.30	2.40	2.50	2.60	2.70	2.80	2.90	3.00
0	.122	.111	.100	.091	.082	.074	.067	.061	.055	.050
1	.380	.355	.331	.308	.287	.267	.249	.231	.215	.199
2	.650	.623	.596	.570	.544	.518	.494	.469	.446	.423
3	.839	.819	.799	.779	.758	.736	.714	.692	.670	.647
4	.938	.928	.916	.904	.891	.877	.863	.848	.832	.815
5	.980	.975	.970	.964	.958	.951	.943	.935	.926	.916
6	.994	.993	.991	.988	.986	.983	.979	.976	.971	.966
7	.999	.998	.997	.997	.996	.995	.993	.992	.990	.988
8	1.000	1.000	.999	.999	.999	.999	.998	.998	.997	.996
9	1.000	1.000	1.000	1.000	1.000	1.000	.999	.999	.999	.999
10	1.000	1.000	1.000	1.000	1.000	1.000	1.000	1.000	1.000	1.000
11	1.000	1.000	1.000	1.000	1.000	1.000	1.000	1.000	1.000	1.000
12	1.000	1.000	1.000	1.000	1.000	1.000	1.000	1.000	1.000	1.000

TABLE 3 (*Continued*)

c	m 3.10	3.20	3.30	3.40	3.50	3.60	3.70	3.80	3.90	4.00
0	.045	.041	.037	.033	.030	.027	.025	.022	.020	.018
1	.185	.171	.159	.147	.136	.126	.116	.107	.099	.092
2	.401	.380	.359	.340	.321	.303	.285	.269	.253	.238
3	.625	.603	.580	.558	.537	.515	.494	.473	.453	.433
4	.798	.781	.763	.744	.725	.706	.687	.668	.648	.629
5	.906	.895	.883	.871	.858	.844	.830	.816	.801	.785
6	.961	.955	.949	.942	.935	.927	.918	.909	.899	.889
7	.986	.983	.980	.977	.973	.969	.965	.960	.955	.949
8	.995	.994	.993	.992	.990	.988	.986	.984	.981	.979
9	.999	.998	.998	.997	.997	.996	.995	.994	.993	.992
10	1.000	1.000	.999	.999	.999	.999	.998	.998	.998	.997
11	1.000	1.000	1.000	1.000	1.000	1.000	1.000	.999	.999	.999
12	1.000	1.000	1.000	1.000	1.000	1.000	1.000	1.000	1.000	1.000
13	1.000	1.000	1.000	1.000	1.000	1.000	1.000	1.000	1.000	1.000
14	1.000	1.000	1.000	1.000	1.000	1.000	1.000	1.000	1.000	1.000

c	m 4.50	5.00	5.50	6.00	6.50	7.00	7.50	8.00	8.50	9.00
0	.011	.007	.004	.002	.002	.001	.001	.000	.000	.000
1	.061	.040	.027	.017	.011	.007	.005	.003	.002	.001
2	.174	.125	.088	.062	.043	.030	.020	.014	.009	.006
3	.342	.265	.202	.151	.112	.082	.059	.042	.030	.021
4	.532	.440	.358	.285	.224	.173	.132	.100	.074	.055
5	.703	.616	.529	.446	.369	.301	.241	.191	.150	.116
6	.831	.762	.686	.606	.527	.450	.378	.313	.256	.207
7	.913	.867	.809	.744	.673	.599	.525	.453	.386	.324
8	.960	.932	.894	.847	.792	.729	.662	.593	.523	.456
9	.983	.968	.946	.916	.877	.830	.776	.717	.653	.587
10	.993	.986	.975	.957	.933	.901	.862	.816	.763	.706
11	.998	.995	.989	.980	.966	.947	.921	.888	.849	.803
12	.999	.998	.996	.991	.984	.973	.957	.936	.909	.876
13	1.000	.999	.998	.996	.993	.987	.978	.966	.949	.926
14	1.000	1.000	.999	.999	.997	.994	.990	.983	.973	.959
15	1.000	1.000	1.000	.999	.999	.998	.995	.992	.986	.978
16	1.000	1.000	1.000	1.000	1.000	.999	.998	.996	.993	.989
17	1.000	1.000	1.000	1.000	1.000	1.000	.999	.998	.997	.995
18	1.000	1.000	1.000	1.000	1.000	1.000	1.000	.999	.999	.998
19	1.000	1.000	1.000	1.000	1.000	1.000	1.000	1.000	.999	.999
20	1.000	1.000	1.000	1.000	1.000	1.000	1.000	1.000	1.000	1.000
21	1.000	1.000	1.000	1.000	1.000	1.000	1.000	1.000	1.000	1.000
22	1.000	1.000	1.000	1.000	1.000	1.000	1.000	1.000	1.000	1.000

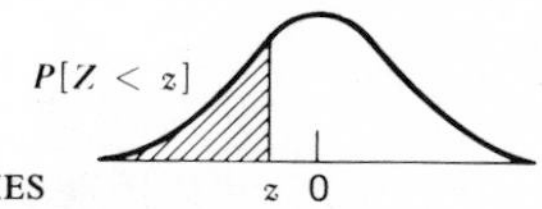

TABLE 4
Standard Normal Probabilities

z	.00	.01	.02	.03	.04	.05	.06	.07	.08	.09
−3.5	.0002	.0002	.0002	.0002	.0002	.0002	.0002	.0002	.0002	.0002
−3.4	.0003	.0003	.0003	.0003	.0003	.0003	.0003	.0003	.0003	.0002
−3.3	.0005	.0005	.0005	.0004	.0004	.0004	.0004	.0004	.0004	.0003
−3.2	.0007	.0007	.0006	.0006	.0006	.0006	.0006	.0005	.0005	.0005
−3.1	.0010	.0009	.0009	.0009	.0008	.0008	.0008	.0008	.0007	.0007
−3.0	.0013	.0013	.0013	.0012	.0012	.0011	.0011	.0011	.0010	.0010
−2.9	.0019	.0018	.0018	.0017	.0016	.0016	.0015	.0015	.0014	.0014
−2.8	.0026	.0025	.0024	.0023	.0023	.0022	.0021	.0021	.0020	.0019
−2.7	.0035	.0034	.0033	.0032	.0031	.0030	.0029	.0028	.0027	.0026
−2.6	.0047	.0045	.0044	.0043	.0041	.0040	.0039	.0038	.0037	.0036
−2.5	.0062	.0060	.0059	.0057	.0055	.0054	.0052	.0051	.0049	.0048
−2.4	.0082	.0080	.0078	.0075	.0073	.0071	.0069	.0068	.0066	.0064
−2.3	.0107	.0104	.0102	.0099	.0096	.0094	.0091	.0089	.0087	.0084
−2.2	.0139	.0136	.0132	.0129	.0125	.0122	.0119	.0116	.0113	.0110
−2.1	.0179	.0174	.0170	.0166	.0162	.0158	.0154	.0150	.0146	.0143
−2.0	.0228	.0222	.0217	.0212	.0207	.0202	.0197	.0192	.0188	.0183
−1.9	.0287	.0281	.0274	.0268	.0262	.0256	.0250	.0244	.0239	.0233
−1.8	.0359	.0351	.0344	.0336	.0329	.0322	.0314	.0307	.0301	.0294
−1.7	.0446	.0436	.0427	.0418	.0409	.0401	.0392	.0384	.0375	.0367
−1.6	.0548	.0537	.0526	.0516	.0505	.0495	.0485	.0475	.0465	.0455
−1.5	.0668	.0655	.0643	.0630	.0618	.0606	.0594	.0582	.0571	.0559
−1.4	.0808	.0793	.0778	.0764	.0749	.0735	.0721	.0708	.0694	.0681
−1.3	.0968	.0951	.0934	.0918	.0901	.0885	.0869	.0853	.0838	.0823
−1.2	.1151	.1131	.1112	.1093	.1075	.1056	.1038	.1020	.1003	.0985
−1.1	.1357	.1335	.1314	.1292	.1271	.1251	.1230	.1210	.1190	.1170
−1.0	.1587	.1562	.1539	.1515	.1492	.1469	.1446	.1423	.1401	.1379
−.9	.1841	.1814	.1788	.1762	.1736	.1711	.1685	.1660	.1635	.1611
−.8	.2119	.2090	.2061	.2033	.2005	.1977	.1949	.1922	.1894	.1867
−.7	.2420	.2389	.2358	.2327	.2297	.2266	.2236	.2206	.2177	.2148
−.6	.2743	.2709	.2676	.2643	.2611	.2578	.2546	.2514	.2483	.2451
−.5	.3085	.3050	.3015	.2981	.2946	.2912	.2877	.2843	.2810	.2776
−.4	.3446	.3409	.3372	.3336	.3300	.3264	.3228	.3192	.3156	.3121
−.3	.3821	.3783	.3745	.3707	.3669	.3632	.3594	.3557	.3520	.3483
−.2	.4207	.4168	.4129	.4090	.4052	.4013	.3974	.3936	.3897	.3859
−.1	.4602	.4562	.4522	.4483	.4443	.4404	.4364	.4325	.4286	.4247
−.0	.5000	.4960	.4920	.4880	.4840	.4801	.4761	.4721	.4681	.4641

TABLE 4 (*Continued*)

z	.00	.01	.02	.03	.04	.05	.06	.07	.08	.09
.0	.5000	.5040	.5080	.5120	.5160	.5199	.5239	.5279	.5319	.5359
.1	.5398	.5438	.5478	.5517	.5557	.5596	.5636	.5675	.5714	.5753
.2	.5793	.5832	.5871	.5910	.5948	.5987	.6026	.6064	.6103	.6141
.3	.6179	.6217	.6255	.6293	.6331	.6368	.6406	.6443	.6480	.6517
.4	.6554	.6591	.6628	.6664	.6700	.6736	.6772	.6808	.6844	.6879
.5	.6915	.6950	.6985	.7019	.7054	.7088	.7123	.7157	.7190	.7224
.6	.7257	.7291	.7324	.7357	.7389	.7422	.7454	.7486	.7517	.7549
.7	.7580	.7611	.7642	.7673	.7703	.7734	.7764	.7794	.7823	.7852
.8	.7881	.7910	.7939	.7967	.7995	.8023	.8051	.8078	.8106	.8133
.9	.8159	.8186	.8212	.8238	.8264	.8289	.8315	.8340	.8365	.8389
1.0	.8413	.8438	.8461	.8485	.8508	.8531	.8554	.8577	.8599	.8621
1.1	.8643	.8665	.8686	.8708	.8729	.8749	.8770	.8790	.8810	.8830
1.2	.8849	.8869	.8888	.8907	.8925	.8944	.8962	.8980	.8997	.9015
1.3	.9032	.9049	.9066	.9082	.9099	.9115	.9131	.9147	.9162	.9177
1.4	.9192	.9207	.9222	.9236	.9251	.9265	.9279	.9292	.9306	.9319
1.5	.9332	.9345	.9357	.9370	.9382	.9394	.9406	.9418	.9429	.9441
1.6	.9452	.9463	.9474	.9484	.9495	.9505	.9515	.9525	.9535	.9545
1.7	.9554	.9564	.9573	.9582	.9591	.9599	.9608	.9616	.9625	.9633
1.8	.9641	.9649	.9656	.9664	.9671	.9678	.9686	.9693	.9699	.9706
1.9	.9713	.9719	.9726	.9732	.9738	.9744	.9750	.9756	.9761	.9767
2.0	.9772	.9778	.9783	.9788	.9793	.9798	.9803	.9808	.9812	.9817
2.1	.9821	.9826	.9830	.9834	.9838	.9842	.9846	.9850	.9854	.9857
2.2	.9861	.9864	.9868	.9871	.9875	.9878	.9881	.9884	.9887	.9890
2.3	.9893	.9896	.9898	.9901	.9904	.9906	.9909	.9911	.9913	.9916
2.4	.9918	.9920	.9922	.9925	.9927	.9929	.9931	.9932	.9934	.9936
2.5	.9938	.9940	.9941	.9943	.9945	.9946	.9948	.9949	.9951	.9952
2.6	.9953	.9955	.9956	.9957	.9959	.9960	.9961	.9962	.9963	.9964
2.7	.9965	.9966	.9967	.9968	.9969	.9970	.9971	.9972	.9973	.9974
2.8	.9974	.9975	.9976	.9977	.9977	.9978	.9979	.9979	.9980	.9981
2.9	.9981	.9982	.9982	.9983	.9984	.9984	.9985	.9985	.9986	.9986
3.0	.9987	.9987	.9987	.9988	.9988	.9989	.9989	.9989	.9990	.9990
3.1	.9990	.9991	.9991	.9991	.9992	.9992	.9992	.9992	.9993	.9993
3.2	.9993	.9993	.9994	.9994	.9994	.9994	.9994	.9995	.9995	.9995
3.3	.9995	.9995	.9995	.9996	.9996	.9996	.9996	.9996	.9996	.9997
3.4	.9997	.9997	.9997	.9997	.9997	.9997	.9997	.9997	.9997	.9998
3.5	.9998	.9998	.9998	.9998	.9998	.9998	.9998	.9998	.9998	.9998

90% CI
1.645 = .95
95% CI
1.96 = .975
98% CI
.99 = 2.327
99% CI
2.575 = .995

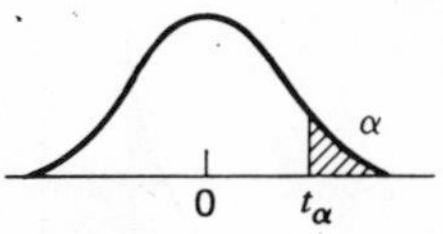

TABLE 5
Percentage Points of t Distributions

d.f.	α .25	.10	.05	.025	.01	.005
1	1.000	3.078	6.314	12.706	31.821	63.657
2	.816	1.886	2.920	4.303	6.965	9.925
3	.765	1.638	2.353	3.182	4.541	5.841
4	.741	1.533	2.132	2.776	3.747	4.604
5	.727	1.476	2.015	2.571	3.365	4.032
6	.718	1.440	1.943	2.447	3.143	3.707
7	.711	1.415	1.895	2.365	2.998	3.499
8	.706	1.397	1.860	2.306	2.896	3.355
9	.703	1.383	1.833	2.262	2.821	3.250
10	.700	1.372	1.812	2.228	2.764	3.169
11	.697	1.363	1.796	2.201	2.718	3.106
12	.695	1.356	1.782	2.179	2.681	3.055
13	.694	1.350	1.771	2.160	2.650	3.012
14	.692	1.345	1.761	2.145	2.624	2.977
15	.691	1.341	1.753	2.131	2.602	2.947
16	.690	1.337	1.746	2.120	2.583	2.921
17	.689	1.333	1.740	2.110	2.567	2.898
18	.688	1.330	1.734	2.101	2.552	2.878
19	.688	1.328	1.729	2.093	2.539	2.861
20	.687	1.325	1.725	2.086	2.528	2.845
21	.686	1.323	1.721	2.080	2.518	2.831
22	.686	1.321	1.717	2.074	2.508	2.819
23	.685	1.319	1.714	2.069	2.500	2.807
24	.685	1.318	1.711	2.064	2.492	2.797
25	.684	1.316	1.708	2.060	2.485	2.787
26	.684	1.315	1.706	2.056	2.479	2.779
27	.684	1.314	1.703	2.052	2.473	2.771
28	.683	1.313	1.701	2.048	2.467	2.763
29	.683	1.311	1.699	2.045	2.462	2.756
30	.683	1.310	1.697	2.042	2.457	2.750
40	.681	1.303	1.684	2.021	2.423	2.704
60	.679	1.296	1.671	2.000	2.390	2.660
120	.677	1.289	1.658	1.980	2.358	2.617
∞	.674	1.282	1.645	1.960	2.326	2.576

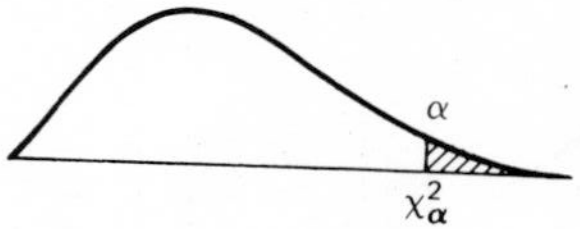

TABLE 6
Percentage Points of χ^2 Distributions

d.f. \ α	.995	.990	.975	.950	.050	.025	.010	.005
1	392704×10^{-10}	157088×10^{-9}	982069×10^{-9}	393214×10^{-8}	3.84146	5.02389	6.63490	7.87944
2	.0100251	.0201007	.0506356	.102587	5.99147	7.37776	9.21034	10.5966
3	.0717212	.114832	.215795	.351846	7.81473	9.34840	11.3449	12.8381
4	.206990	.297110	.484419	.710721	9.48773	11.1433	13.2767	14.8602
5	.411740	.554300	.831211	1.145476	11.0705	12.8325	15.0863	16.7496
6	.675727	.872085	1.237347	1.63539	12.5916	14.4494	16.8119	18.5476
7	.989265	1.239043	1.68987	2.16735	14.0671	16.0128	18.4753	20.2777
8	1.344419	1.646482	2.17973	2.73264	15.5073	17.5346	20.0902	21.9550
9	1.734926	2.087912	2.70039	3.32511	16.9190	19.0228	21.6660	23.5893
10	2.15585	2.55821	3.24697	3.94030	18.3070	20.4831	23.2093	25.1882
11	2.60321	3.05347	3.81575	4.57481	19.6751	21.9200	24.7250	26.7569
12	3.07382	3.57056	4.40379	5.22603	21.0261	23.3367	26.2170	28.2995
13	3.56503	4.10691	5.00874	5.89186	22.3621	24.7356	27.6883	29.8194
14	4.07468	4.66043	5.62872	6.57063	23.6848	26.1190	29.1413	31.3193
15	4.60094	5.22935	6.26214	7.26094	24.9958	27.4884	30.5779	32.8013
16	5.14224	5.81221	6.90766	7.96164	26.2962	28.8454	31.9999	34.2672
17	5.69724	6.40776	7.56418	8.67176	27.5871	30.1910	33.4087	35.7185
18	6.26481	7.01491	8.23075	9.39046	28.8693	31.5264	34.8053	37.1564
19	6.84398	7.63273	8.90655	10.1170	30.1435	32.8523	36.1908	38.5822

20	7.43386	8.26040	9.59083	10.8508	31.4104	34.1696	37.5662	39.9968
21	8.03366	8.89720	10.28293	11.5913	32.6705	35.4789	38.9321	41.4010
22	8.64272	9.54249	10.9823	12.3380	33.9244	36.7807	40.2894	42.7956
23	9.26042	10.19567	11.6885	13.0905	35.1725	38.0757	41.6384	44.1813
24	9.88623	10.8564	12.4011	13.8484	36.4151	39.3641	42.9798	45.5585
25	10.5197	11.5240	13.1197	14.6114	37.6525	40.6465	44.3141	46.9278
26	11.1603	12.1981	13.8439	15.3791	38.8852	41.9232	45.6417	48.2899
27	11.8076	12.8786	14.5733	16.1513	40.1133	43.1944	46.9630	49.6449
28	12.4613	13.5648	15.3079	16.9279	41.3372	44.4607	48.2782	50.9933
29	13.1211	14.2565	16.0471	17.7083	42.5569	45.7222	49.5879	52.3356
30	13.7867	14.9535	16.7908	18.4926	43.7729	46.9792	50.8922	53.6720
40	20.7065	22.1643	24.4331	26.5093	55.7585	59.3417	63.6907	66.7659
50	27.9907	29.7067	32.3574	34.7642	67.5048	71.4202	76.1539	79.4900
60	35.5346	37.4848	40.4817	43.1879	79.0819	83.2976	88.3794	91.9517
70	43.2752	45.4418	48.7576	51.7393	90.5312	95.0231	100.425	104.215
80	51.1720	53.5400	57.1532	60.3915	101.879	106.629	112.329	116.321
90	59.1963	61.7541	65.6466	69.1260	113.145	118.136	124.116	128.299
100	67.3276	70.0648	74.2219	77.9295	124.342	129.561	135.807	140.169

From "*Biometrika Tables for Statisticians*," Vol. 1, (3rd Edition) Cambridge University Press (1966); Edited by E. S. Pearson and H. O. Hartley.

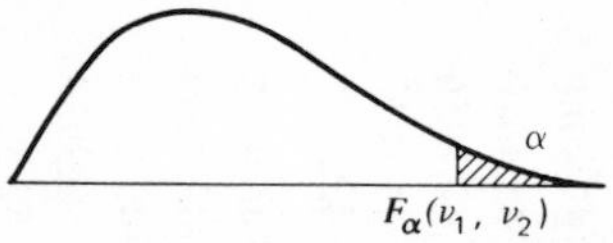

TABLE 7
PERCENTAGE POINTS OF $F(\nu_1, \nu_2)$ DISTRIBUTIONS

$\alpha = .10$

ν_2 \ ν_1	1	2	3	4	5	6	7	8	9
1	39.864	49.500	53.593	55.833	57.241	58.204	58.906	59.439	59.858
2	8.5263	9.0000	9.1618	9.2434	9.2926	9.3255	9.3491	9.3668	9.3805
3	5.5383	5.4624	5.3908	5.3427	5.3092	5.2847	5.2662	5.2517	5.2400
4	4.5448	4.3246	4.1908	4.1073	4.0506	4.0098	3.9790	3.9549	3.9357
5	4.0604	3.7797	3.6195	3.5202	3.4530	3.4045	3.3679	3.3393	3.3163
6	3.7760	3.4633	3.2888	3.1808	3.1075	3.0546	3.0145	2.9830	2.9577
7	3.5894	3.2574	3.0741	2.9605	2.8833	2.8274	2.7849	2.7516	2.7247
8	3.4579	3.1131	2.9238	2.8064	2.7265	2.6683	2.6241	2.5893	2.5612
9	3.3603	3.0065	2.8129	2.6927	2.6106	2.5509	2.5053	2.4694	2.4403
10	3.2850	2.9245	2.7277	2.6053	2.5216	2.4606	2.4140	2.3772	2.3473
11	3.2252	2.8595	2.6602	2.5362	2.4512	2.3891	2.3416	2.3040	2.2735
12	3.1765	2.8068	2.6055	2.4801	2.3940	2.3310	2.2828	2.2446	2.2135
13	3.1362	2.7632	2.5603	2.4337	2.3467	2.2830	2.2341	2.1953	2.1638
14	3.1022	2.7265	2.5222	2.3947	2.3069	2.2426	2.1931	2.1539	2.1220
15	3.0732	2.6952	2.4898	2.3614	2.2730	2.2081	2.1582	2.1185	2.0862
16	3.0481	2.6682	2.4618	2.3327	2.2438	2.1783	2.1280	2.0880	2.0553
17	3.0262	2.6446	2.4374	2.3077	2.2183	2.1524	2.1017	2.0613	2.0284
18	3.0070	2.6239	2.4160	2.2858	2.1958	2.1296	2.0785	2.0379	2.0047
19	2.9899	2.6056	2.3970	2.2663	2.1760	2.1094	2.0580	2.0171	1.9836
20	2.9747	2.5893	2.3801	2.2489	2.1582	2.0913	2.0397	1.9985	1.9649
21	2.9609	2.5746	2.3649	2.2333	2.1423	2.0751	2.0232	1.9819	1.9480
22	2.9486	2.5613	2.3512	2.2193	2.1279	2.0605	2.0084	1.9668	1.9327
23	2.9374	2.5493	2.3387	2.2065	2.1149	2.0472	1.9949	1.9531	1.9189
24	2.9271	2.5383	2.3274	2.1949	2.1030	2.0351	1.9826	1.9407	1.9063
25	2.9177	2.5283	2.3170	2.1843	2.0922	2.0241	1.9714	1.9292	1.8947
26	2.9091	2.5191	2.3075	2.1745	2.0822	2.0139	1.9610	1.9188	1.8841
27	2.9012	2.5106	2.2987	2.1655	2.0730	2.0045	1.9515	1.9091	1.8743
28	2.8939	2.5028	2.2906	2.1571	2.0645	1.9959	1.9427	1.9001	1.8652
29	2.8871	2.4955	2.2831	2.1494	2.0566	1.9878	1.9345	1.8918	1.8560
30	2.8807	2.4887	2.2761	2.1422	2.0492	1.9803	1.9269	1.8841	1.8498
40	2.8354	2.4404	2.2261	2.0909	1.9968	1.9269	1.8725	1.8289	1.7929
60	2.7914	2.3932	2.1774	2.0410	1.9457	1.8747	1.8194	1.7748	1.7380
120	2.7478	2.3473	2.1300	1.9923	1.8959	1.8238	1.7675	1.7220	1.6843
∞	2.7055	2.3026	2.0838	1.9449	1.8473	1.7741	1.7167	1.6702	1.6315

TABLE 7 (*Continued*)

$\alpha = .10$

ν_2 \ ν_1	10	12	15	20	24	30	40	60	120	∞
1	60.195	60.705	61.220	61.740	62.002	62.265	62.529	62.794	63.061	63.328
2	9.3916	9.4081	9.4247	9.4413	9.4496	9.4579	9.4663	9.4746	9.4829	9.4913
3	5.2304	5.2156	5.2003	5.1845	5.1764	5.1681	5.1597	5.1512	5.1425	5.1337
4	3.9199	3.8955	3.8689	3.8443	3.8310	3.8174	3.8036	3.7896	3.7753	3.7607
5	3.2974	3.2682	3.2380	3.2067	3.1905	3.1741	3.1573	3.1402	3.1228	3.1050
6	2.9369	2.9047	2.8712	2.8363	2.8183	2.8000	2.7812	2.7620	2.7423	2.7222
7	2.7025	2.6681	2.6322	2.5947	2.5753	2.5555	2.5351	2.5142	2.4928	2.4708
8	2.5380	2.5020	2.4642	2.4246	2.4041	2.3830	2.3614	2.3391	2.3162	2.2926
9	2.4163	2.3789	2.3396	2.2983	2.2768	2.2547	2.2320	2.2085	2.1843	2.1592
10	2.3226	2.2841	2.2435	2.2007	2.1784	2.1554	2.1317	2.1072	2.0818	2.0554
11	2.2482	2.2087	2.1671	2.1230	2.1000	2.0762	2.0516	2.0261	1.9997	1.9721
12	2.1878	2.1474	2.1049	2.0597	2.0360	2.0115	1.9861	1.9597	1.9323	1.9036
13	2.1376	2.0966	2.0532	2.0070	1.9827	1.9576	1.9315	1.9043	1.8759	1.8462
14	2.0954	2.0537	2.0095	1.9625	1.9377	1.9119	1.8852	1.8572	1.8280	1.7973
15	2.0593	2.0171	1.9722	1.9243	1.8990	1.8728	1.8454	1.8168	1.7867	1.7551
16	2.0281	1.9854	1.9399	1.8913	1.8656	1.8388	1.8108	1.7816	1.7507	1.7182
17	2.0009	1.9577	1.9117	1.8624	1.8362	1.8090	1.7805	1.7506	1.7191	1.6856
18	1.9770	1.9333	1.8868	1.8368	1.8103	1.7827	1.7537	1.7232	1.6910	1.6567
19	1.9557	1.9117	1.8647	1.8142	1.7873	1.7592	1.7298	1.6988	1.6659	1.6308
20	1.9367	1.8924	1.8449	1.7938	1.7667	1.7382	1.7083	1.6768	1.6433	1.6074
21	1.9197	1.8750	1.8272	1.7756	1.7481	1.7193	1.6890	1.6569	1.6228	1.5862
22	1.9043	1.8593	1.8111	1.7590	1.7312	1.7021	1.6714	1.6389	1.6042	1.5668
23	1.8903	1.8450	1.7964	1.7439	1.7159	1.6864	1.6554	1.6224	1.5871	1.5490
24	1.8775	1.8319	1.7831	1.7302	1.7019	1.6721	1.6407	1.6073	1.5715	1.5327
25	1.8658	1.8200	1.7708	1.7175	1.6890	1.6589	1.6272	1.5934	1.5570	1.5176
26	1.8550	1.8090	1.7596	1.7059	1.6771	1.6468	1.6147	1.5805	1.5437	1.5036
27	1.8451	1.7989	1.7492	1.6951	1.6662	1.6356	1.6032	1.5686	1.5313	1.4906
28	1.8359	1.7895	1.7395	1.6852	1.6560	1.6252	1.5925	1.5575	1.5198	1.4784
29	1.8274	1.7808	1.7306	1.6759	1.6465	1.6155	1.5825	1.5472	1.5090	1.4670
30	1.8195	1.7727	1.7223	1.6673	1.6377	1.6065	1.5732	1.5376	1.4989	1.4564
40	1.7627	1.7146	1.6624	1.6052	1.5741	1.5411	1.5056	1.4672	1.4248	1.3769
60	1.7070	1.6574	1.6034	1.5435	1.5107	1.4755	1.4373	1.3952	1.3476	1.2915
120	1.6524	1.6012	1.5450	1.4821	1.4472	1.4094	1.3676	1.3203	1.2646	1.1926
∞	1.5987	1.5458	1.4871	1.4206	1.3832	1.3419	1.2951	1.2400	1.1686	1.0000

TABLE 7 (*Continued*)

$\alpha = .05$

ν_2 \ ν_1	1	2	3	4	5	6	7	8	9
1	161.45	199.50	215.71	224.58	230.16	233.99	236.77	238.88	240.54
2	18.513	19.000	19.164	19.247	19.296	19.330	19.353	19.371	19.385
3	10.128	9.5521	9.2766	9.1172	9.0135	8.9406	8.8868	8.8452	8.8123
4	7.7086	6.9443	6.5914	6.3883	6.2560	6.1631	6.0942	6.0410	5.9988
5	6.6079	5.7861	5.4095	5.1922	5.0503	4.9503	4.8759	4.8183	4.7725
6	5.9874	5.1433	4.7571	4.5337	4.3874	4.2839	4.2066	4.1468	4.0990
7	5.5914	4.7374	4.3468	4.1203	3.9715	3.8660	3.7870	3.7257	3.6767
8	5.3177	4.4590	4.0662	3.8378	3.6875	3.5806	3.5005	3.4381	3.3881
9	5.1174	4.2565	3.8626	3.6331	3.4817	3.3738	3.2927	3.2296	3.1789
10	4.9646	4.1028	3.7083	3.4780	3.3258	3.2172	3.1355	3.0717	3.0204
11	4.8443	3.9823	3.5874	3.3567	3.2039	3.0946	3.0123	2.9480	2.8962
12	4.7472	3.8853	3.4903	3.2592	3.1059	2.9961	2.9134	2.8486	2.7964
13	4.6672	3.8056	3.4105	3.1791	3.0254	2.9153	2.8321	2.7669	2.7144
14	4.6001	3.7389	3.3439	3.1122	2.9582	2.8477	2.7642	2.6987	2.6458
15	4.5431	3.6823	3.2874	3.0556	2.9013	2.7905	2.7066	2.6408	2.5876
16	4.4940	3.6337	3.2389	3.0069	2.8524	2.7413	2.6572	2.5911	2.5377
17	4.4513	3.5915	3.1968	2.9647	2.8100	2.6987	2.6143	2.5480	2.4943
18	4.4139	3.5546	3.1599	2.9277	2.7729	2.6613	2.5767	2.5102	2.4563
19	4.3808	3.5219	3.1274	2.8951	2.7401	2.6283	2.5435	2.4768	2.4227

20	4.3513	3.4928	3.0984	2.8661	2.7109	2.5990	2.5140	2.4471	2.3928
21	4.3248	3.4668	3.0725	2.8401	2.6848	2.5727	2.4876	2.4205	2.3661
22	4.3009	3.4434	3.0491	2.8167	2.6613	2.5491	2.4638	2.3965	2.3419
23	4.2793	3.4221	3.0280	2.7955	2.6400	2.5277	2.4422	2.3748	2.3201
24	4.2597	3.4028	3.0088	2.7763	2.6207	2.5082	2.4226	2.3551	2.3002
25	4.2417	3.3852	2.9912	2.7587	2.6030	2.4904	2.4047	2.3371	2.2821
26	4.2252	3.3690	2.9751	2.7426	2.5868	2.4741	2.3883	2.3205	2.2655
27	4.2100	3.3541	2.9604	2.7278	2.5719	2.4591	2.3732	2.3053	2.2501
28	4.1960	3.3404	2.9467	2.7141	2.5581	2.4453	2.3593	2.2913	2.2360
29	4.1830	3.3277	2.9340	2.7014	2.5454	2.4324	2.3463	2.2782	2.2229
30	4.1709	3.3158	2.9223	2.6896	2.5336	2.4205	2.3343	2.2662	2.2107
40	4.0848	3.2317	2.8387	2.6060	2.4495	2.3359	2.2490	2.1802	2.1240
60	4.0012	3.1504	2.7581	2.5252	2.3683	2.2540	2.1665	2.0970	2.0401
120	3.9201	3.0718	2.6802	2.4472	2.2900	2.1750	2.0867	2.0164	1.9588
∞	3.8415	2.9957	2.6049	2.3719	2.2141	2.0986	2.0096	1.9384	1.8799

TABLE 7 (*Continued*)

$\alpha = .05$

$\nu_2 \backslash \nu_1$	10	12	15	20	24	30	40	60	120	∞
1	241.88	243.91	245.95	248.01	249.05	250.09	251.14	252.20	253.25	254.32
2	19.396	19.413	19.429	19.446	19.454	19.462	19.471	19.479	19.487	19.496
3	8.7855	8.7446	8.7029	8.6602	8.6385	8.6166	8.5944	8.5720	8.5494	8.5265
4	5.9644	5.9117	5.8578	5.8025	5.7744	5.7459	5.7170	5.6878	5.6581	5.6281
5	4.7351	4.6777	4.6188	4.5581	4.5272	4.4957	4.4638	4.4314	4.3984	4.3650
6	4.0600	3.9999	3.9381	3.8742	3.8415	3.8082	3.7743	3.7398	3.7047	3.6688
7	3.6365	3.5747	3.5108	3.4445	3.4105	3.3758	3.3404	3.3043	3.2674	3.2298
8	3.3472	3.2840	3.2184	3.1503	3.1152	3.0794	3.0428	3.0053	2.9669	2.9276
9	3.1373	3.0729	3.0061	2.9365	2.9005	2.8637	2.8259	2.7872	2.7475	2.7067
10	2.9782	2.9130	2.8450	2.7740	2.7372	2.6996	2.6609	2.6211	2.5801	2.5379
11	2.8536	2.7876	2.7186	2.6464	2.6090	2.5705	2.5309	2.4901	2.4480	2.4045
12	2.7534	2.6866	2.6169	2.5436	2.5055	2.4663	2.4259	2.3842	2.3410	2.2962
13	2.6710	2.6037	2.5331	2.4589	2.4202	2.3803	2.3392	2.2966	2.2524	2.2064
14	2.6021	2.5342	2.4630	2.3879	2.3487	2.3082	2.2664	2.2230	2.1778	2.1307
15	2.5437	2.4753	2.4035	2.3275	2.2878	2.2468	2.2043	2.1601	2.1141	2.0658
16	2.4935	2.4247	2.3522	2.2756	2.2354	2.1938	2.1507	2.1058	2.0589	2.0096
17	2.4499	2.3807	2.3077	2.2304	2.1898	2.1477	2.1040	2.0584	2.0107	1.9604
18	2.4117	2.3421	2.2686	2.1906	2.1497	2.1071	2.0629	2.0166	1.9681	1.9168
19	2.3779	2.3080	2.2341	2.1555	2.1141	2.0712	2.0264	1.9796	1.9302	1.8780

20	2.3479	2.2776	2.2033	2.1242	2.0825	2.0391	1.9938	1.9464	1.8963	1.8432
21	2.3210	2.2504	2.1757	2.0960	2.0540	2.0102	1.9645	1.9165	1.8657	1.8117
22	2.2967	2.2258	2.1508	2.0707	2.0283	1.9842	1.9380	1.8895	1.8380	1.7831
23	2.2747	2.2036	2.1282	2.0476	2.0050	1.9605	1.9139	1.8649	1.8128	1.7570
24	2.2547	2.1834	2.1077	2.0267	1.9838	1.9390	1.8920	1.8424	1.7897	1.7331
25	2.2365	2.1649	2.0889	2.0075	1.9643	1.9192	1.8718	1.8217	1.7684	1.7110
26	2.2197	2.1479	2.0716	1.9898	1.9464	1.9010	1.8533	1.8027	1.7488	1.6906
27	2.2043	2.1323	2.0558	1.9736	1.9299	1.8842	1.8361	1.7851	1.7307	1.6717
28	2.1900	2.1179	2.0411	1.9586	1.9147	1.8687	1.8203	1.7689	1.7138	1.6541
29	2.1768	2.1045	2.0275	1.9446	1.9005	1.8543	1.8055	1.7537	1.6981	1.6377
30	2.1646	2.0921	2.0148	1.9317	1.8874	1.8409	1.7918	1.7396	1.6835	1.6223
40	2.0772	2.0035	1.9245	1.8389	1.7929	1.7444	1.6928	1.6373	1.5766	1.5089
60	1.9926	1.9174	1.8364	1.7480	1.7001	1.6491	1.5943	1.5343	1.4673	1.3893
120	1.9105	1.8337	1.7505	1.6587	1.6084	1.5543	1.4952	1.4290	1.3519	1.2539
∞	1.8307	1.7522	1.6664	1.5705	1.5173	1.4591	1.3940	1.3180	1.2214	1.0000

From "*Tables of Percentage Points of the Inverted Beta (F) Distribution,*" Biometrika, Vol. 33 (1943), pages 73–88, by Maxine Merrington and Catherine M. Thompson. By permission of Biometrika.

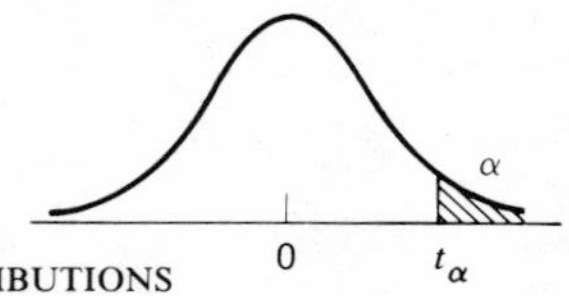

TABLE 8
EXTENDED TABLE OF PERCENTAGE POINTS OF t DISTRIBUTIONS

	α							
d.f.	.0025	.001	.0005	.00025	.0001	.00005	.000025	.00001
1	127.321	318.309	636.619	1,273.239	3,183.099	6,366.198	12,732.395	31,830.989
2	14.089	22.327	31.598	44.705	70.700	99.992	141.416	223.603
3	7.453	10.214	12.924	16.326	22.204	28.000	35.298	47.928
4	5.598	7.173	8.610	10.306	13.034	15.544	18.522	23.332
5	4.773	5.893	6.869	7.976	9.678	11.178	12.893	15.547
6	4.317	5.208	5.959	6.788	8.025	9.082	10.261	12.032
7	4.029	4.785	5.408	6.082	7.063	7.885	8.782	10.103
8	3.833	4.501	5.041	5.618	6.442	7.120	7.851	8.907
9	3.690	4.297	4.781	5.291	6.010	6.594	7.215	8.102
10	3.581	4.144	4.587	5.049	5.694	6.211	6.757	7.527
11	3.497	4.025	4.437	4.863	5.453	5.921	6.412	7.098
12	3.428	3.930	4.318	4.716	5.263	5.694	6.143	6.756
13	3.372	3.852	4.221	4.597	5.111	5.513	5.928	6.501
14	3.326	3.787	4.140	4.499	4.985	5.363	5.753	6.287
15	3.286	3.733	4.073	4.417	4.880	5.239	5.607	6.109
16	3.252	3.686	4.015	4.346	4.791	5.134	5.484	5.960
17	3.223	3.646	3.965	4.286	4.714	5.044	5.379	5.832
18	3.197	3.610	3.922	4.233	4.648	4.966	5.288	5.722
19	3.174	3.579	3.883	4.187	4.590	4.897	5.209	5.627
20	3.153	3.552	3.850	4.146	4.539	4.837	5.139	5.543

21	3.135	3.527	3.819	4.110	4.493	4.784	5.077	5.469
22	3.119	3.505	3.792	4.077	4.452	4.736	5.022	5.402
23	3.104	3.485	3.768	4.048	4.415	4.693	4.972	5.343
24	3.090	3.467	3.745	4.021	4.382	4.654	4.927	5.290
25	3.078	3.450	3.725	3.997	4.352	4.619	4.887	5.241
26	3.067	3.435	3.707	3.974	4.324	4.587	4.850	5.197
27	3.057	3.421	3.690	3.954	4.299	4.558	4.816	5.157
28	3.047	3.408	3.674	3.935	4.275	4.530	4.784	5.120
29	3.038	3.396	3.659	3.918	4.254	4.506	4.756	5.086
30	3.030	3.385	3.646	3.902	4.234	4.482	4.729	5.054
35	2.996	3.340	3.591	3.836	4.153	4.389	4.622	4.927
40	2.971	3.307	3.551	3.788	4.094	4.321	4.544	4.835
45	2.952	3.281	3.520	3.752	4.049	4.269	4.485	4.766
50	2.937	3.261	3.496	3.723	4.014	4.228	4.438	4.711
55	2.925	3.245	3.476	3.700	3.986	4.196	4.401	4.667
60	2.915	3.232	3.460	3.681	3.962	4.169	4.370	4.631
70	2.899	3.211	3.435	3.651	3.926	4.127	4.323	4.576
80	2.887	3.195	3.416	3.629	3.899	4.096	4.288	4.535
90	2.878	3.183	3.402	3.612	3.878	4.072	4.261	4.503
100	2.871	3.174	3.390	3.598	3.862	4.053	4.240	4.478
∞	2.807	3.090	3.291	3.481	3.719	3.891	4.056	4.265

Adapted from Federighi, E. T., "Extended Tables of the Percentage Points of Student's *t*-Distributions," *Journal of American Statistical Association*, (1959) Vol. 54, p. 684–685.

TABLE 9

Selected Tail Probabilities for the Null Distribution of Wilcoxon's Rank-Sum Statistic

$P = P[W_s \geqslant x] = P[W_s \leqslant x^*]$

SMALLER SAMPLE SIZE = 2

LARGER SAMPLE SIZE

3			4			5			6		
x	P	x^*	x	P	x^*	x	P	x^*	x	P	x^*
8	.200	4	10	.133	4	11	.190	5	13	.143	5
9	.100	3	11	.067	3	12	.095	4	14	.071	4
10	0	2	12	0	2	13	.048	3	15	.036	3
						14	0	2	16	0	2

7			8			9			10		
x	P	x^*	x	P	x^*	x	P	x^*	x	P	x^*
15	.111	5	16	.133	6	18	.109	6	19	.136	7
16	.056	4	17	.089	5	19	.073	5	20	.091	6
17	.028	3	18	.044	4	20	.036	4	21	.061	5
18	0	2	19	.022	3	21	.018	3	22	.030	4
			20	0	2	22	0	2	23	.015	3

SMALLER SAMPLE SIZE = 3

LARGER SAMPLE SIZE

3			4			5			6		
x	P	x^*	x	P	x^*	x	P	x^*	x	P	x^*
13	.200	8	16	.114	8	18	.125	9	20	.131	10
14	.100	7	17	.057	7	19	.071	8	21	.083	9
15	.050	6	18	.029	6	20	.036	7	22	.048	8
16	0	5	19	0	5	21	.018	6	23	.024	7
						22	0	5	24	.012	6
									25	0	5

TABLE 9 (*Continued*)

7			8			9			10		
x	P	x^*	x	P	x^*	x	P	x^*	x	P	x^*
22	.133	11	24	.139	12	27	.105	12	29	.108	13
23	.092	10	25	.097	11	28	.073	11	30	.080	12
24	.058	9	26	.067	10	29	.050	10	31	.056	11
25	.033	8	27	.042	9	30	.032	9	32	.038	10
26	.017	7	28	.024	8	31	.018	8	33	.024	9
27	.008	6	29	.012	7	32	.009	7	34	.014	8
28	0	5	30	.006	6				35	.007	7
			31	0	5						

SMALLER SAMPLE SIZE = 4

LARGER SAMPLE SIZE

4			5			6			7		
x	P	x^*	x	P	x^*	x	P	x^*	x	P	x^*
22	.171	14	25	.143	15	28	.129	16	31	.115	17
23	.100	13	26	.095	14	29	.086	15	32	.082	16
24	.057	12	27	.056	13	30	.057	14	33	.055	15
25	.029	11	28	.032	12	31	.033	13	34	.036	14
26	.014	10	29	.016	11	32	.019	12	35	.021	13
27	0	9	30	.008	10	33	.010	11	36	.012	12
			31	0	9				37	.006	11

8			9			10		
x	P	x^*	x	P	x^*	x	P	x^*
34	.107	18	36	.130	20	39	.120	21
35	.077	17	37	.099	19	40	.094	20
36	.055	16	38	.074	18	41	.071	19
37	.036	15	39	.053	17	42	.053	18
38	.024	14	40	.038	16	43	.038	17
39	.014	13	41	.025	15	44	.027	16
40	.008	12	42	.017	14	45	.018	15
			43	.010	13	46	.012	14
						47	.007	13

TABLE 9 (*Continued*)

SMALLER SAMPLE SIZE = 5

LARGER SAMPLE SIZE

5			6			7			8		
x	P	x^*	x	P	x^*	x	P	x^*	x	P	x^*
34	.111	21	37	.123	23	41	.101	24	44	.111	26
35	.075	20	38	.089	22	42	.074	23	45	.085	25
36	.048	19	39	.063	21	43	.053	22	46	.064	24
37	.028	18	40	.041	20	44	.037	21	47	.047	23
38	.016	17	41	.026	19	45	.024	20	48	.033	22
39	.008	16	42	.015	18	46	.015	19	49	.023	21
			43	.009	17	47	.009	18	50	.015	20
									51	.009	19

9			10		
x	P	x^*	x	P	x^*
47	.120	28	51	.103	29
48	.095	27	52	.082	28
19	.073	26	53	.065	27
50	.056	25	54	.050	26
51	.041	24	55	.038	25
52	.030	23	56	.028	24
53	.021	22	57	.020	23
54	.014	21	58	.014	22
55	.009	20	59	.010	21

SMALLER SAMPLE SIZE = 6

LARGER SAMPLE SIZE

6			7			8			9		
x	P	x^*	x	P	x^*	x	P	x^*	x	P	x^*
47	.120	31	51	.117	33	55	.114	35	59	.112	37
48	.090	30	52	.090	32	56	.091	34	60	.091	36
49	.066	29	53	.069	31	57	.071	33	61	.072	35
50	.047	28	54	.051	30	58	.054	32	62	.057	34
51	.032	27	55	.037	29	59	.041	31	63	.044	33
52	.021	26	56	.026	28	60	.030	30	64	.033	32
53	.013	25	57	.017	27	61	.021	29	65	.025	31
54	.008	24	58	.011	26	62	.015	28	66	.018	30
			59	.007	25	63	.010	27	67	.013	29
									68	.009	28

TABLE 9 (*Continued*)

SMALLER SAMPLE SIZE = 6

LARGER SAMPLE SIZE

10		
x	P	x^*
63	.110	39
64	.090	38
65	.074	37
66	.059	36
67	.047	35
68	.036	34
69	.028	33
70	.021	32
71	.016	31
72	.011	30
73	.008	29

SMALLER SAMPLE SIZE = 7

LARGER SAMPLE SIZE

7			8			9			10		
x	P	x^*	x	P	x^*	x	P	x^*	x	P	x^*
63	.104	42	67	.116	45	72	.105	47	76	.115	50
64	.082	41	68	.095	44	73	.087	46	77	.097	49
65	.064	40	69	.076	43	74	.071	45	78	.081	48
66	.049	39	70	.060	42	75	.057	44	79	.067	47
67	.036	38	71	.047	41	76	.045	43	80	.054	46
68	.027	37	72	.036	40	77	.036	42	81	.044	45
69	.019	36	73	.027	39	78	.027	41	82	.035	44
70	.013	35	74	.020	38	79	.021	40	83	.028	43
71	.009	34	75	.014	37	80	.016	39	84	.022	42
			76	.010	36	81	.011	38	85	.017	41
						82	.008	37	86	.012	40
									87	.009	39

TABLE 9 (*Continued*)

SMALLER SAMPLE SIZE = 8

LARGER SAMPLE SIZE 8			9			10		
x	P	x^*	x	P	x^*	x	P	x^*
80	.117	56	86	.100	58	91	.102	61
81	.097	55	87	.084	57	92	.086	60
82	.080	54	88	.069	56	93	.073	59
83	.065	53	89	.057	55	94	.061	58
84	.052	52	90	.046	54	95	.051	57
85	.041	51	91	.037	53	96	.042	56
86	.032	50	92	.030	52	97	.034	55
87	.025	49	93	.023	51	98	.027	54
88	.019	48	94	.018	50	99	.022	53
89	.014	47	95	.014	49	100	.017	52
90	.010	46	96	.010	48	101	.013	51
						102	.010	50

SMALLER SAMPLE SIZE = 9

LARGER SAMPLE SIZE 9			10		
x	P	x^*	x	P	x^*
100	.111	71	106	.106	74
101	.095	70	107	.091	73
102	.081	69	108	.078	72
103	.068	68	109	.067	71
104	.057	67	110	.056	70
105	.047	66	111	.047	69
106	.039	65	112	.039	68
107	.031	64	113	.033	67
108	.025	63	114	.027	66
109	.020	62	115	.022	65
110	.016	61	116	.017	64
111	.012	60	117	.014	63
112	.009	59	118	.011	62
			119	.009	61

SMALLER SAMPLE SIZE = 10

LARGER SAMPLE SIZE 10		
x	P	x^*
122	.109	88
123	.095	87
124	.083	86
125	.072	85
126	.062	84
127	.053	83
128	.045	82
129	.038	81
130	.032	80
131	.026	79
132	.022	78
133	.018	77
134	.014	76
135	.012	75
136	.009	74

Adapted from: Kraft, C., and van Eeden, C., *A Nonparametric Introduction to Statistics*, The Macmillan Company, New York, 1968.

TABLE 10

SELECTED TAIL PROBABILITIES FOR THE NULL DISTRIBUTION OF WILCOXON'S SIGNED-RANK STATISTIC

$P = P[T^+ \geqslant x] = P[T^+ \leqslant x^*]$

n=3			n=4			n=5			n=6		
x	P	x*	x	P	x*	x	P	x*	x	P	x*
5	.250	1	8	.188	2	12	.156	3	17	.109	4
6	.125	0	9	.125	1	13	.094	2	18	.078	3
7	0		10	.062	0	14	.062	1	19	.047	2
			11	0		15	.031	0	20	.031	1
						16	0		21	.016	0
									22	0	

n=7			n=8			n=9			n=10		
x	P	x*	x	P	x*	x	P	x*	x	P	x*
22	.109	6	27	.125	9	34	.102	11	40	.116	15
23	.078	5	28	.098	8	35	.082	10	41	.097	14
24	.055	4	29	.074	7	36	.064	9	42	.080	13
25	.039	3	30	.055	6	37	.049	8	43	.065	12
26	.023	2	31	.039	5	38	.037	7	44	.053	11
27	.016	1	32	.027	4	39	.027	6	45	.042	10
28	.008	0	33	.020	3	40	.020	5	46	.032	9
			34	.012	2	41	.014	4	47	.024	8
			35	.008	1	42	.010	3	48	.019	7
									49	.014	6
									50	.010	5

TABLE 10 (*Continued*)

$n=11$			$n=12$			$n=13$			$n=14$		
x	P	x^*	x	P	x^*	x	P	x^*	x	P	x^*
48	.103	18	56	.102	22	64	.108	27	73	.108	32
49	.087	17	57	.088	21	65	.095	26	74	.097	31
50	.074	16	58	.076	20	66	.084	25	75	.086	30
51	.062	15	59	.065	19	67	.073	24	76	.077	29
52	.051	14	60	.055	18	68	.064	23	77	.068	28
53	.042	13	61	.046	17	69	.055	22	78	.059	27
54	.034	12	62	.039	16	70	.047	21	79	.052	26
55	.027	11	63	.032	15	71	.040	20	80	.045	25
56	.021	10	64	.026	14	72	.034	19	81	.039	24
57	.016	9	65	.021	13	73	.029	18	82	.034	23
58	.012	8	66	.017	12	74	.024	17	83	.029	22
59	.009	7	67	.013	11	75	.020	16	84	.025	21
			68	.010	10	76	.016	15	85	.021	20
						77	.013	14	86	.018	19
						78	.011	13	87	.015	18
						79	.009	12	88	.012	17
									89	.010	16

$n=15$					
x	P	x^*	x	P	x^*
83	.104	37	92	.036	28
84	.094	36	93	.032	27
85	.084	35	94	.028	26
86	.076	34	95	.024	25
87	.068	33	96	.021	24
88	.060	32	97	.018	23
89	.053	31	98	.015	22
90	.047	30	99	.013	21
91	.042	29	100	.011	20
			101	.009	19

Adapted from: Kraft, C., and van Eeden, C., *A Nonparametric Introduction to Statistics*, The Macmillan Company, New York, 1968.

TABLE 11
Confidence Intervals for the Median of a Symmetric Distribution Using the Wilcoxon Signed-Rank Test

Confidence coefficient $= 1 - \alpha$

Confidence interval $= d$th smallest to dth largest of pair averages

n	d	$1-\alpha$	n	d	$1-\alpha$	n	d	$1-\alpha$	n	d	$1-\alpha$
3	1	.750	12	8	.991	18	28	.991	24	62	.990
4	1	.875		9	.988		29	.990		63	.989
5	1	.938		14	.958		41	.952		82	.951
	2	.875		15	.948		42	.946		83	.947
6	1	.969		18	.908		48	.901		92	.905
	2	.937		19	.890		49	.892		93	.899
	3	.906	13	10	.992	19	33	.991	25	69	.990
	4	.844		11	.990		34	.989		70	.989
7	1	.984		18	.952		47	.951		90	.952
	3	.953		19	.943		48	.945		91	.948
	4	.922		22	.906		54	.904		101	.904
	5	.891		23	.890		55	.896		102	.899
8	1	.992	14	13	.991	20	38	.991			
	2	.984		14	.989		39	.989			
	4	.961		22	.951		53	.952			
	5	.945		23	.942		54	.947			
	6	.922		26	.909		61	.903			
	7	.891		27	.896		62	.895			
9	2	.992	15	16	.992	21	43	.991			
	3	.988		17	.990		44	.990			
	6	.961		26	.952		59	.954			
	7	.945		27	.945		60	.950			
	9	.902		31	.905		68	.904			
	10	.871		32	.893		69	.897			
10	4	.990	16	20	.991	22	49	.991			
	5	.986		21	.989		50	.990			
	9	.951		30	.956		66	.954			
	10	.936		31	.949		67	.950			
	11	.916		36	.907		76	.902			
	12	.895		37	.895		77	.895			
11	6	.990	17	24	.991	23	55	.991			
	7	.986		25	.989		56	.990			
	11	.958		35	.955		74	.952			
	12	.946		36	.949		75	.948			
	14	.917		42	.902		84	.902			
	15	.898		43	.891		85	.895			

Adapted from *Introduction to Statistics* by G. E. Noether. Copyright © 1971 by Houghton Mifflin Company. Used by permission of the publisher.

TABLE 12
CONFIDENCE INTERVALS FOR SHIFT USING THE WILCOXON RANK-SUM TEST

Confidence coefficient $= 1 - \alpha$
Confidence interval $=$ dth smallest to dth largest of differences $X_{1i} - X_{2j}$

LARGER SAMPLE SIZE	SMALLER SAMPLE SIZE 3		4		5		6		7		8	
	d	$1-\alpha$	d	$1-\alpha$	d	$1-\alpha$	d	$1-\alpha$	d	$1-\alpha$	d	$1-\alpha$
3	1	.900										
4	1	.943	1	.971								
	2	.886	2	.943								
			3	.886								
					1	.992						
5	1	.964	1	.984	2	.984						
	2	.929	2	.968	3	.968						
	3	.857	3	.937	4	.944						
			4	.889	5	.905						
					6	.849						
					2	.991	3	.991				
6	1	.976	1	.990	3	.983	4	.985				
	2	.952	2	.981	4	.970	6	.959				
	3	.905	3	.962	5	.948	7	.935				
	4	.833	4	.933	6	.918	8	.907				
			5	.886	7	.874	9	.868				
					2	.995	4	.992	5	.993		
7	1	.983	1	.994	3	.990	5	.986	6	.989		
	2	.967	2	.988	6	.952	7	.965	9	.962		
	3	.933	4	.958	7	.927	8	.949	10	.947		
	4	.883	5	.927	8	.894	9	.927	12	.903		
			6	.891			10	.899	13	.872		
8	1	.988	2	.992	3	.994	5	.992	7	.991	8	.993
	3	.952	3	.984	4	.989	6	.987	8	.986	9	.990
	4	.915	5	.952	7	.955	9	.957	11	.960	14	.950
	5	.867	6	.927	8	.935	10	.941	12	.946	15	.935
			7	.891	9	.907	11	.919	14	.906	16	.917
					10	.873	12	.892	15	.879	17	.895
9	1	.991	2	.994	4	.993	6	.992	8	.992	10	.992
	2	.982	3	.989	5	.988	7	.988	9	.988	11	.989
	3	.964	5	.966	8	.958	11	.950	13	.958	16	.954
	4	.936	6	.950	9	.940	12	.934	14	.945	17	.941
	5	.900	7	.924	10	.917	13	.912	16	.909	19	.907
			8	.894	11	.888	14	.887	17	.886	20	.886

TABLE 12 (*Continued*)

Larger Sample Size	Smaller Sample Size 3		4		5		6		7		8	
	d	$1-\alpha$	d	$1-\alpha$	d	$1-\alpha$	d	$1-\alpha$	d	$1-\alpha$	d	$1-\alpha$
10	1	.993	3	.992	5	.992	7	.993	10	.990	12	.991
	2	.986	4	.986	6	.987	8	.989	11	.986	13	.988
	4	.951	6	.964	9	.960	12	.958	15	.957	18	.957
	5	.923	7	.946	10	.945	13	.944	16	.945	19	.945
	6	.888	8	.924	12	.901	15	.907	18	.912	21	.917
			9	.894	13	.871	16	.882	19	.891	22	.899
11	1	.995	3	.994	6	.991	8	.993	11	.992	14	.991
	2	.989	4	.990	7	.987	9	.990	12	.989	15	.988
	4	.962	7	.960	10	.962	14	.952	17	.956	20	.959
	5	.940	8	.944	11	.948	15	.938	18	.944	21	.949
	6	.912	9	.922	13	.910	17	.902	20	.915	24	.909
	7	.874	10	.896	14	.885	18	.878	21	.896	25	.891
12	2	.991	4	.992	7	.991	10	.990	13	.990	16	.990
	3	.982	5	.987	8	.986	11	.987	14	.987	17	.988
	5	.952	8	.958	12	.952	15	.959	19	.955	23	.953
	6	.930	9	.942	13	.936	16	.947	20	.944	24	.943
	7	.899	10	.922	14	.918	18	.917	22	.917	27	.902
			11	.897	15	.896	19	.898	23	.900	28	.885

TABLE 12 (*Continued*)

LARGER SAMPLE SIZE	SMALLER SAMPLE SIZE 9		10		11		12	
	d	$1-\alpha$	d	$1-\alpha$	d	$1-\alpha$	d	$1-\alpha$
9	12	.992						
	13	.989						
	18	.960						
	19	.950						
	22	.906						
	23	.887						
10	14	.992	17	.991				
	15	.990	18	.989				
	21	.957	24	.957				
	22	.947	25	.948				
	25	.905	28	.911				
	26	.887	29	.895				
11	17	.990	19	.992	22	.992		
	18	.988	20	.990	23	.989		
	24	.954	27	.957	31	.953		
	25	.944	28	.949	32	.944		
	28	.905	32	.901	35	.912		
	29	.888	33	.886	36	.899		
12	19	.991	22	.991	25	.991	28	.992
	20	.988	23	.989	26	.989	29	.990
	27	.951	30	.957	34	.956	38	.955
	28	.942	31	.950	35	.949	39	.948
	31	.905	35	.907	39	.909	43	.911
	32	.889	36	.893	40	.896	44	.899

TABLE 13

SELECTED TAIL PROBABILITIES FOR THE NULL DISTRIBUTION OF SPEARMAN'S STATISTIC $\sum_{i=1}^{n} R_i S_i$

$$P = P\left[\sum_{i=1}^{n} R_i S_i \geqslant x\right] = P\left[\sum_{i=1}^{n} R_i S_i \leqslant x^*\right]$$

$n=4$			$n=5$			$n=6$			$n=7$		
x	P	x^*	x	P	x^*	x	P	x^*	x	P	x^*
29	.167	21	52	.117	38	84	.121	63	128	.100	96
30	.042	20	53	.067	37	85	.088	62	129	.083	95
31	0	19	54	.042	36	86	.068	61	130	.069	94
			55	.0083	35	87	.051	60	131	.055	93
						88	.029	59	132	.044	92
						89	.017	58	133	.033	91
						90	.0083	57	134	.024	90
									135	.017	89
									136	.012	88
									137	.0062	87

TABLE 13 (*Continued*)

$n=8$			$n=9$			$n=10$		
x	P	x^*	x	P	x^*	x	P	x^*
183	.108	141	253	.106	197	339	.102	266
184	.098	140	254	.097	196	340	.096	265
185	.085	139	255	.089	195	341	.089	264
186	.076	138	256	.081	194	342	.083	263
187	.066	137	257	.074	193	343	.077	262
188	.057	136	258	.066	192	344	.072	261
189	.048	135	259	.060	191	345	.067	260
190	.042	134	260	.054	190	346	.062	259
191	.035	133	261	.048	189	347	.057	258
192	.029	132	262	.043	188	348	.052	257
193	.023	131	263	.038	187	349	.048	256
194	.018	130	264	.033	186	350	.044	255
195	.014	129	265	.029	185	351	.040	254
196	.011	128	266	.025	184	352	.037	253
197	.0077	127	267	.022	183	353	.033	252
			268	.018	182	354	.030	251
			269	.016	181	355	.027	250
			270	.013	180	356	.024	249
			271	.011	179	357	.022	248
			272	.0086	178	358	.019	247
						359	.017	246
						360	.015	245
						361	.013	244
						362	.012	243
						363	.010	242

Adapted from: Owen, D. B., *Handbook of Statistical Tables*, Addison-Wesley, Boston, 1962. Courtesy of the U.S. Energy Research and Development Administration.

TABLE 14
5000 RANDOM DIGITS

00000	10097	32533	76520	13586	34673	54876	80959	09117	39292	74945
00001	37542	04805	64894	74296	24805	24037	20636	10402	00822	91665
00002	08422	68953	19645	09303	23209	02560	15953	34764	35080	33606
00003	99019	02529	09376	70715	38311	31165	88676	74397	04436	27659
00004	12807	99970	80157	36147	64032	36653	98951	16877	12171	76833
00005	66065	74717	34072	76850	36697	36170	65813	39885	11199	29170
00006	31060	10805	45571	82406	35303	42614	86799	07439	23403	09732
00007	85269	77602	02051	65692	68665	74818	73053	85247	18623	88579
00008	63573	32135	05325	47048	90553	57548	28468	28709	83491	25624
00009	73796	45753	03529	64778	35808	34282	60935	20344	35273	88435
00010	98520	17767	14905	68607	22109	40558	60970	93433	50500	73998
00011	11805	05431	39808	27732	50725	68248	29405	24201	52775	67851
00012	83452	99634	06288	98083	13746	70078	18475	40610	68711	77817
00013	88685	40200	86507	58401	36766	67951	90364	76493	29609	11062
00014	99594	67348	87517	64969	91826	08928	93785	61368	23478	34113
00015	65481	17674	17468	50950	58047	76974	73039	57186	40218	16544
00016	80124	35635	17727	08015	45318	22374	21115	78253	14385	53763
00017	74350	99817	77402	77214	43236	00210	45521	64237	96286	02655
00018	69916	26803	66252	29148	36936	87203	76621	13990	94400	56418
00019	09893	20505	14225	68514	46427	56788	96297	78822	54382	14598
00020	91499	14523	68479	27686	46162	83554	94750	89923	37089	20048
00021	80336	94598	26940	36858	70297	34135	53140	33340	42050	82341
00022	44104	81949	85157	47954	32979	26575	57600	40881	22222	06413
00023	12550	73742	11100	02040	12860	74697	96644	89439	28707	25815
00024	63606	49329	16505	34484	40219	52563	43651	77082	07207	31790
00025	61196	90446	26457	47774	51924	33729	65394	59593	42582	60527
00026	15474	45266	95270	79953	59367	83848	82396	10118	33211	59466
00027	94557	28573	67897	54387	54622	44431	91190	42592	92927	45973
00028	42481	16213	97344	08721	16868	48767	03071	12059	25701	46670
00029	23523	78317	73208	89837	68935	91416	26252	29663	05522	82562
00030	04493	52494	75246	33824	45862	51025	61962	79335	65337	12472
00031	00549	97654	64051	88159	96119	63896	54692	82391	23287	29529
00032	35963	15307	26898	09354	33351	35462	77974	50024	90103	39333
00033	59808	08391	45427	26842	83609	49700	13021	24892	78565	20106
00034	46058	85236	01390	92286	77281	44077	93910	83647	70617	42941
00035	32179	00597	87379	25241	05567	07007	86743	17157	85394	11838
00036	69234	61406	20117	45204	15956	60000	18743	92423	97118	96338
00037	19565	41430	01758	75379	40419	21585	66674	36806	84962	85207
00038	45155	14938	19476	07246	43667	94543	59047	90033	20826	69541
00039	94864	31994	36168	10851	34888	81553	01540	35456	05014	51176
00040	98086	24826	45240	28404	44999	08896	39094	73407	35441	31880
00041	33185	16232	41941	50949	89435	48581	88695	41994	37548	73043
00042	80951	00406	96382	70774	20151	23387	25016	25298	94624	61171
00043	79752	49140	71961	28296	69861	02591	74852	20539	00387	59579
00044	18633	32537	98145	06571	31010	24674	05455	61427	77938	91936
00045	74029	43902	77557	32270	97790	17119	52527	58021	80814	51748
00046	54178	45611	80993	37143	05335	12969	56127	19255	36040	90324
00047	11664	49883	52079	84827	59381	71539	09973	33440	88461	23356
00048	48324	77928	31249	64710	02295	36870	32307	57546	15020	09994
00049	69074	94138	87637	91976	35584	04401	10518	21615	01848	76938
00050	09188	20097	32825	39527	04220	86304	83389	87374	64278	58044
00051	90045	85497	51981	50654	94938	81997	91870	76150	68476	64659
00052	73189	50207	47677	26269	62290	64464	27124	67018	41361	82760
00053	75768	76490	20971	87749	90429	12272	95375	05871	93823	43178
00054	54016	44056	66281	31003	00682	27398	20714	53295	07706	17813

TABLE 14 (*Continued*)

00055	08358	69910	78542	42785	13661	58873	04618	97553	31223	08420
00056	28306	03264	81333	10591	40510	07893	32604	60475	94119	01840
00057	53840	86233	81594	13628	51215	90290	28466	68795	77762	20791
00058	91757	53741	61613	62269	50263	90212	55781	76514	83483	47055
00059	89415	92694	00397	58391	12607	17646	48949	72306	94541	37408
00060	77513	03820	86864	29901	68414	82774	51908	13980	72893	55507
00061	19502	37174	69979	20288	55210	29773	74287	75251	65344	67415
00062	21818	59313	93278	81757	05686	73156	07082	85046	31853	38452
00063	51474	66499	68107	23621	94049	91345	42836	09191	08007	45449
00064	99559	68331	62535	24170	69777	12830	74819	78142	43860	72834
00065	33713	48007	93584	72869	51926	64721	58303	29822	93174	93972
00066	85274	86893	11303	22970	28834	34137	73515	90400	71148	43643
00067	84133	89640	44035	52166	73852	70091	61222	60561	62327	18423
00068	56732	16234	17395	96131	10123	91622	85496	57560	81604	18880
00069	65138	56806	87648	85261	34313	65861	45875	21069	85644	47277
00070	38001	02176	81719	11711	71602	92937	74219	64049	65584	49698
00071	37402	96397	01304	77586	56271	10086	47324	62605	40030	37438
00072	97125	40348	87083	31417	21815	39250	75237	62047	15501	29578
00073	21826	41134	47143	34072	64638	85902	49139	06441	03856	54552
00074	73135	42742	95719	09035	85794	74296	08789	88156	64691	19202
00075	07638	77929	03061	18072	96207	44156	23821	99538	04713	66994
00076	60528	83441	07954	19814	59175	20695	05533	52139	61212	06455
00077	83596	35655	06958	92983	05128	09719	77433	53783	92301	50498
00078	10850	62746	99599	10507	13499	06319	53075	71839	06410	19362
00079	39820	98952	43622	63147	64421	80814	43800	09351	31024	73167
00080	59580	06478	75569	78800	88835	54486	23768	06156	04111	08408
00081	38508	07341	23793	48763	90822	97022	17719	04207	95954	49953
00082	30692	70668	94688	16127	56196	80091	82067	63400	05462	69200
00083	65443	95659	18288	27437	49632	24041	08337	65676	96299	90836
00084	27267	50264	13192	72294	07477	44606	17985	48911	97341	30358
00085	91307	06991	19072	24210	36699	53728	28825	35793	28976	66252
00086	68434	94688	84473	13622	62126	98408	12843	82590	09815	93146
00087	48908	15877	54745	24591	35700	04754	83824	52692	54130	55160
00088	06913	45197	42672	78601	11883	09528	63011	98901	14974	40344
00089	10455	16019	14210	33712	91342	37821	88325	80851	43667	70883
00090	12883	97343	65027	61184	04285	01392	17974	15077	90712	26769
00091	21778	30976	38807	36961	31649	42096	63281	02023	08816	47449
00092	19523	59515	65122	59659	86283	68258	69572	13798	16435	91529
00093	67245	52670	35583	16563	79246	86686	76463	34222	26655	90802
00094	60584	47377	07500	37992	45134	26529	26760	83637	41326	44344
00095	53853	41377	36066	94850	58838	73859	49364	73331	96240	43642
00096	24637	38736	74384	89342	52623	07992	12369	18601	03742	83873
00097	83080	12451	38992	22815	07759	51777	97377	27585	51972	37867
00098	16444	24334	36151	99073	27493	70939	85130	32552	54846	54759
00099	60790	18157	57178	65762	11161	78576	45819	52979	65130	04860

SOURCE: *A Million Random Digits with 100,000 Normal Deviates*, The RAND Corporation, The Free Press, New York, 1955. Reproduced by permission of the publishers.

Answers to Selected Odd-Numbered Exercises

Chapter 2

9. (d) 2, (f) 15
11. (c) $\bar{x}=2.2$, median $=1$, $s^2=3.2$, $s=1.79$, range $=4$
13. (c) $Q_1=22$, $Q_2=24$, $Q_3=30$
17. (a) 146.99, (b) 26.01, (c) 144.8
(d) $Q_1=126.8$, $Q_3=164.0$, (e) 37.2
25. 3

Chapter 3

1(a). In absolute temperature $[0,\infty)$
3(a). $\{0,1,3\}$
5(a). $\{0,1\}$
7(a). $\{(0,0),\ (0,\mathrm{L}),\ (0,\mathrm{R}),\ (0,\mathrm{LR})$
$(\mathrm{L},0),\ (\mathrm{L},\mathrm{L}),\ (\mathrm{L},\mathrm{R}),\ (\mathrm{L},\mathrm{LR})$
$(\mathrm{R},0),\ (\mathrm{R},\mathrm{L}),\ (\mathrm{R},\mathrm{R}),\ (\mathrm{R},\mathrm{LR})$
$(\mathrm{LR},0),\ (\mathrm{LR},\mathrm{L}),\ (\mathrm{LR},\mathrm{R}),\ (\mathrm{LR},\mathrm{LR})\}$
9(b). .712
13(b). $(A\cup B\cup C)^c=\{e_8\}$, $P(e_8)=.1$
15. $2\times 3=6$
19(a). 720
25(c). $\frac{1}{22}=.045$
29(a). $\frac{2}{5}=.4$
31. .0099. Not independent.
35(b). $P(A\cup B)=.65$
39(a). .72
43. .00006
45(b). $\frac{3}{7}=.429$

Chapter 4

1.

	DDD,	DDN,	DND,	NDD,	DNN,	NDN,	NND,	NNN
$x=$	0,	1,	1,	1,	2,	2,	2,	3

5. (b) 3.9, (c) $P[X \geqslant 4]=.6$, $P[2<X \leqslant 4]=.6$

9. (a) 5.2, (b)

y	0	4	16
$f(y)$	.3	.5	.2

11. (a) $m_0=4$
13. (b) Mean $=1$, variance $=.25$
17. (d) Cov $=-.0175$, Corr $=-.054$
21. Smaller. Larger.
23. (c) $-.5$

Chapter 5

3. (a) $(.4)^2=.16$, (c) $3(.4)^2(.6)=.288$
9. $P[X \geqslant 10]=.315$, $E(X)=8.5$
13. (a)

p	.05	.10	.20	.30	.40
$P(A)$	.829	.549	.167	.035	.005

17. $P[X=2]=.424$, $E(X)=1.67$, $sd(X)=.841$
21. (b) $P[X=2]=.16$, $P[X \geqslant 3]=.64$, $E(X)=5$
23. (a) $P[X=3]=.195$, $P[X \leqslant 5]=.785$

Chapter 6

3. (a) .125, (b) .875
7. Rejection region $X \geqslant 14$ has $\alpha=.058$, claim is supported
9. (b)

p	.05	.10	.20	.30	.40
$\gamma(p)$	.171	.451	.833	.965	.995

13. (a) $\alpha=.048$, $\beta(.3)=.851$
17. (b) With $\alpha=.051$, rejection region is $X \geqslant 8$; claim not supported
    (c) .676

Chapter 7

1. (a) .6628, (c) .0455, (e) .8670, (g) .2524
3. (a) $P[-1.15<Z<1.15]=.7498$
5. (a) 115.68
11. \$24.25
15. (b) .0262
19. (a) (i) .904, (b) (i) .8996

21.

Value of $\overline{X}_{(4)}$	1.00	1.25	1.50	1.75	2.00	2.25	2.50	2.75	3.00
Probability	.0081	.0432	.1188	.2064	.2470	.2064	.1188	.0432	.0081

23. (b) (i) .0548
25. (b) Approximately .0668

Chapter 8

1. (b) $\bar{x}=11.8$, estimated S.E. $=.154$
3. (a) 4
7. (b) (124.6, 129.2)
11. (.477, .648)
15. (.346, .414)
19. (a) Observed $t=2.268$ is in rejection region $t>1.895$, model is contradicted
23. (b) smallest $\alpha=.0032$, very strong evidence for rejection
27. Observed $Z=-2.685$ is in rejection region $Z<-2.33$ $(\alpha=.01)$
33. (b) (1.56, 4.15)
35. $n=97$
39. $n=545$

Chapter 9

1. (a) Observed $t=-1.806$, H_0 rejected in favor of H_1: $\mu_1<\mu_2$
   (c) $(-9.09, .69)$
9. Observed $z=2.57$, reject H_0: $\mu_A=\mu_B$ with $\alpha=.01$. 95% confidence interval for $\mu_A-\mu_B$ is $(-3.87, -.52)$
13. (c) $(-1.59, -.51)$ for $\mu_A-\mu_B$
17. Observed $t=1.49$, H_0 is not rejected
23. (b) (.43, 6.77) for left$-$right
29. (a) Observed $z=2.64$, reject H_0: $p_1=p_2$ with $\alpha=.01$
   (b) (.046, .294)
35. Observed $\dfrac{s_1^2}{s_2^2}=.75$, critical region $F\leqslant .315$ or $F\geqslant 3.179$, H_0 not rejected with $\alpha=.10$. Assume populations normal

Chapter 10

5. (a) $-.22, .28, -.02, .08, -.12$
   (b) $SSE=.148$, (c) $s^2=.0493$
7. $\hat{\alpha}=9$, $\hat{\beta}=-1.5$, $SSE=.5$, $s^2=.25$
11. (a) $\hat{y}=112.25+.967x$
   (d) (122.9, 124.9)
13. (a) $\hat{y}=40.35+3.722x$
   (c) (464.4, 732.9)

Chapter 11

3. Yes, Var(Y_i) seems to increase with $\hat{y}$
9. .519
13. $SS_{pe}=.065$, $SS_L=.3113$; $F=1.6$, d.f.$=(6, 2)$. Not significant with $\alpha=.1$
15. $\hat{y}'=-1.161+.0306x$, $SSE=.022$
17. (a) $\hat{y}=-83.46+.353x_1+8.628x_2$
19. $\hat{y}=2.08+.64x-.1x^2$, $SSE=.032$
21. (b) $d=2.09$

Chapter 12

1. $r=-.5$
5. (b) $r=.437$
9. Observed $t=7.2$, H_0 rejected even with $\alpha=.005$
15. Observed $z=.26$, do not reject H_0
17. (.49, .83)
21. (b) $r_1=.66$, (c) $\sqrt{n}\, r_1 = 2.95$, H_0 rejected with $\alpha=.01$
23. $r=.519$

Chapter 13

1. Observed $\chi^2=2.267$, $\chi^2_{.05}=5.991$, model not contradicted
5. (a) Poisson
   (b) Combining last three cells, observed $\chi^2=32.44$; $\chi^2_{.005}=12.84$, model contradicted
7. Observed $\chi^2=.1525$, H_0 not rejected, association very weak
9. .2059
13. Observed $\chi^2=24.61$, $\chi^2_{.05}=9.49$, $\chi^2_{.005}=14.86$, difference is highly significant. Contributions to χ^2—City 1: .74, .17, .75; City 2: 6.15, 3.52, 9.15; City 3: 1.11, .98, 2.04
17. (b) $\chi^2=18.34$, $\chi^2_{.005}=7.88$, H_0 strongly rejected
19. (a)

7	1	8
5	15	20
12	16	28

8	0	8
4	16	20
12	16	28

Probability = .0041 Probability = .0002

(b) $\chi^2=9.11$, $\chi^2_{.05}=3.84$, $\chi^2_{.005}=7.88$

Chapter 14

5. (b) $SS_T=312$, $SS_B=138$, $SSE=32$
9. (a) $t_{.0083}=3.1$. 95% simultaneous confidence intervals are
   $\beta_1-\beta_2$: $(-.055, .231)$
   $\beta_1-\beta_3$: $(.043, .329)$
   $\beta_2-\beta_3$: $(-.045, .241)$

13. (b)

Source	SS	d.f.	MS	F
Powder (A)	6098.2	2	3049.1	51.1
Ingredient (B)	6912.0	1	6912.0	115.8
$A \times B$	228.5	2	114.25	1.91
Residual	358.0	6	59.67	
Total	13,596.7	11		

(c) No

17. (b) Temperature (A) : 10 $\quad A \times B$: 4
pH (B) : 4 $\quad A \times C$: 2
Concentration (C): 8 $\quad B \times C$: 6
(c) $A \times B \times C$: 2

21. (a)

Source	SS	d.f.	MS	F
Row (A)	442.04	1	442.04	2.582
Column (B)	166.33	2	83.17	.486
$A \times B$	580.34	2	290.17	1.695
Residual	3081.25	18	171.18	
Total	4269.96	23		

No effect is significant
(b) Large residuals observed in middle column

Chapter 15

3. Observed $W_s = 25$, reject H_0 since R: $W_s \leqslant 30$ with $\alpha = .051$
9. Observed $S = 7$, significance probability $= .172$, do not reject H_0 at usual levels
11. (b)

Value of T^+	0	1	2	3	4	5	6
Probability	$\frac{1}{8}$	$\frac{1}{8}$	$\frac{1}{8}$	$\frac{2}{8}$	$\frac{1}{8}$	$\frac{1}{8}$	$\frac{1}{8}$

15. For $(1-\alpha) = .905$, $d = 25$, confidence interval [24, 56]
19. Observed $r_{Sp} = .419$, $z = 2.49$; H_0 rejected with $\alpha = .01$
23. $H = 11.01$, adjusted $H = 11.02$; H_0 rejected with $\alpha = .025$

Chapter 16

9. (3.36, 3.54)
11. (.302, .393)
15. $\overline{X}_{st} = 14.21$, error bound .28
17. $n_1 = 345$, $n_2 = 301$, $n_3 = 127$, $n_4 = 227$

Index

Applied Probability and Statistics (Continued)

OTNES and ENOCHSON · Applied Time Series Analysis: Volume I, Basic Techniques
OTNES and ENOCHSON · Digital Time Series Analysis
POLLOCK · The Algebra of Econometrics
PRENTER · Splines and Variational Methods
RAO and MITRA · Generalized Inverse of Matrices and Its Applications
SARD and WEINTRAUB · A Book of Splines
SEAL · Survival Probabilities: The Goal of Risk Theory
SEARLE · Linear Models
SPRINGER · The Algebra of Random Variables
THOMAS · An Introduction to Applied Probability and Random Processes
UPTON · The Analysis of Cross-Tabulated Data
WHITTLE · Optimization under Constraints
WILLIAMS · A Sampler on Sampling
WONNACOTT and WONNACOTT · Econometrics, *Second Edition*
WONNACOTT and WONNACOTT · Introductory Statistics, *Third Edition*
WONNACOTT and WONNACOTT · Introductory Statistics for Business and Economics, *Second Edition*
ZELLNER · An Introduction to Bayesian Inference in Econometrics

Tracts on Probability and Statistics

BARNDORFF-NIELSEN · Information and Exponential Families in Statistical Theory
BHATTACHARYA and RAO · Normal Approximation and Asymptotic Expansions
BIBBY and TOUTENBURG · Prediction and Improved Estimation in Linear Models
BILLINGSLEY · Convergence of Probability Measures
JARDINE and SIBSON · Mathematical Taxonomy
KINGMAN · Regenerative Phenomena
RIORDAN · Combinatorial Identities